THE MECHANICAL UNIVERSE

THE MECHANICAL UNIVERSE

MECHANICS AND HEAT, ADVANCED EDITION

STEVEN C. FRAUTSCHI
PROFESSOR OF THEORETICAL PHYSICS
CALIFORNIA INSTITUTE OF TECHNOLOGY

RICHARD P. OLENICK
ASSISTANT PROFESSOR OF PHYSICS, UNIVERSITY OF DALLAS
VISITING ASSOCIATE, CALIFORNIA INSTITUTE OF TECHNOLOGY

TOM M. APOSTOL
PROFESSOR OF MATHEMATICS, CALIFORNIA INSTITUTE OF TECHNOLOGY

DAVID L. GOODSTEIN
PROFESSOR OF PHYSICS AND APPLIED PHYSICS, CALIFORNIA INSTITUTE OF TECHNOLOGY
AND PROJECT DIRECTOR, *THE MECHANICAL UNIVERSE*

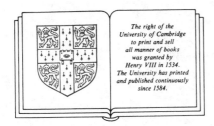

*The right of the
University of Cambridge
to print and sell
all manner of books
was granted by
Henry VIII in 1534.
The University has printed
and published continuously
since 1584.*

CAMBRIDGE UNIVERSITY PRESS
Cambridge
London   New York   New Rochelle
Melbourne   Sydney

Published by the Press Syndicate of the University of Cambridge
The Pitt Building, Trumpington Street, Cambridge CB2 1RP
32 East 57th Street, New York, NY 10022, USA
10 Stamford Road, Oakleigh, Melbourne 3166, Australia

First Published 1986

Printed in the United States of America

Library of Congress Cataloging in Publication Data

The Mechanical universe.

   1. Mechanics. 2. Heat. 3. Physics—Study and
teaching (Higher)—United States—Audio-visual aids.
4. Mechanical universe (Television program)
I. Frautschi, Steven C., 1933–
QC125.M43 1985   531   85-22361
ISBN 0-521-30432-6

British Library Cataloguing in Publication Data

The Mechanical universe: mechanics and heat.—

  Advanced ed.
  1. Mechanics
  I. Frautschi, Steven C.
  531   QC125.2
ISBN 0-521-30432-6

CONTENTS

PREFACE

I GENERAL INTRODUCTION

The Mechanical Universe is a project that encompasses fifty-two half-hour television programs, two textbooks in four volumes (including this one), teacher's manuals, specially edited videotapes for high school use, and much more. It seems safe to say that nothing quite like it has been attempted in physics (or any other subject) before. A few words about how all this came to be seem to be in order.

Caltech's dedication to the teaching of physics began fifty years ago with a popular introductory textbook written by Robert Millikan, Earnest Watson, and Duane Roller. Millikan, whose exploits are celebrated in Chapter 8 of this book, was Caltech's founder, president, first Nobel prizewinner, and all-around patron saint. Earnest Watson was Dean of the Faculty, and both he and Duane Roller were distinguished teachers.

Twenty years ago, the introductory physics courses at Caltech were taught by Richard Feynman, who is not only a scientist of historic proportions, but also a dramatic and highly entertaining lecturer. Feynman's words were lovingly recorded, transcribed, and published in a series of three volumes that have become genuine and indispensable classics of the science literature.

The teaching of physics at Caltech, like the teaching of science courses everywhere, is constantly undergoing transition. Caltech's latest effort to infuse new life in freshman physics was instituted by Professor David Goodstein and eventually led to the creation of *The Mechanical Universe*. Word reached the cloistered Pasadena campus that a fundamental tool of scientific research, the cathode-ray tube, had been adapted to new purposes, and in fact could be found in many private homes. Could it be that a large public might be introduced to the joys of physics by the flickering tube that sells us spray deodorants and light beer?

As the idea of using television to teach physics started to reach serious proportions, a gift was announced by Walter Annenberg, publisher and former U.S. Ambassador to Great Britain, to support the use of broadcast means for teaching at the college level. Ultimately, nearly $6 million of funds from the Annenberg School of Communications to the Corporation for Public Broadcasting through the Annenberg/CPB Project would be spent in support of *The Mechanical Universe*. That, in brief, is the story of how *The Mechanical Universe* came to be.

II PREFACE FOR STUDENTS

Most chapters of this book corresponds to programs in the television series *The Mechanical Universe*, as indicated in the table on pp. xv–xvi. The television series is primarily intended for students who are not preparing (as we assume you are) for careers in science and engineering. However, our experience is that even our most sophisticated professional colleagues greatly enjoy these programs. The purpose of this book is to make it possible for you to enjoy the combined benefits of a rigorous training in physics and the less demanding pleasures of the television programs.

Although many important ideas in this course are presented in the television series, they cannot be learned by simply watching television any more than they can be learned by simply listening to a classroom lecture. Mastering physics requires the active mental and physical effort of asking and answering questions, and especially of working out problems. The examples interspersed through every chapter and the problems at the end of each chapter are intended to play an essential role in the process of learning.

The backbone of *The Mechanical Universe* is a series of events that began in 1543 with a book published by Copernicus and culminated a century and a half later with a view of the universe that, consciously or not, has formed the basis of virtually all human intellectual activity since then. It is our firm view that a thorough understanding of that revolution in human thought is essential to any serious education. Offering that knowledge in an innovative way is the purpose of *The Mechanical Universe*.

III PREFACE FOR INSTRUCTORS AND ADMINISTRATORS

We expect that the ways in which *The Mechanical Universe* television series and textbooks are used will vary widely according to the circumstances and preferences of the institutions that offer it as a college course. The television programs can be viewed at home via broadcast or cable, presented in class, offered for viewing at the student's convenience at campus facilities, or even dispensed with altogether. However, we hope that no institution will imagine that the course can be presented without the services of live, flesh-and-blood college physics teachers. For most students, physics cannot be learned from a book alone, and it cannot be learned from a television screen either.

No laboratory component is offered as a part of *The Mechanical Universe* project. The reason is not that we judge a physics laboratory course to be unimportant or uninteresting, but rather that we judge its presentation by us to be impractical. We expect each institution offering the course to decide how it wishes to handle the laboratory component of learning physics.

The purpose of this volume is to make it possible for students majoring in science and engineering to enjoy the benefits of *The Mechanical Universe* television series. The television programs are intended to supplement lectures and other forms of instruction traditionally used in this kind of course. There are a number of topics, treated as sections within chapters or as entire chapters, which are not covered at all in the television series. Moreover, the order in which subjects are presented in this book does not always correspond to the order of programs intended for broadcast television. The accompanying table indicates the correspondence between chapters and television programs (numbered in the order in which they are intended for broadcast).

Chapter	Program
1. Introduction	1. Introduction to *The Mechanical Universe*
2. The Law of Falling Bodies	2. The Law of Falling Bodies
3. The Language of Nature: Derivatives and Integrals	3. Derivatives 7. Integration
4. Inertia	4. Inertia
5. Vectors	5. Vectors 9. Moving in circles
6. Newton's Laws and Equilibrium	6. Newton's Laws Note: Sections 6.5 and 6.6, on static forces, are not treated in the television series.
7. Universal Gravitation and Circular Motion	8. The Apple and the Moon 9. Moving in Circles
8. Forces	10. Fundamental Forces 11. Gravity, Electricity, Magnetism 12. The Millikan Experiment
9. Forces in Accelerating Reference Frames	No television program for this chapter

10. Energy: Conservation and Conversion	13. Conservation of Energy 14. Potential Energy
11. The Conservation of Momentum	15. Conservation of Momentum
12. Oscillatory Motion	16. Harmonic Motion 17. Resonance 18. Waves Note: Although waves are not specifically treated in this volume, Program 18 is largely a qualitative treatment of that subject.
13. Angular Momentum	19. Angular Momentum
14. Rotational Dynamics for Rigid Bodies	No television program for this chapter
15. Torques and Gyroscopes	20. Torques and Gyroscopes
16. Kepler's Laws and the Conic Sections	21. Kepler's Three Laws
17. Solving the Kepler Problem	22. The Kepler Problem 23. Energy and Eccentricity 25. From Kepler to Einstein
18. Navigating in Space	24. Navigating in Space
19. Temperature and the Gas Law	45. Temperature and the Gas Law
20. Engine of Nature	46. Engine of Nature
21. Entropy and the States of Matter	47. Entropy
22. The Quest for Low Temperature	48. Low Temperature

Note: Parts of Program 25 (From Kepler to Einstein) treat the tides and a brief mention of general relativity, topics not covered in this volume.

As one can see from the table, subjects such as noninertial frames (Chapter 9) and rotation of rigid bodies (Chapter 14) have been added to the material covered by the telecourse because they are considered appropriate to the preparation of science and engineering students. Many other subjects are covered more deeply, and most chapters include more challenging problems than would be presented to other students of *The Mechanical Universe*.

Science and engineering students who take a calculus-based physics course either have some knowledge of elementary calculus or take a separate calculus course concurrently with their physics course. The treatment of derivatives and integrals in Chapter 3 should be useful to both groups. It provides an early introduction of calculus topics that may be needed in physics before they are presented in a concurrent mathematics course, and for those who already know calculus it serves as a review. Care has been taken to make the material compatible with what the student might learn from a course taught by a professional mathematician.

Finally, the chapters on thermal physics, 19–22, have been placed at the end of the book to aid in flexibility of course design. For example, the first term of a three-quarter course might not include these chapters.

Throughout *The Mechanical Universe*, history is used as a means to humanize physics. It should go without saying that we don't expect students to memorize names and dates any more than we expect them to memorize detailed formulas and constants. *The Mechanical Universe* may or may not contribute to the vocational training of any given student. We hope it will contribute to their education.

IV ACKNOWLEDGMENTS

The Mechanical Universe textbooks, like the television series itself, would not have been possible without the cheerful and dedicated work of a long list of people who aided in its realization.

Special mention goes to *The Mechanical Universe* Local Advisory Committee, each member of which read and criticized in detail every chapter of the manuscripts, thus lending the benefit of their very considerable teaching experience: Keith Miller, Professor of Physics, Pasadena City College; Ronald F. Brown, Professor of Physics, California Polytechnic State University, San Luis Obispo; Eldred F. Tubbs, Member of the Technical Staff, Jet Propulsion Laboratory, Caltech; Elizabeth Hodes, Professor of Mathematics, Santa Barbara City College; and Eric J. Woodbury, Chief Scientist (retired), Hughes Aircraft Company.

In addition, parts of the manuscripts have been read and criticized by Margaret Osler (University of Calgary), Judith Goodstein (Caltech), and Robert Westman (UCLA), distinguished historians all; Dave Campbell (Saddleback Community College) and Jim Blinn (Jet Propulsion Laboratory), members of *The Mechanical Universe* team; Theodore Sarachman (Whittier College); and the entire 1983 and 1984 Caltech freshman classes.

In addition to all of these, homework problems were provided by Mark Muldoon and Brian Warr, and problems were checked for accuracy by Mark and Brian as well as by George Siopsis and Milan Mijic, all of Caltech.

Under the splendid direction of Project Secretary Renate Bigalke, the words and equations, mistakes and corrections of the authors were patiently and accurately rendered into the computer memory by Laurie Cornachio, Marcia Goodstein, and Sarb Nam Khalsa. All of the work was watched over anxiously by Hyman Field of the Annenberg/CPB Project (sponsors of *The Mechanical Universe*) and gently prodded along by David Tranah and Peter-John Leone of Cambridge University Press. We are especially pleased that Cambridge, which published Newton's *Principia*, has decided to follow it up with *The Mechanical Universe*. Sally Beaty, Executive Producer of *The Mechanical Universe* television series was present and instrumental at every important juncture in the creation of these books. Geraldine Grant and Richard Harsh supervised an extensive formal evaluation of various components of *The Mechanical Universe* project, including drafts of the chapters from this volume; the results of that effort have had their due effect on the final work. Carol Harrison sniffed out many photos and their sources for us.

Finally, special thanks are due to Don Delson, Project Manager of *The Mechanical Universe*, who, through some miracle of organizational skill, cunning, and compulsive worrying, managed to keep the whole show going.

CHAPTER

INTRODUCTION TO THE MECHANICAL UNIVERSE

In the center of all the celestial bodies rests the sun. For who could in this most beautiful temple place this lamp in another or better place than that from which it can illuminate everything at the same time? Indeed, it is not unsuitable that some have called it the light of the world; others, its minds, and still others, its ruler. Trismegistus calls it the visible God; Sophocles' Electra, the all-seeing. So indeed, as if sitting on a royal throne, the Sun rules the family of the stars which surround it.

Nicolaus Copernicus in *De Revolutionibus Orbium Coelestium* (1543)

1.1 THE COPERNICAN REVOLUTION

We find it difficult to imagine the frame of mind of people who once firmly believed the earth to be the immovable center of the universe, with all the heavenly bodies revolving harmoniously around it. It is ironic that this view, inherited from the Middle Ages and handed down by the Greeks, particularly Greek thought frozen in the writings of Plato and Aristotle, was one designed to illustrate our insignificance amid the grand scheme of the universe – even while we resided at its center.

Aristotle's world consisted of four fundamental elements – fire, air, water, and earth – and each element was inclined to seek its own natural place. Flame leapt through air, bubbles rose in water, rain fell from the heavens, and rocks fell to earth: the world was ordered. Each element strove to return to its sphere surrounding the center of the universe. But even as Aristotle ordered the world, he did not deem it perfect. It was subject to death and decay just as were its inhabitants. Perfection was reserved for the heavens alone, which were serene and immutable.

Above the sphere of fire were the crystalline spheres of the moon, planets, sun, and the stars beyond. Each heavenly body was fixed in its orbiting sphere, traveling across the sky in a circle – the perfect shape Plato declared the ideal path all cosmic bodies should follow. Thus conceived, the universe was so simple that it was fully and adequately represented in great clocks constructed and painted by craftsmen of the Middle Ages. And the motions of the heavenly bodies were like the inner workings of clocks – regular, predictable, and in the mind of man, free of earthly decay.

This scheme, this grand plan, was an effort to describe the environment as it presented itself to the human senses. It was an attempt to find one simple, all-encompassing explanation for natural phenomena. The modern era began when people began to ask questions that Aristotle's world view could not answer.

Our purpose is to understand what has come to be known as classical mechanics – the science that arose to answer those new questions. No discovery in human thought is more important. Through it the stranglehold of Aristotelian thought on Europe was broken and no less than a new view of our place in the universe gradually rose from its ashes to replace it. So before we start to study physics, let's introduce some of the principal heroes of the story that we're going to see unfold.

Nicolaus Copernicus, a timid monk who lived from 1473 to 1543, began the revolution with his book *De Revolutionibus Orbium Coelestium*, or *On the Revolutions of the Celestial Spheres*, published in the year of his death, 1543. In this work, Copernicus froze the sun in the sky and set the earth in motion around it.

Figure 1.1 Nicolaus Copernicus (1473–1573). (Courtesy of the Polish Cultural Institute in London.)

In an effort to simplify the Aristotelian model of the universe with its circles upon circles to describe the complex motion of planets, Copernicus had to place the sun at the center of the universe. In doing so, the earth was reduced to a common planet orbiting the sun just like the five other known planets. This theory so upset the academic world that the word "revolution" has come to be associated with radical change. Thus we have not only the revolution of the earth around the sun but also that of the United States against Great Britain.

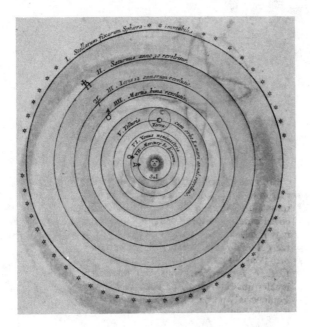

Figure 1.2 Diagram of the sun-centered system from Copernicus's *On Revolutions*. (Courtesy of the Archives, California Institute of Technology.)

Although Copernicus's ideas ultimately changed the Western view of the world, his book was at first largely ignored, and later considered heresy. One of the first scientists to seize upon the monk's revolutionary ideas was Johannes Kepler. Born a generation after the death of Copernicus and living from 1571 to 1630, Kepler ardently believed in the sun-centered system. In a battle to fit the best data on the seemingly wandering motions of the planets to the Copernican model, Kepler arrived at a set of mathematical equations that precisely described the motions. His results were as startling as they were elegant: planets tirelessly move around the sun in elliptical orbits.

At the beginning of the seventeenth century the discovery of the telescope drove the last nail in the coffin of the Aristotelian world view. Galileo Galilei, who lived from 1564 until 1642, explored the sky with the newly invented telescope. There among other unexpected wonders he found moons revolving around Jupiter. This sighting gave direct proof that the earth did not have to be the center of all heavenly motions. Through his fertile experimentation and keen insight into the character of natural phenomena, Galileo's

Figure 1.3 Johannes Kepler (1571–1630). (Courtesy of the Archives, California Institute of Technology.)

Figure 1.4 Galileo Galilei (1564–1642). (SCALA/Art Resource, N.Y.)

genius solidified the work that started with Copernicus's theory and hastened the destruction and replacement of Aristotelian phenomenology with the science of mechanics.

Like the early Greeks, the scientists of the new mechanics were intent on inventing an all-encompassing theory that would explain and describe every aspect of observable phenomena. Towering above all other scientists in the history of mechanics is Sir Isaac Newton. Born in the year Galileo died, 1642, Newton composed the grand synthesis that brought together the laws of the heavens and the laws of the earth. The physics that he established stood unchallenged until the beginning of the twentieth century.

Figure 1.5 Isaac Newton (1642–1727). (Courtesy of the Mansell Collections.)

1.2 UNITS AND DIMENSIONS

One of the themes in the history of science is the great discovery that there's a connection between what goes on in the world – what we can observe – and mathematics. The discovery of this connection was first made by the Pythagoreans – followers of the Greek philosopher Pythagoras, in the fifth century B.C., a time when *The Iliad* and *The Odyssey* were cast into their final form, when Confucius walked the earth, and when the Greeks began to question nature rather than oracles for answers. The connection was preserved through time principally by astronomers who knew that the planets and stars followed courses that could be predicted by mathematical formulas and tables. However, it was believed during all of this time that the laws of the heavens, whatever they were, were in no way connected with the laws that govern the earth. And so even though the Pythagoreans knew that events on earth obeyed mathematical laws, this idea was later forgotten as the views of Aristotle and Plato came to dominate Greek thought – and for that matter, all of Western thought – for nearly 2000 years.

Quantification in the natural sciences, the description of natural phenomena in mathematical terms, began to arise again at the end of the Middle Ages, about the same time as the invention of double-entry bookkeeping, an essential device for keeping track of the blossoming commerce of the period. Scholars still debate which of these two great discoveries inspired the other. No matter which came first, it is certainly true that commerce and science share a common need for standardized units of measure. Throughout history, establishing common units of time, distance, and weight for the sake of orderly agriculture and commerce has been one of the principal responsibilities of government, and the degradation of these standards, usually for the purpose of increasing effective taxes, has been one of the best-known symptoms of corrupt government. The nursery rhyme "Jack and Jill" started as a popular jibe at the inflation of standards of measure (a gill is still a unit of volume; up and down the hill has obvious meaning; keeping the units constant were the responsibility of "the crown," and so on).

Honest or not, units of measure were seldom the same in different political juris-
dictions, and were generally based on some convenient or traditional magnitude. For
example, in the British engineering system, still in use in the United States today, the
mile (from *milia*, thousand) was 1000 standard paces of a Roman legion (the legionnaire's
pace being a double step and consequently about 5 ft in length). The yard was the distance
from someone's nose to his outstretched fingers; the foot is obvious enough; and the inch
was the length of the last joint of someone's thumb.

Example 1
What is the number of inches in 24 yds?

To convert from yards to inches we multiply 24 yd by 1, written in a creative way:

$$24 \text{ yd} = 24 \text{ yd} \times \left[\frac{36 \text{ in}}{\text{yd}} \right] = 864 \text{ in.}$$

The term in brackets is equal to 1 by conversion of units. Upon canceling out the units
like algebraic symbols, we are left with inches as the final unit.

Alternatively, one can treat the units as algebraic symbols and immediately express
the symbols by other equivalents:

$$24 \text{ yd} = 24(36 \text{ in.}) = 864 \text{ in.}$$

Every scientist privately uses one or the other of these two tricks when doing calculations.

One of the legacies of Napoleon's conquests in Europe was the metric system, a new
system not based on tradition and whimsy, but on decimals and cool, precise French
logic. It is, nevertheless, firmly based on human magnitudes, and on the properties of
water, an essential ingredient of life. For example, the central unit of length is the *meter*,
which is roughly a yard. But instead of dividing it into feet and thumbs, it is divided in
tenths (decimeters), hundredths (centimeters), and thousandths (millimeters), as well as
multiplied by thousands (kilometers), and so on. The unit of mass, the gram, is the mass
of 1 cm^3 of water; the unit of volume, the liter, is 1000 cm^3, and so a liter of water has
a mass of 1 kg, and so on. The definitions of these quantities no longer vary from one
country to another; they are fixed by international treaty, and are used almost everywhere
except in the United States. This system is formally known as *Système International
d'Unités*, or SI for short.

Though relatively new, the SI system has already gone through a considerable evolution. Originally the meter was defined to be 1/10,000,000th of the distance between
the earth's equator and the pole. Subsequent redefinitions have kept the household meter
stick unchanged while progressively increasing the precision of the standard for scientific
and technological use. For many years, the meter was defined as the distance between
two lines engraved on a bar of platinum-iridium alloy stored at 0°C at the International
Bureau of Weights and Measures in Sèvres, near Paris. In 1960 it was redefined as
1,650,763.73 wavelengths of orange-red light emitted from a krypton-86 lamp, a standard

that improved the accuracy to four parts in 10^9. Most recently in 1983 the meter was redefined as the distance that light travels in 1/299,792,458th of a second. This latest method of measurement is accurate to two parts in 10^{10}, which permits measurements of the earth's circumference to an accuracy of ± 8 mm.

Another basic physical measurement is that of time. Mastering the flow of time and dividing it into units seems to have been a part of the growth of every civilization on Earth. Astronomer-priests of agricultural societies were responsible for deciding when to begin the annual cycle of tilling, planting, and harvesting. Smaller divisions of time corresponded, at least roughly, to the death and rebirth of the moon (months) and, of course, the daily cycle of light and dark. Intermediate clusters of days, five or ten or, by Roman times, seven days per week, also came up. Dividing time into units smaller than a day proved more difficult, since it involved inventing timekeeping devices rather than mere counting. Hours, minutes, and seconds are relatively recent inventions, as is the idea that these units should have the same duration all year, regardless of the proportion of daylight and darkness in each day. Fortunately, unlike the units of length and mass, the same units of time are used everywhere, even in the United States. Units of one second (s) and longer have the traditional names (minute, hour, day, week, month, year, century) whereas shorter times get metric-style prefixes: milliseconds (ms), microseconds (μs), nanoseconds (ns), etc.

To define a unit of time such as the second, one must identify a recurrent phenomenon that occurs in equal time intervals. At a very low level of accuracy, in the absence of all equipment, the second could be based on the regular beating of the human heart (about 1 s per beat for a healthy person at rest). For centuries it was based on the much more

Table 1.1 Metric Prefixes and Abbreviations

Multiple	Prefix	Abbreviation
10^{12}	tera	T
10^{9}	giga	G
10^{6}	mega	M
10^{3}	kilo	k
10^{-2}	centi	c
10^{-3}	milli	m
10^{-6}	micro	μ
10^{-9}	nano	n
10^{-12}	pico	p
10^{-15}	femto	f

regular rotation of the earth (the second was defined as 1/86,400th of a day, with refinements added later to take account of the not-quite-uniformly circular movement of
the earth around the sun, and the slow lengthening of the year by about half a second
per century). In recent years the second has been based on the even more precise move-
ments of atomic electrons (beginning in 1967, when the second was redefined in terms
of an atomic clock controlled by one of the characteristic frequencies associated with
atoms of the isotope cesium 133).

Quantities like seconds, grams, and meters not only have units, which may vary
from one jurisdiction to another, but also dimensions, meaning respectively, time, mass,
and distance. Quantities with the same *dimensions*, but different *units*, can easily be
compared (see Example 1): 1 in. is bigger than 2 cm, but less than a light-year. But
quantities with different dimensions cannot be compared at all, regardless of their units:
a kilogram is neither bigger nor smaller than an hour or a yard.

For many years after the beginning of the quantification of nature, all quantification
consisted of comparing quantities of the same dimension. The idea of compounding
quantities of different dimensions – say, dividing a distance by a time to form a velocity
– is a rather recent but richly useful invention. A compound quantity like velocity has
a unique dimension (distance divided by time) but various units (cm/s, furlongs/fortnight,
etc.). Average velocity is defined as

$$\bar{v} = \frac{\text{change in position}}{\text{time elapsed}}.$$

(1.1)

The bar over the v reminds us that we are talking about an average velocity during that
time interval. If distance is measured in feet or meters, and time is measured in seconds,
the units of velocity are feet per second, written ft/s, or meters per second, written m/s.
Velocity can be positive or negative; its absolute value is called speed.

Another physical quantity which we will encounter is acceleration, a term which will
be used to denote increase as well as decrease of velocity. We often hear about cars
which can speed up, say from 0 to 60 mi/h, in so many seconds. That's an example of
acceleration. The average acceleration during a time interval is the change in velocity
during that interval divided by time,

$$\bar{a} = \frac{\text{change in velocity}}{\text{time elapsed}}.$$

(1.2)

Just as velocity can be measured in ft/s or m/s, the units of acceleration can be (ft/s)/s
(read feet per second per second), usually abbreviated as ft/s^2 (read feet per second
squared), or m/s/s, abbreviated as m/s^2.

Appendix A contains a list of U.S. customary and metric units, conversions between
them, standardized abbreviations such as ft for foot and cm for centimeter, and the metric
prefixes.

In this text we shall use U.S. customary and metric units interchangeably at first,
but mainly metric units later on. While engineers continue to use the customary units in
America, science is mainly conducted in the metric system for several reasons:

(i) Conversions between units in the metric system involve only factors of 10, while
conversions in the U.S. customary system from 1 mi to 1760 yd, 1 yd to 3 ft, 1 ft
to 12 in., etc., involve a potpourri of bothersome factors.

(ii) When we come to weights, the English word pound will prove confusing because its use grew up before the scientific distinction between weight and mass was clarified. The metric units, having been established much later, are not subject to this confusion.

Example 2

What is the speed of a car traveling 60 mi/hr in kilometers per hour? In feet per second?

Using the fact that there are 1.6 km in 1 mi, we can convert

$$60 \, \frac{\text{mi}}{\text{h}} = 60 \, \frac{(1.6 \text{ km})}{\text{h}} = 96 \, \frac{\text{km}}{\text{h}} \, .$$

Similarly we find

$$60 \, \frac{\text{mi}}{\text{h}} = 60 \, \frac{(5280 \text{ ft})}{(60 \text{ min})} = 60 \, \frac{(5280 \text{ ft})}{60(60 \text{ s})} = 88 \, \frac{\text{ft}}{\text{s}} \, .$$

Always remember, most physical quantities have dimensions, and therefore are characterized by both a number and a unit. It is meaningless to describe a physical quantity by a number alone without the unit. A change in the unit requires a corresponding change in the number.

1.3 A FINAL WORD

The rise of modern science grows out of quantification. But quantification means much more than expressing observations in mathematical form. It is also a turning away from natural philosophy – grand schemes based on aesthetic preference – to detailed and precise observation and measurement. In other words, it is a turning to the accumulation of knowledge by small, detailed increments. ''I would rather learn a single fact, no matter how ordinary,'' wrote Galileo, ''than discourse endlessly on Great Issues.'' Before Copernicus, many speculated on a sun-centered universe but Copernicus took the trouble to do the detailed calculations and produce the astronomical tables that made his system a serious competitor for the very successful Ptolemaic system that came before it. Others before Galileo speculated on the properties of matter in motion, but Galileo based his arguments on detailed observations. His experiments with balls rolling down smooth inclined planes led to the law of falling bodies (Chapter 2), the law of inertia (Chapter 4), and ultimately to the law of conservation of energy (Chapter 10). Before clocks as we know them were invented, he devised means of timing his experiments by weighing how much water flowed through a specially constructed pipe. These measurements were accurate to a tenth of a second. The ultimate result of this kind of careful attention to detail, together with ingenuity, was no less than a new view of the universe.

Quantification of physics tends to condense its ideas into mathematical formulas. As Galileo so delightfully expressed it, the great book of nature lies ever open before our eyes, but it is written in mathematical characters. History teaches us that mathematics helps to advance physics, but it also shows that, like the tides, ideas flow in both directions. New discoveries in one field often lead to improvements in the other. For example, early in the seventeenth century the French mathematician Pierre de Fermat devised a crude method for drawing a tangent line to a curve. This gave Newton a hint for determining the velocity of a moving point, and this, in turn, led to Newton's version of differential calculus. The mathematics in this book is developed in the same spirit. When new mathematical concepts are introduced, such as derivatives, integrals, and vectors, they arise naturally from physical problems. And then these new concepts help us to read the great book of nature and to write new chapters.

After Copernicus and his revolution and the events that led Europe through the years of Kepler and Galileo and finally Newton, our view of the universe was, and still is today, that we live on a speck of dust in a lost corner, somewhere in the universe. Aristotle and Plato tried to teach us humility by placing us in a lowly sphere, isolated from the serene perfection of the heavens, but never in their wildest dreams could they have imagined the impact of the psychological change that occurred in the human race when we first realized that we were not at the center of the universe.

When we study history, we learn about kings and princes, about social problems, wars, economics, and so on. All of these things come and go. And when they're gone the world is pretty much the same as it was before. If you walk through an ancient Roman town today, say the town of Herculaneum in Italy, you easily understand where you are and what the people were doing, and why the town was built. In its essentials, the human condition has not changed very much in the past 2000 years. But there is one profound change that has altered the human race forever. That is the discovery of our real place in the universe. Studying history, we learn about the Renaissance and Reformation, the Counter-Reformation and the Thirty Years' War – events that dominated the history of Europe during the time of Copernicus, Kepler, Galileo, and Newton. But those events were minor readjustments in the social fabric compared to this one monumental change that was occurring – new ideas that changed the human race absolutely forever. Our job in this volume is to study that story, to see exactly how and why it happened that we found our real place in the universe.

Problems

1. Determine which of the following distances is greatest:

   (a) 560 yd,   **(b)** 0.3 mi,   **(c)** 0.5 km,   **(d)** 52,300 cm.

2. (a) If a baseball pitcher's fastball travels 100 mi/h, what is its speed in feet per second?

   (b) If a sprinter runs 100 meters in 10 seconds, what is his speed in miles per hour?

   Convert the following quantities into feet per second:

   (c) 23 km/h,   **(d)** 88 nm/ms,   **(e)** 40 cm/yr.

3. Determine which of the following bizarre quantities have units of distance, velocity, or acceleration:

   (a) $2.3 \dfrac{\text{m yr}}{\text{s}}$,   **(b)** $144 \dfrac{\text{m yd s}}{\text{mi yr}^2}$,   **(c)** $4 \dfrac{\text{cm}}{\text{s century}}$.

4. Convert each of the quantities in Problem 3 into metric units of distance, velocity, or acceleration.

CHAPTER  2

THE LAW OF
FALLING BODIES

(Natural) philosophy is written in this enormous book, which is continuously
open before our eyes (I mean the universe) but it cannot be understood
without first learning to understand the language, and to recognize the
characters in which it is written. It is written in a mathematical language. . . .

Galileo Galilei, *Il Saggiatore* (1623)

2.1 HISTORICAL BACKGROUND

Before we learned to read Galileo's mathematical book of the universe, our descriptions
of nature were qualitative and verbal. For centuries, only words were used to describe
the motion of objects, and those words were largely based upon the writings of Aristotle
from the fourth century B.C. Aristotle's description of motion centered on the idea of a
"natural place." In his view earth had the lowest natural place, so naturally if you

dropped a heavy object made of earth, it fell. Heavy bodies, containing more earth, fell faster than light ones. And any object, when released, initially gained speed toward its natural place.

Aristotle's ideas present a qualitative (and apparently correct) description of motion. He believed that mathematics and precise quantitative measurement were of little value in describing the world around him.

Nevertheless, Aristotle's doctrine did leave open the question, how does a falling body initially speed up? What rule does it follow? And so over a period of several centuries in the Middle Ages, various sages offered suggestions on this tricky question. Galileo himself started out following this tradition, but soon his genius for incisive observation and experiment, and his instinct for precise mathematical description led him to a wholly new understanding, quite outside of Aristotelian doctrine.

We shall state Galileo's result, and then go back to see in detail how it compared to previous ideas and how he established it.

2.2 GALILEO'S LAW

Galileo's law of falling bodies can be written

In a vacuum, all bodies fall with the same constant acceleration.

This innocent-sounding statement makes three separate points:

   (i) Gravity has the *same* effect on *all* bodies, regardless of their size or weight. From Galileo to Isaac Newton, right down to Albert Einstein, that fact has been one of the central mysteries in all of physics. We shall return to it at several levels as we proceed through the course.
  (ii) The result of gravity on the earth's surface is *constant acceleration.* To understand what this means we shall introduce a new mathematical invention called the derivative. It will turn out to be a crucial mathematical tool for understanding the world.
 (iii) Points (i) and (ii), profound and important as they are, violate our simplest intuition, because they happen only in a vacuum, not in the world of air and water we're familiar with. In the presence of air resistance, feathers do fall more slowly than stones, but this turns out to be a complicated distraction, not the essence of the phenomenon of falling.

2.3 DO HEAVY BODIES FALL FASTER THAN LIGHT ONES?

In Aristotelian physics, heavy bodies fall faster than light ones, and of course this is the behavior we observe in the atmosphere. But in Galileo's 1638 publication *Dialogues Concerning Two New Sciences*, he considered bodies falling without air resistance.

This was a brilliant insight because Galileo could not produce a vacuum; he imagined one. He realized that the question of whether the vacuum could be produced was not crucial. What was important was that to understand falling bodies, the effects of air resistance should be ignored.

The argument he used was this: Consider a heavy rock connected to a light one by a string. According to Aristotle, when they are released, the heavy one pulls the lighter one down and tries to make it fall faster than it would if unattached. The light one, on the other hand, tends to slow down the heavier one. Thus the combined body must fall faster than the lighter one alone, and slower than the heavy one. But the combined body is heavier than the heavy one. It should therefore fall faster than the heavy one. The Aristotelian view thus leads to a contradiction if we ignore atmospheric effects such as air resistance.

It was arguments like this that led Aristotle himself to the belief that a vacuum was impossible – it would lead to absurd results. Galileo concluded instead that if there were a vacuum, all bodies would fall at the same rate.

Galileo's ability to test this conclusion experimentally was very limited. He had no vacuum. The best he could do was roll balls of differing mass down the same gently inclined plane and verify that they rolled at equal rates (when bodies move slowly, air resistance is not important).

Today we can readily verify Galileo's law. Innumerable students have witnessed a feather and a penny fall at the same rate in an evacuated tube as a classroom demonstration. When astronaut David R. Scott of the Apollo 15 mission found himself on the airless surface of the moon, he could not resist repeating this classic experiment for the whole world to see. As he said then, Galileo's discovery was a major step toward getting to the place he was standing.

Figure 2.1 Apollo astronaut David R. Scott dropping a falcon feather and hammer on the moon.

2.4 MEDIEVAL LAWS OF FALLING BODIES

As we have mentioned, a number of medieval scholars tried to describe the falling motion of a heavy body. One of the earliest attempts is credited to the fourteenth-century scholar Albert of Saxony. In trying to answer the question, ''In what way does a body get faster as it falls?'' Albert argued that the speed of a body is proportional to the distance it has fallen. For centuries most people who thought about it at all assumed Albert was right. Another fourteenth-century scholar, Nicole Oresme, had a different conjecture. From a mathematical study of various possible types of motion, he found a law which he at one time suspected could describe falling bodies, although he was not much concerned with anything so messy as the real world. What he suggested was a relationship between speed and time rather than distance: the speed is directly proportional to the time spent falling.

These two ideas focused on how the speed of a falling body changes. Leonardo da Vinci, in the fifteenth century, formulated a different kind of law. He expressed his law in terms of quantities that are easy to measure: intervals of distance and time. He proposed that the distances fallen in successive equal intervals of time are proportional to the consecutive integers (1, 2, 3, . . .). In other words, if a body falls one unit of distance in one unit of time, then in the second unit of time it falls two units of distance, three in the third, and so on.

Galileo stated his ideas about falling bodies in a number of different ways. One of these resembled Leonardo's formulation. He said that the distance in successive time intervals is proportional to the odd numbers 1, 3, 5, ... In other words, if a body falls one unit of distance in one unit of time, then in the second unit of time it falls *three* units of distance, *five* in the third, and so on.

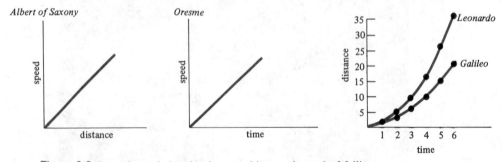

Figure 2.2 Laws interrelating the time, position, and speed of falling bodies, as proposed by Albert of Saxony, Oresme, Leonardo da Vinci, and Galileo.

We've now encountered four different attempts to describe falling motion: speed is proportional to distance or to time; distance per time interval is proportional to the integers or to the odd numbers. And none of them sounds very much like what we already know to be the correct answer: constant acceleration!

One reason for the profusion of laws was that some of the descriptions proposed for falling bodies were answers to *different* questions concerning their motion: how does speed depend on distance? how does speed depend on time? and so forth. It was not clear just how the different questions were related to one another, or whether one was more basic than the others. A second reason was the absence of any attempt to test the laws. The Aristotelian cast of mind did not make it seem important to perform experiments. And, in the absence of good clocks, measurements would have been prohibitively difficult in any case. Bodies fall rapidly once they get going. Even if Galileo had dropped a cannonball from the Leaning Tower of Pisa (he probably did not), it would have taken only about two seconds to reach the ground – too short a time to make any detailed measurements given the technology available to him.

2.5 EXPERIMENTAL DETERMINATION OF THE LAW OF FALLING BODIES

Faced with these difficulties, Galileo found a way to slow down falling motion so that he could measure it. He used (among other techniques) a water clock to time balls rolling slowly down very gently inclined slopes. He found that if he increased the slope, the ball would roll faster, but the nature of the motion was always the same: the total distance the ball went was proportional to the square of the elapsed time. In other words, it would go four times as far in 2 s as in 1 s, nine times as far in 3 s, and so on. He guessed that the same would happen if his slope were vertical – in other words, in free fall.

This is a statement that can be written in the form of an equation. We shall call the distance fallen s, and write it as $s(t)$, where t is the time. This $s(t)$ is read "s of t"; it represents s as a function of t, or in other words, how s depends on t. The answer to how s depends on t we write on the other side of the equation:

$$s(t) = ct^2 \tag{2.1}$$

where c is constant.

Since $s(t)$ has units of length, and t has units of time, the constant c must also have units. These units must be length divided by time squared; only in this way will the time-squared units cancel to yield units of length for $s(t)$.

When $t = 1$, $s(1) = c$, so c is numerically equal to the distance any object will fall in the first unit of time. The value of c can be determined by measurement of an actual falling body. In U.S. customary units its value turns out to be 16 ft/s^2. In metric units c has the value 4.9 m/s^2. And it has the same value for every object falling in a vacuum near the surface of the earth.

But what has this to do with the other laws? The connection to Galileo's odd-number rule has to do with a mathematical fact that was known all the way back in the time of the Pythagoreans (in the sixth century B.C.): the intervals between the square numbers (1, 4, 9, 16, 25, . . .) are the odd numbers (3, 5, 7, 9, . . .). Thus if the *total* distance fallen in successive units of time increases as

$$0^2 = 0, \qquad 1^2 = 1, \qquad 2^2 = 4, \qquad 3^2 = 9, \qquad 4^2 = 16,$$

then the *incremental* distance fallen in successive units of time increases as 1, 3, 5, 7, . . . , since

$$1^2 - 0^2 = 1, \qquad 2^2 - 1^2 = 3, \qquad 3^2 - 2^2 = 5, \qquad 4^2 - 3^2 = 7.$$

Galileo's two laws are equivalent descriptions of falling bodies, whereas Leonardo's differing law for time intervals disagrees with experiment.*

---

*See, however, Problem 5.

With modern clocks we can study free fall directly and verify Galileo's results for ourselves. The following picture shows a falling ball in successive instants of time. The picture was taken by a camera with an open shutter, while the ball was illuminated by an instrument called a stroboscope. The stroboscope emits flashes of light at regular intervals of time. It is our clock in this experiment. The same interval of time has elapsed between each successive image of the ball in our picture. A scale behind the ball allows us to measure the distance fallen.

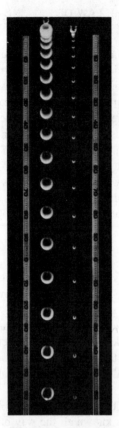

Figure 2.3 Stroboscopic picture of a falling ball. (*PSSC Physics*, 2nd ed., 1965. D.C. Heath and Co. with Education Development Center, Inc., Newton, Mass.)

Data of this kind do not directly test Albert's law or Oresme's law for the speed of a falling body, because we have measured not the speed, but the position of the body. On the other hand, these data are ideal for choosing between Leonardo's law and Galileo's law for the distance fallen in successive intervals of time.

Before we try to choose, we should understand that these are real experimental data. Like all real experimental data, they are imperfect. For example, we can't tell exactly where the ball is because the images are a bit fuzzy, and some of them overlap. Besides, there really is air resistance, which Galileo (but not Leonardo) wanted to ignore. There are other possible sources of error as well. There is experimental error in all experiments, no matter how ingenious or carefully done.

We examine our data in the following way: we measure the distance fallen between the first and second flashes (the first flash illuminates the ball at the instant of release). This distance we call c. According to Leonardo's law, the next distance should be $2c$, the one after that should be $3c$, and so on. Figure 2.4 shows our picture with lines drawn on it to indicate where each image should be, constructed in this way.

Now we reconstruct the predictions of Galileo's law of odd numbers. If c is the distance between the first and second flashes, then the next distance should be $3c$, the one after that should be $5c$, and so on (remember that Galileo's law is the same as the time-squared law, which we have just used). The result is shown in Fig. 2.4.

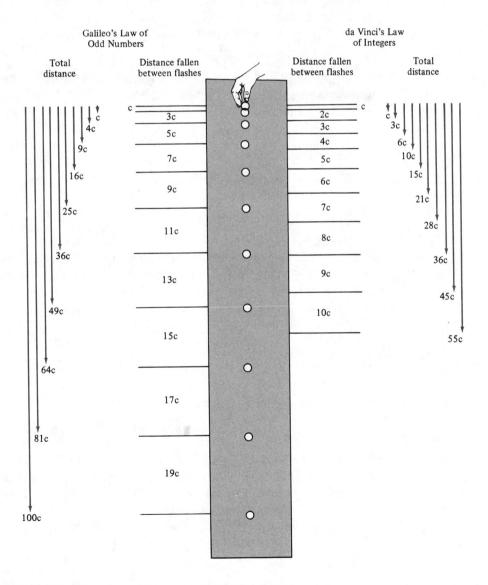

Figure 2.4 Comparison of Leonardo's and Galileo's predictions for falling bodies.

In spite of all the experimental uncertainties, the comparison between the two predictions is decisive. It is Galileo's law that we choose. Measurements of this type can
also determine the value of the constant in the equation for free fall, $c = 16$ ft/s^2.

2.6 THE AVERAGE VELOCITY OF A FALLING BODY

Now that we know how the position changes with time, can we find how fast the body
is falling? We must be very careful to frame our question correctly. How fast it's moving
means how far it travels in a specified amount of time. We need to specify the time
interval during which the object drops.

If we do specify the time interval, the free-fall equation will allow us to determine
the average velocity during that time, as follows. First we determine $s(t)$ at some time t
using $s(t) = ct^2$. Next we calculate the distance fallen after a later time, say t_1. That is,
we calculate $s(t_1)$ from our equation. The length of the time interval is $t_1 - t$, and the
change in position during this time interval is $s(t_1) - s(t)$. The quotient is the average
velocity for that interval:

$$\overline{v} = \frac{\text{change in position}}{\text{travel time}} = \frac{s(t_1) - s(t)}{t_1 - t}. \tag{2.2}$$

For example, an object falls $s(1) - s(0) = 4.9$ m in the first second. Its average
velocity during the first second is 4.9 m/s. After 2 s, $s(2) = 4.9 \times 2^2 = 4.9 \times 4 =$
19.6 m, for an average velocity of 9.8 m/s over the first 2 s. The average velocity keeps
on increasing – that's to be expected – but there's a problem. In our example, the object
is moving slower than 9.8 m/s at the start of the 2-s interval and faster at the end of it.
What we would really like to know is the velocity, not on the average, but at each instant
of time. But the velocity at some instant means that there is no time interval. If we try
to take the change of position in no time interval and divide by that zero time, we end
up trying to divide zero by zero, which is a kind of mathematical disaster. Yet we have
an intuitive feeling that at every instant a falling body does have a definite velocity.

This problem plagued mathematicians for thousands of years, and it was not until
calculus was invented, a generation after the death of Galileo, that the solution was found.

2.7 INSTANTANEOUS VELOCITY: THE DERIVATIVE

Our goal in this section is to find the velocity of a falling body at any instant. Our strategy
will be to calculate the average velocity for a certain time interval, then, at the very end
of the calculation, allow the time interval to shrink to zero. By waiting until the end
before shrinking the time interval to zero, we will be able to avoid the meaningless result
of dividing zero by zero. This simple idea leads to one of the most important inventions
in the history of mathematics. It is called the *derivative*. The derivative lies at the heart
of calculus, and it solves the problem of finding instantaneous velocity.

The distance the body has fallen after time t is

$$s(t) = ct^2. \tag{2.1}$$

Suppose we ask what distance the body has fallen a short time interval h later. This small time interval can be one second or one-millionth of a second; right now its value is unimportant. According to Eq. (2.1), the distance fallen after $t + h$ seconds is

$$s(t + h) = c(t + h)^2. \tag{2.3}$$

[Note that the parentheses indicate functional dependence on the left, but multiplication of c by $(t + h)^2$ on the right. Which usage of parentheses is intended will normally be clear from the context.]

At this point we will rewrite our last result by explicitly multiplying out the square. We obtain

$$s(t + h) = c(t^2 + 2th + h^2). \tag{2.4}$$

The average velocity $\bar{v}$ over the small time interval h is, by Eq.(2.2),

$$\bar{v} = \frac{s(t + h) - s(t)}{h}, \tag{2.5}$$

or in other words,

$$\bar{v} = \frac{c(t^2 + 2th + h^2) - ct^2}{h}. \tag{2.6}$$

In this last expression the two terms ct^2 will cancel, so we perform that cancelation first:

$$\bar{v} = \frac{2cth + ch^2}{h}. \tag{2.7}$$

Now we can cancel the h in the denominator with an h in each of the terms in the numerator. Our result is

$$\bar{v} = 2ct + ch. \tag{2.8}$$

This is the average velocity of the falling body for any time interval h after time t.

But something quite extraordinary has happened. If we now try to find the precise velocity at any instant by taking $h = 0$, we're no longer dividing by zero! Instead we obtain a simple equation for the instantaneous velocity as a function of time, $v(t)$:

$$v(t) = 2ct \quad \text{when } h = 0 \tag{2.9}$$

[the bar is no longer written over $v(t)$ because it is no longer an average velocity but an instantaneous velocity, the limiting value of the average velocity as the time interval shrinks to zero].

This is an important result. For one thing, we've answered one of the questions posed earlier: Galileo's law $s(t) = ct^2$ is the same as Oresme's law: the speed is indeed proportional to the time (the law of Albert of Saxony, that speed is proportional to distance, turns out to be incorrect; we will find in Example 1 how speed actually does depend on distance).

But in the long run, the argument leading to Eq. (2.9) represents something much more important: it is the basis of the differential calculus. Here is the general idea. We start by calculating the average rate at which something is changing. Then we ask, what happens if we compute that average over a smaller interval? Usually, as the interval gets smaller and smaller, the average rate of change doesn't jitter wildly or blow up. Instead it smoothly approaches some definite result. If that's true in the limit as we let the interval shrink to zero, then the result is called the *derivative* or the *instantaneous rate of change* of the original function. The derivative is the fundamental building block of calculus.

Velocity $v(t)$ is the derivative of displacement $s(t)$ with respect to time. The process of obtaining the velocity from the displacement is called *differentiation*. To emphasize that $v(t)$ comes from $s(t)$ we write

$$v(t) = s'(t). \qquad (2.10)$$

The right-hand side reads "*s* prime of *t*," where the prime indicates the derivative of *s* with respect to *t*.

2.8 ACCELERATION

We next ask, "How rapidly does the velocity of a falling body change?" The answer is obtained by differentiating $v(t)$.

We first want the average rate of change of $v(t)$ between t and $t + h$. The rate of change of velocity is called acceleration; we are calling average acceleration $\bar{a}$. Then

$$\bar{a} = \frac{v(t + h) - v(t)}{h} . \qquad (2.11)$$

This is exactly what we did with $s(t)$. We eventually want to let h become zero, but not yet. First substitute in $v(t) = 2ct$ and $v(t + h) = 2c(t + h)$:

$$\bar{a} = \frac{2c(t + h) - 2ct}{h} = \frac{2ch}{h} = 2c. \qquad (2.12)$$

This is the simplest result yet. It turns out that $\bar{a}$ doesn't depend on t; it doesn't even depend on h! No matter what we do, even if we let h shrink to zero, we get the same constant result, $2c$. That means the instantaneous acceleration is also $2c$:

$$a(t) = v'(t) = 2c. \qquad (2.13)$$

But that's exactly the expression of the law of falling bodies we gave at the very beginning: all bodies fall with the same constant acceleration.

That constant acceleration is so important in all of mechanics that we give it a name of its own. It's called g (for gravity); that is, all the formulas we now have are usually written with g replacing $2c$:

$$s(t) = \tfrac{1}{2}gt^2, \qquad (2.14)$$

$$v(t) = s'(t) = gt, \qquad (2.15)$$

$$a(t) = v'(t) = g. \qquad (2.16)$$

The value of g at the surface of the earth is about 9.8 m/s^2 (32 ft/s^2). On another planet or moon, g may have a different value, but the form of the law of falling bodies will not change. For example, if you are on the earth's moon, objects fall slower: g is only 1.6

m/s^2, but you can still use $s(t) = \frac{1}{2}gt^2$ to find the distance fallen. In fact Eqs. (2.14)–(2.16) apply, with appropriate values of g, to any problem involving constant acceleration, even if the source of the acceleration is not gravity.

To express the fact that the acceleration is obtained from the distance by differentiating twice, we often write

$$a(t) = v'(t) = s''(t). \tag{2.17}$$

The notation s'' means $(s')'$; it is called the *second derivative* of s with respect to t.

The three equations (2.14)–(2.16) contain what we need to know about falling bodies. Each of them correctly describes the motion. They are interrelated by the concept of differentiation. Among them, the statement that acceleration is constant will turn out to be the most important, the clue to the underlying physics, when we study Newton's laws in later chapters. A fourth statement, Galileo's law of odd numbers, is also correct but far less useful. The suggestions by Albert of Saxony and Leonardo da Vinci were wrong.

Around 1350, long before the discovery of the derivative, Nicole Oresme managed to figure out all of these relationships, using essentially geometrical arguments. Galileo used almost identical arguments 250 years later to figure it out all over again. He added the crucial ingredient of experimental test, and realized that it was not just a theoretical kind of motion, but the way falling bodies actually behave, or at least would behave if they were falling in a vacuum. The derivative was invented yet another generation later, by Isaac Newton and Gottfried Wilhelm Leibniz.

For Oresme and Galileo, discovering the properties of uniformly accelerated motion was an extraordinary intellectual feat. The problem is that the simple and constant aspect of the motion is not where the body is, nor even how fast it's going, but rather how fast it's getting faster. That's hard even to think about, just using words. It's difficult using geometry. But we have done it much more easily with the aid of the differential calculus, which is ideally suited to this kind of analysis.

Example 1

A flower pot falls from a window ledge 36 ft above the pavement. What is the velocity of the flower pot just before it strikes the ground?

From Eq. (2.15) we have $v(t) = gt$. The time t is not specified, but solving the free-fall Eq. (2.14) for t, we find that it is related to the distance fallen by $t = \sqrt{2s/g}$. Thus the relation between velocity and distance fallen is

$$v = gt = g\sqrt{\frac{2s}{g}} = \sqrt{2gs},$$

and the velocity after falling 36 ft is

$$v = \sqrt{2 \times 32\,\frac{\text{ft}}{\text{s}^2} \times 36\text{ ft}} = 48\,\frac{\text{ft}}{\text{s}}.$$

Example 2

The flower pot in Example 1 falls past a window partway down with a speed of 24 ft/s. How far above this window did the flower pot originate?

The relation obtained above by eliminating t from Eqs. (2.14) and (2.15), $v = \sqrt{2gs}$, can be used here as well in the form

$$s = \frac{v^2}{2g} = \frac{(24 \text{ ft/s})^2}{2 \times 32 \text{ ft/s}^2} = 9 \text{ ft.}$$

Note that in these examples we put in numbers only at the last step, after first solving for s algebraically. This is usually the best way to proceed, both because the algebraic answer is so informative (e.g., in the present case it gives s for *any* v and g) and because when numerical values are inserted at an early stage, carrying them through subsequent steps is a cumbersome and error-prone process.

2.9 A FINAL WORD

One of the jobs of physics is to find simple, economical underlying principles that explain the complicated world we live in. We have done that here. If we drop an object, it falls. As it falls, its motion is opposed, with varying degrees of success, by the air through which it moves. If we can imagine disposing of the air and describing the effect of gravity alone, we discover a dramatic and surprising fact: all bodies fall at the same rate.

We could be satisfied with that fact. After all, discovering it was quite an impressive accomplishment. But of course, we are not satisfied. We want to know, why is it true? What is the nature of gravity that leads to such strange behavior? That question has turned out to be one of the most profound in the history of physics. It has persisted even into our own century. It was the starting point from which Albert Einstein built his celebrated general theory of relativity.

But we are getting ahead of our story. Once we knew there was one law for all falling bodies, the job was then to express that law with precision. We have done that too. We learned that all bodies fall with the same constant acceleration.

Acceleration is the rate of change of velocity. Constant acceleration follows from the fact that the velocity of a falling body released from rest increases in proportion to the time spent falling. Velocity is the rate of change of position. The proportionality of velocity to time follows from the proportionality of distance to the square of the time. So we have in fact three precise, mathematical statements of the law of falling bodies. They are all true, and they are connected by one of the great and crucial discoveries in mathematics: the differential calculus.

Galileo Galilei discovered and expressed the law of falling bodies. He was a brilliant and arrogant man who managed so much to offend the ecclesiastical authorities of his time that he spent the last eight years of his life a prisoner, under house arrest, at his estate near Florence.

The calculus was discovered by Isaac Newton and Gottfried Wilhelm Leibniz. It was a mighty triumph, the most important event in mathematics in thousands of years. But

Newton and Leibniz sacrificed the joy of their discovery in a bitter, acrimonious dispute over who deserved credit for discovering it first.

On the other hand, Albert Einstein became a folk hero to a whole world that never pretended to understand what he had done, or why he deserved honor.

All of these are threads in the story we are going to see unfold. We shall see how Newton's laws of mechanics give acceleration a special role, and how his universal law of gravitation causes gravitational acceleration to take on different values on the moon, in space, and even to some slight extent over the earth's surface. We shall learn how rotation of the earth and other bodies affects the perceived acceleration. And we shall suggest how complicated corrections such as air resistance can be taken into account.

Problems

Medieval Laws of Falling Bodies

1. In his work *Two New Sciences*, Galileo asserts that the law of Albert of Saxony cannot be correct because when an object is released, it has not fallen any distance and therefore, by this law, has no speed. Having no speed, of course, it would not begin to fall. Consequently, he argues, if this law were correct, the body would never fall. Are you convinced by this argument? Why?

2. Construct graphs of average velocity, defined as $[s(t) - s(0)]/t$, as a function of time for Leonardo's and Galileo's laws.

3. Followers of the sixth-century B.C. Greek philosopher Pythagoras believed that numbers had shapes. One shape was a square. The first square number was 1 and you could draw a little square around one dot like this: ⊡. The next square number was 4, and that can be constructed from your original square by adding more dots to form a larger square like this: ⊞

   The next square number was 9, and you can continue to enlarge your square by adding more dots. (You can also do this with squares on graph paper instead of dots.) Examine this procedure for 1, 4, 9, 16, and 25, by considering how many dots you are adding to form each successive square. Galileo knew of this Pythagorean discovery. How is it related to his law of falling bodies?

4. Refer to Problem 3. There are also *triangular* numbers in the Pythagorean system:

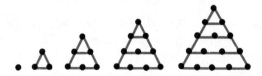

   The picture for the *n*th triangular number is made by adding a row of *n* dots to the bottom of the previous triangular number.

   (a) What do these numbers have to do with Leonardo's law of falling bodies? Write the law in the form $s = s(t)$.

(b) What is the instantaneous velocity for this law?

(c) What is the instantaneous acceleration?

5. Show that if Eq. (2.1), $s(t) = ct^2$, is evaluated at successive units of time $\frac{1}{2}$, $\frac{3}{2}$, $\frac{5}{2}$, $\frac{7}{2}$, . . . , instead of the usual times 0, 1, 2, 3, . . . , the incremental distance fallen in these successive units of time increases in a manner equivalent to Leonardo's law.

Galileo's Law of Falling Bodies

6. An object falls according to the time-squared law. If it falls 16 ft in 1 s, how long will it take to fall 144 ft?

7. A cliff diver is in the air for 2 s before hitting the water. Calculate the following for his dive:

(a) The height from which he fell.

(b) His average velocity over the entire dive.

(c) His average velocity over the last half second.

8. According to an ad for children's safety seats, "a car collision at only 30 mph has the same crushing effect as falling from a three-story building." Assuming the ad writer means that one acquires a velocity of 30 mi/h in falling three stories, find the exact height of the three-story building.

9. A ball is dropped from a very high tower. One second later another ball is dropped.

(a) Does the distance between the balls increase, decrease, or remain the same as they fall? Explain your reasoning.

(b) Does the ratio of their velocities (the velocity of the first one dropped compared to the second) increase, decrease, or remain the same as they fall? Explain.

(c) What is the distance between the two balls after the first one has fallen for 3 s?

10. (a) Calculate the time required for a falling body to reach a velocity of 80 km/s.

(b) If the body in (a) reaches that velocity just before hitting the ground, from how high up was it dropped?

11. (a) Determine how long it would take an object to fall 65 m.

(b) Find the velocity of the object when it has fallen 65 m.

12. A 5-ft 6-in.-tall geologist was walking in a cave. She saw a drop of water fall past her face and splash on the ground. Struck by inspiration, she waited and timed the next drop; it took 0.1 s to fall past from the top of her head to the ground. How high up was the ceiling?

13. In a certain fountain two drops of water splash over the edge of a bowl and fall to a pool below. The drop from the left falls directly into the pool. The drop from the right falls 1 m, hits a projection and instantaneously loses all its speed, and then begins falling again. After it has fallen another meter it hits another projection, . . . , and so on in equal stages until it hits the pool. How much longer does it take the drop on the right to reach the water? Express your answer in terms of the distance fallen on the left, and the number of stages on the right.

14. One afternoon Chuang Tzu awoke after dreaming he was a butterfly. He wondered if he was really a butterfly dreaming of being a person. Let's suppose he dreamed this: Chuang Tzu was an exceptionally fast and agile butterfly. One day he looked up and saw an anvil poised 2 m above his head. Just as the anvil began to fall, Chuang Tzu flew up toward it at 20 m/s (a very fast butterfly). When he reached the anvil he instantly changed direction and flew toward the ground at the same speed. Upon reaching the ground he instantly changed direction and flew toward the anvil, . . . , and so forth, until he awoke just as the dream anvil crushed the dream butterfly. How far did he fly?

Constant Acceleration

15. An experiment performed on the moon finds that a feather falls 20 m in 5 s. Find the value of g on the moon.

16. A freight train, starting from rest, acquires a velocity of 12 mi/h while moving 3 mi under constant acceleration.

  (a) Find its acceleration in miles per hour squared.
  (b) Express this in feet per second squared.

17. Suppose that the distance traveled as a function of time for falling bodies had turned out to be $s(t) = kt$, where k is a constant.

  (a) What would be the velocity of a falling object at any instant?
  (b) What is the instantaneous acceleration of the object?

18. To find the depth of a well, a stone is dropped into it. It is heard to strike the bottom 3 s after release.

  (a) Compute the depth of the well, assuming sound travels infinitely fast.
  (b) Compute the depth, assuming the speed of sound to be 1000 ft/s.

19. An astronaut is carrying equipment up a narrow, 30-m-long rock chimney on Mars. Encumbered as she is, she can only crawl up or down the tunnel at the same steady speed. Suddenly she sees a stone 10 m above the chimney begin to fall toward her. She instantly realizes that she can just barely escape by crawling out *either* end of the chimney. How far up into it has she climbed?

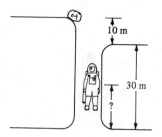

CHAPTER

THE LANGUAGE OF NATURE: DERIVATIVES AND INTEGRALS

It is most useful that the true origins of memorable inventions be known, especially of those which were conceived not by accident but by an effort of meditation. . . . One of the noblest inventions of our time has been a new kind of mathematical analysis, known as the differential calculus.

Gottfried Wilhelm Leibniz, *Historia et origo calculi differentialis* (1714)

3.1 THE DEVELOPMENT OF DIFFERENTIAL CALCULUS

After the advent of algebra in the sixteenth century, a flood of mathematical discoveries swept through Europe. The most important were *differential calculus* and *integral calculus*, bold new methods for attacking a host of problems that had challenged the world's best minds for more than 2000 years. Differential calculus deals with ideas such as *speed*, *rate of growth*, *tangent lines*, and *curvature*, whereas integral calculus treats topics such as *area*, *volume*, *arc length*, and *centroids*.

Work begun by Archimedes in the third century B.C. led ultimately to the birth of integral calculus in the seventeenth century A.D. This development has a long and fascinating history which we will explore in more detail later.

Differential calculus has a relatively short history. The concept of derivative was first formulated early in the seventeenth century when the French mathematician Pierre de Fermat tried to devise a way of finding the smallest and largest values of a given function. He imagined the graph of a function having, at each of its points, a direction given by a tangent line, as suggested by the points labeled in Fig. 3.1. He noted that the tangent line is horizontal at points like A and C where the function has a maximum or minimum. This observation made it seem worthwhile to have a general method for finding the slope of the tangent line at an arbitrary point. That method turned out to be differentiation. Fermat found the method in 1629, communicated it to Descartes in 1638, and published it in 1642.

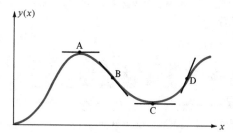

Figure 3.1 A curve with its tangent lines at various points.

3.2 DERIVATIVES AND SLOPES OF TANGENT LINES

To have a concrete example in mind, imagine that x in Fig. 3.1 represents the horizontal distance from your home, and $y(x)$ represents elevation above your home. The steepness of the curve is highly noticeable if you bicycle to school. We wish to express this steepness by a number called the slope of the curve.

The slope of a straight line joining two points (x_1, y_1) and (x_2, y_2) is

$$\text{slope} = \frac{y_2 - y_1}{x_2 - x_1}.$$

For the more general case of a curve, an average value of the slope between two points can be found by approximating the curve between the two points by a straight line as in Fig. 3.2. Let's call the vertical distance between the two points Δy, and the horizontal distance Δx. The slope of that line joining the two points (formally known as the chord) is given by

$$\text{average slope} = \frac{\Delta y}{\Delta x} = \frac{y_2 - y_1}{x_2 - x_1}. \tag{3.1}$$

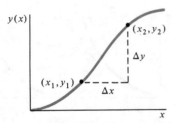

Figure 3.2 Graph of a function with points indicated for determining the slope of a chord.

But a bicyclist wants to know more than the average slope between, say, his home and school. Especially when he's going up a hill, the slope right where he is seems more important than the average; it's the local slope that determines how hard he has to pedal.

We encountered a similar problem in the study of falling bodies when we wanted to know not only the average velocity of the falling object but also the velocity at one instant. By starting with the average velocity and allowing the time interval to shrink to zero we found the instantaneous velocity, which was a derivative. We will use an analogous strategy to find the slope of the hill at each particular point.

To be specific, suppose the hill has the shape of a parabola, $y(x) = x^2$. For the two points we'll use to calculate the slope we choose $x_1 = x$ and $x_2 = x + h$. We are using h as an alternative symbol for Δx, as was done in Chapter 2. The corresponding values of y will be $y_1 = y(x)$ and $y_2 = y(x + h)$. The slope of the line between the points is

$$\text{average slope} = \frac{\Delta y}{\Delta x} = \frac{y(x + h) - y(x)}{(x + h) - x}. \tag{3.2}$$

Substituting $y(x) = x^2$ and $y(x + h) = (x + h)^2$ we find

$$\text{average slope} = \frac{\Delta y}{\Delta x} = \frac{(x + h)^2 - x^2}{(x + h) - x}$$

$$= \frac{2xh + h^2}{h}$$

$$= 2x + h. \tag{3.3}$$

Because h has canceled out of the denominator, we can now safely let h shrink to zero, obtaining the slope at any particular point along the parabola:

$$\text{slope} = 2x.$$

This result may look familiar, and indeed it is: we have just calculated the derivative $y'(x)$ of the function $y(x) = x^2$ with respect to x [just as we calculated the derivative of the distance function $s(t) = ct^2$ with respect to t in Chapter 2 and found $s'(t) = 2ct$]. But now we have a geometrical interpretation, shown in the sequence in Fig. 3.3. As h shrinks to zero, $x + h$ approaches x and the chord moves closer and closer to a line through (x, y) with slope $2x$. This line is called the *tangent line* at (x, y), and the number $2x$ is its slope.

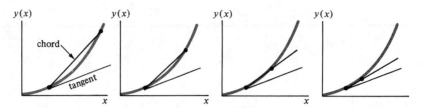

Figure 3.3 The tangent line is the limiting position of the chord as one end point approaches the other.

In other words, the line through (x, y) with slope $y'(x)$ is, by definition, the tangent line to the curve at (x, y). Geometrically, then, differentiation gives us the slope of the tangent line at each point of the curve. This is the connection which Fermat sought and which gives us a geometric view of differentiation. At the special points A and C in Fig. 3.1, where the function has a local maximum or minimum, the slope $y'(x) = 0$.

If the hill is just a straight line of slope m, then $y(x) = mx + b$ and the average slope between any two points is

$$\frac{\Delta y}{\Delta x} = \frac{m(x + h) + b - (mx + b)}{h} = m,$$

so the slope at each point of the line is also m: $y'(x) = m$.

Example 1

For the function shown below, can you roughly sketch a graph of its derivative?

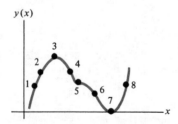

Using the idea that the derivative of a function at a point is the slope of the tangent line at that point, we can estimate the slopes of the tangents at points 1–8. At point 1, the slope is steep, which means it's large and positive, at 2 it is still positive but smaller, at 3 it is zero, at 4 it is negative, and so forth. Plotting these points and smoothly connecting them, we obtain the following graph:

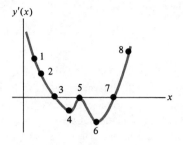

Example 2
Sketch a rough graph of the second derivative of the function in Example 1.

As we saw in Chapter 2, the second derivative is the derivative of the derivative. Therefore it is the slope of the tangent line of the *first* derivative at each point. Using this idea we estimate that $y''(x)$ has the form shown here.

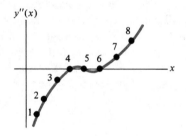

negative slope — negative #'s

This example illustrates three possible circumstances under which $y'(x) = 0$:

a *maximum* $[y''(x) < 0$ as at point 3],

a *minimum* $[y''(x) > 0$ as at 7],

a "*point of inflection*" $[y''(x) = 0$ as at 5].

3.3 LEIBNIZ'S NOTATION. ANALYTIC DEFINITION OF THE DERIVATIVE

Another way of writing the derivative, instead of $y'(x)$, is dy/dx. This is the notation that Leibniz invented, and here we see where it comes from. The derivative dy/dx denotes the limit* of $\Delta y/\Delta x$ when Δx shrinks to zero:

*In this book the limit concept is used in an intuitive way, which suffices for our purposes. A more systematic treatment of limits can be found in most calculus texts.

$$\frac{dy}{dx} = \lim_{\Delta x \to 0} \frac{\Delta y}{\Delta x} = \lim_{h \to 0} \frac{y(x + h) - y(x)}{h} .$$

(3.4)

Equation (3.4) is the analytic definition of the derivative, and can be used to calculate the derivative of any function you will encounter in physics. You can verify that it corresponds to the steps we have gone through in calculating each example thus far. The results for the parabola $y(x) = x^2$ and for the straight line $y(x) = mx + b$ are written in Leibniz notation as follows:

$$\frac{d}{dx}(x^2) = 2x, \quad \frac{d}{dx}(mx + b) = m.$$

Leibniz thought of the derivative dy/dx as a quotient of "infinitesimal" quantities dy and dx, which he regarded as unimaginably small, but still not quite exactly equal to zero. This concept provoked a great deal of philosophical controversy during the early decades of the development of calculus because neither Liebniz nor his followers could give a satisfactory definition of infinitesimals. Eventually the controversy was resolved with the introduction of the theory of limits, which treated dy/dx as a single symbol rather than as a ratio of infinitesimals. Nevertheless, Leibniz's idea of treating dy/dx as if it were a fraction, dy divided by dx, is still popular today because it often leads quickly to correct results that would be much harder to obtain without the use of infinitesimals.

Sometimes the functional notation $y'(x)$ is preferable to the Leibniz notation dy/dx because it specifies precisely where the derivative is wanted. For example, if $y'(x) = 2x$ then $y'(4) = 8$. In the Leibniz notation the latter equation would be written in the less convenient form

$$\left. \frac{dy}{dx} \right|_{x=4} = 8.$$

The vertical stroke next to the symbol dy/dx with $x = 4$ near the bottom specifies the point where the derivative is evaluated. We will not use this awkward symbol in this book, but it does occur in the literature, so the reader should be aware of its existence.

3.4 RULES OF DIFFERENTIATION AND DERIVATIVES OF SPECIAL FUNCTIONS

While we have given simple examples where the derivative is useful, the power of the newly developed calculus of Newton and Leibniz lies in its wide range of applicability to more complicated functions. Central to this power are a few simple rules of differentiation. Newton and Leibniz realized that complicated functions are often composed of simple functions which are added, subtracted, multiplied, or divided. Consequently, the rules of differentiation concern these ways in which functions are built.

Consider first the derivative of the sum of two functions, $y(x) + z(x)$. Physically $y(x)$ might represent the amount of water in one barrel as a function of the rainfall x, and $z(x)$ the amount of water in another barrel of different shape. Intuition tells us that the change in the total amount of water in the barrels with a change in rainfall, $[y(x) +$

1. sum of derivatives

$z(x)]'$, is simply the sum of the changes in each, $y'(x) + z'(x)$. To prove this, apply the definition of the derivative in Eq. (3.4) to the sum:

$$\frac{d}{dx}(y + z) = \lim_{h \to 0} \frac{[y(x + h) + z(x + h)] - [y(x) + z(x)]}{h}$$

$$= \lim_{h \to 0} \frac{y(x + h) - y(x)}{h} + \lim_{h \to 0} \frac{z(x + h) - z(x)}{h}$$

$$= \frac{dy}{dx} + \frac{dz}{dx} . \tag{3.5}$$

A second rule concerns the derivative of a product of two functions, $d(yz)/dx$. For example, if a rectangle has sides of length $y(x)$ and $z(x)$ that expand as temperature x rises, then $d(yz)/dx$ is the rate of change of the area with respect to temperature. We can work this out as Newton did, using geometric intuition.

Figure 3.4a shows a rectangle with sides y and z and area yz. If y grows by a small amount Δy and z by a small amount Δz, the area increases as shown in Fig. 3.4b.

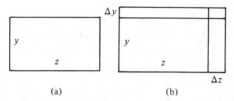

(a)   (b)

Figure 3.4 (a) Rectangle of sides y and z and area yz. (b) Increases in the sides and area.

The change in area comes from three extra pieces:

In other words, the change in the product yz is y times the change in z, plus z times the change in y, and a tiny piece in the corner of area $\Delta y\, \Delta z$ which is very small compared to the other two pieces.

This led Newton to announce the following product rule: *the rate of change of yz is y times the rate of change of z, plus z times the rate of change of y.* In symbols,

$$\frac{d}{dx}(yz) = y\frac{dz}{dx} + z\frac{dy}{dx} . \tag{3.6}$$

What happened to the little piece in the corner? Newton chose to ignore it. He knew that he was right, but offered no proof.

The use of limits shows why Eq. (3.6) is correct. Applying the definition of derivative we have

$$\frac{d}{dx}(yz) = \lim_{\Delta x \to 0} \frac{(y + \Delta y)(z + \Delta z) - yz}{(x + \Delta x) - x}$$

$$= \lim_{\Delta x \to 0} \frac{y\,\Delta z + z\,\Delta y + \Delta y\,\Delta z}{\Delta x}$$

$$= \lim_{\Delta x \to 0} \left(y\frac{\Delta z}{\Delta x} + z\frac{\Delta y}{\Delta x} + \frac{\Delta y\,\Delta z}{\Delta x} \right)$$

$$= y\frac{dz}{dx} + z\frac{dy}{dx} + \lim_{\Delta x \to 0} \frac{\Delta y\,\Delta z}{\Delta x}. \tag{3.7}$$

The last term corresponds to the little piece in the corner. We can write it as

$$\lim_{\Delta x \to 0} \Delta y \left(\frac{\Delta z}{\Delta x} \right) \quad \text{or as} \quad \lim_{\Delta x \to 0} \Delta z \left(\frac{\Delta y}{\Delta x} \right).$$

Since both $\Delta y \to 0$ and $\Delta z \to 0$ as $\Delta x \to 0$, the extra term is the product of a quantity that tends to 0 and a quantity that tends to a derivative. In other words, the last limit is 0, and we get Eq. (3.6). That is why Newton was correct to ignore the piece in the corner.

Both Newton and Liebniz applied the product rule to find the derivative of the power function $y(x) = x^n$, where n is an integer. We illustrate with $y(x) = x^3$, which we write as $x \cdot x^2$. We already know that $dx/dx = 1$ and $d(x^2)/dx = 2x$, so the product rule gives us

$$\frac{d}{dx}(x^3) = \frac{d}{dx}(x \cdot x^2) = x(2x) + (x^2)1 = 3x^2.$$

The same argument shows that $d(x^4)/dx = 4x^3$, $d(x^5)/dx = 5x^4$, and so forth. The pattern that emerges is the power rule:

$$\frac{d}{dx}(x^n) = nx^{n-1}. \tag{3.8}$$

This holds for any positive integer $n = 1, 2, 3, \ldots$, and it also holds for $n = 0$ since the derivative of a constant is zero. Newton was able to show that the power rule applies to negative exponents as well. In fact, Eq. (3.8) holds for all values of n – positive or negative, integer or not.

Another rule of differentiation concerns the derivative of a function which is itself a function of another function. Suppose for example that we call the distance your car goes y, and the amount of fuel it uses x. The distance you go depends on how much fuel you use; in other words, y is a function of x. And as you drive, you consume fuel, so x changes with time: x is a function of t. Now if you are told that your car gets 25 mi/gal, and is using fuel at a rate of 2 gal/h, you readily calculate that its speed is 50 mi/h:

$$25 \, \frac{\text{mi}}{\text{gal}} \times 2 \, \frac{\text{gal}}{\text{h}} = 50 \, \frac{\text{mi}}{\text{h}} \, . \tag{3.9}$$

In the language of functions and derivatives, 25 mi/gal $= dy/dx$, 2 gal/h $= dx/dt$, and 50 mi/h $= dy/dt$, so

$$\frac{dy}{dx} \frac{dx}{dt} = \frac{dy}{dt} \, . \qquad\qquad chain \; rule \tag{3.10}$$

Equation (3.10), which we have written on the basis of one especially simple case, is generally valid. This result is called the *chain rule*.

The reason the rule works is that dy/dx is the limit of $\Delta y/\Delta x$ as Δx gets small and dx/dt is the limit of $\Delta x/\Delta t$ as Δt gets small. If Δt and Δx are not zero we can form the product of the ordinary fractions:

$$\frac{\Delta y}{\Delta x} \frac{\Delta x}{\Delta t} = \frac{\Delta y}{\Delta t} \, . \tag{3.11}$$

As Δt gets small so does Δx since $\Delta x = (\Delta x/\Delta t)\Delta t$. Passing to the limit in Eq. (3.11) we find

$$\frac{dy}{dx} \frac{dx}{dt} = \frac{dy}{dt} \, . \tag{3.12}$$

In this equation you can arrive at the right answer by treating dy/dx as if it were an ordinary fraction and canceling the dx in dy/dx and dx/dt.

In addition to power functions, several other functions occur frequently in physics. Phenomena which are cyclic or repeat their motion can conveniently be described by functions which are periodic – the trigonometric functions – sine, cosine, and their combinations. Therefore we should know their derivatives. The results, which we state without proof, are as follows:

$$\frac{d}{d\theta} (\sin \theta) = \cos \theta \tag{3.13}$$

and

$$\frac{d}{d\theta} (\cos \theta) = - \sin \theta \tag{3.14}$$

where θ is measured in radians. Equation (3.13) is illustrated in Fig. 3.5. The slope at each point of the sine curve is equal to the height at the corresponding point on the cosine curve. Proofs can be found in any standard calculus text.

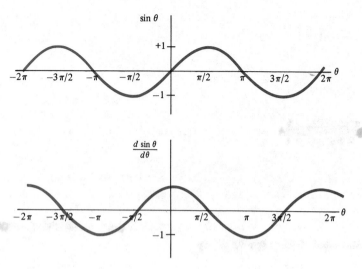

Figure 3.5 Geometric meaning of the derivative of sin θ.

In many applications θ is a function of another variable, for example, time t. The chain rule then gives us

$$\frac{d}{dt}\sin\theta(t) = \frac{d}{d\theta}\sin\theta(t)\frac{d\theta}{dt} = \cos\theta(t)\frac{d\theta}{dt}.$$

In particular, if $\theta(t) = bt$, where b is a constant, we have $d\theta/dt = b$, so

$$\frac{d}{dt}\sin(bt) = b\cos(bt).$$

We also state (without proof) the derivatives of two other commonly encountered functions:

$$\frac{d}{dx}(e^x) = e^x \tag{3.15}$$

and

$$\frac{d}{dx}(\ln x) = \frac{d}{dx}(\log_e x) = \frac{1}{x}, \quad x > 0. \tag{3.16}$$

These, together with the chain rule, give the more general formulas

$$\frac{d}{dx}(e^u) = e^u\frac{du}{dx}$$

and

$$\frac{d}{dx}(\ln u) = \frac{1}{u}\frac{du}{dx}, \quad u > 0,$$

where u is a function of x.

The exponential function e^x is especially noteworthy insofar as the principal reason for its wide application to diverse phenomena involving the growth of physical quantities – population, money in the bank, etc. – is the behavior of its derivative. Any quantity which grows at a rate proportional to itself will be described by an exponential. The symbol $\ln x$ or $\log_e x$ in (3.16) is the natural logarithm or logarithm to the base $e = 2.71828 \ldots .*$

By careful application of the rules of differentiation, and the derivatives of a few special functions such as x^n, $\sin \theta$, and e^x, most functions you will see in physics can be differentiated easily and efficiently: you cannot and need not memorize every derivative in the world – most can be calculated. After some practice the process of calculating derivatives will become second nature. We summarize below the rules we have learned and the derivatives of all the special functions that will be used in this book.

SUM RULE

$$\frac{d}{dx}(y + z) = \frac{dy}{dx} + \frac{dz}{dx} \qquad (3.5)$$

PRODUCT RULE

$$\frac{d}{dx}(yz) = y\frac{dz}{dx} + z\frac{dy}{dx} \qquad (3.6)$$

CHAIN RULE

$$\frac{dy}{dt} = \frac{dy}{dx}\frac{dx}{dt} \qquad (3.12)$$

Table 3.1 Derivatives of Special Functions Used in Physics (b Represents a Constant)

$y(x)$	$y'(x) = dy/dx$
x^n	nx^{n-1}
$\sin bx$	$b \cos bx$
$\cos bx$	$-b \sin bx$
e^{bx}	be^{bx}
$\log_e x = \ln x$	x^{-1} $(x > 0)$

*The reader should be cautioned that calculus books and mathematical tables often write $\log x$ for $\ln x$, whereas some engineering books reserve the symbol $\log x$ for logarithms to the base 10.

3.5   ANTIDIFFERENTIATION, THE REVERSE OF DIFFERENTIATION

In discussing falling bodies in Chapter 2 we started with a knowledge of the distance function $s(t)$ (how far a body falls in time t) then took its derivative to find the velocity $v(t) = ds/dt$ (how fast it is falling), then took the derivative of the velocity to find its acceleration $a(t) = dv/dt$ (how fast it was getting faster).

But we are not always given the position as a function of time. As we have stressed for the law of falling bodies, the most fundamental statement of the law is in terms of the acceleration:

$$a = g. \tag{3.17}$$

Suppose we start with this fundamental law and try to determine the velocity v and the position s at time t. Because $a = dv/dt$, the fundamental law states that

$$\frac{dv}{dt} = g$$

and we want to find v. Finding v is therefore simply a matter of reversing dv/dt, or, in other words, finding a function $v(t)$ whose derivative with respect to t is the constant g. This is typical of many problems in physics where we try to determine something from its rate of change. The process of finding a function or quantity whose derivative is known is called *antidifferentiation*.*

For the problem at hand we know one such function, namely, $v(t) = gt$. Another is $v(t) = 3 + gt$, and yet another is $v(t) = -5 + gt$. In fact, if we take $v(t) = C + gt$ for any constant C then $dv/dt = g$, so we see that there are many such functions, one for each value of C.

This is typical of the process of antidifferentiation. A function is not uniquely determined by its derivative because there can be many functions having the same derivative. But it is easy to see that any two of them can differ only by a constant. In fact, if two functions $g(t)$ and $f(t)$ have the same derivative, $g'(t) = f'(t)$, their difference $g(t) - f(t)$ has derivative 0, which means the difference doesn't change, so the difference $g(t) - f(t) = C$, where C is a constant. Hence $g(t) = f(t) + C$. In other words, if $f(t)$ is one antiderivative of $f'(t)$, then *all* antiderivatives are $f(t) + C$, where C is an arbitrary constant.

In our velocity problem, the constant C has a specific physical meaning. It represents the velocity when $t = 0$, that is, the initial velocity v_0. Thus, among all possible functions $v(t) = C + gt$ with derivative g, we choose that one for which $C = v_0$ and get

$$v = v_0 + gt.$$

Now we repeat the process. Knowing that $v = v_0 + gt$ and that

$$v = \frac{ds}{dt}$$

---

*Antidifferentiation is also called *integration*, a term that is also used for an entirely different concept, the process of calculating areas. The connection between the two processes is developed in Section 3.6.

we want to find $s(t)$, an antiderivative of v. Again we know (or can guess) one such antiderivative, namely $s = v_0t + \frac{1}{2}gt^2$, because its derivative is $v_0 + gt$. But we also know that all antiderivatives must be equal to this one plus some constant C', so

$$s = C' + v_0t + \tfrac{1}{2}gt^2.$$

In this case, the constant C' represents the initial position at time $t = 0$ which we call s_0. Thus we get the required position function

$$s = s_0 + v_0t + \tfrac{1}{2}gt^2.$$

Here the initial position $s_0 = s(0)$ is measured from whatever point we choose as the origin of our coordinates, and $v_0 = v(0)$ is the initial velocity imparted to the body at time $t = 0$. If the body starts at rest, then $v_0 = 0$.

Success in this method depends on being able to find antiderivatives. By transposing the columns in Table 3.1 we obtain (with minor changes in notation) the table of antiderivatives shown below. The derivative of each entry on the right is the corresponding entry on the left.

3.6 ANTIDIFFERENTIATION AND QUADRATURE

A famous problem from antiquity that challenged the world's best minds for nearly 2000 years was that of *quadrature*: given a region with curved boundaries, find a square having the same area. One of the most important events in the history of mathematics was the discovery by Newton and Leibniz that the ancient problem of quadrature could be solved with the help of antidifferentiation.

Table 3.2 Antiderivatives Used in Physics (b and C Denote Constants)

Function	Antiderivative	
bx^n	$b\dfrac{x^{n+1}}{n+1} + C$	$(n \neq -1)$
$\cos bx$	$\dfrac{1}{b}\sin bx + C$	$(b \neq 0)$
$\sin bx$	$-\dfrac{1}{b}\cos bx + C$	$(b \neq 0)$
e^{bx}	$\dfrac{e^{bx}}{b} + C$	$(b \neq 0)$
$\dfrac{1}{x}$	$\ln x + C$	$(x > 0)$

It is not at all obvious that quadrature and differentiation are related. Quadrature deals with area, while differentiation deals with rate of change. We can discover the relationship between the two by looking at some simple familiar examples.

Figure 3.6 shows the graph of the constant acceleration of a falling body: $a = g$.

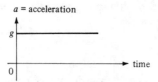

Figure 3.6 Graph of constant acceleration.

Let's calculate the area of the region between this graph and the time axis. The area depends, of course, on where we start and where we stop. Suppose we start at time 0 and stop at time t. Then the region is the shaded rectangle in Fig. 3.7, and its area is simply gt, the product of base times height. If we think of t not as a fixed number but as a variable, then the area gt is a function of t, which we can call the area function.

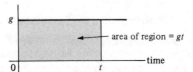

Figure 3.7 Area of the region under the graph from 0 to t.

This function has derivative g, so it is an antiderivative of the constant function $a(t) = g$ that we started with. But an antiderivative of acceleration is velocity, hence the area function gt represents the velocity of a body falling from rest:

velocity $= v(t) = gt$.

Now let's draw the graph of the velocity function, a line of slope g shown in Fig. 3.8a, and calculate the area of the region under its graph from time 0 to any time t. This region, shown shaded in Fig. 3.8b, is a triangle of base t and height gt, so *its* area (one-half base times height) is $\frac{1}{2}gt^2$, a new function of t.

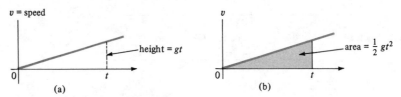

Figure 3.8 Velocity curve of a body falling from rest. (a) Graph of velocity $v(t) = gt$. (b) Area of region under its graph.

Again, we see that this area function is an antiderivative of the curve we started with. The area $\frac{1}{2}gt^2$ is equal to the distance fallen at any time t.

The relationship between area and antidifferentiation revealed in these two examples is not merely an accident. It is the idea underlying the stunning discovery made by Newton and Leibniz.

To explore this idea further, let's try to calculate the area of the parabolic segment shown in Fig. 3.9. The curve (part of a parabola) is the graph of the function $y = x^2$, and we want to find the area of the region between the curve and the x axis, from $x = 0$ to $x = t$. This region (shaded in Fig. 3.9) is called a parabolic segment of base t and altitude t^2.

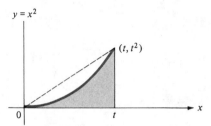

Figure 3.9 Parabolic segment of base t and altitude t^2.

From Fig. 3.9 it is clear that the area of the parabolic segment is less than that of a triangle with the same base and the same altitude. This triangle has area $\frac{1}{2}t(t^2) = \frac{1}{2}t^3$, so the area of the parabolic segment is smaller than $\frac{1}{2}t^3$. How much smaller is it?

By an ingenious geometric argument, Archimedes (287–212 B.C.) showed* that the area is exactly $\frac{1}{3}t^3$. He was the first to solve the quadrature problem for a parabolic segment. We will now solve the same problem by antidifferentiation.

Let $A(t)$ denote the area of the parabolic segment. This is the function we are trying to determine by antidifferentiation, so let's try to find its derivative dA/dt. Recall that the derivative dA/dt is the limit of the quotient

$$\frac{A(t + h) - A(t)}{h}$$

as h shrinks to 0. The numerator of this quotient is the difference of two areas, the area of a parabolic segment of base $t + h$ and the area of a parabolic segment of base t. This difference is shown in Fig. 3.10c.

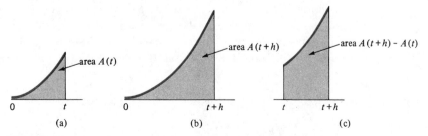

Figure 3.10 $A(t + h) - A(t)$ is the difference of the areas of two parabolic segments.

*See A. Rosenthal, "The History of Calculus," *American Mathematical Monthly*, vol. 58 (1951), pp. 75–86, for a description of Archimedes' argument.

In Fig. 3.11 we have sketched a rectangle with the same area as the region in Fig. 3.10c. It has base h and height x^2 for some (unknown) value of x between t and $t + h$.

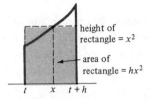

height of
rectangle $= x^2$

area of
rectangle $= hx^2$

$t \quad x \quad t + h$

Figure 3.11 A rectangle having the same area as $A(t + h) - A(t)$.

The area of the rectangle is hx^2, hence we have

$$A(t + h) - A(t) = hx^2, \tag{3.18}$$

or, dividing by h,

$$\frac{A(t + h) - A(t)}{h} = x^2$$

for some x between t and $t + h$. Now we let h shrink to 0 and see what happens to this equation. The left-hand side becomes dA/dt, and x^2 becomes t^2. In other words we have shown that

$$\frac{dA}{dt} = t^2,$$

the derivative of $A(t)$ is t^2. Therefore $A(t)$ must be an antiderivative of t^2. But we know all antiderivatives of t^2 have the form $\frac{1}{3}t^3 + C$ for some constant C, so

$$A(t) = \tfrac{1}{3}t^3 + C.$$

But when $t = 0$ the area is 0 and hence $C = 0$, so we find $A(t) = \frac{1}{3}t^3$. We have obtained Archimedes' quadrature formula using antidifferentiation!

The reason this discovery is important is not that Newton and Leibniz solved the quadrature problem for a parabolic segment. After all, Archimedes had already done it nearly 2000 years earlier. The importance of their discovery is that exactly the same method is applicable when the parabola is replaced by *any* continuous curve. The argument goes as follows: Take a function $f(x)$ whose graph lies above the x axis as shown in Fig. 3.12, and let $A(t)$ denote the area of the shaded region from $x = a$ to $x = t$. The difference $A(t + h) - A(t)$ is the area of the shaded region in Fig. 3.13a, and Fig. 3.13b shows a rectangle of height $f(x)$ with the same area. Therefore

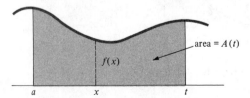

area $= A(t)$

$f(x)$

$a \qquad x \qquad t$

Figure 3.12 Area $A(t)$ of the region under the curve $y = f(x)$ from $x = a$ to $x = t$.

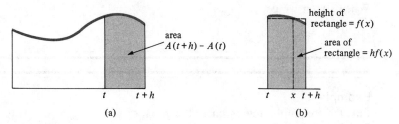

Figure 3.13 (a) A region with area $A(t + h) - A(t)$. (b) A rectangle with the same area.

$$A(t + h) - A(t) = hf(x).$$

This is the same result obtained in (3.18) except that x^2 is now replaced by $f(x)$. Dividing by h and letting $h \to 0$ we find

$$\frac{dA}{dt} = f(t). \qquad (3.19)$$

Consequently, if $P(t)$ is any antiderivative of $f(t)$ we have

$$A(t) = P(t) + C \qquad (3.20)$$

for some constant C. But since the area $A(t)$ is zero when $t = a$ we find $C = A(a) - P(a) = -P(a)$ and hence Eq. (3.20) becomes

$$A(t) = P(t) - P(a). \qquad (3.21)$$

This formula gives us a straightforward recipe for finding areas. If you want to find the area of the region under a curve $y = f(x)$ from $x = a$ to $x = t$, first find an anti-derivative of the given function, any function $P(x)$ whose derivative is $f(x)$. Then the required area is simply $P(t)$ minus $P(a)$.

Example 3

Calculate the area of the region under one arch of the curve $y = \sin x$ (shown shaded in the figure).

Let $A(t)$ denote the area from $x = 0$ to $x = t$. We want $A(\pi)$, but it requires no more effort to find $A(t)$ for all t between 0 and π. All we need is an antiderivative of $\sin x$. The function $P(x) = -\cos x$ is an antiderivative, because $P'(x) = \sin x$. Therefore, by formula (3.21), we have

$$A(t) = P(t) - P(0) = -\cos t + \cos 0 = 1 - \cos t.$$

When $t = \pi$ we get

$$A(\pi) = 1 - \cos \pi = 1 - (-1) = 2.$$

Example 4

Find the area of the region under the graph of $y = x^n$ from $x = a$ to $x = t$, where n is a positive integer and $0 \le a \le t$. (The graph is sometimes called a *generalized parabola*.)

Again, all we need is an antiderivative of x^n. One such antiderivative is $P(x) = x^{n+1}/(n + 1)$, so the area $A(t)$ from $x = a$ to $x = t$ is equal to

$$A(t) = P(t) - P(a) = \frac{t^{n+1} - a^{n+1}}{n + 1}.$$

This quadrature formula for the generalized parabola had been obtained prior to Newton and Leibniz by Cavalieri, Fermat, Pascal, and Roberval, but they did not use antiderivatives.

3.7   THE LEIBNIZ INTEGRAL NOTATION

Leibniz introduced a special notation for representing the area $A(t)$ of the region under the graph of a function $f(x)$ from $x = a$ to $x = t$ (the region shown in Fig. 3.12). He denoted this area symbolically as

$$A(t) = \int_a^t f(x)\, dx.$$

This is read "$A(t)$ is the integral from a to t of $f(x)\, dx$." The symbol $\int$ (an elongated S) is called an *integral sign*. You can see it today in copies of the original Declaration of Independence and the Constitution. Leibniz introduced it to mathematics in 1675.

The function $f(x)$ under the integral sign is called the *integrand*, and the interval from a to t is called the *interval of integration*, with the numbers a and t being the *limits of integration*.

Leibniz's symbol for the integral was readily accepted by many early mathematicians because they liked to think of the region under the graph as being composed of many thin rectangles of height $f(x)$ and base dx, as suggested in Fig. 3.14. The symbol $\int_a^t f(x)\, dx$ represented the process of summing the areas of all these thin rectangles.

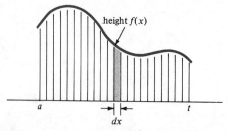

Figure 3.14 The region under the curve conceived as being filled with thin rectangles of height $f(x)$ and base dx.

All the results we obtained earlier concerning the area function $A(t)$ can now be expressed in Leibniz's integral notation. For example, the formula for quadrature of the parabolic segment becomes

$$\int_0^t x^2 \, dx = \tfrac{1}{3} t^3.$$

Equation (3.19), which tells us that the derivative of the area function $A(t)$ is the function $f(t)$ we started with, now becomes

$$\frac{d}{dt} \int_a^t f(x) \, dx = f(t). \qquad (3.22)$$

This charming result, which relates the derivative and the integral, is known as the *first fundamental theorem of calculus*. It says that the derivative of an integral $\int_a^t f(x) \, dx$ with respect to the upper limit t is equal to the integrand evaluated at t.

The first to notice this connection between derivatives and integrals was Isaac Barrow, a Cambridge professor of Greek, and later of mathematics and theology. He was also Newton's teacher, and later resigned his prestigious chair in favor of Newton. But Barrow never realized the importance of his discovery. Both Newton and Leibniz, however, appreciated the significance of the result and exploited it to develop their powerful technique for solving quadrature problems by antidifferentiation, as described above by Eq. (3.21). In integral notation, Eq. (3.21) can be stated as follows:

If $P(x)$ is any antiderivative of $f(x)$, then

$$\int_a^t f(x) \, dx = P(t) - P(a). \qquad (3.23)$$

This is known as the *second fundamental theorem of calculus*. It can also be written entirely in terms of P. Since $P'(x) = f(x)$ it states that

$$\int_a^t P'(x) \, dx = P(t) - P(a). \qquad (3.24)$$

In other words, if you integrate the derivative $P'(x)$ of some function $P(x)$ from $x = a$ to $x = t$, the value of the integral is $P(t) - P(a)$, the difference of the function values at the endpoints.

Sometimes the special symbol

$$P(x) \Big|_a^t$$

is used to designate the operation of evaluating $P(x)$ first for $x = t$ and then for $x = a$ and subtracting. With this symbol, the second fundamental theorem, as stated in Eq. (3.23), can be written as follows:

$$\int_a^t f(x)\ dx = P(x)\Big|_a^t = P(t) - P(a).$$

Note that the value of the integral, $P(t) - P(a)$, depends only on the endpoints a and t and not on the running variable which varies from a to t. Because the result doesn't depend on x, we call x a *dummy variable*.

In any integral $\int_a^t f(x)\ dx$ we can replace the dummy variable x by any other convenient symbol, for example, y, u, or τ (the Greek letter tau). Thus we write

$$\int_a^t f(x)\ dx = \int_a^t f(y)\ dy = \int_a^t f(u)\ du = \int_a^t f(\tau)\ d\tau.$$

In choosing a dummy variable it is best to avoid letters that are already used for other purposes. Thus it's not a good idea to write $\int_a^t f(t)\ dt$ since the t attached to the integral sign is supposed to represent an endpoint of the interval, whereas the dummy variable is supposed to run through all values in the interval. A good compromise is to write

$$\int_a^t f(t')\ dt'.$$

Leibniz adapted his integral notation to introduce a special symbol for antiderivatives. He wrote

$$\int f(x)\ dx$$

without any limits attached to the integral sign to denote any antiderivative of $f(x)$. In this notation, an equation like

$$\int f(x)\ dx = P(x) + C$$

simply means that $P'(x) = f(x)$. For example, since $d/dx(\sin x) = \cos x$ we can write

$$\int \cos x\ dx = \sin x + C.$$

The symbol C represents an arbitrary constant. Thus we could write

$$\int \cos x\ dx = \sin x + 5, \quad \text{or} \quad \int \cos x\ dx = \sin x - 7.$$

All of these are correct because the derivative of each right-hand side is $\cos x$.

Despite similarity in appearance, the symbol $\int f(x)\ dx$ is conceptually distinct from the symbol $\int_a^x f(t)\ dt$. The symbols originate from two different processes – the first from antidifferentiation, the second from quadrature. But they are related to each other because of the first and second fundamental theorems. Each represents a function whose derivative is $f(x)$. Therefore they differ only by a constant, so we can write

$$\int f(x)\ dx = \int_a^x f(t)\ dt + C \qquad\qquad (3.25)$$

for some constant C.

Because of long historical usage, the symbol $\int f(x)\,dx$ is often referred to as an "indefinite integral" rather than as an antiderivative. By contrast, $\int_a^x f(t)\,dt$ is called a "definite integral." This is justified, in part, by Eq. (3.25), which tells us that $\int f(x)\,dx$ is an integral from some unspecified point a to x, plus some unspecified constant C. Handbooks of mathematical tables often contain extensive lists of formulas labeled "tables of indefinite integrals" that, in reality, are "tables of antiderivatives." Our skill in calculating integrals depends on our ability to find antiderivatives, so such tables are very useful. Any systematic method for finding antiderivatives is called a "technique of integration." When one is asked to "integrate $\int f(x)\,dx$" what is really wanted is the most general antiderivative of $f(x)$.

Here is how our list of antiderivatives in Table 3.2 would appear in the Liebniz notation:

Table 3.3 Indefinite Integrals (Antiderivatives) Used in Physics

$\int bx^n\,dx = b\dfrac{x^{n+1}}{n+1} + C$	$(n \neq -1)$
$\int \cos bx\,dx = \dfrac{1}{b}\sin bx + C$	$(b \neq 0)$
$\int \sin bx\,dx = -\dfrac{1}{b}\cos bx + C$	$(b \neq 0)$
$\int e^{bx}\,dx = \dfrac{e^{bx}}{b} + C$	$(b \neq 0)$
$\int \dfrac{b}{x}\,dx = b\ln x + C$	$(x > 0)$

In defining the integral in terms of area, we implicitly assumed that the integrand $f(x)$ was nonnegative, so its graph never went below the x axis. If $f(x)$ takes both positive and negative values, as shown in Fig. 3.15, we define the integral to be the algebraic sum of the areas of the regions above the axis, minus the sum of the areas of the regions below the axis. Areas above the axis are added, whereas those under the axis are subtracted. Under this extended definition the second fundamental theorem is still valid.

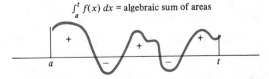

$\int_a^t f(x)\,dx$ = algebraic sum of areas

Figure 3.15 The integral is the sum of the areas of the regions above the axis minus the sum of those below.

One more remark about the definition of the integral. In writing $\int_a^t f(x)\,dx$ we have always assumed that the lower limit a is less than the upper limit t. If $a < t$, we also define

$$\int_t^a f(x)\,dx = -\int_a^t f(x)\,dx.$$

In other words, switching the limits around changes the sign of the integral. Finally, if $a = t$ we define $\int_a^a f(x)\,dx = 0$. This is consistent with the previous equation when $t = a$.

3.8 APPLICATIONS OF THE SECOND FUNDAMENTAL THEOREM TO PHYSICS

The second fundamental theorem of calculus states that

$$\int_a^t P'(x)\,dx = P(t) - P(a),\tag{3.24}$$

or

$$P(t) = \int_a^t P'(x)\,dx + P(a).\tag{3.26}$$

This formula tells us how to recover a function $P(t)$ from its derivative: integrate the derivative $P'(x)$ from $x = a$ to $x = t$ and then add $P(a)$. We now have a mathematical prescription for what we did earlier in this chapter when we found the velocity and position of a falling body from the acceleration. Let's cast those results into our new language of integrals.

We can recover the velocity $v(t)$ of a body at any instant from a knowledge of the acceleration function $a(\tau)$ by applying (3.26):

$$v(t) = v(0) + \int_0^t a(\tau)\,d\tau.\tag{3.27}$$

(We're using τ for the dummy variable because τ suggests time.) When a body is dropped from rest at $t = 0$ then $a(\tau) = g$ and $v(0) = 0$, so (3.27) gives us

$$v(t) = gt,$$

the same result we found earlier.

Similarly, the position function $s(t)$ can be recovered from the velocity by integration:

$$s(t) = s(0) + \int_0^t v(\tau)\,d\tau.\tag{3.28}$$

For example, if $v(\tau) = g\tau$, the velocity of a body falling from rest, and if the initial position $s(0) = 0$, we find the distance fallen in time t is

$$s(t) = \int_0^t g\tau\,d\tau = \frac{1}{2}gt^2,$$

as obtained earlier.

It is also interesting to know the vertical motion of a body that is thrown up or down with initial velocity $v(0)$ at $t = 0$. The first thing to decide is which direction to take as positive. We shall follow the usual custom of taking height y *upward* from a given level of reference as *positive*. The acceleration is then *negative*, equal to $-g$. With this convention the basic form of the law of falling bodies is

$$a(t) = \frac{dv}{dt} = -g.$$

Integrating, we find

$$v(t) = v(0) + \int_0^t a(\tau)\, d\tau = v_0 - gt, \tag{3.29}$$

where $v_0 = v(0)$.

The velocity is the derivative of the height so we can find y by integrating v. Writing y_0 for $y(0)$ we obtain

$$y(t) = y_0 + \int_0^t v(\tau)\, d\tau = y_0 + v_0 t - \tfrac{1}{2} g t^2. \tag{3.30}$$

This formula applies not only to falling bodies but to any problem with constant acceleration $-g$.

What velocity does a body acquire in falling through a given distance? Neither Eq. (3.29) nor (3.30) alone provides the answer. But solving for t in (3.29) we obtain

$$t = \frac{v_0 - v}{g}$$

and when this is substituted into (3.30) we find

$$y - y_0 = v_0 \frac{v_0 - v}{g} - \tfrac{1}{2} g \frac{(v_0 - v)^2}{g^2} = \frac{v_0^2 - v^2}{2g}$$

so the velocity acquired satisfies

$$v^2 = v_0^2 - 2g(y - y_0). \tag{3.31}$$

Example 5

A ball is hit straight up with an initial velocity of 96 ft/s.

(a) What is the velocity after 4 s?
(b) How high will it be after 4 s?
(c) How long does it take to reach its peak, and how high will it rise?
(d) How long will it stay up in the air?
(e) What is the velocity after 50 ft of rise?

Since we are taking y positive upward, v_0 is positive in the present problem and the acceleration is $-g = -32$ ft/s². The resulting acceleration, velocity and height of the ball are shown as functions of time in the graphs below.

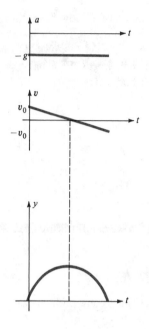

(a) To answer question (a), use Eq. (3.29):

$$v(4\ \text{s}) = 96\ \frac{\text{ft}}{\text{s}} - 32\ \frac{\text{ft}}{\text{s}^2} \times 4\ \text{s} = -32\ \frac{\text{ft}}{\text{s}}.$$

The negative result implies that the ball has reached its peak and turned downward.

(b) Equation (3.30) provides the answer to (b). We have $y_0 = 0$, and

$$y(4\ \text{s}) = 96\ \frac{\text{ft}}{\text{s}} \times 4\ \text{s} - \frac{1}{2} \times 32\ \frac{\text{ft}}{\text{s}^2} \times 16\ \text{s}^2 = 128\ \text{ft}.$$

(c) When the ball reaches its peak, the slope of the tangent to $y(t)$ vanishes, so $v = 0$ (see figure). Then from Eq. (3.29), $0 = v_0 - gt$ or

$$t = \frac{v_0}{g} = \frac{96\ \text{ft/s}}{32\ \text{ft/s}} = 3\ \text{s}.$$

The peak height is determined by Eq. (3.30) evaluated at this time:

$$y\left(\frac{v_0}{g}\right) = v_0\left(\frac{v_0}{g}\right) - \frac{1}{2}g\left(\frac{v_0}{g}\right)^2 = \frac{v_0^2}{2g} = \frac{(96\ \text{ft/s})^2}{2 \times 32\ \text{ft/s}^2} = 144\ \text{ft}.$$

(d) When the ball touches the ground, $y = y_0 = 0$. Equation (3.30) then reads

$$0 = v_0 t - \tfrac{1}{2}gt^2,$$

which has two solutions, $t = 0$ and $t = 2v_0/g$. The first solution is just the starting time when the ball left the ground. The second solution is the one we want,

$$t = \frac{2 \times 96 \text{ ft/s}}{32 \text{ ft/s}^2} = 6 \text{ s}$$

for the time when the ball returns. As we found earlier, the ball took 3 s to reach its peak, so the total flight time is twice the time spent rising, as we might have expected.

(e) Use Eq. (3.31) to find

$$v^2 = (96 \text{ ft/s})^2 - 2 \times (32 \text{ ft/s}^2) \times 50 \text{ ft},$$

$$v^2 = (9216 - 3200) \text{ ft}^2/\text{s}^2 = 6016 \text{ ft}^2/\text{s}^2,$$

$$v = 77.6 \text{ ft/s}.$$

3.9 A FINAL WORD

More than 2000 years ago Greek mathematicians developed a method to determine the areas of various geometric figures with curved boundaries. The method of exhaustion, as it is known, consisted of inscribing polygons with an increasing number of edges into a region whose area is to be determined. Archimedes (287–212 B.C.) successfully used this method to calculate the area of a circle and some other special figures.

By the seventeenth century, when algebraic symbols and manipulations had become standard techniques, the method of exhaustion was slowly transformed into what is known today as the process of *integration*, a systematic method for calculating areas and volumes. Many mathematicians in various countries had developed special techniques to treat special problems. In Germany, Kepler found formulas for the volumes of barrels. In Italy, Cavalieri formulated a comparison principle to determine when two solids cut by parallel planes have equal volumes. In France, Descartes, Fermat, and Pascal calculated areas of special regions, as did Wallis in England and Guldin in Switzerland. The air of seventeenth-century Europe was swarming with ideas of differential and integral calculus.

Yet Newton and Leibniz are considered the founders of calculus. Why? They provided an important missing ingredient. Newton and Leibniz recognized that *differentiation and integration are inverse processes*. A few mathematicians had come close to this knowledge, but it was Newton and Liebniz who independently realized the significance of this discovery. Moreover, they exploited this relation to create a systematic and satisfactory calculus which treated whole classes of problems by routine operations, to be executed strictly by rules, without using geometric arguments.

At the ripe age of 23, during the plague year of 1665–6, Isaac Newton undertook an incredible program of study executed in solitude at his family residence in Lincolnshire. The result of his concentration was the genesis of differential and integral calculus. But Newton wasn't much of an extrovert. Being a reserved, uncommunicative individual, he delayed publication of his work for 20 years. As a result, Leibniz's publications of his calculus preceded those of Newton. Each had his own notation, but the ideas and methods were the same. Both of these lofty scholars were conscious of the profound power of calculus to such an extent that they waged a bitter battle over the priority of discovery. Newton's followers charged Leibniz with plagiarism, suggesting that during an exchange of letters with Newton he had learned crucial ideas and later used them in his published work without giving credit to Newton.

In an attempt to settle the dispute, Leibniz appealed to the Royal Society of London, of which he was a member. This was an unfortunate step. Newton was president of the Society. Embittered over the controversy, Newton stage-managed the final report, which was backed by the investigating committee but unsigned. It ruled that Leibniz was essentially guilty as charged.

The controversy was based on nationalistic rivalry rather than scholarship. Today, historians realize that the important discoveries of Newton and Leibniz were made independently, and that both men deserve to be recognized as the founders of calculus.

Ironically, the temporary English victory had a deplorable effect on British mathematics for over a century. Blinded by patriotic loyalty to Newton, British mathematicians refused to adopt Leibniz's superior notation and cut themselves off from the spectacular advances in eighteenth-century mathematics and physics that came from continental mathematicians using Liebniz's symbols rather than Newton's.

Problems

Derivatives and Slopes

1. (a) Roughly sketch the graph of the derivative for each of the functions shown below using knowledge of the connection between the derivative and the tangent line.
   (b) Sketch the graph of the second derivative of the function in the first (upper left) figure.

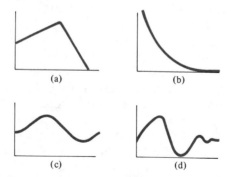

2. Which of the following functions has the largest slope at $x = 2$? At $x = -2$?

    (a) $y(x) = x$,   **(b)** $y(x) = x^2 + 1$,
    (c) $y(x) = 5$,   **(d)** $y(x) = 3x^2 - 2$.

Rules of Differentiation

3. Your car gets 20 mi/gal at 70 mi/h. If you are driving at 70 mi/h, how much fuel do you use in 20 min?

4. Suppose gas is pumped into a spherical balloon at a rate that makes the volume increase by 250 cm^3/s. How fast is the radius of the balloon changing when the radius is 10 cm?

5. Each edge of a cube is expanding at the rate of 1 cm/s. How fast is the volume changing when the length of the edge is (a) x cm, (b) 5 cm?

6. Suppose your income y is directly proportional to the number x of hours you work: $y = cx$, where c is a constant. In addition, suppose you're a big spender and the money z you spend varies with income as $z(y) = a + by^2$, where a and b are also constants. (a) How does the amount of money you spend change with a change in the number of hours worked? (b) For your income to be larger than your expenses, what conditions on y, a, and b must be met?

7. A cylindrical stream of water flows at a speed v_0 out of a pipe. The amount of water flowing through a cross section of the stream is the same at all distances from the pipe; that is, if five liters (5 L) flow out of the pipe in 1 s, then 5 L also flow past a point 1 m below the end of the pipe. Since the falling water accelerates downward the stream must get narrower the farther it gets from the end of the pipe.

   (a) Find an equation for the change dr/dy in the radius r of the stream with respect to distance y below the end of the pipe.

   (b) If the pipe is 10 cm across and the water flows out at 2.0 m/s, at which distance beneath it will the radius of the stream be 2.5 cm?

8. A droplet of perspiration falls off a steelworker who is 40 m above a vat of hot metal. Throughout the problem, assume the droplet is spherical.

   (a) Suppose that as it falls the droplet evaporates at a constant rate,

$$\frac{d(\text{volume})}{dt} = -c$$

   where c is a positive constant. Find an expression for dr/dh, the rate of change of the droplet's radius r with respect to height h above the vat.

(b) Suppose instead that the rate of evaporation is proportional to the surface area:

$$\frac{d(\text{volume})}{dt} = -b(\text{area})$$

where b is a positive constant. Again, find dr/dh.

(c) In case (a), is there a value of c, the rate of evaporation, such that the drop disappears just as it hits the vat? For case (b), is there a choice of b for which the droplet evaporates away just as it reaches the vat? If these constants exist, find them.

Derivatives of Special Functions

9. (a) Let $y(t) = Ce^{kt}$, where C and k are constants. Show that

$$\frac{dy}{dt} = ky.$$

In other words, the rate of change of y with respect to t is proportional to y, with constant of proportionality k.

(b) Now let $u(t)$ be any function of t with the same property:

$$\frac{du}{dt} = ku.$$

where k is a constant. Prove that $u(t)$ must have the form $u(t) = Ce^{kt}$ for some constant C. [*Hint.* Let $f(t) = u(t)e^{-kt}$ and calculate the derivative $f'(t)$.]

10. Let N be the number of human beings at time t. In the absence of limiting factors such as wars, plagues, and food shortages, the change in the population with time is $dN/dt = kN$. Draw on your knowledge of biology, sociology, psychology, and literature to estimate k. Then estimate how long it would have taken to get from Adam and Eve to the present world population in the absence of limiting factors.

11. We are told that in 1626 Dutch settlers bought the island of Manhattan from the Manhattan Indians for about $24 worth of beads and fabric. Since the island and the things on it are now worth about half a trillion dollars the Indians are usually considered to have made a bad deal. But suppose the Indians had invested their $24. At what annual interest rate would their money have had to grow to make its current value equal to that of Manhattan?

12. The radioactive isotope ^{11}C (carbon 11) decays in such a way that the number of atoms $N(t)$ at any instant is given by $N(t) = N_0e^{-kt}$ where N_0 is the initial number of atoms and $k = 0.035$ min^{-1}.

(a) What fraction of radioactive ^{11}C atoms remain after 14 min?

(b) In how many minutes does the number of ^{11}C atoms decrease to one-tenth the original amount?

(c) After how many minutes does the number of decays in a minute equal 1% of the original number of ^{11}C atoms?

13. Neutrons striking the atmosphere sometimes react with nitrogen to produce radio-active carbon, ^{14}C. Because living things use carbon from the environment, the ratio of ^{14}C in animals and plants is the same as it is in the world. When an animal dies it stops taking in new radiocarbon and the ^{14}C already in it begins to decay with a half-life of 5568 years. (The half-life is the time it takes for half the sample to decay.) Charcoal made from modern wood gives 15.3 decays per minute per gram (dpm/g). If we assume that the concentration of radiocarbon in the environment has been constant for many thousands of years we can estimate the age of ancient material by the number of radiocarbon decays it produces.

   (a) Charcoal from the linen wrapper of the copy of the Book of Isaiah found with the Dead Sea scrolls gave 12.0 dpm/g. When was the scroll made?
   (b) A bit of wood from one of the deck planks of the funerary ship of Sesostris III produced charcoal that gave 9.7 dpm/g. When did Sesostris rule? This radio-carbon date is in fair agreement with dates established from other records.
   (c) Charcoal from the cave at Altamira, Spain, gave 2.8 dpm/g. Use this to estimate the age of the amazing paintings on the roof of the cave.

Antidifferentiation

14. A body has a velocity described by $v(t) = ct^2$, where $c = 2$ m/s^3.

   (a) What is the acceleration of the body at any instant?
   (b) How far does the body travel during the period from $t = 0$ to $t = 10$ s?

15. A body has an acceleration $a(t) = a_0 \sin \omega t$, where a_0 and ω are constants. Find the velocity $v(t)$ of the object and the distance $s(t)$, assuming it started with $s(0) = 0$ and $v(0) = 0$.

16. As a racecar starts along a course, its acceleration is described by $a(t) = 0.7ge^{-ct}$, where $c = 0.07$/s and t is in seconds. (a) If the car started from rest, how fast is it traveling after 5 s? (b) Find the distance traveled after time t.

Applications of Integration

17. A car is waiting at a stoplight, and when the light turns green the car accelerates uniformly for 6 s at 2 m/s^2 and then moves with uniform velocity. At the instant the car started, a truck moving in the same direction with a constant velocity of 10 m/s passed it.

   (a) Sketch graphs for the motion of the car and the truck on the same axes.
   (b) When will the car catch up with the truck?
   (c) How far will the car have traveled before it catches the truck?

18. A stone is dropped from a diving board which is 5 m above the water surface. After hitting the water, assume it sinks to the bottom, 3 m below the surface, with the same velocity with which it struck the water.

(a) How long is the stone in the air (ignoring air resistance)?
(b) What is the velocity of the stone when it hits the water?
(c) How long does it take the stone to reach the bottom of the pool?

19. Suppose you are in a stopped elevator when suddenly both the cable and the brakes fail and the elevator starts falling.

(a) If, just as the elevator begins to fall, you drop your keys from one meter above the floor, how long will it be before they land on the floor?
(b) Now suppose that after the elevator has fallen three floors (10 m) the brakes catch and slow the elevator with a constant acceleration. If the elevator stops after one more second, how far had it fallen when the keys hit the floor?

20. Yeast in warm bread dough reproduces about as fast as it can. When the yeast is growing this quickly it's a good approximation to say

$$\frac{dN}{dt} = \alpha N$$

where α is a constant and N is the number of yeast cells.

(a) Assuming you know the initial number of yeast cells $N(0)$, find $N(t)$, the number of yeast cells at time t.
(b) In the course of using its food, yeast produces carbon dioxide as a waste gas; this is why yeast makes bread rise. We can approximate the rate of gas production in the dough by saying that each cell puts out gas at a steady rate:

$$\frac{d(\text{volume})}{dt} = \beta N(t)$$

where β is the rate of gas production per cell per unit time. When you make bread you mix the dough and set it aside to rise. After about 90 min it has doubled in bulk. You then punch and knead the dough so all the gas goes out of it and it returns to its original size (but now with more yeast cells than before). You then set it aside and wait for it to again double in bulk. If it also takes 90 min for the yeast population to double, how long should the second rise take?

CHAPTER

INERTIA

It has been observed that missiles, that is to say, projectiles follow some kind of curved path, but that it is a parabola no one has shown. I will show that it is, together with other things, neither few in number nor less worth knowing, and what I hold to be even more important, they open the door to a vast and crucial science of which these our researches will constitute the elements; other geniuses more acute than mine will penetrate its hidden recesses.

Galileo Galilei, *Two New Sciences*, Third Day (1638)

4.1 IF THE EARTH MOVES: ARISTOTELIAN OBJECTIONS

In 1543 Nicolaus Copernicus's book *De Revolutionibus Orbium Coelestium* (*On the Revolutions of the Heavenly Spheres*) appeared in print. Copernicus, mindful of his personal safety, had waited until his deathbed to publish his ideas. Within the pages of *De Revolutionibus* Copernicus set the earth spinning on its axis and revolving around the sun. In his attempt to return the heavens to their simple beauty, Copernicus had to make the sun, not the earth, the center of the universe, and in doing so he tore the heart out of the Aristotelian world. Without the solid, immovable earth at the center of the universe, there could be no Aristotelian laws of motion. And without these laws, there were none at all, for Copernicus had no laws to replace those he destroyed.

For nearly half a century Copernicus's system was largely ignored. Then, with the rise of Galileo's mechanics and Kepler's developments in astronomy, the Copernican system received serious consideration. The science of mechanics arose as an answer to the challenges of the post-Copernican Aristotelians and completely, irrevocably replaced the old physics of Aristotle.

Why had educated people before Galileo clung to Aristotle's conceptions? Aristotle's scheme had described how everyday objects move: rocks fall, smoke rises, and an apple cart stops when no longer pulled. All motion required a cause, and depending on the nature of its cause was either natural or violent. A heavy rock falling toward its natural resting place on earth was in natural motion because its cause was the internal nature of the rock, its heaviness. On the other hand, Aristotle knew that a horse was needed to pull a cart, and if the horse stopped, the cart also stopped. The motion of the cart he classified as violent motion, motion caused by some agent or force outside the object. If the force were removed, violent motion ceased.

But why should a rock fall toward the center of the earth if the earth itself is spinning dizzily through space? If the sun and not the earth is the center of the universe, Aristotle's scheme of natural places and natural motions crumbles. And without the solid stationary earth beneath our feet, what do we have left? Even granting that there is something called gravity that keeps us from flying off into space, how do we begin to understand the motion of other objects?

If the earth rotated from west to east, the post-Copernican Aristotelians argued, then things like clouds not attached to the earth would always seem to move from east to west, just like the sun. Even birds would suffer a similar fate because, with the earth rotating once every 24 h, the speed of the earth at the equator would be 1000 mph, and no bird could fly swiftly enough to keep up with it. Lacking a constraining force to push them along with the earth, birds should always appear to fly from east to west as the earth rapidly revolves underneath.

And the Aristotelians argued further. On a motionless earth, everyone knew that a stone dropped from a tower always landed at the foot of the tower. But if the earth were indeed rotating under a falling stone, then the stone should strike the ground west of the foot of the tower. They made their argument more persuasive by pointing out the size of the effect: If the tower were only 16 ft high, then in its 1-s fall, the earth would move 1500 ft!

A falling stone on a spinning earth should behave just like an object dropped on a moving ship. If dropped from the mast, the object lands not at the foot of the mast, but at some distance behind it, according to their description. The distance corresponds exactly to how far the ship moved during the fall. Everyone knew this, so they thought. The Aristotelians never bothered to test their predictions: they preferred words and logic to experiments. Figure 4.1 shows how they compared motions of falling bodies on a moving earth and on a moving ship.

(a)   (b)

Figure 4.1 Aristotelian description of objects in free fall on (a) a
rotating earth and (b) a moving ship.

Precisely because birds are seen flying in all directions and stones land directly below
the point where they are dropped, the seventeenth-century Aristotelians thought they had
a sound case against the spinning world of Copernicus. Familiar observations attested to
Aristotle's laws of motion and to the stationary earth.

4.2   THE EARTH MOVES: GALILEO'S LAW OF INERTIA

Galileo was fascinated by the Copernican world view and became a staunch defender of it. Besides, he liked a good fight. By his investigations he hoped to convince the world that the heliocentric system was not merely a fiction that simplified calculations of astronomers, but that it embodied physical truth about the universe. In his *Dialogue Concerning the Two Chief World Systems*, published in 1632, Galileo summarized his proofs for the Copernican system. He based his arguments in this book on experiments, reasoning, and penetrating insights.

One key experiment involved balls allowed to roll down and then up inclined planes. He chose balls because he wanted to minimize friction and focus purely on the motion. Figure 4.2 illustrates how Galileo set up his experiment. As he changed the plane on the right from position A to B, he noticed that the ball rolls further along the plane but always to the same height as it was released from.

From this observation Galileo imagined what would happen if the second plane were horizontal. In that case the ball could never reach its original height; it would keep on going forever with the speed that it had acquired in rolling down the inclined plane. Galileo concluded that any object in motion, if not obstructed, will continue to move with a constant speed along a horizontal line. That was Galileo's version of inertia. Inertia resists changes, not only from rest, but also from motion with a constant speed in the horizontal direction.

Although Galileo's arguments were brilliant, they were not quite right. To him a horizontal surface was one which was everywhere perpendicular to the direction pointing toward the center of the earth. His law of inertia then means that in the absence of any outside forces an object set into motion would continue moving in a circle around the earth, forever. That's where Galileo was wrong.

Objects set into motion tend to continue moving with a constant speed in a *straight line*, not in a circle. The law of inertia is

> *A body will remain at rest or continue to move with a constant speed in a straight line unless acted upon by an outside force.*

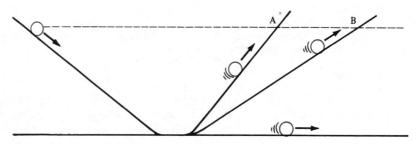

Figure 4.2 Illustration of Galileo's experiment with rolling balls which led him to the idea of inertia.

It is an outside force, the gravitational attraction of the earth, which makes satellites and the moon deviate from straight-line motion and circle the earth. In this instance even Galileo was unable to free himself completely from his earthbound perspective and imagine the natural motion in a place where there was no earth.

Nonetheless, Galileo was able to take one more step. He realized that a small portion of a very large circle such as the circumference of the earth is very nearly a straight line. Therefore objects on the earth will appear to move in straight lines. And thinking that this was only approximately true, Galileo perfectly described the law of inertia and worked out many of its consequences. He realized that all arguments against a rotating earth were due to a failure to understand the law of inertia.

To counter his critics, Galileo shrewdly used precisely the examples proposed by his opponents. Considering a stone dropped from the crow's nest of a ship he pointed out that if the ship is at rest, the stone lands at the foot of the mast. No one disagrees with that. But if the ship were moving with a constant velocity, Galileo claimed that the stone would still land at the foot of the mast. It makes no difference if the ship is at rest or moving with a constant speed in a straight line, the stone will land at the same place on the ship. But why?

Galileo realized that before the stone is released it is moving along as part of the ship. This means that it has the same velocity as the ship along the straight line. When it's released, the stone's inertia keeps it moving with the same velocity in the horizontal direction. So the stone and ship continue to move together. But in addition, gravity pulls the stone vertically downward. Galileo conjectured that this vertical motion doesn't interfere with the stone's horizontal motion. So as the stone falls, it continues to move horizontally with the same constant velocity as the ship, and the stone lands right at the foot of the mast as illustrated in Figure 4.3.

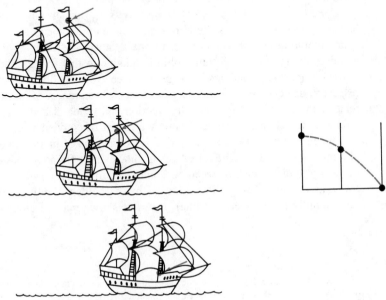

Figure 4.3 Path of a stone falling from a ship as seen from the shore.

Galileo was right on target with his prediction. A sailor on board would see the stone fall in the very same way in a moving ship as in one at rest. It's hard to say whether the Aristotelians would have been convinced of their errors by seeing this demonstrated, but in any case it wasn't until 1640 that a French philosopher, Pierre Gassendi, carried out the experiment and showed that Galileo was right.

Using the law of inertia, Galileo was able to explain the natural motion of falling objects on a rotating earth and answer the arguments of his critics. Dropping the stone from the mast of a moving boat is just like dropping a rock from a tower on the moving surface of the earth. Because the rock continues to share the horizontal motion of the earth as it falls, the stone lands at the base of the tower. An observer on Earth could not tell whether or not the earth were moving by looking at the stone.* Motion, Galileo realized, is relative.

But Galileo grasped even more. The natural state of motion is motion with a constant speed in a straight line, the simplest law one can imagine. In the absence of propulsion the tendency of an object was not, as Aristotle thought, to reach a state of rest, but to continue moving. For Galileo, if there were no friction and a horse pulling a cart suddenly stopped and indulgently stepped aside, the cart would continue to move; its inertia would keep it going.

4.3 RELATIVE MOTION

Galileo argued that by watching a stone fall, you cannot tell whether the earth is at rest or moving. You might wonder whether there is any way to tell.

In one sense, all motion is relative. Someone on the moon would think he saw the earth moving, but to you it is the man on the moon who's moving. To describe motion you need to specify a frame of reference relative to which the motion is measured. We habitually use nearby walls, buildings, trees, and other objects attached to the solid earth – and therefore at rest with respect to one another – as defining a frame of reference, and describe the motion of other objects relative to this frame. But on a boat, we tend to use our local surroundings in the boat – once again, a collection of objects at rest with respect to one another – as defining our frame of reference, even though the boat moves relative to the earth. Kinematically, motion can be described relative to any frame of reference; the choice of frame is simply a matter of convenience.

When one turns from mere kinematic description of motion to dynamics – the laws of motion – one still has great freedom in choosing the frame of reference, but it becomes necessary to distinguish between different categories of frames. Let us define an *inertial frame* as one in which Galileo's law of inertia holds (a body acted upon by no forces moves uniformly in a straight line). It can be shown that if frame S is inertial, then any

---

*Strictly speaking this would be true only if the earth moved exactly in a straight line and did not rotate on its axis. Later we shall find that the earth's rotation gives rise to a small deviation of the stone's trajectory called the Coriolis effect.

frame S' with respect to which S moves with constant speed in a straight line is also inertial. To take the simplest example, if a body has a velocity v in the x direction in frame S, and S moves in the x direction at velocity v_0 with respect to S', then the body has velocity

$$v' = v + v_0$$

in the x direction in frame S'. If v and v_0 are constant, then so is v', and the body will obey the law of inertia in frame S' as well as S, even though its precise location and velocity will be different in the two frames. Thus the two frames are equivalent for discussing the law of inertia. The equivalence still holds for objects and reference frames moving in different directions as long as their velocities are constant, as we shall show in Chapter 9. Any inertial frame is suitable to describe the motion; none is preferred over any other. In this sense there is no way to tell whether the earth or any other body is in a state of absolute rest.

However, not all frames of reference move in straight lines. The earth, for example, rotates daily and revolves around the sun. As a consequence, a frame fixed in the earth is *not* an inertial reference frame, and it turns out that motions of objects with respect to the earth's surface do give some evidence of the earth's rotation. We shall discuss the evidence later in Chapters 9 and 14: so-called centrifugal forces and Coriolis forces. So one *can* tell that the earth rotates after all.

But the moving surface of the earth, because it is so large, behaves *approximately* like a reference frame moving with constant speed in a straight line. Thus, confined to the earth's surface, and with the limited measurements available to them, the best minds in medieval Europe could not easily discover that the earth is spinning.

Although seemingly trivial, the choice of reference frame is of great significance in physics. In defending the idea of shifting the center of the universe from the earth to the sun, Galileo discovered the broader principle that all inertial reference frames are equivalent for describing motion. Not only is that idea important for understanding mechanics, it lies at the very heart of Einstein's theory of relativity.

4.4 PROJECTILE MOTION: A CONSEQUENCE OF INERTIA

When Galileo was a professor at the University of Padua, he managed to earn a few extra scudi by giving military advice to the Venetian government. His scientific work had potential relevance to military affairs because, using his ideas about inertia and falling bodies, he had for the first time figured out the correct trajectory a projectile, such as an arrow or a cannonball, would have in the absence of air resistance.

The motion of projectiles had always been a weak spot for Aristotle. He couldn't successfully fit this motion into his scheme of natural and constrained motions because there is no mover of an arrow once it leaves the bow.

In the fourteenth century the idea arose that a projectile was endowed with a sort of internal force called *impetus*. The bowstring gave the arrow a certain quantity of impetus. According to this theory, an arrow, once shot, should fly along until its impetus ran out,

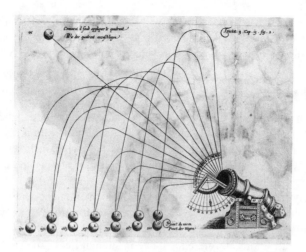

Figure 4.4 A 1621 illustration of the path of a cannonball according to the impetus theory. (Department of Special Collections, Stanford University Libraries.)

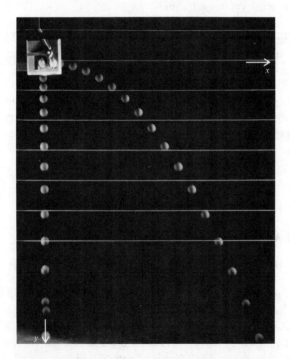

Figure 4.5 Stroboscope picture of two balls, one projected horizontally at the same instant the other is dropped. The distance between lines was 6 in. and the time between flashes was 1/30th s. (*PSSC Physics*, 2nd ed., 1965. D.C. Heath and Co. with Education Development Center, Inc., Newton, Mass.)

and then fall straight down to the earth. Figure 4.4 is an illustration of the theory from a book published in 1621.

Galileo correctly understood that impetus is not needed; there is no horizontal propelling force. Instead, inertia is the key to explaining projectile motion. He saw the motion of a cannonball as a combination of two motions, one horizontal and the other vertical. Moreover, he realized that these motions don't interfere with one another; each motion behaves as if the other is not present. This means that the cannonball falls vertically under the force of gravity while its inertia keeps it moving in the horizontal direction (where there is no force) with a constant velocity. This is the same idea he had applied to a stone falling on a moving ship.

Figure 4.5 is a photograph of the paths of two balls taken with a stroboscope, which illuminated the balls every 1/30th of a second. One ball is projected horizontally at the same instant the other is dropped. The picture reveals that the two balls are at identical vertical positions at the same instant; they are both falling vertically according to the law of falling bodies. In addition, the projected ball moves the same horizontal distance between each flash, so the horizontal velocity is constant. Galileo was correct!

4.5 CALCULATING A PARTICULAR TRAJECTORY

With a combined knowledge of the law of falling bodies and the principle of inertia, we can find the shape of a projectile's trajectory. Since different motions occur along the horizontal and vertical directions, these are the natural directions for coordinate axes. We'll choose the x axis along the horizontal and the y axis pointing vertically upward with the origin at the starting point of the trajectory as shown in Fig. 4.6.

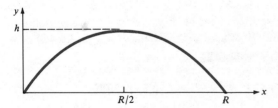

Figure 4.6 Flight of a projectile.

If we start the projectile off with initial velocity component v_{x0} in the x direction and v_{y0} in the y direction, the initial conditions at $t = 0$ can be written $x(0) = y(0) = 0$, $v_x(0) = v_{x0}$, and $v_y(0) = v_{y0}$. The principle of inertia governs motion in the x direction and states that $v_x(t)$ is constant. The horizontal change in position is therefore proportional to the time:

$$x(t) = v_{x0}t. \qquad (4.1)$$

Motion in the y direction is governed by the constant downward acceleration $d^2y/dt^2 = -g$. The change in height $y(t)$ of a vertically thrown projectile was worked out in Eq. (3.30):

$$y = v_{y0}t - \tfrac{1}{2}gt^2.$$ (3.30)

To find the trajectory we eliminate t from (3.30) and (4.1). Solving Eq. (4.1) for t, we find $t = x/v_{x0}$, and substituting into Eq. (3.30), we obtain

$$y = \frac{v_{y0}x}{v_{x0}} - \frac{gx^2}{2v_{x0}^2}.$$ (4.2)

By completing the square we can rewrite (4.2) in the form

$$y = \frac{-g}{2v_{x0}^2}\left(x - \frac{v_{x0}v_{y0}}{g}\right)^2 + \frac{v_{y0}^2}{2g},$$ (4.3)

which is the equation of a parabola, $y - h = a(x - b)^2$ with a, b, h constants. The trajectory of any projectile (ignoring air resistance) is a parabola called, even today, a Galilean parabola.

The height h at the peak of the trajectory is

$$h = \text{maximum } y = \frac{v_{y0}^2}{2g}.$$ (4.4)

At peak height the projectile has moved $x = v_{x0}v_{y0}/g$ horizontally. The *range R*, the distance x reached when the projectile returns to the ground at $y = 0$, is

$$R = \frac{2v_{x0}v_{y0}}{g}.$$ (4.5)

[This is most easily obtained by solving Eq. (4.2) for x when $y = 0$; the other solution, $x = 0$, refers to the starting point of the trajectory.] Note that the range is twice the horizontal distance traveled at peak height; the projectile travels equal distances horizontally during ascent and descent.

The time of flight can be obtained from Eq. (3.30). The y coordinate vanishes at two points, the starting time and the time when the projectile returns to the ground. Hence Eq. (3.30) has two solutions: $t = 0$ and

$$t = \frac{2v_{y0}}{g}.$$ (4.6)

The latter is the time of flight.

Example 1
A ski jumper leaves the end of the slide with a purely horizontal velocity of $v_0 = 20$ m/s. At this point he is 4 m above the ground, which falls off below him at a slope of 45°.

(a)   How far does he jump horizontally?
(b)   How long is he in the air?
(c)   What is his vertical velocity just before landing?

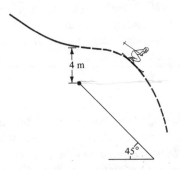

Let $x = 0$, $y = y_0$ ($= 4$ m) at the beginning of the jump. Since the jumper has no vertical velocity initially, his position as a function of time is, by Eqs. (4.1) and (3.30),

$$x = v_0 t, \qquad y = y_0 - \tfrac{1}{2} g t^2.$$

Eliminating t, we find for this trajectory

$$y = y_0 - \frac{g x^2}{2 v_0^2}.$$

This trajectory intersects the ground when $y = -x$:

$$-x = y_0 - \frac{g x^2}{2 v_0^2}.$$

The two solutions to this quadratic equation are, by the quadratic formula,

$$x_{\pm} = \frac{v_0^2}{g} \left(1 \pm \sqrt{1 + \frac{2 g y_0}{v_0^2}} \right).$$

Only the positive root is of physical interest. (The negative root gives the other intersection of the parabola and the 45° slope if they are continued *uphill* to the left of the jumpoff point.) Numerically the horizontal distance jumped is

$$x_{+} = \frac{(20 \text{ m/s})^2}{9.8 \text{ m/s}^2} \left(1 + \sqrt{1 + \frac{2(9.8)(4)}{(20)^2}} \right) = 85.4 \text{ m}.$$

The jump takes

$$t = \frac{x_{+}}{v_0} = \frac{85.4 \text{ m}}{20 \text{ m/s}} = 4.27 \text{ s}.$$

The vertical velocity just before landing is

$$v_y = -gt = -9.8 \frac{m}{s^2}(4.27\ s) \approx -42 \frac{m}{s},$$

the minus sign indicating that y is decreasing.

In practice the distance quoted for a ski jump is not x_+ but $\sqrt{x_+^2 + (y - y_0)^2}$, the length of the straight line from takeoff to landing; you are invited to work this out for our case. Also note that real ski jumps are appreciably modified by air resistance.

4.6   A FINAL WORD

Neither Galileo nor artillery gunners would claim that real projectiles move in precisely parabolic trajectories, or travel equal horizontal distances in ascent and descent. For example, it has been estimated that air resistance reduces the distance a well-hit baseball travels by as much as 40%.* Artillery gunners must take such effects into account. In Galileo's time they did so simply by watching where the shells landed and adjusting their aim accordingly. In modern times they commonly do so with the aid of computers.

When air resistance is included, the trajectory of a projectile may resemble the impetus theory prediction (Fig. 4.4) more closely than Galileo's parabola. Why then should we prefer Galileo's theory for everyday applications in the earth's atmosphere?

One reason is that even without constructing a vacuum, one can find projectiles of high density, low cross-sectional area, and relatively low speed that follow Galileo's prediction very well. A shot-put follows a much more nearly parabolic orbit than a beach ball.

A second reason concerns the distinction between technology and science. For seventeenth-century technology, unable as it was to compute the effects of air resistance, Galileo's prediction offered little immediate advance in the practice of gunnery. But for science, Galileo's ideas ushered in modern thought and opened the way to Newton's great discoveries.

Problems

Motion of the Earth

1. Using the fact that the circumference of the earth is approximately 25,000 mi (40,000 km), substantiate the statement in the text that the earth's surface moves 1500 ft in 1 s (460 m/s) at the equator. For comparison calculate the speed of the earth around the sun [the distance from Earth to sun is about 93,000,000 mi (150,000,000 km)].

Inertia

2. When a car suddenly brakes to a screeching stop, you lurch forward. Why? If your car is stopped at a stop sign and another car strikes you from the rear, your head snaps back relative to your car. Why?

*P. Brancazio, *Scientific American*, Vol. 248 (April 1983), p. 76.

3. Discuss how the law of inertia could explain why a sharp jerk to a dusty coat succeeds in removing the dust.

4. Imagine you are Galileo. How would you answer an Aristotelian who said that a cannonball fired horizontally westward would travel further than one fired horizontally eastward on the rotating earth?

5. A train moves along a horizontal track with a constant speed of 20 m/s. A passenger drops a ball from the ceiling, which is 2.5 m above the floor of the train.

   (a) According to the passenger, how long does it take the ball to reach the floor?
   (b) According to the passenger, where does the ball hit the floor?
   (c) According to someone standing by the tracks, how far does the train move while the ball is in the air?
   (d) What horizontal distance does a person by the tracks see the ball move while it falls?

Relative Motion

6. Suppose you are a passenger on a very smoothly riding car that is driving along a straight, level road at a constant speed. If you close your eyes and plug your ears, can you tell that you are moving? If the car rounds a curve, can you tell with your eyes closed? Why?

7. Imagine a spaceship in deep space cruising along with a constant velocity. A mechanical arm reaches into a garbage bag and pulls out a load of garbage, then releases it outside the spaceship. Describe the motion of the garbage as seen by a passenger on the ship.

8. Suppose you are speeding along at 120 km/h on an interstate highway and pass a Ferrari parked at a roadside restaurant. Twenty minutes later the Ferrari leaves the restaurant and follows you at 150 km/h.

   (a) How fast is the Ferrari moving relative to you?
   (b) How fast are you moving relative to the Ferrari?
   (c) How long does it take the Ferrari to catch you?

9. While a train is stopped at a station, Mike, at the front of a car, and Ann, at the rear of the car, roll a ball back and forth. Each can roll the ball with a speed of 10 m/s.

   (a) What is the speed of the ball relative to Mike? to Ann?
   (b) If the train is moving forward at 10 m/s, what then is the speed of the ball relative to Mike? to Ann?
   (c) While the train is moving as in (b), what is the speed of the ball according to someone on the ground as it goes from Mike to Ann? from Ann to Mike?

10. A ball of speed v strikes a massive, stationary wall at normal incidence, and rebounds with a final speed v in the opposite direction. This is an example of an elastic collision. In a second situation, the only change is that now the wall is also moving with speed u toward the ball. If the ball's initial speed is again v directed normally to the wall, and it rebounds elastically from the moving wall, what is the ball's final speed?

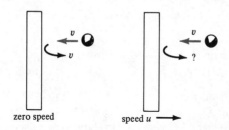

Projectile Motion

(In the following problems ignore air resistance.)

11. To strike a weapons factory, a bomber should release its bombs

    (a) when it is directly over the target,
    (b) before it is over the target,
    (c) after passing over the target.

If the bomber continues to move with constant velocity, what is the path the pilot sees for a falling bomb? What kind of path does a witness on the ground see?

12. A plane dropping bales of hay to snow-stranded cattle is flying horizontally at an altitude of 200 m with a speed of 65 m/s.

    (a) How long does it take the bale of hay to reach the ground?
    (b) How fast is the bale falling just before it hits the ground?
    (c) What is the velocity component of the bale in the horizontal direction just before it strikes the ground?
    (d) How far does the bale move horizontally while it is in the air?

13. Find the time when a projectile reaches the highest point on its trajectory. What is the relation of this time to the total time of flight given by Eq. (4.6)?

14. A rock is swung on a string in a circle. When the string is 45° from the vertical, it is cut. What trajectory does the rock follow? If the rock is initially moving at 5

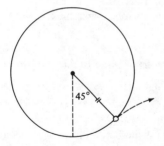

m/s, how far does it fly horizontally before falling back to the same height it had when the string was cut?

15. An airplane in horizontal flight with given speed v and altitude h prepares to bomb a target T a distance d ahead. The bombardier sights the target through the bombsight at an angle $\theta = \tan^{-1}(h/d)$ below the horizontal. At what value of θ (neglecting air resistance) should the bomb be released to hit the target? (*Note.* An automatic bombsight sets the correct θ on the basis of h and v supplied to it via connections with the altimeter and tachometer, with further adjustments for air resistance, wind, vertical component of flight, etc.)

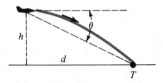

16. A divebomber drops its bomb from an altitude of 500 m while diving toward the ground at a speed of 600 km/h and an angle of 45°.

(a) How long does it take the bomb to reach the ground?

(b) How far does the bomb travel horizontally between the moment of release and the moment of striking the ground?

17. (a) A Mongol inscription from the time of Genghis Khan records a championship archery shot of 500 m horizontally. [O. Lattimore, *Nomads and Commissars* (Oxford University, London, 1962), p. 22.] If the arrow was shot at the optimal angle with air resistance neglected, what was its initial speed?

(b) A favorite Mongol tactic was to shoot toward the enemy while riding toward them, thus gaining the extra velocity of the horse, then wheel and ride back. If the shot has initial speed V_0 and initial angle θ with respect to the horse and if the horse moves with horizontal velocity V_H relative to the ground, find the ratio of the distance shot from a moving horse to the distance shot with the same θ and V_0 from a stationary position. (Neglect the height of the horse.)

18. A world-class sprinter can run 100 m in about 10 s (men) or 11 s (women). What ratio would you expect between the men's and women's records in the long jump? Compare this to the August 1983 world records, 29 ft 2.5 in. for men and 24 ft 4.5 in. for women.

19. A juggler is in a room with a ceiling 3 m above her hands.

    (a) What initial vertical velocity keeps a ball in the air longest?

    (b) If she tosses a ball up every 0.7 s, with each ball spending 0.5 s in her hands between the time it is caught and the time it is tossed, how many balls can she juggle? How many could she juggle in a similar room on the moon?

    (c) While she juggles, her hands are about 35 cm apart. What is the ratio of the horizontal velocity of balls thrown on the moon (always assuming the same height) as compared to those thrown on Earth?

20. One of the limits on how many objects a person can juggle is the need for accurate throws.

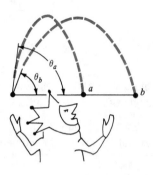

    (a) Assume that all throws reach the same height h. Consider a ball thrown from one hand at $x = 0$. To be caught it must come down somewhere near the other hand, in the range $a < x < b$. Find an expression relating the initial angle of the throw, θ, to the position where the ball is caught. You should get an expression in terms of h and the lateral position.

    (b) In practice, a and b are of order 0.2 m and 0.4 m, respectively. For a world-class juggler who tosses balls very rapidly, the required heights (in meters) for juggling 3, 7, and 13 balls are as follows:

	h	a	b	θ_a	θ_b	$\theta_a - \theta_b$
3 balls	0.35	0.2	0.4			
7 balls	3.25	0.2	0.4			
13 objects	13	0.2	0.4			

    Fill in the rest of the table. (The world record is 13 objects. Note how severe the requirement for angular accuracy has become in this case.)

21. A kind of firework called a spider looks like a large ball of glowing streamers. The streamers are trails left by phosphorus burning on bits of an exploding shell. The shell is launched from a mortar and typically rises to 250 m. If the shell were at rest when it exploded, the phosphorus would all have burned off in about 1 s, leaving a fireball 45 m in radius. For a shell that is launched at 5° from the vertical, draw trajectories for several bits of the shell in each of the following cases:

(a) The shell explodes at the peak of its arc.

(b) The shell explodes on the way up, at 200 m.

(c) Show that, for an observer moving with the shell, the fireball is a sphere.

CHAPTER 5

VECTORS

If I wished to attract the student of any of these sciences to an algebra for vectors, I should tell him that the fundamental notions of this algebra were exactly those with which he was daily conversant. . . . In fact, I should tell him that the notions which we use in vector analysis are those which he who reads between the lines will meet on every page of the great masters of analysis, or of those who have probed the deepest secrets of nature. . . .

J. W. Gibbs (in *Nature*, 16 March 1893)

5.1 COORDINATE SYSTEMS

Galileo discovered through the law of inertia that there is no single preferred reference frame. To use this discovery most efficiently, we need to discuss some geometrical ideas. The first kind of geometrical construction we need is a way of describing where things are.

If the world were only one-dimensional, everything would be on a single line. To describe where something is on that line we would first pick a point on the line as a point of reference, the origin. Then we would pick a direction along the line as the positive direction – let's say to the right of the starting point. And having made those choices we would only need to give one number – call it the x coordinate – to specify the location of a point.

On a two-dimensional surface, like this page, we need a slightly more complicated procedure to describe the location of a point. We can pick one axis, and then a second axis perpendicular to it as in Fig. 5.1 (it is not essential that the axes be perpendicular, but it is convenient). The intersection of the axes is the origin. We have to pick a positive direction by convention, so we say everything to the right of the origin is positive (call this axis x), and everything above the origin is positive (call this axis y). The position of a point can be described by rectangular coordinates x_0 and y_0 as in Fig. 5.1a. Or equivalently it can be described by polar coordinates r and θ, the distance from the origin and the angle the radial line makes with the x axis, as in Fig. 5.1b. In any case, when any system of coordinates is chosen it takes two numbers to specify where a point is in two-dimensional space.

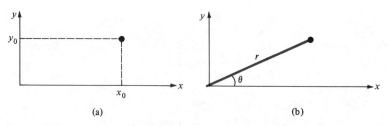

Figure 5.1 Two-dimensional coordinate system with point described by (a) rectangular coordinates x_0 and y_0, (b) polar coordinates r and θ.

In the real three-dimensional world it takes three numbers to specify the location of a point. The coordinate system in most common use is a rectangular grid called a Cartesian coordinate system in honor of the French mathematician and philosopher René Descartes. A Cartesian coordinate system in three dimensions starts with three mutually perpendicular axes labeled x, y, and z, like the ones shown in Fig. 5.2. Such a coordinate system is called right-handed because of the way the axes are arranged: if you point your right hand in the direction of the positive x axis and curl your fingers toward the positive y axis, your thumb will point in the direction of the positive z axis.

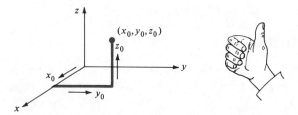

Figure 5.2 Right-handed coordinate system in three dimensions with a point described by rectangular coordinates (x_0, y_0, z_0).

To specify a point relative to this coordinate system, we specify the distance x_0 we would move out from the origin along the x axis to come abreast of it, then the distance y_0 moved parallel to the y axis to come directly under or above it, and then the distance z_0 moved parallel to the z axis to reach it (Fig. 5.2).

5.2 VECTORS

A vector is a mathematical object which has both *magnitude* and *direction* in space. Geometrically a vector is represented by an arrow whose length corresponds to the magnitude and whose direction, from the tail to the head of the arrow, corresponds to the direction of the vector.

Although a vector has a magnitude and direction, it has no fixed position in space. You can take a vector, like the one shown in Fig. 5.3, and slide it parallel to itself anywhere, and you still have the same vector. Two vectors having the same magnitude and direction when placed one on top of the other are considered equal.

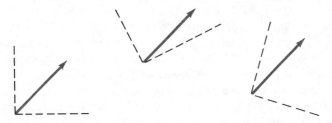

Figure 5.3 The same vector shown at different positions in space and in different coordinate systems.

Precisely because vectors don't have a definite position, they exist independent of any coordinate system. For this reason they are useful in mechanics. Why? Before Copernicus, there was only one conceivable reference frame, centered at the center of the earth. The essence of the Copernican revolution was to move the origin of the coordinate system describing all physics from the center of the earth to the center of the sun. Then, in devising a new mechanics, Galileo discovered through the law of inertia that there is no preferred reference frame. There is nothing unique about coordinate systems; no inertial frame is better than any other. Consequently coordinate systems (reference frames) can be anywhere in space, oriented in any way. Likewise vectors can be anywhere in space. Vectors are natural devices to describe physical quantities with complete generality and independence from particular reference frames.

Historically the use of vector analysis did not become standard until late in the nineteenth century when Josiah Willard Gibbs, a Yale physicist, and Oliver Heaviside, a self-taught British scientist, showed that it simplifies the equations and notation describing electricity and magnetism. But once accepted for electromagnetism, the elegant vector analysis also proved to be the method of choice for mechanics and other branches of physics.

Because a vector has a magnitude and a direction, it is more than a single number, and as such should not be represented by the same kind of symbol that we use to represent a single number. Vectors require a special kind of symbol. Physicists often write a vector as a letter with an arrow over it: $\vec{A}$. Others like to use the printer's symbol for boldface, a letter with a squiggle under it: $\underset{\sim}{A}$. You can use whatever you wish, but we'll use Gibbs's notation: boldface letters like **A** will denote vectors.

The magnitude or length of a vector **A** is denoted by $|A|$ or simply by A.

Magnitude of vector $\mathbf{A} = |\mathbf{A}|$ or A.

To emphasize the difference between vectors and ordinary numbers, one uses the word "scalar" for a quantity that has magnitude but not spatial direction. Scalars can be ordinary numbers like 2, -5, 144, and π, or physical quantities such as mass and time. Like vectors, they are independent of the coordinate system. The magnitude of a vector is also a scalar (but is never negative). Distance and speed are examples of magnitudes of vectors.

In physics, different names are often used to distinguish vector quantities from their scalar magnitudes. For example, the velocity **v** of a moving object is a vector represented by an arrow pointing in the direction of the motion. The length of the velocity vector, $v = |\mathbf{v}|$, is called the *speed*; it tells us how fast the object is moving. Another example is displacement versus distance. If a particle moves a distance s from one point to another along a straight line, the vector **s** joining the two points is called the displacement; its magnitude $s = |\mathbf{s}|$ is a scalar. Table 5.1 lists common scalars and vectors in mechanics.

Although vectors are defined independently of coordinate systems, it is often convenient to describe a vector in a particular coordinate system. One way this can be done is by sliding the tail of the vector to the origin. The head of the vector is now at some point whose coordinates provide instructions telling us how to get from the tail to the head of the vector. For example, in Cartesian coordinates the head of a vector **A** will have three rectangular coordinates which we call the *components* of the vector and which we denote by (A_x, A_y, A_z) as illustrated in Fig. 5.4.

Table 5.1 Common Scalar and Vector Quantities in Mechanics

Scalars	Vectors
distance s	displacement **s**
speed v	velocity **v**
acceleration a	acceleration **a**
force F	force **F**
time t	
mass m	

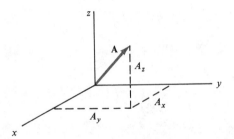

Figure 5.4 Cartesian components (A_x, A_y, A_z) of the vector **A**.

Since there are two equivalent ways to specify a vector – either by a magnitude and a direction, or by components – there should exist some way of finding components when the magnitude and direction are known, and vice versa. This process of finding components is known as resolving a vector into its components. Many of the applications of vector analysis will involve the resolution of vectors in a plane into their components; therefore, let's examine just how to resolve vectors in a plane.

If we know the magnitude and direction of a vector **A** which lies in the *xy* plane, we can easily find its components. Applying trigonometry to Fig. 5.5, we relate the components A_x and A_y to the magnitude A and angle θ as follows:

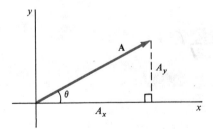

Figure 5.5 Resolution of a vector into its components in a plane.

$$A_y = A \sin \theta \qquad (5.1)$$

and

$$A_x = A \cos \theta. \qquad (5.2)$$

In some cases you might know the components of a vector but want its magnitude and direction. The change from components to magnitude and direction is made by reversing the process just described. Suppose you know the components A_x and A_y of a vector as shown in Fig. 5.5. The vector **A** and the edges labeled A_x and A_y form a right triangle. By the Pythagorean theorem, the hypotenuse, which is the magnitude A, is

$$A = \sqrt{A_x^2 + A_y^2}. \qquad (5.3)$$

The angle θ as defined in Fig. 5.5 can be calculated in many ways; the easiest is to use the definition of tangent:

$$\tan \theta = \frac{A_y}{A_x},$$

so

$$\theta = \tan^{-1}\left(\frac{A_y}{A_x}\right). \tag{5.4}$$

Most calculators have an inverse tangent key, $\tan^{-1}$, so when you know the components you can easily calculate the angle.

Example 1

For a given initial speed $v_0 = \sqrt{v_{x0}^2 + v_{y0}^2}$, find the maximum horizontal distance a projectile can move.

In Chapter 4 we derived Eq. (4.5) for the range of a projectile which lands at the same height it was launched from:

$$R = \frac{2v_{x0}v_{y0}}{g}.$$

In terms of θ, the angle the initial motion makes with the horizontal (Fig. 4.6), we can write the initial velocity components as $v_{x0} = v_0 \cos \theta$ and $v_{y0} = v_0 \sin \theta$. The range is then

$$R = \frac{2v_0^2 \sin \theta \cos \theta}{g} = \frac{v_0^2 \sin 2\theta}{g},$$

which reaches its maximum when $\sin 2\theta = 1$,

$$\max R = \frac{v_0^2}{g},$$

i.e., at $\theta = 45°$. Maximum range is achieved by launching the projectile at an angle of $45°$.

5.3   ADDITION AND SUBTRACTION OF VECTORS, AND MULTIPLICATION BY A SCALAR

Because vectors are new objects, algebraic operations involving vectors must be defined. We will define addition, subtraction, and three kinds of multiplication.

The simplest operation is *multiplication of a vector by a scalar*. If you have a vector **B** and multiply it by the scalar 3, then the new vector is three times longer than **B** and is written 3**B**. More generally, multiplying **B** by a scalar c leads to another vector, written c**B** (possibly longer or shorter) with the same direction if c is positive, and with the opposite direction if c is negative. The magnitude of c**B** is $|c|\,|\mathbf{B}|$, the absolute value of c times the magnitude of **B**.

Division of a vector by a scalar presents no new problem because we can always regard division by a scalar c as multiplication by $1/c$ if $c \neq 0$.

A vector **A**, multiplied by the scalar -1 is called the negative of the vector, $-$**A**. The result has the same magnitude but points in the opposite direction.

Multiplication by a scalar that has dimensions, such as mass or time, changes a vector into another vector which is parallel to it but has a different dimension and physical significance. For example, an object moving with constant velocity **v** for time t undergoes a displacement

$$\mathbf{s} = t\mathbf{v}. \tag{5.5}$$

Displacement is a vector with dimension of length; velocity is a vector with dimension of length over time.

Just as zero is an exception among numbers, the zero vector is also an exception among vectors. The zero vector is the result of multiplying any vector by the scalar zero. It has a magnitude (of zero) but no direction. In vector equations we will encounter, the zero vector will simply be written as **0**.

To define *addition* of vectors, let's suppose we have two vectors **A** and **B** like those shown in Fig. 5.6a. To add the two vectors, begin the second vector where the first one ends (Fig. 5.6b). Remember that you can always slide the vectors around as long as you keep them pointing in the same direction, so you can perform this tail-to-head construction. The vector sum, also called the resultant vector, forms a new vector $\mathbf{C} = \mathbf{A} + \mathbf{B}$ which points from the beginning of the first vector to the end of the second vector; it is "as the crow flies." Once you've constructed the resultant vector, you can use a ruler to measure its length and determine its magnitude. The angle from some reference line can be found easily with a protractor. Then you're done because you have two things which specify the resultant vector – its magnitude and direction.

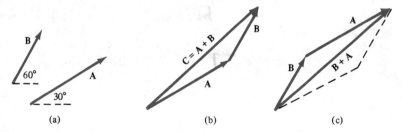

Figure 5.6 (a) Two vectors **A** and **B**. (b) The vector sum **C** = **A** + **B**, found by the tail-to-head method. (c) The vector sum **B** + **A**.

Figure 5.6c shows a useful feature of vector addition. In the figure, we start with **B** and add **A** to it; that is, we find the sum **B** + **A**. It turns out to be the same vector **C**. This means that the order in which you add two vectors is unimportant because you'll always get the same result. This property of vector addition is called commutativity:

Commutativity of vector addition:

$$\mathbf{A} + \mathbf{B} = \mathbf{B} + \mathbf{A}. \tag{5.6}$$

Note that the sum **A** + **B** is a diagonal of the parallelogram determined by **A** and **B** as shown in Fig. 5.6c. For this reason, vector addition is sometimes called addition by the parallelogram law.

The graphical technique of adding vectors can be extended to determine the sum of more than two vectors. For example, suppose you want the sum of three vectors **A**, **B**, and **C**. You can take (**A** + **B**) and add it to **C**, as shown in Fig. 5.7, or you can add **A** to (**B** + **C**). Both results are the same. This property is known as associativity:

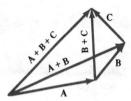

Figure 5.7 Associative law of vector addition.

Associativity of vector addition:

$$\mathbf{A} + (\mathbf{B} + \mathbf{C}) = (\mathbf{A} + \mathbf{B}) + \mathbf{C}. \tag{5.7}$$

Because of this we can write the sum as **A** + **B** + **C** without indicating which way they are grouped together.

All these results we have derived by the graphical method for addition of vectors in a plane also apply to vectors in space.

Vectors can be related to their components by vector addition with the help of unit vectors. A unit vector is a vector whose length is equal to one – not 1 cm, or 1 m, just 1; it is dimensionless. We use the symbol $\hat{\mathbf{i}}$ to represent the unit vector along the positive x axis. A hat (ˆ) over a boldface letter is used to remind us that the vector under the hat is a unit vector. Similarly, the unit vector $\hat{\mathbf{j}}$ points in the positive y direction, and $\hat{\mathbf{k}}$ points in the positive z direction. Just like the coordinate axes, the unit vectors $\hat{\mathbf{i}}$, $\hat{\mathbf{j}}$, $\hat{\mathbf{k}}$ are mutually perpendicular. The three unit vectors $\hat{\mathbf{i}}$, $\hat{\mathbf{j}}$, $\hat{\mathbf{k}}$ are shown in Fig. 5.8.

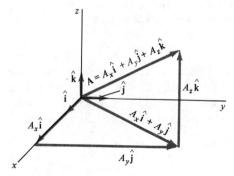

Figure 5.8 A vector **A** expressed in terms of its components.

Using these unit vectors we can express any vector $\mathbf{A}$ algebraically in terms of its components. The relation, illustrated in Fig. 5.8, is

$$\mathbf{A} = A_x\hat{\mathbf{i}} + A_y\hat{\mathbf{j}} + A_z\hat{\mathbf{k}}. \tag{5.8}$$

You can think of the displacement from the origin to the tip of $\mathbf{A}$ as a vector sum of three displacements: $A_x\hat{\mathbf{i}}$ followed by $A_y\hat{\mathbf{j}}$ followed by $A_z\hat{\mathbf{k}}$.

As an example, consider the trajectory of a projectile as discussed in Chapter 4. In its horizontal motion, following the law of inertia, a cannonball moves with constant speed, given it by the explosion of gunpowder. At the same time and quite independently, in the vertical direction, the ball falls with constant acceleration, obeying the law of falling bodies. What we did when we analyzed this motion was to break the motion down into components.

In Fig. 5.9 the horizontal component of the ball's displacement $\mathbf{s}$ is $s_x = v_x t$, whereas the vertical component is given by the law of falling bodies, $s_y = -\frac{1}{2}gt^2$. The total displacement can be written as

$$\mathbf{s}(t) = v_x t\hat{\mathbf{i}} - \tfrac{1}{2}gt^2\hat{\mathbf{j}}. \tag{5.9}$$

The arrows in the figure show this vector at different times.

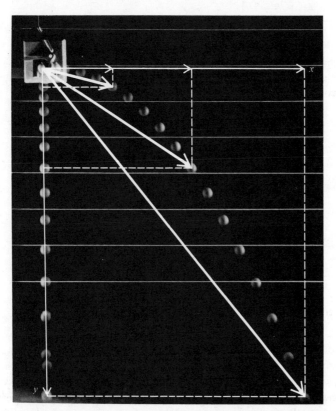

Figure 5.9 Displacement vector of a horizontally projected ball.
(*PSSC Physics*, 2nd ed., 1965. D.C. Heath and Co. with Educational Development Center, Inc., Newton, Mass.)

In terms of components, if

$$\mathbf{B} = B_x\hat{\mathbf{i}} + B_y\hat{\mathbf{j}} + B_z\hat{\mathbf{k}} \tag{5.10}$$

and c is any scalar, then, multiplying both sides of Eq. (5.10) by c, we find

$$c\mathbf{B} = cB_x\hat{\mathbf{i}} + cB_y\hat{\mathbf{j}} + cB_z\hat{\mathbf{k}}. \tag{5.11}$$

Thus the components of $c\mathbf{B}$ are cB_x, cB_y, cB_z.

Vector addition can also be expressed in terms of components. If we add the vectors $\mathbf{A} = A_x\hat{\mathbf{i}} + A_y\hat{\mathbf{j}} + A_z\hat{\mathbf{k}}$ and $\mathbf{B} = B_x\hat{\mathbf{i}} + B_y\hat{\mathbf{j}} + B_z\hat{\mathbf{k}}$ we find

$$\mathbf{A} + \mathbf{B} = (A_x + B_x)\hat{\mathbf{i}} + (A_y + B_y)\hat{\mathbf{j}} + (A_z + B_z)\hat{\mathbf{k}}. \tag{5.12}$$

Hence the components of the sum $\mathbf{C} = \mathbf{A} + \mathbf{B}$ are obtained by adding corresponding components of $\mathbf{A}$ and $\mathbf{B}$:

$$
\boxed{
\begin{aligned}
C_x &= A_x + B_x, \\
C_y &= A_y + B_y, \\
C_z &= A_z + B_z.
\end{aligned}
}
$$

$$\tag{5.13}$$

The simple-looking vector equation $\mathbf{C} = \mathbf{A} + \mathbf{B}$ is shorthand for three scalar equations. This is one of the powerful advantages of vector algebra. Figure 5.10 illustrates addition of vector components for two vectors in the xy plane.

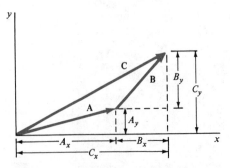

Figure 5.10 Vector addition relating geometric construction and addition of components.

Both of the two methods we have given for working with vectors, geometric and analytic, are useful. The geometric method provides a physical picture and is sometimes more elegant, avoiding as it does any specification of the coordinate system. Analysis by components is perhaps cumbersome but offers an absolutely reliable way of manipulating vectors. Typically geometry is better for seeing relationships but analysis by components is needed for obtaining numerical results.

Example 2

A ship sails 60 mi in a direction 30° north of east, then 30 mi due east, then 40 mi 30° west of north. Where is the ship with respect to its starting point?

Labeling the three vectors representing the three legs of the trip as **A**, **B**, and **C**, we can construct the vector sum **R** = **A** + **B** + **C** geometrically by adding the vectors head to tail:

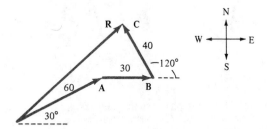

If a scale of 1 cm = 22.5 mi is used, the length of **R** can be measured to be approximately 4 cm, which corresponds to 90 mi. The direction from the starting point (as measured with a protractor) is approximately 46° N of E.

Alternatively we can work algebraically; **A**, **B**, and **C** can be written in components as

$$\mathbf{A} = 60(\cos 30°)\hat{\mathbf{i}} + 60(\sin 30°)\hat{\mathbf{j}} = 51.9\hat{\mathbf{i}} + 30\hat{\mathbf{j}},$$

$$\mathbf{B} = 30\hat{\mathbf{i}},$$

$$\mathbf{C} = 40(\cos 120°)\hat{\mathbf{i}} + 40(\sin 120°)\hat{\mathbf{j}} = -20\hat{\mathbf{i}} + 34.6\hat{\mathbf{j}}.$$

The algebraic sum is

$$\mathbf{R} = \mathbf{A} + \mathbf{B} + \mathbf{C} = (51.9 + 30 - 20)\hat{\mathbf{i}} + (30 + 0 + 34.6)\hat{\mathbf{j}}$$

$$= 61.9\hat{\mathbf{i}} + 64.6\hat{\mathbf{j}},$$

which has magnitude

$$R = \sqrt{(61.9)^2 + (64.6)^2} = 89.5 \text{ mi}$$

and direction

$$\theta = \tan^{-1}\left(\frac{64.6}{61.9}\right) = 46.2°.$$

Once we know how to add vectors, *subtraction* is straightforward because we can always treat subtraction as the addition of a negative vector. As we saw earlier the negative of a vector is a vector with the same magnitude but opposite direction. So if you had two vectors, **G** and **H**, and you wanted **G** − **H**, you would first form −**H** then add it tail to head to **G**. The vector from the tail of **G** to the head of $-$**H** is **G** $-$ **H**, as shown in Fig. 5.11b. In summary,

$$\mathbf{G} - \mathbf{H} = \mathbf{G} + (-\mathbf{H}) \qquad (5.14)$$

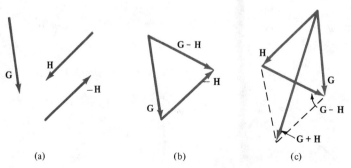

(a)   (b)   (c)

Figure 5.11 (a) Vectors **G** and **H**. (b) Graphical subtraction of vectors to form **G** $-$ **H**. (c) The vector sum **G** $+$ **H** and difference **G** $-$ **H** shown as diagonals of the same parallelogram.

It was apparent earlier in Fig. 5.6c that the sum of the two vectors **G** and **H** represents one diagonal of the parallelogram determined by **G** and **H**. The difference **G** $-$ **H** represents the other diagonal from the tip of **H** to the tip of **G**, as shown in Fig. 5.11c. The difference in the other order, **H** $-$ **G**, is, of course, just the negative of this, the diagonal from the tip of **G** to the tip of **H**.

A swimmer in a stream and an airplane in a wind are good physical illustrations of vector addition and subtraction. The swimmer, for example, makes a certain progress with respect to the water he's in, but he's also swept along in some other direction by the current. What is his actual velocity as seen by someone standing on the bank of the stream? It's not hard to see that it is simply the vector sum of the swimmer's velocity in still water plus the velocity of the stream itself. The reason is that each velocity vector is equal to the corresponding displacement vector in one unit of time, and since displacements add vectorially, the same is true for velocities.

Example 3

A weekend pilot whose plane has a speed of 173 km/h relative to the air tries to fly due north. But there is a strong wind blowing due east. As a result the airplane, as seen from the ground, moves 30° east of north with a speed of 200 km/h. What are the speed and direction of the wind?

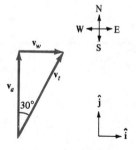

Let's call the plane's velocity relative to the air $\mathbf{v_a}$, the wind's velocity $\mathbf{v_w}$, and the plane's total velocity with respect to the ground $\mathbf{v_t}$. These velocity vectors are related by the vector equation

$$\mathbf{v_t} = \mathbf{v_a} + \mathbf{v_w},$$

which can be solved for the wind's velocity:

$$\mathbf{v_w} = \mathbf{v_t} - \mathbf{v_a}.$$

In terms of vector components we have

$$\mathbf{v_a} = 173\hat{\mathbf{j}},$$

$$\mathbf{v_t} = 200[(\cos 60°)\hat{\mathbf{i}} + (\sin 60°)\hat{\mathbf{j}}] = 100\hat{\mathbf{i}} + 173\hat{\mathbf{j}},$$

so the wind's velocity is

$$\mathbf{v_w} = (100 - 0)\hat{\mathbf{i}} + (173 - 173)\hat{\mathbf{j}} = 100\hat{\mathbf{i}} \text{ km/h}.$$

The wind is blowing 100 km/h due east.

Example 4
A pier is situated on a riverbank at point P, and a raft is anchored a distance D away at point Q. A boy swims straight from P to Q and back again, moving with constant speed V relative to still water. Calculate the total time for the round trip if the river current flows with speed v and the raft is

(a) directly downstream from the pier,
(b) directly offshore (perpendicular to the current).

(a)   The boy's speed relative to the pier is $V + v$ on the way out, and $V - v$ on the return. Since the time on each section of the trip is D/speed, the total time taken is

$$t_a = \frac{D}{V + v} + \frac{D}{V - v} = \frac{2DV}{V^2 - v^2}.$$

(b)   Before setting this case to algebra, one must carefully visualize the scene!

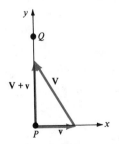

If $\mathbf{v}$ lies along the x axis, and the offshore direction along the y axis, the boy's total velocity $\mathbf{V} + \mathbf{v}$ relative to the pier lies straight along the y axis. But to maintain this course the boy's velocity $\mathbf{V}$ relative to the water must have an upstream component $-\mathbf{v}$ sufficient to just cancel the current. In vector component notation the vectors are

$$\mathbf{v} = v\hat{\mathbf{i}}, \qquad \mathbf{V} = -v\hat{\mathbf{i}} + \sqrt{V^2 - v^2}\,\hat{\mathbf{j}}$$

where the $\hat{\mathbf{j}}$ component of $\mathbf{V}$ follows from Pythagoras' theorem. The boy's net velocity is therefore

$$\mathbf{V} + \mathbf{v} = \sqrt{V^2 - v^2}\,\hat{\mathbf{j}}$$

on the way out, and minus this on the return. The total time for the round trip is

$$t_b = \frac{2D}{\sqrt{V^2 - v^2}}.$$

Note that in both cases (a) and (b) the round-trip time is increased by the existence of the current.

5.4 THE SCALAR PRODUCT OF VECTORS

Thus far we have defined addition and subtraction for vectors and multiplication of a vector by a scalar. Next we wish to define two different ways of multiplying vectors, one of which produces a scalar result, the other a vector. We'll examine the *scalar product* first.

The scalar product is also called the dot product, written $\mathbf{A} \cdot \mathbf{B}$, with a dot between the vectors, and is defined as the scalar

$$\boxed{\mathbf{A} \cdot \mathbf{B} = AB \cos \theta} \tag{5.15}$$

where θ is the angle between the two vectors (measured so that $0 \leq \theta \leq 180°$). In other words, to compute the scalar product of two vectors, multiply their magnitudes and the cosine of the angle between them. Note that the scalar product is commutative:

Commutativity of the scalar product:

$$\mathbf{A} \cdot \mathbf{B} = \mathbf{B} \cdot \mathbf{A}. \tag{5.16}$$

The scalar quantity

$$b = B \cos \theta \tag{5.17}$$

is called the component of $\mathbf{B}$ along $\mathbf{A}$ and is shown in Fig. 5.12a. The dot product

$$\mathbf{A} \cdot \mathbf{B} = AB \cos \theta = Ab \tag{5.18}$$

is obtained by multiplying the magnitude A with the component of **B** along **A**. It is also equal to the product of B and the component of **A** along **B**, as shown in Fig. 5.12b.

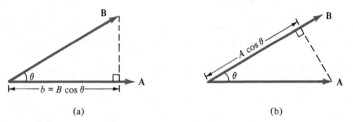

(a)
(b)

Figure 5.12 (a) $B \cos \theta$ is the component of **B** along **A**. (b) $A \cos \theta$ is the component of **A** along **B**.

If **A** and **B** are perpendicular, $\cos \theta$ is zero and

$$\mathbf{A} \cdot \mathbf{B} = 0. \tag{5.19}$$

Conversely, if the scalar product is zero, at least one of the vectors is zero or else the two are perpendicular.

If **A** and **B** have the same direction, $\cos \theta = 1$ and $\mathbf{A} \cdot \mathbf{B}$ is simply the product of the magnitudes of the two vectors. In particular,

$$\mathbf{A} \cdot \mathbf{A} = A^2, \tag{5.20}$$

the square of the magnitude of **A**.

In Fig. 5.13, b and c are the components of **B** and **C** along **A**. It is clear that $b + c$ is the component of $\mathbf{B} + \mathbf{C}$ along **A**. But

$$A(b + c) = Ab + Ac, \tag{5.21}$$

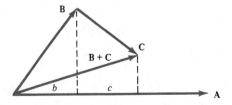

Figure 5.13 Illustration of the distributive law $\mathbf{A} \cdot (\mathbf{B} + \mathbf{C}) = \mathbf{A} \cdot \mathbf{B} + \mathbf{A} \cdot \mathbf{C}$.

which implies that the dot product is distributive:

Distributivity of the dot product:

$$\mathbf{A} \cdot (\mathbf{B} + \mathbf{C}) = \mathbf{A} \cdot \mathbf{B} + \mathbf{A} \cdot \mathbf{C}. \tag{5.22}$$

Let us turn now to the algebraic representation of vectors in terms of components and unit vectors. We recall that the unit vectors $\hat{\mathbf{i}}$, $\hat{\mathbf{j}}$, $\hat{\mathbf{k}}$ are mutually perpendicular. In terms of dot products, this means that

$$\hat{\mathbf{i}} \cdot \hat{\mathbf{j}} = \hat{\mathbf{j}} \cdot \hat{\mathbf{k}} = \hat{\mathbf{k}} \cdot \hat{\mathbf{i}} = 0. \tag{5.23}$$

These relations give us an algebraic way to determine the components of $\mathbf{A} = A_x\hat{\mathbf{i}} + A_y\hat{\mathbf{j}} + A_z\hat{\mathbf{k}}$. Just take the dot product of $\mathbf{A}$ with $\hat{\mathbf{i}}$ to get

$$\mathbf{A} \cdot \hat{\mathbf{i}} = A_x\hat{\mathbf{i}} \cdot \hat{\mathbf{i}} + A_y\hat{\mathbf{j}} \cdot \hat{\mathbf{i}} + A_z\hat{\mathbf{k}} \cdot \hat{\mathbf{i}} = A_x, \tag{5.24}$$

because $\hat{\mathbf{i}} \cdot \hat{\mathbf{i}} = 1$ and $\hat{\mathbf{j}} \cdot \hat{\mathbf{i}} = \hat{\mathbf{k}} \cdot \hat{\mathbf{i}} = 0$. Similarly we find

$$\mathbf{A} \cdot \hat{\mathbf{j}} = A_y \qquad \text{and} \qquad \mathbf{A} \cdot \hat{\mathbf{k}} = A_z. \tag{5.25}$$

It is useful to express the scalar product of two vectors $\mathbf{A}$ and $\mathbf{B}$ in terms of components. The dot product $\mathbf{A} \cdot \mathbf{B}$ is given by

$$\mathbf{A} \cdot \mathbf{B} = (A_x\hat{\mathbf{i}} + A_y\hat{\mathbf{j}} + A_z\hat{\mathbf{k}}) \cdot (B_x\hat{\mathbf{i}} + B_y\hat{\mathbf{j}} + B_z\hat{\mathbf{k}}). \tag{5.26}$$

Using the distributive law (5.22), we multiply out the expression on the right and get

$$\mathbf{A} \cdot \mathbf{B} = A_xB_x\hat{\mathbf{i}} \cdot \hat{\mathbf{i}} + A_xB_y\hat{\mathbf{i}} \cdot \hat{\mathbf{j}} + A_xB_z\hat{\mathbf{i}} \cdot \hat{\mathbf{k}} + A_yB_x\hat{\mathbf{j}} \cdot \hat{\mathbf{i}} + A_yB_y\hat{\mathbf{j}} \cdot \hat{\mathbf{j}}$$

$$+ A_yB_z\hat{\mathbf{j}} \cdot \hat{\mathbf{k}} + A_zB_x\hat{\mathbf{k}} \cdot \hat{\mathbf{i}} + A_zB_y\hat{\mathbf{k}} \cdot \hat{\mathbf{j}} + A_zB_z\hat{\mathbf{k}} \cdot \hat{\mathbf{k}}. \tag{5.27}$$

But this isn't as bad as it looks because all the terms involving the dot product of a unit vector with a different unit vector are zero. Only the three terms with $\hat{\mathbf{i}} \cdot \hat{\mathbf{i}}$, $\hat{\mathbf{j}} \cdot \hat{\mathbf{j}}$, and $\hat{\mathbf{k}} \cdot \hat{\mathbf{k}}$ survive. So we're left with

$$\boxed{\mathbf{A} \cdot \mathbf{B} = A_xB_x + A_yB_y + A_zB_z.} \tag{5.28}$$

In other words, to calculate the dot product we simply add the products of corresponding components.

When we take the dot product of a vector with itself, we find through Eq. (5.28)

$$\mathbf{A} \cdot \mathbf{A} = A_x^2 + A_y^2 + A_z^2. \tag{5.29}$$

This says that the square of the length of $\mathbf{A}$ is the sum of the squares of its components. For vectors in the xy plane this is just the Pythagorean theorem, so we've found an extension of the Pythagorean theorem to three-dimensional space.

Example 5

(a) Calculate the dot product of $\mathbf{A} = 3\hat{\mathbf{i}} + 2\hat{\mathbf{j}}$ and $\mathbf{B} = \hat{\mathbf{i}} - \hat{\mathbf{j}}$.
(b) What is the angle between these two vectors?

(a) Using Eq. (5.28) one easily finds

$$\mathbf{A} \cdot \mathbf{B} = (3)(1) + (2)(-1) + (0)(0) = 1.$$

(b) At first glance, it may appear that we need to draw a graph of the two vectors and then measure the angle between them. But instead we can find the angle by combining the result we just found, $\mathbf{A}\cdot\mathbf{B} = 1$, with our other expression for the dot product, Eq. (5.15):

$\mathbf{A}\cdot\mathbf{B} = AB \cos\theta.$

The magnitudes A and B are easily found by Eq. (5.29), i.e., the Pythagorean theorem:

$$A = \sqrt{\mathbf{A}\cdot\mathbf{A}} = \sqrt{A_x^2 + A_y^2 + A_z^2} = \sqrt{3^2 + 2^2 + 0^2} = \sqrt{13},$$

$$B = \sqrt{1^2 + (-1)^2 + 0^2} = \sqrt{2}.$$

Putting everything together, we get

$$\cos\theta = \frac{\mathbf{A}\cdot\mathbf{B}}{AB} = \frac{1}{\sqrt{13}\sqrt{2}} = 0.20$$

which tells us that $\theta = 79°$.

Example 6

The ski jumper of Example 1 in Chapter 4 has velocity $\mathbf{v} = (20\hat{\mathbf{i}} - 42\hat{\mathbf{j}})$ m/s just before landing on a 45° slope. Find v_s, the component of $\mathbf{v}$ parallel to the slope, and v_p, the component of $\mathbf{v}$ perpendicular to the slope. (These components are of interest because v_s will be his speed in the direction of the slope just after landing, whereas the jolt he receives on landing will be proportional to v_p.)

We note that $v_s = \mathbf{v}\cdot\hat{\mathbf{u}}_s$, where $\hat{\mathbf{u}}_s$ is a unit vector directed along the 45° slope. Since

$$\hat{\mathbf{u}}_s = \frac{\hat{\mathbf{i}} - \hat{\mathbf{j}}}{\sqrt{2}}$$

we find

$$v_s = (20\hat{\mathbf{i}} - 42\hat{\mathbf{j}}) \cdot \frac{\hat{\mathbf{i}} - \hat{\mathbf{j}}}{\sqrt{2}}$$

$$= \frac{20}{\sqrt{2}} + \frac{42}{\sqrt{2}} = 44 \frac{m}{s}.$$

Similarly, $v_p = \mathbf{v} \cdot \hat{\mathbf{u}}_p$, where $\hat{\mathbf{u}}_p$ is a unit vector perpendicular to the slope. Since

$$\hat{\mathbf{u}}_p = \frac{-\hat{\mathbf{i}} - \hat{\mathbf{j}}}{\sqrt{2}}$$

we obtain

$$v_p = (20\hat{\mathbf{i}} - 42\hat{\mathbf{j}}) \cdot \frac{(-\hat{\mathbf{i}} - \hat{\mathbf{j}})}{\sqrt{2}} = \frac{-20}{\sqrt{2}} + \frac{42}{\sqrt{2}} = 16 \frac{\text{m}}{\text{s}}.$$

An alternative solution can be given by using methods discussed earlier in this chapter. Just before landing the jumper has speed

$$v = \sqrt{v_x^2 + v_y^2} = 46.5 \text{ m/s}$$

at an angle $\theta = \cos^{-1}(v_x/v) = 64.5°$ below the horizontal, or $19.5°$ below the $45°$ slope.

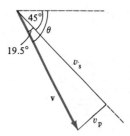

The right triangle with edges v_s and v_p has hypotenuse v so

$$v_s = v \cos 19.5° = 44 \text{ m/s}$$

and

$$v_p = v \sin 19.5° = 16 \text{ m/s}.$$

Our result for v_p, obtained for an imaginary hill that drops off at constant slope from the takeoff, is too large for a safe landing. To keep v_p down to a safe level, real ski jumps are designed so that the hill has a curved profile nearly matching the trajectory of a typical jump, making it probable that the jumper will land on a slope nearly parallel to his trajectory. For jumps ending between points A and B in the figure below, v_p is small enough for safe landings.

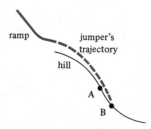

Beyond point B the hill flattens out and v_p becomes dangerously large. For safe operation, when snow or wind conditions produce jumps much beyond point B, the jumpers are restarted further down the ramp to reduce their takeoff speed. At the "70-meter" ski jump built for the 1980 Olympics at Lake Placid, New York, 70 m is the distance from takeoff to A, and the distance from takeoff to B is 86 m.

5.5 THE CROSS PRODUCT OF VECTORS

We turn now to the second way of multiplying two vectors together. To describe many phenomena in physics, it is helpful to have a method for constructing a vector perpendicular to each of two given vectors. The product of two vectors, say **A** and **B**, which produces a vector **C** perpendicular to both **A** and **B**, is called the *vector product* or *cross product*, written as

$$\mathbf{A} \times \mathbf{B} = \mathbf{C}, \tag{5.30}$$

and read "A cross B."

The direction of **C** is related to **A** and **B** by the right-hand rule. If you place your fingers in the direction of the first vector, **A**, then curl them through the smaller angle toward **B**, your thumb points in the direction of **C**, as shown in Fig. 5.14a. Note that the cross product **B** × **A** produces a vector in the opposite direction, as Fig. 5.14b illustrates. Mathematically, the cross product is said to be *anticommutative*:

$$\mathbf{A} \times \mathbf{B} = -\mathbf{B} \times \mathbf{A}. \tag{5.31}$$

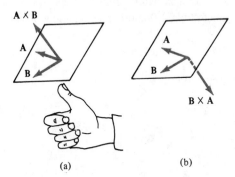

(a)   (b)

Figure 5.14 (a) Right-hand rule used to find the direction of **A** × **B**.
(b) Right-hand rule used to find the direction of **B** × **A**.

This property sharply distinguishes vectors from ordinary numbers.

The magnitude of the cross product is defined to be

$$|\mathbf{A} \times \mathbf{B}| = |\mathbf{A}|\,|\mathbf{B}|\,\sin\theta = AB\sin\theta \tag{5.32}$$

where θ is the angle between **A** and **B** measured so that $0 \leq \theta \leq 180°$. As illustrated in Fig. 5.15a, $B \sin \theta$ is the component of **B** perpendicular to **A**, sometimes written as $B_\perp$. In this notation, the magnitude of the cross product is $|\mathbf{A} \times \mathbf{B}| = AB_\perp$.

Equivalently, the magnitude can be thought of as B times the component of **A** perpendicular to **B**, which would be called $A_\perp$, so $|\mathbf{A} \times \mathbf{B}| = A_\perp B$, as Fig. 5.15b shows. Geometrically, $|\mathbf{A} \times \mathbf{B}|$ represents the area of the parallelogram spanned by **A** and **B** (shaded in Fig. 5.15).

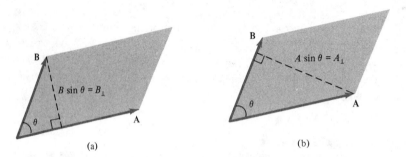

Figure 5.15 (a) $|\mathbf{A} \times \mathbf{B}| = AB_\perp = A(B \sin \theta)$. (b) $|\mathbf{A} \times \mathbf{B}| = A_\perp B = (A \sin \theta)B$.

Example 7

Two vectors lie in the xy plane; vector **A** has magnitude 1.5 and makes an angle of 30° with the x axis. Vector **B** has magnitude 2.0 and makes an angle of 100° with the x axis. Find **A** × **B**.

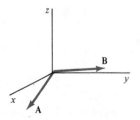

The direction of **A** × **B** is found from the right-hand rule: placing the fingers of your right hand in the direction of **A** and curling them toward **B**, you find that your thumb points along the positive z direction. To find the magnitude, we need to know the angle θ between **A** and **B**. We have $\theta = 100° - 30° = 70°$. Therefore the magnitude of the cross product is

$$|\mathbf{A} \times \mathbf{B}| = AB \sin \theta = 1.5(2.0) \sin 70° = 2.8.$$

Using the definition of the length of the cross product, Eq. (5.32), we see that the cross product of two parallel vectors is zero (because sin 0 = 0 and sin 180° = 0). For the unit vectors $\hat{\mathbf{i}}, \hat{\mathbf{j}}, \hat{\mathbf{k}}$, we find that

$$\hat{\mathbf{i}} \times \hat{\mathbf{i}} = \hat{\mathbf{j}} \times \hat{\mathbf{j}} = \hat{\mathbf{k}} \times \hat{\mathbf{k}} = \mathbf{0}, \tag{5.33}$$

whereas by the right-hand rule

$$\hat{\mathbf{i}} \times \hat{\mathbf{j}} = \hat{\mathbf{k}}, \qquad \hat{\mathbf{j}} \times \hat{\mathbf{k}} = \hat{\mathbf{i}}, \qquad \hat{\mathbf{k}} \times \hat{\mathbf{i}} = \hat{\mathbf{j}}. \tag{5.34}$$

For any two vectors $\mathbf{A}$ and $\mathbf{B}$ the cross product is

$$\mathbf{A} \times \mathbf{B} = (A_x\hat{\mathbf{i}} + A_y\hat{\mathbf{j}} + A_z\hat{\mathbf{k}}) \times (B_x\hat{\mathbf{i}} + B_y\hat{\mathbf{j}} + B_z\hat{\mathbf{k}}). \tag{5.35}$$

Multiplying out the terms and using the properties of the cross product between unit vectors in (5.34), we find

$$\boxed{\mathbf{A} \times \mathbf{B} = (A_yB_z - A_zB_y)\hat{\mathbf{i}} + (A_zB_x - A_xB_z)\hat{\mathbf{j}} + (A_xB_y - A_yB_x)\hat{\mathbf{k}}.} \tag{5.36}$$

Equation (5.36) provides a useful way to calculate the cross product of two vectors when the components are known. (Though it would not have been the most efficient way to find $\mathbf{A} \times \mathbf{B}$ in Example 7!)

If you are familiar with determinants, there is a handy mnemonic device for remembering the cross product:

$$\mathbf{A} \times \mathbf{B} = \begin{vmatrix} \hat{\mathbf{i}} & \hat{\mathbf{j}} & \hat{\mathbf{k}} \\ A_x & A_y & A_z \\ B_x & B_y & B_z \end{vmatrix}.$$

When this is "expanded" along the first row we obtain (5.36). We shall have occasion to use the cross product soon in discussing torque and rotational motion.

One sometimes needs to know the "vector triple product" $\mathbf{A} \times (\mathbf{B} \times \mathbf{C})$ of three vectors. The result has the form

$$\mathbf{A} \times (\mathbf{B} \times \mathbf{C}) = (\mathbf{C} \cdot \mathbf{A})\mathbf{B} - (\mathbf{B} \cdot \mathbf{A})\mathbf{C}. \tag{5.37}$$

This can be verified by taking $\mathbf{A} = \hat{\mathbf{i}}, \hat{\mathbf{j}}, \hat{\mathbf{k}}$, respectively, and combining terms using components to obtain (5.37). The fact that the right side of (5.37) is a linear combination of vectors $\mathbf{B}$ and $\mathbf{C}$ has a geometrical interpretation. Because $\mathbf{B} \times \mathbf{C}$ is perpendicular to the plane of $\mathbf{B}$ and $\mathbf{C}$, the vector $\mathbf{A} \times (\mathbf{B} \times \mathbf{C})$ must lie in this plane, so it is a linear combination of $\mathbf{B}$ and $\mathbf{C}$.

Example 8

Find the cross product of the vectors $\mathbf{A} = \hat{\mathbf{i}} - \hat{\mathbf{j}} + 3\hat{\mathbf{k}}$ and $\mathbf{B} = \hat{\mathbf{i}} - 5\hat{\mathbf{j}} - 2\hat{\mathbf{k}}$, and the angle θ between them.

Using Eq. (5.36), we calculate the cross product to be

$$\mathbf{A} \times \mathbf{B} = [-1(-2) - 3(-5)]\hat{\mathbf{i}} + [3(1) - 1(-2)]\hat{\mathbf{j}} + [1(-5) - (-1)(1)]\hat{\mathbf{k}},$$

$$\mathbf{A} \times \mathbf{B} = 17\hat{\mathbf{i}} + 5\hat{\mathbf{j}} - 4\hat{\mathbf{k}}.$$

To find θ let's make use of the magnitude of the cross product, Eq. (5.32):

$$|\mathbf{A} \times \mathbf{B}| = AB \sin \theta.$$

If we know all the magnitudes, we can use this relation to find $\sin \theta$ and consequently θ. We easily calculate

$$A = \sqrt{1^2 + (-1)^2 + 3^2} = \sqrt{11},$$
$$B = \sqrt{1^2 + (-5)^2 + (-2)^2} = \sqrt{30},$$
$$|\mathbf{A} \times \mathbf{B}| = \sqrt{17^2 + 5^2 + (-4)^2} = \sqrt{330}.$$

Therefore we find

$$\sin \theta = \frac{|\mathbf{A} \times \mathbf{B}|}{AB} = \frac{\sqrt{330}}{\sqrt{11}\sqrt{30}} = 1,$$

which tells us that the angle between $\mathbf{A}$ and $\mathbf{B}$ is $90°$.

Alternatively we could have concluded that $\theta = 90°$ by noting that $\mathbf{A} \cdot \mathbf{B} = 0$.

5.6 DERIVATIVES OF VECTOR FUNCTIONS IN A FIXED COORDINATE SYSTEM

A final topic we wish to discuss in this chapter is uniform circular motion. Before doing so, we need to know what is meant by the derivative of a vector function. If a particle moves along a curve in space, its position coordinates (x, y, z) at time t can be specified by three scalar equations expressing each of x, y, and z as a function of t:

$$x = x(t), \qquad y = y(t), \qquad z = z(t).$$

In vector notation, the position vector is

$$\mathbf{r}(t) = x(t)\hat{\mathbf{i}} + y(t)\hat{\mathbf{j}} + z(t)\hat{\mathbf{k}}, \tag{5.38}$$

and is illustrated in Fig. 5.16. This is an example of a vector function of a real variable, which here is time.

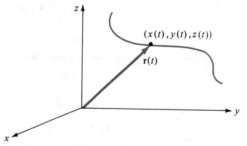

Figure 5.16 Position vector tracing out a curve in space.

As t varies through some interval, the vector $\mathbf{r}(t)$ can change both its magnitude and direction. To study this change we introduce the idea of the derivative of a vector function. As for ordinary functions (which we encountered in Chapter 3), we consider the position vector $\mathbf{r}(t)$ and at some time $t + h$ later, $\mathbf{r}(t + h)$. The difference in position, shown in Fig. 5.17a, is the vector $\mathbf{r}(t + h) - \mathbf{r}(t)$. The average velocity over that time is

$$\bar{\mathbf{v}} = \frac{\mathbf{r}(t + h) - \mathbf{r}(t)}{h}, \tag{5.39}$$

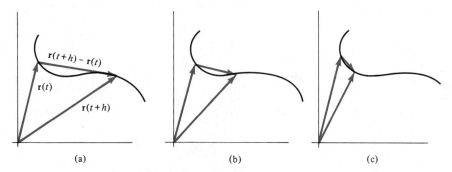

(a)   (b)   (c)

Figure 5.17 The shrinking of $\mathbf{r}(t + h) - \mathbf{r}(t)$.

and is parallel to $\mathbf{r}(t + h) - \mathbf{r}(t)$. If we allow h to shrink to zero, the two positions become closer together, and in the limit we obtain the derivative of $\mathbf{r}(t)$, which we call the velocity vector $\mathbf{v}(t)$:

$$\mathbf{v}(t) = \frac{d\mathbf{r}}{dt} = \lim_{h \to 0} \frac{\mathbf{r}(t + h) - \mathbf{r}(t)}{h}. \tag{5.40}$$

As Figs. 5.17b and 5.17c illustrate, in the limit the velocity vector $\mathbf{v}(t)$ is tangent to the curve. Expressing the difference quotient in terms of components, we find*

$$\frac{\mathbf{r}(t + h) - \mathbf{r}(t)}{h} = \frac{x(t + h) - x(t)}{h}\hat{\mathbf{i}} + \frac{y(t + h) - y(t)}{h}\hat{\mathbf{j}}$$
$$+ \frac{z(t + h) - z(t)}{h}\hat{\mathbf{k}}. \tag{5.41}$$

As h tends to zero, the components on the right tend to dx/dt, dy/dt, dz/dt and hence it is natural to define the limit of the difference quotient on the left to be the vector

$$\frac{d\mathbf{r}}{dt} = \frac{dx}{dt}\hat{\mathbf{i}} + \frac{dy}{dt}\hat{\mathbf{j}} + \frac{dz}{dt}\hat{\mathbf{k}}. \tag{5.42}$$

If we know the Cartesian components of the position vector, the velocity vector is obtained by the prescription of Eq. (5.42), which contains the three scalar equations for its components:

*We assume here that although the particle under consideration moves along a curve, the coordinate system is fixed (or moves along a straight line), so the unit vectors $\hat{\mathbf{i}}$, $\hat{\mathbf{j}}$, and $\hat{\mathbf{k}}$ do not change with time.

$$v_x = \frac{dx}{dt}, \qquad v_y = \frac{dy}{dt}, \qquad v_z = \frac{dz}{dt}. \tag{5.43}$$

The velocity $\mathbf{v}(t)$ is itself a vector function, and often we want to know its derivative $\mathbf{a}(t)$, the acceleration. Figure 5.18 shows the velocity vectors $\mathbf{v}(t)$ and $\mathbf{v}(t + h)$, a short time later. The change in velocity $\Delta\mathbf{v} = \mathbf{v}(t + h) - \mathbf{v}(t)$, so the average acceleration is

$$\bar{\mathbf{a}} = \frac{\mathbf{v}(t + h) - \mathbf{v}(t)}{h}. \tag{5.44}$$

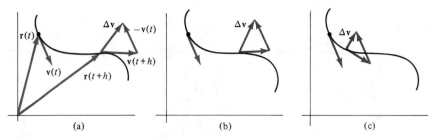

$$\text{(a)} \qquad\qquad\qquad \text{(b)} \qquad\qquad\qquad \text{(c)}$$

Figure 5.18 The change in the velocity vector $\Delta\mathbf{v} = \mathbf{v}(t + h) - \mathbf{v}(t)$ as the time interval shrinks to zero.

In the limit as the time interval shrinks to zero, as Fig. 5.18b and 5.18c indicate, the average acceleration vector becomes the instantaneous acceleration, the derivative of velocity:

$$\mathbf{a}(t) = \lim_{h \to 0} \frac{\mathbf{v}(t + h) - \mathbf{v}(t)}{h} = \frac{d\mathbf{v}}{dt}. \tag{5.45}$$

In terms of components, we have

$$\mathbf{a}(t) = \frac{d\mathbf{v}}{dt} = \frac{dv_x}{dt}\hat{\mathbf{i}} + \frac{dv_y}{dt}\hat{\mathbf{j}} + \frac{dv_z}{dt}\hat{\mathbf{k}}. \tag{5.46}$$

Unlike the velocity vector, the acceleration vector is not necessarily tangent to the curve $\mathbf{r}(t)$.

Since the acceleration vector is obtained from the position vector by differentiating it twice, we often use the notation

$$\mathbf{a}(t) = \frac{d^2\mathbf{r}}{dt^2} = \frac{d^2x}{dt^2}\hat{\mathbf{i}} + \frac{d^2y}{dt^2}\hat{\mathbf{j}} + \frac{d^2z}{dt^2}\hat{\mathbf{k}} \tag{5.47}$$

to indicate this relationship. The notation $d^2\mathbf{r}/dt^2$ means $d(d\mathbf{r}/dt)/dt$, the derivative of $d\mathbf{r}/dt$.

The same rules of differentiation hold for vector functions as for scalar functions. The derivative of the sum of two vector functions $\mathbf{A}(t)$ and $\mathbf{B}(t)$ is the sum of the derivatives:

$$\frac{d}{dt}(\mathbf{A}(t) + \mathbf{B}(t)) = \frac{d\mathbf{A}}{dt} + \frac{d\mathbf{B}}{dt}. \tag{5.48}$$

The derivative of the dot product of two vector functions follows from the product rule for scalar differentiation:

$$\frac{d}{dt}\mathbf{A}\cdot\mathbf{B} = \mathbf{A}\cdot\frac{d\mathbf{B}}{dt} + \frac{d\mathbf{A}}{dt}\cdot\mathbf{B}. \tag{5.49}$$

Similarly,

$$\frac{d}{dt}\mathbf{A}\times\mathbf{B} = \mathbf{A}\times\frac{d\mathbf{B}}{dt} + \frac{d\mathbf{A}}{dt}\times\mathbf{B}.$$

Finally, if $\mathbf{A}(t)$ is a vector function of t and t is a scalar function of another variable, say $t = t(u)$, then for $\mathbf{A}[t(u)]$, the chain rule becomes

$$\frac{d}{du}\mathbf{A}[t(u)] = \frac{d\mathbf{A}}{dt}\frac{dt}{du}. \tag{5.50}$$

Example 9
A wire helix of radius R is oriented vertically along the z axis. A frictionless bead slides down along the wire. Its position vector varies with time as

$$\mathbf{r}(t) = (R\cos t^2)\hat{\mathbf{i}} + (R\sin t^2)\hat{\mathbf{j}} - \tfrac{1}{2}bt^2\hat{\mathbf{k}}.$$

Find $\mathbf{v}(t)$ and $\mathbf{a}(t)$.

Using Eq. (5.42) and the chain rule we form the derivatives and find

$$\mathbf{v}(t) = \frac{d\mathbf{r}}{dt} = (-2tR\sin t^2)\hat{\mathbf{i}} + (2tR\cos t^2)\hat{\mathbf{j}} - bt\hat{\mathbf{k}}.$$

Note that the horizontal component of velocity, $\sqrt{v_x^2 + v_y^2}$, grows rapidly with t. Equation (5.46) gives the acceleration

$$\mathbf{a}(t) = \frac{d\mathbf{v}}{dt} = (-2R\sin t^2 - 4t^2R\cos t^2)\hat{\mathbf{i}}$$

$$+ (2R\cos t^2 - 4t^2R\sin t^2)\hat{\mathbf{j}} - b\hat{\mathbf{k}}.$$

The z component of acceleration is constant as expected, but the horizontal component of acceleration grows rapidly with time. We will gain some understanding of why acceleration grows as the circular component of motion speeds up when we analyze the simpler case of uniform circular motion in Section 5.8.

5.7 POSITION VECTOR EXPRESSED IN POLAR COORDINATES

Let

$$\mathbf{r} = x\hat{\mathbf{i}} + y\hat{\mathbf{j}} \tag{5.51}$$

denote the position vector from the origin to a point in the plane with rectangular coordinates (x, y). If $(x, y) \neq (0, 0)$ the position vector can also be described by specifying its length r and the angle θ the vector makes with the x axis, as illustrated in Fig. 5.19. The two numbers r and θ (with θ measured in radians) are called *polar coordinates* and they are related to the rectangular coordinates by the equations

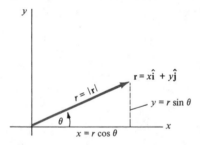

Figure 5.19 Polar coordinates r and θ of a point (x, y).

$$x = r \cos \theta, \qquad y = r \sin \theta. \tag{5.52}$$

These equations can be used to find x and y if r and θ are given. Conversely, if x and y are given we can find r by the theorem of Pythagoras: $r^2 = x^2 + y^2$, so

$$r = \sqrt{x^2 + y^2}. \tag{5.53}$$

The angle θ is related to x and y by the equations

$$\cos \theta = \frac{x}{r}, \qquad \sin \theta = \frac{y}{r}, \qquad \tan \theta = \frac{y}{x}. \tag{5.54}$$

We usually require that θ lie in the interval $0 \leq \theta < 2\pi$ and call θ the *polar angle* of the position vector. Expressed in polar coordinates, the position vector is

$$\mathbf{r} = (r \cos \theta)\hat{\mathbf{i}} + (r \sin \theta)\hat{\mathbf{j}}. \tag{5.55}$$

5.8   UNIFORM CIRCULAR MOTION

Circular motion, whether it be of a planet, Ferris wheel, or any other object, is an important type of motion in physics, and to describe it we use polar coordinates with the origin placed at the center of the circle, as shown in Fig. 5.20.

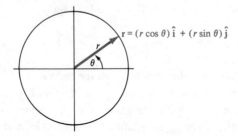

Figure 5.20 Circular motion about the origin: $r = $ const.

Ancient cultures discovered very early that a circle can be constructed by driving a stake into the ground, attaching a rope to it in such a way that the rope is free to move around the stake, and drawing a figure with the other end of the taut rope while walking around the stake.

In our mathematical language, this means that the length of the vector **r** describing a circle is the same everywhere. The vector **r** is *not* constant, since its direction changes, but its magnitude is constant:

$$|\mathbf{r}| = \text{const} = r.$$

So far we've described what we mean by circular motion, but the motion we really want to discuss is *uniform* circular motion. This means that the polar angle θ changes at a constant rate. In other words, the derivative $d\theta/dt$ is constant. We call that constant ω (omega) so

$$\frac{d\theta}{dt} = \omega, \tag{5.56}$$

from which we get, by integration, $\theta = \omega t + \alpha$, where α is the value of θ when $t = 0$. If the motion starts on the positive x axis, then $\alpha = 0$ and we have

$$\theta = \omega t. \tag{5.57}$$

If θ increases with time the quantity ω is positive and is known as the *angular speed* of the object. When θ is measured in radians and t in seconds, ω has units of radians per second (rad/s).

The angular speed ω is related to the time T it takes to complete one revolution. This time T is called the *period*. Since θ moves through 2π radians in one revolution we find from (5.57) that $2\pi = \omega T$, so the angular speed is also given by

$$\omega = 2\pi/T. \qquad\qquad \textit{say } T = 1\ \text{mpr} \tag{5.58}$$

Taking $\theta = \omega t$ in Eq. (5.55) we find the position vector (also called the *radius vector*) of uniform circular motion is given by

$$\mathbf{r} = (r \cos \omega t)\hat{\mathbf{i}} + (r \sin \omega t)\hat{\mathbf{j}} \tag{5.59}$$

where r and ω are constant.

The velocity of the object at any instant is the derivative of the radius vector. We can easily compute this derivative using Eq. (5.42) and obtain

$$\mathbf{v}(t) = \frac{d\mathbf{r}}{dt} = (-r\omega \sin \omega t)\hat{\mathbf{i}} + (r\omega \cos \omega t)\hat{\mathbf{j}}. \tag{5.60}$$

In calculating the derivative we used our knowledge of the derivatives of sine and cosine and the chain rule, along with the fact that r and the unit vectors $\hat{\mathbf{i}}, \hat{\mathbf{j}}$ are constant.

The velocity vector is not in the same direction as $\mathbf{r}$. As shown in Fig. 5.21, the velocity vector is perpendicular to $\mathbf{r}(t)$, being tangent to the circle at the instantaneous position of the particle. This can also be seen by taking the scalar product of Eqs. (5.55) and (5.60) to obtain

$$\mathbf{r}\cdot\mathbf{v} = (r \cos \omega t)(-r\omega \sin \omega t) + (r \sin \omega t)(r\omega \cos \omega t) = 0.$$

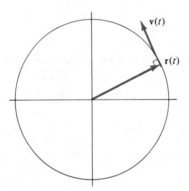

Figure 5.21 Directions of $\mathbf{r}(t)$ and $\mathbf{v}(t)$ for circular motion.

This result implies that $\mathbf{r}(t)$ and $\mathbf{v}(t)$ are perpendicular.

Example 10
Prove analytically that for any circular motion, uniform or not, the vectors $\mathbf{r}(t)$ and $\mathbf{v}(t)$ are perpendicular.

We will, in fact, prove a more general result: *Any vector function* $\mathbf{F}(t)$ *of constant length is perpendicular to its derivative.*

To prove this, consider the scalar product $\mathbf{F}\cdot\mathbf{F} = F^2$ and take the derivative of both sides of this equation, Since F^2 is constant the derivative of the right-hand side is zero. Using the product rule to take the derivative of the left-hand side, we have

$$\frac{d}{dt}(\mathbf{F}\cdot\mathbf{F}) = \mathbf{F}\cdot\frac{d\mathbf{F}}{dt} + \mathbf{F}\cdot\frac{d\mathbf{F}}{dt} = 2\mathbf{F}\cdot\frac{d\mathbf{F}}{dt}.$$

Therefore we have $2\mathbf{F}\cdot d\mathbf{F}/dt = 0$, so $\mathbf{F}(t)$ is perpendicular to $\mathbf{F}'(t)$.

This property is easily understood geometrically, as well. If we place the tail of each vector $\mathbf{F}(t)$ at the origin, as shown in the accompanying figure, constant length means the head of $\mathbf{F}(t)$ lies on a circle of radius F. For small h the vector $\mathbf{F}(t + h) - \mathbf{F}(t)$ is along a chord of the circle, as is $[\mathbf{F}(t + h) - \mathbf{F}(t)]/h$. As $h \to 0$ this chord approaches the tangent line to the circle, a line perpendicular to the radius vector $\mathbf{F}(t)$, so $\mathbf{F}'(t)$ is perpendicular to $\mathbf{F}(t)$.

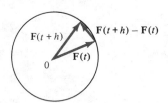

Using Eq. (5.60) and the Pythagorean theorem we can find the speed for uniform circular motion:

$$v = \sqrt{v_x^2 + v_y^2}$$

$$= \sqrt{(-r\omega \sin \omega t)^2 + (r\omega \cos \omega t)^2}$$

$$= \sqrt{r^2\omega^2(\sin^2 \omega t + \cos^2 \omega t)}.$$

Because $\sin^2 \theta + \cos^2 \theta = 1$, we have the simple result

$$v = r\omega. \tag{5.61}$$

This tells us that the *speed* is constant as the particle moves uniformly along the circle. Although the velocity is changing in direction, the magnitude of the velocity is constant. And since the velocity has constant length, its derivative (the acceleration) must be perpendicular to the velocity, by the principle proved in Example 10.

We can easily calculate the acceleration by differentiating the velocity given in Eq. (5.60):

$$\mathbf{a}(t) = \frac{d\mathbf{v}}{dt} = \frac{d}{dt}[(-r\omega \sin \omega t)\hat{\mathbf{i}} + (r\omega \cos \omega t)\hat{\mathbf{j}}]$$

$$= (-r\omega^2 \cos \omega t)\hat{\mathbf{i}} - (r\omega^2 \sin \omega t)\hat{\mathbf{j}}. \tag{5.62}$$

Factoring out $-\omega^2$,

$$\mathbf{a}(t) = -\omega^2[(r \cos \omega t)\hat{\mathbf{i}} + (r \sin \omega t)\hat{\mathbf{j}}], \tag{5.63}$$

and recalling from Eq. (5.55) that $\mathbf{r}(t) = (r \cos \omega t)\hat{\mathbf{i}} + (r \sin \omega t)\hat{\mathbf{j}}$, we cast our result into a simpler form,

$$\mathbf{a}(t) = -\omega^2 \mathbf{r}. \tag{5.64}$$

This tells us that **a** *rotates around with* **r**, *always pointing radially inward*, and is an example of a *centripetal acceleration*.

The magnitude of the acceleration is simply

$$a = \omega^2 r. \tag{5.65}$$

Using Eq. (5.61),

$$v = r\omega, \tag{5.61}$$

we obtain the alternative form

$$a = v^2/r. \tag{5.66}$$

Any object moving uniformly in a circle has a centripetal acceleration, which is directed radially inward and constant in magnitude, as shown in Fig. 5.22.

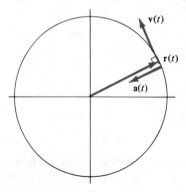

Figure 5.22 Directions of $\mathbf{r}(t)$, $\mathbf{v}(t)$, and $\mathbf{a}(t)$ for uniform circular motion.

Example 11

A space shuttle orbits the earth (radius 6400 km) at 240 km above the surface, making one revolution in 90 min. What is the acceleration of the shuttle?

From Eq. (5.65) we know that the acceleration has magnitude

$$a = \omega^2 r.$$

Since $\omega = 2\pi/T$ by Eq. (5.58), we can write the acceleration as

$$a = 4\pi^2 r/T^2.$$

The period T is 90 min = 5400 s, and the radius r is 6400 km + 240 km = 6640 km, so we obtain

$$a = 4\pi^2(6.64 \times 10^6 \text{ m})/(5400 \text{ s})^2 = 8.99 \text{ m/s}^2.$$

Summary of Vector Algebra

COMPONENTS: $\mathbf{A} = A_x\hat{\mathbf{i}} + A_y\hat{\mathbf{j}} + A_z\hat{\mathbf{k}}, \quad \mathbf{B} = B_x\hat{\mathbf{i}} + B_y\hat{\mathbf{j}} + B_z\hat{\mathbf{k}}$

LENGTH: $A = |\mathbf{A}| = \sqrt{A_x^2 + A_y^2 + A_z^2}$

DOT PRODUCT: $\mathbf{A} \cdot \mathbf{B} = AB\cos\theta$, where θ is the angle from $\mathbf{A}$ to $\mathbf{B}$,
$\quad 0 \le \theta \le 180°$
$\mathbf{A} \cdot \mathbf{B} = A_xB_x + A_yB_y + A_zB_z$
$\mathbf{A} \cdot \mathbf{B} = 0$ if and only if $\mathbf{A} = \mathbf{0}$, $\mathbf{B} = \mathbf{0}$, or
$\quad \mathbf{A}$ perpendicular to $\mathbf{B}$
$\mathbf{A} \cdot \mathbf{A} = A^2$
$\mathbf{A} \cdot \hat{\mathbf{i}} = A_x, \quad \mathbf{A} \cdot \hat{\mathbf{j}} = A_y, \quad \mathbf{A} \cdot \hat{\mathbf{k}} = A_z$

CROSS PRODUCT:
$$\mathbf{A} \times \mathbf{B} = \begin{vmatrix} \hat{\mathbf{i}} & \hat{\mathbf{j}} & \hat{\mathbf{k}} \\ A_x & A_y & A_z \\ B_x & B_y & B_z \end{vmatrix}.$$
$\mathbf{A} \times \mathbf{B} = (A_yB_z - A_zB_y)\hat{\mathbf{i}} + (A_zB_x - A_xB_z)\hat{\mathbf{j}}$
$\quad + (A_xB_y - A_yB_x)\hat{\mathbf{k}}.$
$\mathbf{A} \times \mathbf{B} = -(\mathbf{B} \times \mathbf{A}).$

DIRECTION: $\mathbf{A} \times \mathbf{B}$ is perpendicular to both $\mathbf{A}$ and $\mathbf{B}$, in the direction
$\quad$ determined by the right-hand rule

LENGTH: $|\mathbf{A} \times \mathbf{B}| = AB\sin\theta$, where θ is the angle from $\mathbf{A}$ to $\mathbf{B}$,
$\quad 0 \le \theta \le 180°$
$|\mathbf{A} \times \mathbf{B}| = $ area of parallelogram spanned by $\mathbf{A}$ and $\mathbf{B}$
$|\mathbf{A} \times \mathbf{B}|^2 = A^2B^2 - (\mathbf{A} \cdot \mathbf{B})^2$
$\mathbf{A} \times \mathbf{B} = \mathbf{0}$ if and only if $\mathbf{A} = \mathbf{0}$, $\mathbf{B} = \mathbf{0}$, or
$\quad \mathbf{A}$ is parallel to $\mathbf{B}$

5.9 A FINAL WORD

Before Copernicus, the center of the earth was the center of the universe. That was the starting point for describing something's location in the Aristotelian world. Any other idea about place had no meaning. Then Copernicus came along. He described the motion of the planets in a different coordinate system, and changed the world forever. That's very significant, but somewhat misleading. It suggests that the really important thing in science is to have the right coordinate system. There is an element of truth in this; for a particular problem one often finds that one coordinate system is more convenient or useful than others. But the deeper truth is exactly the opposite. What we finally learned is that all coordinate systems *are equally valid*. Copernicus said the origin is in the sun; the United States Coast Guard, pinpointing the position of a distressed craft, says that the origin is in an airbase – each choice is convenient for the problem at hand, and *both choices are correct!* That's a very valuable lesson, and as a statement, it's even more profound. What it really means is that the laws of physics are the same everywhere in the universe. The laws that Newton gave us work as well in the Crab Nebula as in Kansas City. Because we believe that's true, we need a mathematical device for expressing those laws in a way that's the same in all coordinate systems.

That device is the vector. The idea of a vector is a little disconcerting, because it has a size and a direction, but not a place, unless we find it convenient to give it one. But that makes it a perfect tool for expressing laws, such as Newton's, that work equally well everywhere. The next thing we'll study, Newton's second law, is a vector equation that lies at the heart of our understanding of the world.

Problems

Vectors

1. In Table 5.1, time is listed as a scalar, but often we hear about the "arrow of time," implying that time has a direction from the past toward the future. Then shouldn't time be considered a vector?

2. Sketch the curve for the equation $R = (v_0^2 \sin 2\theta)/g$ of Example 1, expressing R as a function of θ. Rederive the result that maximum range is obtained at $\theta = 45°$ by setting $dR/d\theta = 0$.

3. Determine the components A_x and A_y if **A** represents

   (a) a velocity of 30 m/s along the x direction,
   (b) a displacement of 15 m at an angle of 120° from the x axis,
   (c) an acceleration of 20 m/s² directed 90° from the x axis.

4. A sailboat moves along the axis of its keel. Assume the sail lies in a vertical plane making an angle α with the line of motion, and that the wind makes an angle β as shown in the figure.

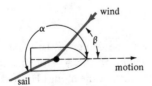

Only the component of the wind normal to the sail exerts a force, call it $\mathbf{F}_w$, on the boat. And only the component of $\mathbf{F}_w$ parallel to the keel is effective in moving the sailboat. (The component normal to the keel is opposed so effectively by water resistance that sideways motion can be neglected.)

(a) If the boat heads into the wind as shown in the figure, for what values of α (in terms of β) will the boat go forward?

(b) What is the optimum angle α (in terms of β) that makes the boat go as fast as possible?

(c) If the wind is directly against the direction we wish to go, so that $\beta = 0$, we must "tack," i.e., reorient the boat by an angle $\Delta\beta$ relative to the wind, and crisscross back and forth with orientations $+\Delta\beta$, $-\Delta\beta$, $+\Delta\beta$, ... , to make a net progress in the desired direction. Assuming the speed of the boat is proportional to the force component you found along the keel, find a formula for the time taken to go a distance L in the desired direction, and show that the optimum angle for tacking in this case is $\Delta\beta = 60°$.

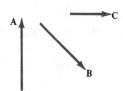

Vector Addition and Subtraction

5. If three vectors add up to zero, what geometric shape do they form when added head to tail? Must three vectors be coplanar (lie in the same plane) in order to add up to zero?

6. For the three vectors shown below, demonstrate associativity of vector addition by constructing geometrically

$$(\mathbf{A} + \mathbf{B}) + \mathbf{C} \qquad \text{and} \qquad \mathbf{A} + (\mathbf{B} + \mathbf{C}).$$

7. Find the magnitude and direction of each of the following vectors:

(a) $\mathbf{s} = (10 \text{ m})\hat{\mathbf{i}} + (30 \text{ m})\hat{\mathbf{j}}$,

(b) $\mathbf{B} = -3\hat{\mathbf{i}} - 4\hat{\mathbf{j}}$,

(c) $\mathbf{v} = (8 \text{ m/s})\hat{\mathbf{i}} - (6 \text{ m/s})\hat{\mathbf{j}}$.

8. Using your combined knowledge of the law of falling bodies and the law of inertia, write down the velocity vector of the projected ball in Fig. 5.9. What is the acceleration vector?

9. A pilot originally started to fly due north at an air speed of 250 km/h, but a strong wind out of the east results in the plane traveling at 289 km/h 30° west of north. What is the velocity of the wind?

10. Raindrops fall vertically with constant speed v. A man runs through the rain with constant speed V. At what angle should he tilt his umbrella forward to keep dry? With what speed do the raindrops strike his umbrella?

11. Consider the following four vectors:

$$\mathbf{A} = 3\hat{\mathbf{i}} + 5\hat{\mathbf{j}}, \quad \mathbf{B} = -2\hat{\mathbf{j}} + \hat{\mathbf{k}}, \quad \mathbf{C} = \hat{\mathbf{i}} + \hat{\mathbf{k}}, \quad \mathbf{D} = 6\hat{\mathbf{i}} + 4\hat{\mathbf{j}} + 3\hat{\mathbf{k}}.$$

    (a) Do **A**, **B**, and **C** lie in the same plane?
    (b) Find a, b such that $a\mathbf{A} + b\mathbf{B} + \mathbf{D} = \mathbf{0}$.

12. In Example 4, find how long the boy's round trip takes if the current is at an angle θ to the line from P to Q. For what value of θ is the round-trip travel time a minimum?

13. If vectors **A** and **B** are consecutive sides of a regular hexagon, determine (in terms of **A** and **B**) the vectors forming the other four sides.

14. A lost dog runs 20 m due north, then 10 m 60° south of east across a field, then 30 m along a road which runs northwest, before stopping.

    (a) Choose an appropriate set of coordinate axes and express the dog's three displacements in terms of components.
    (b) What is the dog's resultant displacement?
    (c) How far is the dog from where he started?

15. Three vectors **A**, **B**, and **C** of magnitude 1, 2, and 3, respectively, lie along the diagonals of the faces of a cube which meet at a corner.

    (a) Choose a coordinate system with the origin at the corner and express each vector in terms of $\hat{\mathbf{i}}$, $\hat{\mathbf{j}}$, $\hat{\mathbf{k}}$.
    (b) Find the components and the magnitude of the resultant $\mathbf{A} + \mathbf{B} + \mathbf{C}$.

16. In chess a knight makes L-shaped moves, two squares in one direction (horizontally or vertically) followed by one square in a perpendicular direction. Use vector methods to determine whether or not a knight can eventually reach every square on an 8 × 8 chess board.

17. A tourist agency wants to start a boat service on the Danube at Budapest. The city is made of two older towns, Buda on the west bank and Pest on the east. Magrit Island is in the river between them. The tour boats go 2.2 m/s in still water, and the Danube flows at 0.9 m/s.

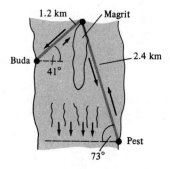

(a) If the route is as shown and the agency wants to have a boat arrive every 20 min, how many boats do they need?

(b) If the boats arrive every 5 min, on which part of the route are successive boats proceeding in the same direction closest together?

Scalar Product

18. Find the scalar product of the vectors **A** and **B** shown below. The magnitude of **A** is 10 and the magnitude of **B** is 12.

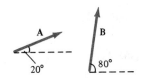

19. (a) If $\mathbf{A} \cdot \mathbf{B} = \mathbf{A} \cdot \mathbf{C}$ does it follow that $\mathbf{B} = \mathbf{C}$?

(b) Can the scalar product of two vectors be negative? If so, what condition must hold?

20. Using the vectors shown below which have magnitudes $A = 3$, $B = 5$, and $C = 6$, calculate the following:

(a) $\mathbf{A} \cdot \mathbf{B}$,   **(b)** $\mathbf{A} \cdot \mathbf{C}$,   **(c)** $(\mathbf{A} + \mathbf{B}) \cdot \mathbf{C}$,

(d) Compare $\mathbf{A} \cdot \mathbf{C} + \mathbf{B} \cdot \mathbf{C}$ with $(\mathbf{A} + \mathbf{B}) \cdot \mathbf{C}$ and verify the distributive law.

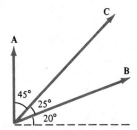

21. **(a)** Calculate the dot product of $\mathbf{A} = 6\hat{\mathbf{i}} + 4\hat{\mathbf{j}} - 5\hat{\mathbf{k}}$ and $\mathbf{B} = \hat{\mathbf{i}} - 2\hat{\mathbf{j}} + \hat{\mathbf{k}}$.

    (b) What is the angle between the vectors $\mathbf{A} = 3\hat{\mathbf{j}} - \hat{\mathbf{k}}$ and $\mathbf{B} = 2\hat{\mathbf{i}} + 2\hat{\mathbf{k}}$?

22. Calculate the value of α such that the vector $\mathbf{A} = \alpha\hat{\mathbf{i}} - 6\hat{\mathbf{j}}$ is perpendicular to the vector $\mathbf{D} = 3\hat{\mathbf{i}} + 2\hat{\mathbf{j}}$.

23. The accompanying diagram shows three vectors $\mathbf{A}$, $\mathbf{B}$, $\mathbf{C}$ arranged to form a triangle. Use properties of vector addition and the scalar product to prove the *law of cosines*:

$$C^2 = A^2 + B^2 - 2AB \cos \gamma.$$

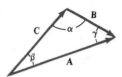

Vector Product

24. Two vectors, $\mathbf{A}$ and $\mathbf{B}$, lie in the xy plane and have magnitude 2.8 and 3.2, respectively, whereas their directions are 210° and 45° measured from the x axis. Find $\mathbf{A} \times \mathbf{B}$.

25. Two vectors are given by $\mathbf{A} = 3\hat{\mathbf{i}} + 5\hat{\mathbf{j}}$ and $\mathbf{B} = -1\hat{\mathbf{i}} + 2\hat{\mathbf{j}} - 3\hat{\mathbf{k}}$. Find $\mathbf{A} \times \mathbf{B}$ and the angle between the vectors.

26. Supply the steps leading from Eq. (5.35) to (5.36).

27. Show that $|\mathbf{A} \times \mathbf{B}|^2 = |\mathbf{A}|^2 |\mathbf{B}|^2 - (\mathbf{A}\cdot\mathbf{B})^2$.

28. Refer to the diagram of Problem 23 and use the cross product to prove the *law of sines*:

$$\frac{A}{\sin \alpha} = \frac{B}{\sin \beta} = \frac{C}{\sin \gamma}.$$

29. The dot product $\mathbf{A} \cdot (\mathbf{B} \times \mathbf{C})$ is a scalar called the *scalar triple product* of $\mathbf{A}$, $\mathbf{B}$, $\mathbf{C}$.

    (a) Prove that its absolute value represents the volume of the parallelepiped spanned by $\mathbf{A}$, $\mathbf{B}$, $\mathbf{C}$, as shown in the figure.

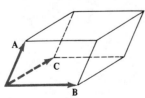

    (b) For the vectors shown in the figure, show by geometrical arguments that

$$\mathbf{A} \cdot (\mathbf{B} \times \mathbf{C}) = \mathbf{B} \cdot (\mathbf{C} \times \mathbf{A}) = \mathbf{C} \cdot (\mathbf{A} \times \mathbf{B}).$$

Derivatives of Vector Functions

30. A particle moves according to $\mathbf{r}(t) = 2t^4\hat{\mathbf{i}} + 5t\hat{\mathbf{j}} - t^3\hat{\mathbf{k}}$, where t is in seconds and r is in meters.

    (a) Find the speed of the particle at $t = 2$ s.
    (b) Calculate the particle's acceleration at $t = 2$ s.

31. The path of an object is described by $\mathbf{r}(t) = 4t\hat{\mathbf{i}} - 2t^2\hat{\mathbf{j}} + 3\hat{\mathbf{k}}$, where t is in seconds and r in meters.

    (a) When is the object at rest?
    (b) Is the acceleration of the object constant?

Uniform Circular Motion

32. A particle moves uniformly in a circle of radius 0.25 m, completing one loop every 20 s. Find

    (a) its speed,
    (b) the magnitude of its acceleration.

33. A jet airplane can withstand an acceleration of 5 g, that is, an acceleration which is five times the acceleration due to gravity. What is the radius of a circle that the jet can safely follow at a speed of 220 m/s?

34. The laws of magnetism are such that an electron traveling 8.0×10^7 cm/s in a direction perpendicular to the earth's magnetic field is forced into a circle 9 cm in radius. An electron traveling parallel to the field is unaffected.

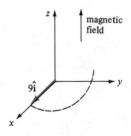

    (a) Using the coordinate system shown, describe the motion of an electron which, at time $t = 0$, has

$$\mathbf{r} = (9 \text{ cm})\hat{\mathbf{i}},$$

$$\mathbf{v} = (8.0 \times 10^7 \text{ cm/s})\hat{\mathbf{j}} + (7.0 \times 10^6 \text{ cm/s})\hat{\mathbf{k}}.$$

    (b) Write an equation for the position $\mathbf{r}(t)$.
    (c) Find the velocity $\mathbf{v}(t)$ and the acceleration $\mathbf{a}(t)$, and show that $\mathbf{v}(t)$ is always perpendicular to $\mathbf{a}(t)$.

CHAPTER

NEWTON'S LAWS AND EQUILIBRIUM

Then from these forces, by other propositions which are also mathematical, I deduce the motions of the planets, the comets, the moon, and the sea. I wish we could derive the rest of the phenomena of Nature by the same kind of reasoning from mechanical principles, for I am induced by many reasons to suspect that they may all depend upon certain forces by which the particles of bodies, by some cause hitherto unknown, are either mutually impelled towards one another, and cohere in regular figures, or are repelled and recede from one another. These forces being unknown, philosophers have hitherto attempted the search of nature in vain; but I hope the principles here laid down will afford some light either to this or some truer method of philosophy.

Isaac Newton, *Principia* (1686)

6.1 THE END OF THE CONFUSION

In 1543 Copernicus published his book, and a tremor rocked the foundations of the Aristotelian world. A century later the Aristotelian world lay in ruins, but nothing had risen to replace it. Galileo and Kepler had made mighty discoveries, but there was no central principle that could organize the world. The unified harmony of the Aristotelian view had been replaced by buzzing confusion.

Galileo was concerned not with the causes of motion but instead with its description. The branch of mechanics he reared is known as *kinematics*; it is a mathematically descriptive account of motion without concern for its causes. Central to Galileo's arguments was the law of inertia, which we discussed in Chapter 4. Armed with this principle, Galileo could neutralize Aristotelian arguments against a moving earth, but the reconstruction of a new mechanics which he promised in his final book had scarcely begun.

The honor of creating a new order fell to Isaac Newton. Newton adopted a statement of the law of inertia as his first law of motion. In his second law he added an explicit rule describing how an impressed force alters a body's motion. By this understanding, he capped and completed Galileo's kinematics with *dynamics* – a theory of the *causes* of motion.

6.2 NEWTON'S LAWS OF MOTION

Newton had hit upon the essence of his second law as a young man, but it was not until 20 years later that the rest of his dynamics fell into place during an intense burst of activity in the autumn and winter of 1684–5. The *Principia*, detailing all his work on the motion of bodies, was published in 1686 when Newton was 44. The style of the *Principia* is reflective of its author: cold and rigid; its pages are laden with diagrams and geometric proofs. Although Newton undoubtedly arrived at his results by using his newly developed calculus, in the *Principia* he presented geometric proofs – the language of physics in the seventeenth century. At Newton's own Cambridge University, a stately institution not given to undue haste, the *Principia* was used as a textbook right into the twentieth century.

Newton inherited from Galileo and Descartes the essential idea that motion along a straight line with a constant speed was the natural state of any body, needing no further explanation. This is Newton's first law, the law of inertia. Stated in his own words,

First Law: *Every body continues in its state of rest, or of uniform motion in a*
             straight line, unless it is compelled to change that state by forces
             impressed upon it.

The essence of the first law is the principle of inertia.

Newton, like Galileo before him, realized that an object's inertia was somehow connected to its mass. The greater the mass of an object, the more difficult it is to prevent it from continuing in motion with a constant velocity. This idea led to his second law. In the *Principia* it is modestly stated as

Second Law: *The change of motion of an object is proportional to the force*
             impressed; and is made in the direction of the straight line in
             which the force is impressed.

What Newton meant by *motion* involved not only a body's velocity, but also its mass. It is the quantity we call *momentum*, the product of mass m and velocity $\mathbf{v}$. Stated as an equation, the second law is

$$\mathbf{F} = \frac{d}{dt}(m\mathbf{v}). \qquad (6.1)$$

The equation can be read ''Force is equal to the rate of change of momentum.'' Like the first law, it holds in general in any inertial frame of reference.

In situations where m is constant, $d(m\mathbf{v})/dt = m \, d\mathbf{v}/dt$ by the rules of differentiation. But we know that the instantaneous rate of change of velocity is acceleration, $\mathbf{a} = d\mathbf{v}/dt$. So for an object (or collection of objects) whose mass doesn't change, Newton's second law tells us that acceleration is caused by forces. It is usually written as

$$\boxed{\text{Second Law: } \mathbf{F} = m\mathbf{a}.} \qquad (6.2)$$

The essence of the second law is that force equals mass times acceleration.

This form was first presented by the Swiss mathematician Leonhard Euler 65 years after the publication of the *Principia*. It is probably the most useful equation in all of physics.

But what is force? In everyday language force is associated with a push or pull. When you push on something, you can feel yourself exerting a force. Once armed with that sensation, you look around and find countless examples of things exerting forces on other things. Pushes, pulls, gravity, tension in a string, and friction are all examples of forces which enter Newton's second law. But these forces must originate *outside* the object whose motion we're trying to describe. In other words, only *external* forces acting on an object can change its motion. Through applications to specific problems, as given later in this and following chapters, we shall flesh out these rather vague statements and see how to use forces in $\mathbf{F} = m\mathbf{a}$ to understand the world.

The force $\mathbf{F}$ need not be just one force acting on the body. It is the *vector sum* of *all* external forces acting on the object. Even though the mathematics of vectors hadn't been invented yet, Newton knew that forces have both a magnitude and a direction. Whenever we write $\mathbf{F} = m\mathbf{a}$, $\mathbf{F}$ symbolizes the *vector sum* of external forces acting on a body. The net acceleration of an object is a result of the total force acting on it. We write $\Sigma \, \mathbf{F}$ to symbolize the vector sum.

The second law, being a vector equation, is shorthand for three equations involving the Cartesian components of $\mathbf{F}$ and $m\mathbf{a}$:

$$\sum F_x = ma_x, \qquad (6.3a)$$

$$\sum F_y = ma_y, \qquad (6.3b)$$

$$\sum F_z = ma_z. \qquad (6.3c)$$

It is this form which is more useful for solving problems. Often, for example, the mass and external forces are known, while the acceleration and motion are to be determined. In such a case we use the known components of external force in this set of equations to solve for the components of the acceleration. From the acceleration, we can then mathematically reconstruct the motion of the body.

Newton added one additional law to express what happens when several bodies interact with one another. His third law is:

Third Law: *To every action there is always opposed an equal reaction; or, the mutual actions of two bodies upon each other are always equal, and directed to contrary parts.*

The essence of the third law is action and reaction.

When you push on anything – a door, a pencil – it pushes back on you with a force equal in magnitude but in the opposite direction. In other words, you can't touch without being touched. That's the essence of the third law – a law of interactions. As illustrated in Fig. 6.1, if Body 1 exerts a force $\mathbf{F}_{12}$ on Body 2, then Body 2 exerts a force $\mathbf{F}_{21}$ on Body 1 such that $\mathbf{F}_{12} = -\mathbf{F}_{21}$.

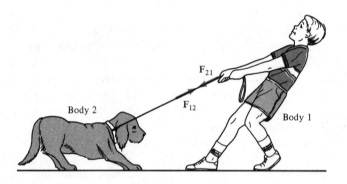

Figure 6.1   An illustration of Newton's third law.

Sometimes it is difficult to isolate the action–reaction pairs of forces in Newton's third law. As a guide, remember that they always act on *different* bodies, never on the same body. If you know one force, for example, you pull on a rope, you can find the reaction force by turning the sentence around: the rope pulls on you. As with the second law, the third law is best understood through applications.

Example 1

A farmer urges an Aristotelian horse to pull his wagon, but the horse refuses to try. In his defense, the horse cites Newton's third law and claims, "If I pull on the wagon, the wagon pulls equally back on me. I can never exert a greater force on the wagon than it exerts on me, so I could never start it moving." What advice would you give the farmer to counter this argument?

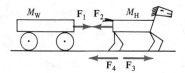

The essential horizontal forces in the problem (neglecting the friction of the wagon wheels) are indicated in the diagram and listed below:

Action Force	*Reaction Force*
$\mathbf{F}_1$ horse pulls on wagon	$\mathbf{F}_2 = -\mathbf{F}_1$ wagon pulls on horse
$\mathbf{F}_4$ horse's feet push back earth	$\mathbf{F}_3 = -\mathbf{F}_4$ earth pushes horse's feet

The only external force on the wagon is $\mathbf{F}_1$ ($\mathbf{F}_2$ doesn't count here!) so its equation of motion is

$$\mathbf{F}_1 = M_W \mathbf{a}.$$

The external forces on the horse are $\mathbf{F}_2 (= -\mathbf{F}_1)$ and $\mathbf{F}_3$, so his equation of motion is

$$\mathbf{F}_3 - \mathbf{F}_1 = M_H \mathbf{a}$$

(horse and wagon accelerate at the same rate, except for transient jerks which we'll ignore).

The equation of motion for the combined horse and wagon system is

$$\mathbf{F}_3 = (M_W + M_H) \mathbf{a}.$$

There are two possible ways to derive the external force $\mathbf{F}_3$ in this case. One is to write the external force on the combined horse and wagon as the sum of the external forces on each, $(\mathbf{F}_3 - \mathbf{F}_1) + \mathbf{F}_1 = \mathbf{F}_3$. Since the horse–wagon reaction pair are internal, they cancel. The second way is to consider, from the beginning, only forces external to the combined system. Here one sees immediately that only the external force exerted by the earth ($\mathbf{F}_3$, reduced in practice by friction) contributes. This second method is preferable, especially for complex systems.

Finally, the earth experiences an external force $\mathbf{F}_4 (= -\mathbf{F}_3)$, so its equation of motion is

$$-\mathbf{F}_3 = (M\mathbf{a})_{\text{earth}}.$$

Of course the earth's acceleration is very small because its mass is so large.

6.3 UNITS OF MASS, MOMENTUM, AND FORCE

Mass, length, and time form the basic physical quantities used in mechanics. The unit of mass in the metric system (SI) is the kilogram, abbreviated kg. The standard kilogram is a platinum–iridium cylinder kept in a vault at the International Bureau of Weights and Measures in Sèvres, near Paris. It is the only SI unit still defined by such an artifact (the standard meter having been recently redefined in terms of properties of light, as described in Sec. 2.4). Secondary standards are housed all over the world. With an equal-arm balance, these standards can determine the mass of objects to a precision of two parts in 100,000. The kilogram is also equal to 1000 grams (1000 g); the gram is the unit of mass in the cgs (centimeter, gram, second) system. Other SI prefixes can be used with the gram, such as milligram (1 mg $= 10^{-3}$ g) and microgram (1 μg $= 10^{-6}$ g). The unit of mass in the British engineering system is the slug; one slug is equal to 14.6 kg.

Momentum, being a product of mass and velocity, is a derived physical quantity. In the metric system, the basic unit is the kilogram-meter per second. In the British system, momentum comes in units of slug-feet per second.

Through Newton's second law, force should have the same units as mass times acceleration. Therefore, the SI unit of force is the kilogram-meter per second squared, or *newton*, abbreviated N. One newton is the force required to accelerate a 1-kg mass at 1 m/s^2:

$$1 \text{ N} = 1 \text{ kg m/s}^2. \tag{6.4}$$

The unit of force in the cgs system is the *dyne* and is the force that will accelerate a 1-g mass at an acceleration of 1 cm/s^2. Using 1 kg $= 10^3$ g and 1 m/s$^2 = 10^2$ cm/s^2, you can show that 1 N $= 10^5$ dyn.

In the SI and cgs systems mass, length, and time are the fundamental quantities. Force is a derived unit. But in the British system, the standard quantities are force, length, and time. The unit of force in the British system is the pound, abbreviated lb, which is 1 slug ft/s^2. By working out the conversion of units, you can show that 1 lb $=$ 4.45 N.

When an object is in free fall, gravity accelerates it downward with a constant acceleration g. Newton's second law tells us that the force must be

$$\mathbf{F} = m\mathbf{g} \tag{6.5}$$

where the direction of the force is vertically downward, toward the center of the earth. This is what we mean by the weight of an object; weight is the force of gravity acting on an object (whether it is falling or not). Being a force, weight is a vector. If we call the magnitude of the vector W, then $W = mg$ near the surface of the earth.

In countries that still use the British system, people are often confused between kilograms and pounds. These units refer to different physical quantities. Yet labels list the weight of an item in pounds along with its mass in kilograms and do not specify that one is weight and the other is mass.* Unlike the mass of a body, which is an intrinsic property of the body, the weight of a body depends on its location. If you know the mass of an object, you can find its weight if you also know the acceleration of gravity at that location. Moving an object around on the surface of the earth doesn't change its weight very much, but moving it to the moon does change its weight considerably without changing its mass. In Chapter 8 we'll find out why weight varies with location when we discuss Newton's universal law of gravity.

Table 6.1 summarizes the units and conversions between the three systems of units.

---

*To be sure, it *is* possible to define a pound of mass (2.2 of which equal a kilogram) with a weight of one pound on the earth's surface. The 2.2 comes from the fact that a kilogram of mass weighs 2.2 lb.

Table 6.1 Units of Mass and Force in the SI, cgs, and British Systems

	Units of mass		
	kilogram	gram	slug
1 kg	1	10^3	0.0685
1 g	10^{-3}	1	6.85×10^{-5}
1 slug	14.6	1.46×10^4	1
	Units of force		
	newton	dyne	pound
1 N	1	10^5	0.225
1 dyn	10^{-5}	1	2.25×10^{-6}
1 lb	4.45	4.45×10^5	1

6.4 PROJECTILE MOTION AS AN APPLICATION OF NEWTON'S SECOND LAW

Galileo was the first to describe the motion of a projectile correctly. Using his laws of inertia and of falling bodies, he showed that projectiles follow parabolic trajectories and was able to deduce many other properties of their motion. Newton's second law does not change Galileo's description of particle trajectories, but it puts the several separate insights that Galileo required into a unified framework, generalizes them, and expresses them in a remarkably concise way. Let us reexamine a number of the properties of projectile motion in the light of Newton's second law.

One famous conclusion of Galileo was that the flight of a projectile is independent of its mass if air resistance is neglected. In Newtonian mechanics one starts with $\mathbf{F} = m\mathbf{a}$ and uses this to deduce the trajectory of the projectile. In this case we don't know at the outset what the force of gravity is but we can calculate it from the law of falling bodies. We learned in Chapter 2 that any body in a vacuum, be it a penny, feather, or cannonball, falls with constant acceleration $\mathbf{a}$ which we can write as

$$\mathbf{a} = -g\hat{\mathbf{k}},$$

$\hat{\mathbf{k}}$ being an upward-pointing unit vector. Since $\mathbf{F} = m\mathbf{a}$ it follows that the gravitational force near the earth's surface is

$$\mathbf{F} = -mg\hat{\mathbf{k}}. \tag{6.6}$$

We shall explore this special property of the gravitational force further in the next chapter.

In Galileo's discussion of projectile motion, the law of falling bodies governed the vertical motion and the law of inertia governed the horizontal motion. The two laws were unrelated. In Newton's formulation, the first law concerning inertia appears as simply a special case of the second law, applying when no force is present. And since $\mathbf{F} = m\mathbf{a}$ is a vector equation, actually expressing three scalar equations, it includes Galileo's case of a force confined to the vertical direction, with motion in the horizontal direction governed by the law of inertia.

But how can a single equation, $\mathbf{F} = m\mathbf{a}$, tell us about the movement of all projectiles? Surely they do not all have the same trajectory! Might we not be better off knowing a little something about the cannon that fires the cannonball? Such as where it was, what direction it was aimed in, and how much powder was used?

The last point, the force momentarily exerted on the cannonball by the explosion of the gunpowder, is an interesting physical question in its own right and determines the initial velocity, but for our present purposes we need only to know the initial velocity, not how it was obtained.

The main point is that Newton's law $\mathbf{F} = m\mathbf{a}$ can be regarded as a *differential equation*,

$$\mathbf{F} = m\frac{d^2\mathbf{r}}{dt^2}, \tag{6.7}$$

that is, an equation involving derivatives of an unknown function, in this case, the position function $\mathbf{r}(t)$. Any function $\mathbf{r}(t)$ satisfying (6.7) is called a solution of the differential equation. Laws of physics which involve rates of change can be expressed as differential equations, and solving a physical problem often requires finding solutions of a differential equation. There are general methods known for solving many types of differential equations, but we do not presuppose a knowledge of these methods. In this book we shall encounter only a few simple differential equations and we will learn how to solve each one as it occurs.

One fact we learned early is that a differential equation can have several solutions. For example, we have just seen that Newton's equation (6.7) with $\mathbf{F} = -mg\hat{\mathbf{k}}$ gives a whole family of parabolic trajectories as solutions. The initial conditions [that is, the initial position $\mathbf{r}(0)$ and the initial velocity $\mathbf{r}'(0)$] determine the particular parabola that solves our problem.

Formulas for particular solutions of differential equations need not be memorized; understanding how they were obtained is more important. The famous twentieth-century physicist Enrico Fermi once remarked that if he had had a good memory he would have been a biologist. In physics you need not memorize detailed equations; if you understand the basic laws and how to solve them, then you have solutions for particular cases at your disposal.

6.5 EQUILIBRIUM: BALANCE OF FORCES

One simple but nonetheless important application of Newton's laws is the treatment of bodies in equilibrium. This subject is called *statics*, implying absence of any motion, and is typically applied to objects at rest on the earth's surface. However, equilibrium does not necessarily mean a state of rest; in fact motion depends on the reference frame, as we discussed in the chapter on inertia, and there is no such thing as absolute rest. A more general and better definition of equilibrium is *absence of acceleration*.

If absence of acceleration defines equilibrium, Newton's second law immediately tells us that the net force acting on a body in equilibrium vanishes:

$$\sum \mathbf{F} = \mathbf{0}. \tag{6.8}$$

Forces may be present, but they cancel in this vector sum.

In order to apply Eq. (6.8) usefully, we need a working knowledge of the properties of several kinds of force. Some of the simplest forces to deal with, apart from gravity, are those in thin ropes, rods, etc. We shall introduce these simple forces here, and return to a more comprehensive discussion of forces in Chapter 8.

Nonrigid elements such as ropes, wires, cables, or chains can only pull, not push. The pull of a rope on an object is directed along its length. By Newton's third law the object pulls back on the rope with an equal and opposite force, putting the rope under *tension*. The tension is transmitted to the other end of the rope.

Example 2

A strongman suspends a weight *mg* by a light rope (of negligible mass) at point P, so that the rope makes an angle θ on both sides as shown.

    (a)   Calculate the tension in the rope on either side of P.

    (b)   Could the strongman support the weight by pulling the rope horizontally?

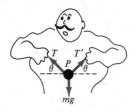

Point P is in equilibrium under the weight *mg* and the tensions in each direction along the rope. The horizontal and vertical components of the condition for equilibrium of forces read

$$T \cos \theta - T' \cos \theta = 0,$$

$$T \sin \theta + T' \sin \theta - mg = 0.$$

The first equation shows that the tension is the same on either side, $T = T'$, and the second equation gives the magnitude,

$$T = \frac{mg}{2 \sin \theta}.$$

If the strongman makes θ smaller and smaller, the tension will grow until either the rope breaks or the strongman reaches the limit of his strength at some small but nonzero angle. For the same reason the main cable supporting the weight of a suspension bridge, for example, must be hung with substantial curvature. If it were stretched straight across, the tension might be great enough to break the cable.

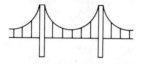

A rigid element such as a rod can either pull or push on an object. By Newton's third law the object pulls or pushes back on the rod with an equal and opposite force, putting the rod under tension (pulls) or *compression* (pushes).

While the tension in a rope is directed along its length, a rigid rod can also support transverse forces. For example, a thick plank crossing a ditch horizontally must be supported by transverse upward forces to hold up its weight (Fig. 6.2a).

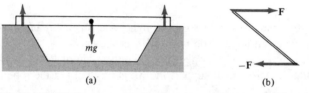

(a)   (b)

Figure 6.2 (a) Transverse forces on a heavy plank in equilibrium. (b) These forces on a light rod, though equal and opposite, produce rotation rather than equilibrium.

One frequently encounters, however, special cases where, to be in equilibrium, the forces applied to a thin rod of negligible weight cannot have a transverse component. In Fig. 6.2b, for example, the equal and opposite forces applied to the ends of the rod tend to rotate it. This shows that although the condition $\Sigma \mathbf{F} = \mathbf{0}$ is necessary for equilibrium, it does not ensure it when the forces are applied to different points. The complete specification of the extra conditions required to ensure equilibrium will be studied in the next section, but in the simple case of Fig. 6.2b one sees that rotation is avoided only if the forces act along the rod.

Example 3

A weight *mg* is held up by a system of rigid but lightweight rods (approximated as weightless) as shown in the figure. At each joint (A, B, C, D) the rods are held by a pin. Find the forces when the system is in equilibrium.

Since the rods are weightless, the only external forces applied to each rod are those exerted by the pins at either of its ends. These must be equal and opposite to satisfy the equilibrium condition (6.8) for the rod. To avoid producing rotation as in Fig. 6.2b, the forces applied to the ends of one of our thin weightless rods must be directed *along* the rod.

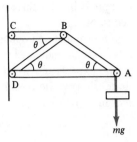

The reaction forces exerted by each rod on its neighboring pins are similarly back to back along the rod, as indicated in the next figure, where the external forces exerted on the pins by the hanging weight (mg) and the wall ($\mathbf{F}_1$ and $\mathbf{F}_2$) are also indicated. To aid in visualization of the forces on each pin, you may find it helpful to cut each rod and replace it by the appropriate force, and to remove the wall support and replace it by a force. In this way you obtain a force diagram for each pin, as suggested by the isolating dashed line around each pin in the figure.

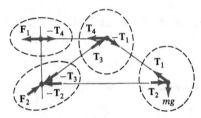

Forces exerted on the pins.

To find the forces, apply $\Sigma\,\mathbf{F} = \mathbf{0}$ at each pin. Starting with pin A at the lower right and writing

$$\mathbf{T}_1 = T_{1x}\hat{\mathbf{i}} + T_{1y}\hat{\mathbf{j}},$$

we find the component force relations

$$T_{1y} - mg = T_1 \sin\theta - mg = 0 \qquad \text{(upward forces)}$$

and

$$T_{2x} + T_{1x} = T_2 - T_1 \cos\theta = 0 \qquad \text{(rightward forces)}$$

with solutions

$$T_1 = mg/\sin\theta, \qquad T_2 = T_1 \cos\theta = mg \cot\theta.$$

For the forces on the upper right-hand pin, B, we find

$$-T_1 \sin\theta + T_3 \sin\theta = 0,$$

$$T_1 \cos\theta + T_3 \cos\theta - T_4 = 0,$$

with solutions

$$T_3 = T_1 = mg/\sin\theta, \qquad T_4 = 2T_1 \cos\theta = 2mg \cot\theta.$$

(Note that since the two equations of equilibrium for pin B involve three initially unknown forces, this pin would not have been a good place to begin.)

By action–reaction the pins exert forces on the rods equal and opposite to the forces we have just found on the pins. These forces on the rods are indicated in the next figure. Note that the directions correspond to intuition: one expects the weight to pull on the upper and upper right-hand bars (putting them in tension) while pushing the lower and lower left-hand bars toward the wall (putting them in compression).

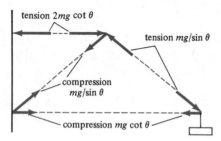

Forces exerted on the rods by the pins.

Finally, it is also easy to determine the forces exerted by the wall by balancing the forces on the upper left-hand pin C:

$$F_{1x} = -T_{4x} = -2mg \cot \theta, \qquad F_{1y} = 0,$$

and by balancing the forces at the lower left-hand pin D, which leads to the result

$$F_{2x} = 2mg \cot \theta, \qquad F_{2y} = mg.$$

The steps we have gone through in Examples 2 and 3 are useful not only for these specific situations, but for many other problems as well. Let us try to list the essential steps in solving a problem in equilibrium mechanics.

Figure 6.3 (a) Gymnast suspended from a rope. (b) Free-body diagram for gymnast.

1. Draw a force diagram or *free-body diagram* for each object you need to analyze. In such a diagram the object is separated from all others and all inessential details, no matter how relevant for other purposes, are eliminated. The objective is to focus your attention on the external forces acting on the body, which must be clearly indicated by arrows which start or end on the body (Fig. 6.3).

2. Label all external forces acting on the objects. In labeling, use the fact that action–reaction pairs of forces are equal and opposite (e.g., in Example 3 keeping track of action–reaction pairs enabled us to relate the forces acting on opposite ends of each rod).

3. Choose a coordinate system for each object under consideration. To simplify the ensuing algebra, choose axes such that the forces whose direction is known *a priori* (such as rope tensions and gravity) have as few nonzero components as possible.

4. Apply the equilibrium condition in component form,

$$\sum F_x = 0, \qquad \sum F_y = 0, \qquad \sum F_z = 0,$$

to *each* object. This requires resolving forces into components.

5. By now you should have as many equations as unknown quantities. If not, see if consideration of further bodies, together with Newton's third law, provides more information [e.g., in Example 3 the tension T_4 in the fourth rod cannot be determined merely by balancing forces on the pins at its two endpoints B and C (try it!). To determine T_4 the forces on pin A must be balanced (even though pin A is not in direct contact with the fourth rod), and then Newton's third law used to relate the force of rod 1 on pin A to its force on pin B, as we did in the solution].

6. Solve the equations. Whenever possible it is best to first solve for the unknowns algebraically, before substituting numbers and units to obtain quantitative answers. This allows you to check your work more easily and reduces errors in calculations.

6.6 EQUILIBRIUM: BALANCE OF TORQUES

Figure 6.2b demonstrated that equal and opposite forces can produce not equilibrium, but a tendency to rotate when applied to different points of an extended body. To establish the full conditions for equilibrium of a rigid body we need a measure of this tendency.

Consider the seesaw pictured in Fig. 6.4. The tendency to rotate counterclockwise increases not only with the weight of the child on the left, but also with his distance from the pivot. Moreover, a force of given magnitude is more effective if applied at right angles to the seat than if applied with a component along the seat toward the pivot.

To express these familiar facts, we call the tendency of a force to produce rotation *torque*, and define its magnitude as the product of the magnitude of a force and the perpendicular distance between its line of application and the axis of rotation. For example, in Fig. 6.5, the torque τ about an axis passing through point P normal to the page is

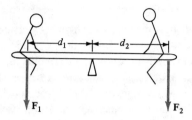

Figure 6.4 Children on a seesaw.

$$\tau = Fd. \tag{6.9}$$

You can think of the distance d as a lever arm. The same force $\mathbf{F}$ applied anywhere along the dotted line would produce the same torque. Equivalently the torque about P can be expressed in terms of r, the distance from the axis of rotation at P to the point of application of the force, and θ, the angle defined in Fig. 6.5, by using $d = r\sin\theta$:

$$\tau = Fr\sin\theta. \tag{6.10}$$

It is important to note that a torque refers to a specific axis of rotation; the same force can exert different torques about different axes of rotation.

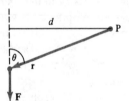

Figure 6.5 Torque $\tau = Fd = Fr\sin\theta$.

Returning once again to the seesaw, we note that equilibrium requires the counterclockwise torque exerted by the child on the left to be balanced by the clockwise torque exerted by the child on the right:

$$F_1d_1 - F_2d_2 = 0. \tag{6.11}$$

The convention is to call the counterclockwise torques positive and clockwise torques negative when drawn on the page and viewed from above.

Example 4
Verify that the external forces applied to the system of rods and pins in Example 3 do not rotate it.

In Example 3 we considered the forces on each individual pin and on each rod, establishing equilibrium and therefore the absence of rotation. But as a check it is illuminating to consider the assemblage of rods and pins (everything contained within the dotted line in the figure) as a unit. All internal forces cancel, and the external forces are as indicated. They add up to zero, a partial check that we have found the right forces for equilibrium. To complete the check, consider the torques about some possible axis of rotation, for example, an axis passing through the lower left pin D perpendicular to the page. Force $\mathbf{F}_1$ exerts a counterclockwise torque $(2mg \cot \theta)L$, and the weight exerts a clockwise torque $mg(2L \cot \theta)$, so the torques indeed balance. We will learn later that there is no torque about *any* axis. You may find it instructive to check that the torques about the upper left-hand pin likewise cancel. This is most easily done by treating the components of $\mathbf{F}_2$ as two separate forces.

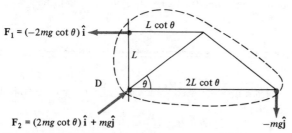

$$\mathbf{F}_1 = (-2mg \cot \theta)\,\hat{\mathbf{i}}$$

$$\mathbf{F}_2 = (2mg \cot \theta)\,\hat{\mathbf{i}} + mg\hat{\mathbf{j}}$$

$$-mg\hat{\mathbf{j}}$$

Forces on the system of rods of Example 3, with relative lengths corresponding to the angle θ.

In dealing with a set of forces, the concept of *resultant* is sometimes helpful. We have already used this term to refer to the sum of vectors. We now extend the meaning of the term and say that the resultant is a force equal to the sum of a set of forces, applied at a point such that it gives the *same torque* as the set of forces.*

For example, gravity pulls downward on all parts of a body. The weight of a body is the resultant of these downward pulls; the point where it acts is called the *center of gravity*. For symmetric bodies of uniform density, the case we shall normally consider, the center of gravity is at the geometric center. In more complicated cases there are methods for determining the center of gravity which we won't describe here; they are very similar to the closely related determination of the center of the mass presented in a later chapter. However, it is worth mentioning an empirical method for determining the center of gravity of a plane lamina. First suspend the lamina from an arbitrary point P and draw the vertical line passing through P. The center of gravity must lie somewhere on this line, since the lamina is in equilibrium under the downward pull of gravity and the upward pull at the point of suspension (Fig. 6.6). Next suspend the lamina from a second point P′ and draw the vertical through P′. The intersection of the two lines is at the center of gravity.

---

*Although, as stated in Chapter 5, vectors can be moved around freely for the purposes of mathematical operations, in physical applications vectors are often fixed at a particular point (e.g., the forces on the pins in Example 3) or along a particular line (e.g., the tensions along the rods in Example 3). The resultant vector in its present extended meaning is an example of this.

We should also acknowledge that it is not always possible to find a resultant. The set of forces $\mathbf{F}_i$ has no resultant when the $\mathbf{F}_i$ give nonzero net torque with $\sum \mathbf{F}_i = \mathbf{0}$ (see, for example, Problem 6.23).

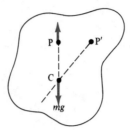

Figure 6.6 Method for determining the center of gravity C of a plane lamina.

Example 5

A horizontal rod is acted upon by two forces as indicated by the solid lines in the figure below. Find the resultant.

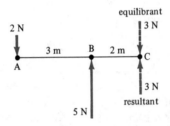

The two forces exert a net force $5 \text{ N} - 2 \text{ N} = 3 \text{ N}$ acting upward, and a net counterclockwise torque of $5 \text{ N} \times 3 \text{ m} = 15 \text{ N m}$ about an axis passing through point A normal to the page. Thus the resultant is an upward-directed force of 3 N acting at point C five meters to the right of point A ($3 \text{ N} \times 5 \text{ m} = 15 \text{ N m}$, the correct torque).

It is important to note that the resultant also gives the same torque as the original forces about an axis through point B, or indeed any other point, as it should to be fully equivalent to the original pair of forces. Thus one method of finding a third force that puts the system in equilibrium is to find the resultant; then the negative of the resultant (called the *equilibrant*) applied at the same point balances the forces and torques.

Example 6

A beam balance consists of a scale, a rigid cross bar and pointer pivoted at P, and weighing pans which hang vertically from each end of the cross bar. The pans are of equal weight W_1, and P is centered as well as possible. The position of the pointer when the pans are empty is marked zero on the scale. The rigid cross bar–pointer system has weight W_2 acting through its center of gravity at point C, a distance b below the pivot P.

When a small weight W_3 is placed in the right-hand pan, the cross bar tilts through an angle α. The new equilibrium condition for torques about the pivot axis through P is

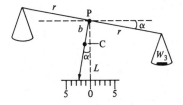

$$W_1 r \cos \alpha + W_2 b \sin \alpha - (W_1 + W_3) r \cos \alpha = 0$$

where the three terms represent the torques due to the left-hand pan, the cross bar–pointer system, and the weighted right-hand pan, respectively. Solving for $\tan \alpha$, we find

$$\tan \alpha = \frac{W_3 r}{W_2 b}.$$

The deflection on the scale, $L \tan \alpha$, equals W_3 times the known quantity $Lr/W_2 b$. The larger this quantity, the more sensitive the scale for measuring W_3. A sensitive scale requires a large but light rigid system (Lr large, W_2 small) with the pointer lighter than the cross bar (b small). The sensitivity is ultimately limited by the conflict between making the cross bar long but light and the need to keep it rigid. Note that the pointer not only points, it also plays an essential role as a counterweight to keep the tilted scale balanced.

Thus far we have defined the torque, due to a force $\mathbf{F}$ acting at distance r from the axis of rotation, as a quantity with magnitude

$$\tau = Fr \sin \theta, \tag{6.10}$$

θ being the angle between $\mathbf{r}$ and $\mathbf{F}$ as drawn in Fig. 6.5. If we look at Eq. (6.10) from a mathematical point of view, we recognize that the right side is the magnitude of the vector cross product $\mathbf{r} \times \mathbf{F}$ defined in the previous chapter. This suggests that we introduce a more general definition of torque as

$$\boxed{\tau = \mathbf{r} \times \mathbf{F}.} \tag{6.12}$$

In this more general definition torque is a vector.

Does such a definition make physical sense? First of all, note that the direction of the vector τ is perpendicular to the plane of $\mathbf{r}$ and $\mathbf{F}$, or, in other words, along the axis of rotation. Thus the direction does have a physical significance related to the rotation the torque would produce. Second, the axis of rotation of a body can point in any direction, so it is reasonable to represent the torque producing the rotation by a vector that can point in any direction.

As shown in Fig. 6.7, the $\mathbf{r}$ in Eq. (6.12) is a vector with its tail at some reference point P and its head at the point of application of $\mathbf{F}$. In speaking of the torque vector, one says it is taken with reference to point P. The axis of rotation runs through P and

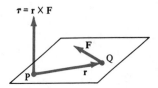

Figure 6.7 Torque about point P due to a force applied at point Q.

has the direction specified by $\mathbf{r} \times \mathbf{F}$, whereas in the scalar definition of torque as $Fr \sin \theta$ we had to specify the direction of the axis separately. Note how neatly the cross product $\mathbf{r} \times \mathbf{F}$, by vanishing when $\mathbf{F}$ is parallel to $\mathbf{r}$, represents the physical fact that a force directed radially outward from P cannot induce rotation about P.

Equation (6.12) ties the convention introduced earlier for positive and negative torques to the right-hand convention for coordinate systems. Consider Fig. 6.8. Since $\mathbf{F}_1$ exerts a counterclockwise torque about the z axis passing through O, its torque is positive according to our convention. The corresponding directional property of the same torque $\boldsymbol{\tau}_1$ as determined by Eq. (6.12) is that $\mathbf{r}_1 \times \mathbf{F}_1$ lies along the $+\hat{\mathbf{k}}$ axis when $\hat{\mathbf{i}}, \hat{\mathbf{j}}, \hat{\mathbf{k}}$ are oriented according to the right-hand rule. Similarly $\mathbf{F}_2$ exerts a clockwise torque, negative according to our convention. Equation (6.12) places $\mathbf{r}_2 \times \mathbf{F}_2$ along the $-\hat{\mathbf{k}}$ axis; in this formulation the sign of $\boldsymbol{\tau}_2$ is reversed from $\boldsymbol{\tau}_1$ because the direction of $\mathbf{r}_2$ is reversed from $\mathbf{r}_1$.

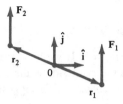

Figure 6.8 Positive and negative torques about O.

Now that we have recognized that both forces and torques must balance to ensure equilibrium, and have identified torque as a vector, we have as the full conditions for equilibrium of a rigid body

$$\sum \mathbf{F} = \mathbf{0},$$ (6.8)

$$\sum \boldsymbol{\tau} = \mathbf{0}.$$ (6.13)

Throughout this chapter we confine ourselves to the special case of forces in a plane, say the xy plane. Such forces produce torques only about the z axis (or an axis parallel to the z axis). Thus the equilibrium conditions for motion in a plane simplify to

$$\sum F_x = 0,$$
$$\sum F_y = 0,$$
$$\sum \tau_z = 0.$$

(6.14)

Conditions (6.14) are sufficient to handle many problems.

For equilibrium to hold, $\sum \tau = 0$ must hold about *all* points. It turns out that *if* $\sum \tau$ *around one point vanishes and if* $\sum F = 0$, *then* τ *vanishes around all points*. The student is urged to check this in Example 4, trying for instance, the point that F_1 or $-mg\hat{j}$ acts upon. The result is physically reasonable because if, for example, a body remains at rest and does not rotate about its center under the action of a set of forces (implying that $\sum F = 0$ and $\sum \tau = 0$ about the center), it certainly does not rotate about any other point either. A general proof proceeds as follows. Suppose a system is in equilibrium, with $\sum F_i = 0$ and $\tau = \sum(r_i \times F_i) = 0$ around some axis. The torque about a second axis displaced by r_0 from the first one is

$$\tau' = \sum_i (r_i + r_0) \times F_i$$

$$= \sum_i (r_i \times F_i) + r_0 \times \sum F_i = 0.$$

(6.15)

The vector τ' equals zero for any r_0 because the first term vanishes by assumption, the displacement r_0 is a common factor that can be taken outside the sum, and $\sum F_i$ vanishes by assumption.

In solving equilibrium problems for motion in a plane when torques as well as forces are involved, the free-body diagrams and other steps recommended in Section 6.5 are again useful. To treat the torques properly, it is essential to indicate carefully on the free-body diagram the point of application of each force as well as its direction. The equilibrium conditions to be used in step 4 are now those of Eq. (6.14). The algebraic expression for the torque can usually be simplified by taking the torque about an axis where a force acts, which is permitted because, in equilibrium, $\sum \tau = 0$ about any axis.

Example 7

The diagram on the next page shows the drawbridge of a medieval castle. The drawbridge has mass M and is supported by two ropes. Assume that the mass of the drawbridge is distributed uniformly, that the tension is the same on each rope, and that the drawbridge cannot go below horizontal.

(a)   What is the minimum breaking tension of the ropes (in units of Mg) that will allow safe operation of the drawbridge?

(b)   If the breaking tension of each rope is $\frac{1}{4}Mg$, at what angle θ will the ropes break if the drawbridge is slowly lowered from the vertical?

(a)   As we'll see later, the tension T in each rope increases as the drawbridge is lowered. So the tension is greatest when the drawbridge is horizontal; let's examine this geometrically simple case first.

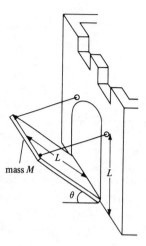

The forces on the horizontal drawbridge are shown in the free-body diagram below. They include $2\mathbf{T} = 2T(\hat{\mathbf{i}} + \hat{\mathbf{j}})/\sqrt{2}$ (doubled because there are two ropes), the weight $-Mg\hat{\mathbf{j}}$, and the unknown force $\mathbf{F}$ exerted by the hinge at the pivot.

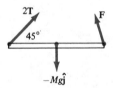

The equilibrium conditions (6.14) are

$$\sum F_x = \sqrt{2}T + F_x = 0,$$

$$\sum F_y = \sqrt{2}T + F_y - Mg = 0,$$

$$\sum \tau \text{ (about P)} = Mg\frac{L}{2} - 2T\frac{L}{\sqrt{2}} = 0.$$

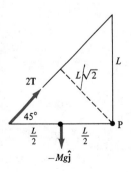

Here the torque condition has been simplified substantially, eliminating the $\mathbf{F}$ dependence, by considering torques about the pivot at P where $\mathbf{F}$ acts.

We have three equations, making it possible to determine all three unknowns T, F_x, and F_y. However, in our case, where we are only interested in T and where $\mathbf{F}$ does not appear in the expression for torque about the pivot axis, we need only to solve the torque condition for the tension:

$$T = \tfrac{1}{4}Mg\sqrt{2}.$$

Since the tension at all higher positions is less than in the horizontal position, the bridge will operate safely if the breaking tension of each rope is greater than $\tfrac{1}{4}Mg\sqrt{2}$.

(b)   To study the tension when the drawbridge is at arbitrary angle θ, it is again simplest to take the torque about the pivot axis. The torque balance equation is now

$$\sum \tau \text{ (about P)} = \frac{1}{2}MgL \cos\theta - 2TL \cos\alpha = 0,$$

giving

$$T = \frac{Mg\cos\theta}{4\cos\alpha}.$$

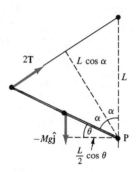

But $2\alpha + \theta = 90°$, so $\cos\theta = \cos(90° - 2\alpha) = \sin 2\alpha = 2\sin\alpha\cos\alpha$. Therefore the formula for T becomes

$$T = \tfrac{1}{2}Mg \sin\alpha.$$

As θ decreases from $90°$ (vertical) to $0°$ (horizontal), α increases from $0°$ to $45°$ and the tension T increases monotonically from $T = 0$ (at $\alpha = 0°$) to $T = \tfrac{1}{4}Mg\sqrt{2}$ (at $\alpha = 45°$). Such an increase is physically reasonable because as the drawbridge is lowered the lever arm of the drawbridge's weight increases, necessitating increased tension to hold it. With regard to part (a), this confirms that our result for $\theta = 0°$ in (a) was the largest tension encountered in operation of the drawbridge. With regard to part (b), the breaking point $T = \tfrac{1}{4}Mg$ for this weaker rope is reached when $\sin\alpha = \tfrac{1}{2}$, i.e., at $\alpha = 30°$, $\theta = 30°$.

6.7   A FINAL WORD

$\mathbf{F} = m\mathbf{a}$ is probably the most useful equation in all of physics. But what does it mean? To understand Newton's second law, we need to answer the questions: What is force? What is mass? Both of these quantities appear in the equation along with acceleration, which we already understand. In the years since Newton, philosophers have debated these questions at great length.

Even though Newton explained what he meant by mass, some people think that mass is a quantity which is defined by this equation. If you know the magnitude of the force acting on an object, measure the magnitude of the resulting acceleration; then the mass of the object is $m = F/a$. Other people think that the second law is a definition of force. Knowing the mass of an object, measure its acceleration, and the applied force can be found through Eq. (6.1). Now, if the second law is a definition of either mass or force, it is not a deep discovery in physics; it is simply a definition. But if the equation is used to define *two* of the three quantities which appear in it, then it has no meaning at all. Can it be that Newton's famous second law is meaningless?

The way to understand what $\mathbf{F} = m\mathbf{a}$ means is by making use of it. Through applications to specific problems, such as those given throughout this chapter, we can see how it works and understand how it organizes the world. Sometimes it is used to determine a mass. Once determined, the response of that mass to various different forces can be studied. Sometimes $\mathbf{F} = m\mathbf{a}$ is used to determine a force, whose effect on various different masses is then studied. In many cases m and $\mathbf{F}$ are known independently and $\mathbf{F} = m\mathbf{a}$ successfully predicts the motion. Though the precise logical status of Newton's second law may pose a recondite philosophical question, in practice simple, workable determinations of $\mathbf{F}$ and m exist which lead to consistent results over the enormous range of phenomena described by classical physics.

Beyond its usefulness, Newton's second law changed the whole nature of physics. Before the time of Copernicus and Galileo, physics had been descriptive. Aristotle provided a qualitative description of motion on and near the earth's surface, and in the heavens a complex set of "epicycles" describing the motions of stars and planets had come down from the Greeks via Ptolemy. Copernicus and Galileo shattered the old framework without fully replacing it. Newton's second law introduced not only a new order, but also a new focus which has characterized physics ever since. It has a mass factor, leading to the questions: What is it that has a mass in the universe? What does matter ultimately consist of? And it has a force term, leading to the questions: What forces operate on matter? What are the fundamental forces of nature? These were the central questions suggested by Newton's second law, and they are still the central questions of physics even today.

Problems

Newton's Laws of Motion

1. Suppose you have two identical cans, one filled with lead and the other empty, which are in an orbiting spacecraft where everything is weightless. How can you tell which can is empty without looking inside?

2. What physical principles are behind the reasoning for making wrecking cranes with massive weights at the end of a cable?

3. A train consisting of an engine and three boxcars moves down the track with a constant acceleration. Between which two cars is the tension in the coupling the greatest? the least? why? If the boxcars have equal mass, what are the ratios among the tensions between the three cars?

4. While you are driving along the freeway, a bug splatters on your windshield. Which experiences the greater force, the bug or the windshield? Which experiences the greatest acceleration?

5. Discuss whether the following pairs of forces are action–reaction forces:

   (a) An athlete standing on a scale pushes down on it; the scale pushes up on the athlete.
   (b) The earth attracts a stone; the stone attracts the earth.
   (c) The tires of a car push on the road; the earth pulls down on the tires.
   (d) A chair pushes down on the floor; gravity pulls down on the chair.

6. A fan is mounted on a cart as shown below. If the fan is turned on, does the cart move? If so, in which direction?

   Suppose a sail were added to the cart. What would be the motion of the cart if the fan were now turned on?

Mass and Weight; Units

7. Determine whether the following combinations of units are units of mass, momentum, or force:

   (a) N s,   (b) dyn s^2/cm,
   (c) slug ft/s,   (d) lb s.

8. (a) How many newtons does a typical 160-lb man weigh?
   (b) What is the mass of a 0.75-lb can of beans?

9. A rock weighs 60 N on the moon, where the acceleration due to gravity is one-sixth that on Earth. What is the mass of this rock on the earth?

10. Suppose you hand an object to each of two people and ask them to guess its weight. One holds it still and guesses; the other hoists it up and down before guessing. One is estimating the mass and the other the weight. Which is which? Explain.

11. A jar of lightning bugs is tightly capped. Does it weigh more, less, or the same when the bugs are flying around compared to when they are at rest?

Projectile Motion

12. Suppose you made a movie of an arrow flying through the air, then played it backward. The arrow would be seen to move in reverse. What would be the direction of the arrow's acceleration? Is it also reversed?

13. A monkey hunter sits on the ground armed with a tranquilizer dart gun. He aims at a monkey hanging from a tree. Startled by the noise of the gun, the monkey lets go of the branch at the same instant the dart leaves the gun. Explain why the dart will strike the monkey as he falls to the ground. How does the initial speed of the dart affect where it will hit the monkey?

14. Suppose air resistance is a force in the direction opposite to the body's motion and proportional to the body's speed.

    (a) How can this be expressed by using Newton's equations? Write three differential equations for the components of the motion.
    (b) For a body which feels no forces other than air resistance and which has an initial velocity $\mathbf{v}_0$, solve the equations from (a).

15. Two identical cannons, A and B, are aimed at each other as shown. Cannonballs from each are fired simultaneously and at the same speeds.

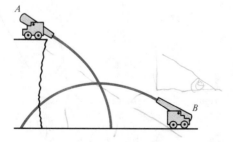

Discuss which of the following statements about the subsequent motion is correct:

    (a) Cannonball A hits the ground first.
    (b) Cannonball A is in the air longer.
    (c) The two cannonballs collide in midair.
    (d) They reach the ground at the same time.

Equilibrium: Balance of Forces

16. A wire 70 cm long is stretched loosely across the back of a picture weighing 3 kg, with the two ends of the wire placed 50 cm apart. If the picture is hung from the center of the wire, how much tension must the wire be able to support?

17. (a) A 5-kg bunch of sausages is suspended from the ceiling using a massless rope. What is the tension on the rope?

    (b) The same sausages are suspended using a 1-m-long chain weighing 2 kg. What is the tension on the lower end of the chain? On the upper end?

    (c) What is the tension at height h above the bottom of the chain?

18. A weathered 60-cm, 50-g cord supports a 1-kg bird feeder. If the cord will snap when the tension in it exceeds 20 N, where will it break when a 1-kg squirrel climbs onto the feeder? Give your answer as a distance above the feeder.

19. The pulleys in the picture below are frictionless and weightless. Find the weight W and the tension in each rope such that the system is in equilibrium, and find the downward pull of the support at A on the ceiling.

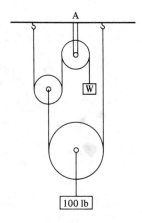

20. After a snowy Wisconsin winter one sometimes sees buildings whose walls have been pushed out by the weight of the roof and the snow.

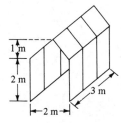

    (a) For the building shown, what is the horizontal force that each of the eight rafters exerts on the walls? The roof has mass 100 kg and the rafters share the load equally.

(b) Slushy snow has a density of 0.1 g/cm^3. How much does the force on the walls increase because of a 60-cm snowfall? Compare this to the force in (a).

21. A 70-kg window washer is sitting on a 25-kg boatswain's chair. With one hand he holds a rope which supports him and the chair in equilibrium. With the other hand he uses a squeegee on the windows. The rope's weight is negligible compared to that of the chair and the man.

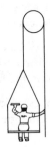

(a) Draw separate force diagrams of the window washer and the chair showing the forces acting on each.

(b) What is the tension in the rope? (In reality there would be a block-and-tackle arrangement, so continuous application of such strength would not be required!)

(c) How large is the force the chair seat exerts on the window washer?

Equilibrium: Balance of Forces and Torques

22. A mobile sculpture is made of four weights and three lightweight 2-m sticks, as in the diagram.

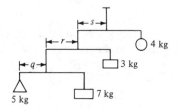

(a) Draw a force diagram of the lowest stick showing the forces exerted on it by the 5- and 7-kg weights.

(b) What is the resultant of the forces shown in (a)? What is the relation between the force exerted by the cord and the equilibrant of the forces shown in (a)?

(c) How should the lengths q, r, and s be chosen so that the sculpture will hang as shown?

23. (a) What is the total force on the bar shown?

(b) What is the net torque about the point a?

(c) This configuration of forces is called a "couple." Is it possible to find a resultant for a couple?

(d) Show that the torque around any point on the bar is the same as the torque around the point a.

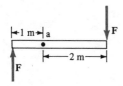

24. The weight of the 400-lb boom in the derrick pictured below is uniformly distributed along its length. By considering the equilibrium conditions for the boom, find the tension in the cable running from A to B (which need not be the same as the tension in the cable below B) and the horizontal and vertical forces exerted on the boom at point C by the hinge of the derrick. By considering the equilibrium conditions for point B, calculate the force exerted by the boom on the cable at point B.

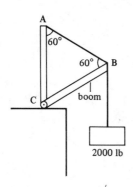

25. Two clowns, Orsene and Waldo, support a 3-m-long, 10-kg plank while a third, Bobo, rides a unicycle back and forth between the two ends at a steady speed. Bobo and the unicycle together come to 55 kg. If Orsene can't hold masses over 40 kg for more than 5 s, how fast should Bobo ride?

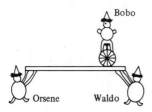

CHAPTER 7

UNIVERSAL GRAVITATION AND CIRCULAR MOTION

Hitherto we have explained the phenomena of the heavens and of our sea by the power of gravity, but have not yet assigned the cause of this power. This is certain, that it must proceed from a cause that penetrates to the very centres of the sun and planets, without suffering the least diminution of its force; that operates not according to the quantity of the surfaces of the particles upon which it acts (as mechanical causes used to do), but according to the quantity of the solid matter which they contain, and propagates its virtue on all sides to immense distances, decreasing always as the inverse square of the distances.

Isaac Newton, *Principia* (1686)

7.1 THE GENESIS OF AN IDEA

The year was 1665; the month was August; and England was besieged by bubonic plague. Isaac Newton, then a 23-year-old Cambridge University student, retired to the solitude of his family's farm in Lincolnshire until the plague subsided and the university reopened. Not given to inactivity, Newton composed 22 questions for himself ranging from geometric constructions to Galileo's new mechanics to Kepler's planetary laws. During the next 18 months, he immersed himself in the search for answers and along the way discovered calculus, the laws of motion, and the universal law of gravity.

There is a myth that Isaac Newton was inspired one of those plague days in his Lincolnshire orchard, by the fall of an apple, to consider whether gravity was responsible for the motion of the moon as well. Newton himself never wrote about that day in the orchard, but he did reminisce about it to friends some 50 years later. He must have had a definite image in mind when he compared the way the moon "falls" with gravity on Earth, and there is every reason to believe that the fall of an apple gave rise to it.

But the story of the apple reduces one of the greatest discoveries of mankind to a simple bright idea – a flash of insight. In fact the universal law of gravitation did not yield even to the great Newton at his first effort. He battled with it and struggled with questions on the behavior of gravity. In what way must gravity depend on distance to account for Kepler's third law, which relates the period and radius of a planet's orbit? What other physical quantities could this force depend on? And how is Galileo's law of falling bodies – gravity on the earth – related to gravity in the heavens?

The secret of the heavens was Newton's for nearly 20 years. In 1684 he amazed a trusted friend, Edmund Halley, when he calmly stated that a force law which decreases inversely as the square of the distance leads to orbits which are conic sections (ellipses, circles, parabolas, and hyperbolas). At Halley's entreaty, Newton wrote a nine-page paper, "On the Motion of Bodies in Orbit," which divulged his secret of universal gravitation to the world and later grew into the *Principia*. Halley recognized that Newton's short paper embodied an immense step forward: the physics of the heavens became the same as the physics of the earth.

7.2 THE LAW OF UNIVERSAL GRAVITATION

Newton had been struggling to find an explanation for the basic rules of planetary motion, which had been laid down by Johannes Kepler half a century earlier. What he perhaps realized that day in Lincolnshire was that the explanation of Kepler's orbits would also explain why an apple falls to the earth. But the answer, if he could find it, would also have to resolve the riddle of why all bodies fall at the same rate regardless of their mass.

From his study of Kepler's orbits, Newton already had an inkling of what he needed. The force between any two bodies in the universe would have to diminish as the bodies moved further apart. The force, he said, would be inversely proportional to the square of the distance between the two bodies. This relationship would satisfy Kepler's empirical law relating the radius of an orbit to its period.

To complete his law of universal gravitation, Newton said that the force of gravity is proportional to the mass of each of the two bodies involved. If m_1 and m_2 are two masses and r is the distance between them, Newton's law of universal gravitation may be expressed as follows:

$$F = G\frac{m_1 m_2}{r^2} .$$

(7.1)

The constant G is a universal constant, having the same value for any two bodies in the universe. This constant should not be confused with g – the acceleration of a body on the surface of the earth due to the earth pulling on it. The value of G is 6.67×10^{-11} N m^2/kg^2; in Chapter 8 we'll find out how it was first measured.

Force is a vector quantity but the relation above only states its magnitude. Since gravity tends to pull objects directly toward one another, the direction of the gravitational force between any two bodies is along the lines that join them. We signify the direction by using the unit vector $\hat{\mathbf{r}}$, pointing, as shown in Fig. 7.1, in the direction from one mass

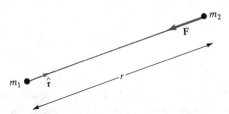

Figure 7.1 Vector quantities for the force exerted by m_1 on m_2 in Newton's law of universal gravitation.

(say m_1) to the mass (say m_2) on which you want to know the force. Like the unit vectors we encountered in Chapter 5, $\hat{\mathbf{r}}$ is dimensionless with length one ($\hat{\mathbf{r}} \cdot \hat{\mathbf{r}} = 1$). The force on m_2, being attractive, is in the direction opposite to $\hat{\mathbf{r}}$. Therefore the vector equation for the universal law of gravity is

$$\mathbf{F} = -G\frac{m_1 m_2}{r^2}\hat{\mathbf{r}}. \qquad (7.2)$$

Equation (7.2) tells us the force between two *point masses*, which are idealized objects having all their mass concentrated at one point. If there are several masses, like m_1, m_2, and m_3 as shown in Fig. 7.2, how would we calculate the gravitational force on

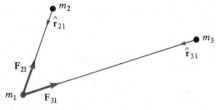

Figure 7.2 Gravitational force on one point mass due to two other point masses.

one of them, say m_1? If only m_1 and m_2 were present, the force from m_2 on m_1 would be

$$\mathbf{F}_{21} = -G\frac{m_2 m_1}{r_{21}^2}\hat{\mathbf{r}}_{21}, \qquad (7.3)$$

where r_{21} is the distance between m_1 and m_2 and $\hat{\mathbf{r}}_{21}$ is the unit vector pointing from m_2 to m_1, as shown in Fig. 7.2. Similarly, if only m_1 and m_3 were present, the force from m_3 on m_1 would be

$$\mathbf{F}_{31} = -G\frac{m_3 m_1}{r_{31}^2}\hat{\mathbf{r}}_{31}. \tag{7.4}$$

If both m_2 and m_3 are attracting m_1, the total force on m_1 is the vector sum of $\mathbf{F}_{21}$ and $\mathbf{F}_{31}$:

$$\mathbf{F}_1 = \mathbf{F}_{21} + \mathbf{F}_{31}$$

$$= -G\frac{m_2 m_1}{r_{21}^2}\hat{\mathbf{r}}_{21} - G\frac{m_3 m_1}{r_{31}^2}\hat{\mathbf{r}}_{31}. \tag{7.5}$$

The *superposition principle* we have used here – that *the resultant force on a mass is the vector sum of the individual forces* – is an important aspect of Newton's second law which allows us to calculate the gravitational force on an object from any number of bodies.

Example 1

Locate the point between two fixed masses $m_3 = 50.0$ kg and $m_2 = 80.0$ kg, which are separated by 1.0 m, where a third mass $m_1 = 10.0$ kg feels no force.

Let's call x the distance from m_1 to m_2. Then the distance from m_1 to m_3 is $1 - x$, as shown below.

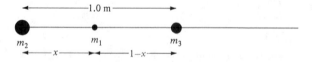

The forces $\mathbf{F}_{21}$ and $\mathbf{F}_{31}$ are in opposite directions, so the magnitude of the total force on m_1 is

$$F_1 = G\frac{m_3 m_1}{(1 - x)^2} - G\frac{m_2 m_1}{x^2}.$$

Setting this force equal to zero and solving for x, we obtain (after a dose of algebra)

$$x = \frac{1}{1 \pm \sqrt{m_3/m_2}}.$$

The $\pm$ comes from taking square roots. Since the distance x must be positive and less than 1 m, the physically acceptable solution is the one with the $+$ sign. Therefore our answer is

$$x = \frac{1}{1 + \sqrt{50/80}} = 0.56 \text{ m}.$$

A good technique for checking the answer to any physics problem that involves several parameters, such as m_2 and m_3 in the present problem, is to look for limiting cases where the problem simplifies, allowing the answer to be tested against common sense or against a familiar result of a simpler problem. For example, in the present problem you can guess the answer without doing any algebra in the limiting cases $m_3/m_2 \to 0$, $m_3 = m_2$, and $m_2/m_3 \to 0$. And indeed the general answer $x = (1 + \sqrt{m_3/m_2})^{-1}$ approaches the expected values 1 m, 0.5 m, and 0 in these three cases.

Now consider an apple of mass m plummeting to the earth. What is the r characterizing the "distance from the earth," and what is the direction $\hat{\mathbf{r}}$? The gravitational law says that each bit of apple is attracted by each bit of the earth; the forces $\mathbf{F}_1$ and $\mathbf{F}_2$ in Fig. 7.3 are two such forces of attraction. To find the total force on the apple we would use the superposition principle, vectorially adding all the forces between each bit of the apple and each bit of the earth. In essence, the calculation requires integration.

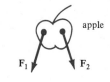

Figure 7.3 Calculating the gravitational force on an apple.

Newton seems to have grasped the solution of this problem immediately, but apparently had difficulty in proving it mathematically. He was a man of rigor and would not consider exposing his ideas in print before he had completely satisfied himself of their soundness. But eventually Newton, with the power of his integral calculus, proved that when two spherically symmetric objects are not touching, each acts as if all its mass were concentrated at its center. This important realization allows us to consider the earth as a point mass having all its mass concentrated at its center. It is a property unique to the $1/r^2$ force law.

Newton's proof can be carried out rigorously by a straightforward, if somewhat lengthy, integration. But we shall content ourselves with a nonrigorous physical argument that emphasizes the unique nature of a $1/r^2$ force law.

To begin with, we remark on how a $1/r^2$ law arises in nature. For example, consider a physical quantity seemingly unrelated to gravitational force: the light from a spherical source. If there is no medium to absorb the light, the total amount of light does not diminish at all with distance r from the source. But the intensity (the amount of light per unit area) falls as r^{-2}. This is easily seen by drawing a concentric sphere around the sources and noting that at the surface of the sphere

$$\text{Intensity} = \frac{\text{Amount}}{\text{Area}} = \frac{\text{Amount}}{4\pi r^2} . \tag{7.6}$$

So a $1/r^2$ law occurs when something is being conserved as it spreads out radially in all directions (in the case of light, the number of individual particles called photons is the "amount" that is being conserved).

This suggests not a wild-eyed identification of light with gravity, but a physical picture of gravity in terms of conserved "lines of constant force" spreading out radially and symmetrically from each point-mass source. The number of lines of force is proportional to the mass of the source.* Lines of force from different sources do not bend or interact with one another. Each object pierced by parallel lines of force feels a gravitational attraction proportional to the number of lines piercing it, and along their direction, irrespective of how far away the lines of force originated. The forces due to nonparallel lines add vectorially, and the outward-spreading lines of force do not terminate anywhere in space. It is this last "conservation property" which suggests that the density of lines, and thus the force, from a point source decreases as $1/r^2$.

Now suppose we introduce an observational sphere concentrically surrounding the mass point which is the source of lines of force (Fig. 7.4a). If we move the source off center, the density of force lines changes at certain portions of the sphere. For example,

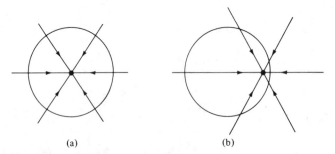

(a)   (b)

Figure 7.4 Force lines due to a mass point (a) at the center of an observational sphere (b) off center.

in Fig. 7.4b there are five lines piercing the right hemisphere and only one piercing the left hemisphere. This means that the force is greater on the right, close to the off-center mass point, than on the left. But we make a new and crucial observation: moving the

*The lines of force concept presented above was developed for analogous problems in the theory of electric fields by the great nineteenth-century English scientist Michael Faraday. Lacking the benefits of formal schooling, he was left a lifelong illiterate in mathematics, but nevertheless made the crucial discoveries of his time in electricity and magnetism with the aid of physical concepts such as lines of force. The concept can be made rigorous with the aid of a mathematical result called Gauss's theorem, and it has been extremely fruitful. Handed down to the next generation of physicists, lines of force were transmuted into that most subtle and powerful of all ideas of modern physics, the idea of the field of force.

source off center does not change the total number of force lines piercing the sphere. That number is a conserved quantity, proportional to the mass.

As a particular application, let us arrange a large number of mass points into a spherically symmetric mass, inside and concentric with our observational sphere. Although the force from an individual off-center mass point on the right is stronger on the right side of the observational sphere, as depicted in Fig. 7.4b, the force from an off-center mass point on the left is correspondingly stronger on the left side. The result is that the total force, i.e., the superposition of the forces from all the individual mass points, is uniform and radially inward everywhere on the observational sphere by symmetry (Fig. 7.5a). And – crucial property – the total number of lines piercing the observational sphere is proportional to the total mass of the source independent of the configuration of the mass. The number of lines of force remains unchanged if the source is compressed to a point at its geometric center, as pictured in Fig. 7.5b. So the force exerted at the observational sphere by a spherically symmetric mass is the same as if all the mass were concentrated at its center. Using the special properties of the $1/r^2$ law, we can see why Newton's result is reasonable without explicit integration!

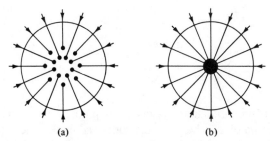

(a)　　　　　　　　　　　(b)

Figure 7.5 Force lines due to (a) a spherically symmetric array of mass points, concentric with an observational sphere; (b) the same total mass as in (a), concentrated at the center.

Our argument clearly applies to spherical shells as well as solid spheres; the gravitational force from a mass distributed symmetrically over a spherical shell is the same as if the mass were concentrated at the center, if you're at a point outside the shell. Newton realized that spherical shells have an additional intriguing property: the force of gravity from a mass distributed symmetrically over a spherical shell is *zero* anywhere *inside* the shell. To understand this result, we present an argument due to Newton.

Newton considered first a thin, uniformly dense spherical shell. To find the gravitational force on a small mass m located at point P inside this shell, he constructed a narrow double cone with the apex at P intersecting the areas A_1 and A_2 on the shell as shown in Fig. 7.6a. Area A_1 is a distance r_1 from P, and A_2 is a distance r_2 from P. The mass contained in each of these areas pulls on m; they pull in opposite directions so there is a cancelation.

To show that the cancelation is exact, Newton needed to know how much mass was contained in each area. Let ρ represent the mass density (the mass per unit volume). Because the density of the shell is constant, the mass m_1 in the area A_1 is the product of

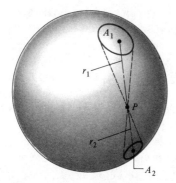

Figure 7.6 (a) Determination of the force on a point mass inside a spherical shell. (b) A straight line intersecting a sphere makes the same angle θ with respect to the normal to the sphere at both intersections.

the density ρ, the area A_1, and the shell thickness t. In other words, $m_1 = \rho A_1 t$. Similarly, $m_2 = \rho A_2 t$. Now we can write the magnitude of the force on m:

$$F = G\frac{mm_1}{r_1^2} - G\frac{mm_2}{r_2^2}$$

$$= Gm\rho t\left(\frac{A_1}{r_1^2} - \frac{A_2}{r_2^2}\right).$$

But the area A_1 is proportional to r_1^2, and A_2 is proportional to r_2^2. The coefficients of proportionality depend on the shapes of the cones and on the orientations of the cones with respect to the intersected surfaces. Because the upper and lower cones have the same shape, and because their midline intersects A_1 and A_2 with the same orientation as indicated in Fig. 7.6b, the coefficients of proportionality are the same for A_1 and A_2, and the force on mass m due to A_1 and A_2 cancels completely. This can be repeated with double cones covering the entire shell. Thus, because the gravitational force decreases inversely with distance squared, the force on a mass inside a thin, uniform spherical shell is zero.

By adding together the forces from concentric thin spherical shells, Newton was able to generalize this result to an arbitrary thick spherical shell. This construction shows that the gravitational force vanishes inside a spherical shell even if the density varies from one thin component shell to the next.

Newton's result for the force inside a spherical shell can be used to estimate the gravitational force acting on an object inside the earth.

Example 2

Suppose that a tunnel could be drilled completely through the earth, passing through its center. Assuming (not very accurately!) that the earth is a uniformly dense sphere, find the gravitational force on a mass m inside the tunnel.

We know that the gravitational force on a body located a distance r from the center of the earth is due entirely to the amount of matter within a sphere of radius r. The shell of matter outside of the object exerts no force on it. Let's call the earth's density ρ. Then the mass M inside a sphere of radius r is the product of ρ and the volume of the sphere, $\frac{4}{3}\pi r^3$, that is, $M = \frac{4}{3}\pi\rho r^3$. This mass can be treated as if it were concentrated at the center of the earth. Therefore the force on the mass m is (in magnitude)

$$F = G\frac{Mm}{r^2} = G\frac{\frac{4}{3}\rho\pi r^3 m}{r^2} = kr$$

where $k = \frac{4}{3}\pi G\rho m$, and is directed toward the center of the earth.

As shown in Example 2, if the earth is approximated by a uniformly dense sphere, the force of gravity decreases linearly as you penetrate the earth. To be sure, the force contains a factor $1/r^2$ that increases as you move closer to the center, but this is more than compensated by the r^3 decrease that occurs because spherical layers of the earth outside a particle do not contribute any force on it. We know that the gravitational force also decreases, like $1/r^2$, as you move outward from the surface. Thus one finds that the gravitational force exerted on a particle by the earth is maximal at the earth's surface. Figure 7.7 shows a graph of the magnitude of the gravitational force as a function of distance from the earth's center.

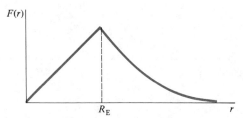

Figure 7.7 Graph of gravitational force versus distance from the center of a uniformly dense earth.

7.3 ACCELERATION OF GRAVITY ON THE EARTH

We can use the law of universal gravitation, together with Newton's second law, to derive Galileo's law of falling bodies and understand g.

When you employ the law of universal gravitation near the earth, you can think of the earth as having all its mass concentrated at its center. Let's apply this law to an apple falling near the surface of the earth. By Eq. (7.1), if m is the mass of the apple and M_E is the mass of the earth, the magnitude of the force of gravity on the apple is

$$F = G\frac{mM_E}{r^2}. \qquad (7.1)$$

The direction of this force is along the line from the center of the apple to the center of the earth and r is the distance from the center of the apple to the center of the earth. As shown in Fig. 7.8, r is equal to the radius of the earth R_E plus the height h of the apple above the surface of the earth. So the force of gravity can be expressed as

$$F = G\frac{mM_E}{(R_E + h)^2}. \qquad (7.7)$$

We can gain greater insight into this result by a simple example. The radius of the earth is about 6000 km or 4000 mi. The tallest building is about 350 m high. So if we

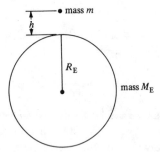

Figure 7.8 Quantities pertaining to gravity on the surface of the earth.

drop something from the tallest building, the distance from the center of the earth to the object changes from

6000 km + 350 m

to 6000 km. The relative change in distance is $(350\ \text{m})/(6 \times 10^6\ \text{m})$ or about 6/100,000, a negligible amount for most purposes. Therefore a very good approximation is to ignore h completely in the force, and write

$$F = G\frac{mM_E}{R_E^2}. \qquad (7.8)$$

According to Newton's second law, $\mathbf{F} = m\mathbf{a}$, we can calculate the acceleration of a falling body by substituting F from the gravitational law. Doing just that, we write

$$G\frac{mM_E}{R_E^2} = ma. \qquad (7.9)$$

The mass of the object appears on both sides of the equation, so it cancels out, leaving

$$a = G\frac{M_E}{R_E^2}.$$

(7.10)

This is a remarkable result. Everything on the right-hand side of the equation is a constant on the earth, yet the left-hand side refers to the acceleration of an object falling near the surface of the earth. We've found the reason for Galileo's law of falling bodies. All objects near the earth fall with the same constant acceleration because the force of gravity has little spatial variation near the earth's surface, and because it depends linearly on the mass of the falling object. A more massive object feels a stronger force, and so, by $a = F/m$, the acceleration is the same. This constant acceleration we called g, and now we see how it is related to the characteristics of the earth (or any other planet):

$$g = G\frac{M_E}{R_E^2}.$$

(7.11)

Example 3

Knowing that the mass of Mars is 0.1 the mass of the earth and that its radius is 0.5 that of the earth, what is the acceleration due to gravity on the surface of Mars?

From Eq. (7.11) we can set up a ratio of g_M to g on the earth:

$$\frac{g_M}{g} = \frac{GM_M/R_M^2}{GM_E/R_E^2} = \frac{M_M}{M_E}\left(\frac{R_E}{R_M}\right)^2,$$

and inserting values of the ratios, we have

$$g_M = g(0.1)(1/0.5)^2 = 0.4g = 3.9 \text{ m/s}^2.$$

7.4 WHY THE MOON DOESN'T FALL TO THE EARTH

Though at first Newton didn't name the force explicitly, he knew that something had to attract the moon if it was to remain in orbit, for the law of inertia stated that the moon would tend to travel in a straight line unless some force acted on it. He coined the word *centripetal force* for any force which is directed inward, toward the center of an object's motion (centripetal means "center seeking"). In the case of the moon, gravity is the centripetal force that holds it in orbit.

Before he compared the force of gravity on a falling apple and on the moon, Newton realized that any satellite (the moon for example) is a projectile. He considered projecting a stone horizontally from a high mountain. Figure 7.9 is an illustration of this idea taken from the *Principia*. If the stone is given a small initial velocity, it doesn't travel far horizontally before hitting the earth, following the path from V to D in the figure. As we discussed in Chapter 4, its inertia keeps it moving horizontally with a constant speed while at the same time it is falling under the influence of gravity. Now if the stone is projected with a greater speed, it travels further before it is at last brought to the ground, as path VE illustrates.

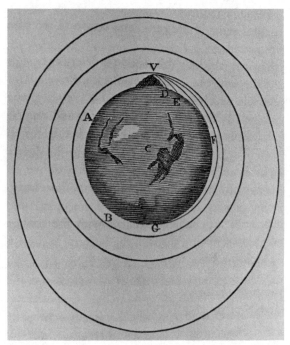

Figure 7.9 Paths of a stone, projected horizontally with different speeds from a tall mountain, leading to orbital motion (from the *Principia*).

The greater the initial speed of the projectile, the further it goes before it falls to the earth. But we must also remember that as the projectile falls, the earth curves away from under it. In the absence of air resistance, the stone could be projected so fast that it would follow path VBA and return to the mountaintop. The stone is continually falling, but because of the curvature of the earth it never reaches the ground. Instead, the stone orbits the earth. The moon, Newton realized, has just the right speed to orbit the earth as it does; it is always falling toward the earth, but never reaching it. This, too, is how satellites and space shuttles orbit the earth.

Once Newton understood that the moon is falling, he could use his theory of gravitation to predict how much it should fall in one second in comparison to an apple falling near the surface of the earth, and then check this prediction against the amount it actually falls in a second. We already know that the distance s_a an apple falls is described by

$$s_a = \tfrac{1}{2}gt^2. \tag{2.14}$$

Equation (7.11),

$$g = G\frac{M_E}{R_E^2}, \tag{7.11}$$

relates g to the earth's mass and radius. The moon is far above the surface of the earth and consequently it falls with an acceleraton less than that of Eq. (7.11). Using the universal law of gravity to calculate its acceleration, we find

$$a = G\frac{M_E}{r_m^2}, \tag{7.12}$$

where r_m is the distance from the center of the earth to the center of the moon. The moon, therefore, should fall according to

$$s_m = \tfrac{1}{2}at^2. \tag{7.13}$$

On the earth, an apple falls 16 ft (4.9 m) in one second. How far, Newton asked, does the moon fall? Using Eqs. (2.14) and (7.11)–(7.13), we find for the ratio of s_m to s_a

$$\frac{s_m}{s_a} = \frac{\tfrac{1}{2}at^2}{\tfrac{1}{2}gt^2} = \frac{a}{g} = \frac{GM_E/r_m^2}{GM_E/R_E^2} = \frac{R_E^2}{r_m^2}. \tag{7.14}$$

The distance the moon falls in one second depends on the square of the ratio of the radius of the earth to the distance to the moon.

By analyzing eclipses of the moon, ancient Greek astronomers had figured out that the distance to the moon is about 60 times the radius of the earth. In his *Principia*, Newton cited values of this distance from Ptolemy, Kepler, Tycho, and Copernicus and used the value of 60 earth radii. Thus the ratio is

$$\frac{s_m}{s_a} = \frac{1}{60^2} = \frac{1}{3600}. \tag{7.15}$$

This means that if an apple falls 16 ft in 1 s, the moon in 1 s falls 1/20th of an inch:

$$s_m = \frac{16 \text{ ft}}{3600} = 0.05 \text{ in.} \tag{7.16}$$

That is the prediction. It remained for Newton to check it against the actual movement of the moon. The distance the moon falls in one second is pictured in Fig. 7.10. The distance d is the horizontal distance the moon moves in the same amount of time. From the theorem of Pythagoras we have $(r_m + s_m)^2 = r_m^2 + d^2$, so

$$r_m^2 + 2r_m s_m + s_m^2 = r_m^2 + d^2. \tag{7.17}$$

Canceling r_m^2 we get $2r_m s_m + s_m^2 = d^2$. Now s_m^2 is a quantity so much smaller than r_m and d that we can safely neglect it to get

$$2r_m s_m = d^2, \tag{7.18}$$

or

$$s_m = \frac{d^2}{2r_m}. \tag{7.19}$$

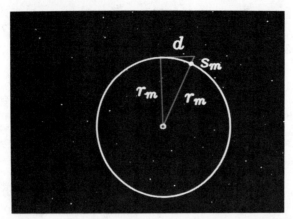

Figure 7.10 Geometry to determine the distance the moon falls in one second.

To find d we observe that the moon travels a distance $2\pi r_m$ (the circumference of the orbit) in one month, so the distance it travels in one second is in the ratio

$$\frac{d}{2\pi r_m} = \frac{1 \text{ s}}{1 \text{ month}} = 4.2 \times 10^{-7}. \tag{7.20}$$

Therefore

$$d = 2\pi r_m (4.2 \times 10^{-7}) = 2\pi(240,000 \text{ mi})(4.2 \times 10^{-7}) = 0.63 \text{ mi}. \tag{7.21}$$

Substituting this into Eq. (7.19) we obtain

$$s_m = (0.63 \text{ mi})^2/(2 \times 240,000 \text{ mi}) = 8.3 \times 10^{-7} \text{ mi} = 0.05 \text{ in}. \tag{7.22}$$

The actual motion of the moon, Eq. (7.22), agrees beautifully with the theoretically expected value in Eq. (7.16)!

The temple of Aristotelian physics had been crumbling for more than a century. But all through that time, no one imagined that an experiment done on the earth could reveal the laws of the heavens. The moment when Isaac Newton realized that the moon falls 1/20th of an inch every second, just as his theory of universal gravitation predicted, was the magic moment in human history when the physics of the heavens and physics on Earth became united in one coherent science.

7.5 CIRCULAR ORBITS

As we have seen, Newton showed that the moon stays in orbit because it is always falling as a result of gravity. We'll now give another description, closely related but more useful for solving many other problems, of how the moon, or any other heavenly body, can stay in a circular orbit. The new description begins with the mathematics of uniform circular motion, studied in Section 5.8. In uniform circular motion we found that the body is not only constantly changing its position as it goes around, but is also constantly changing its velocity, which means that it has an acceleration. We showed that the acceleration always points toward the center, giving rise to the name centripetal, and has magnitude v^2/r:

$$\mathbf{a} = \frac{-v^2}{r}\hat{\mathbf{r}}. \tag{7.23}$$

According to Newton's second law, if a body has a centripetal acceleration there must be a force inducing the acceleration. In the case of orbits it's the force of gravity that supplies the centripetal acceleration. The force of gravity on the moon is

$$\mathbf{F} = \frac{-GM_M M_E}{r^2}\hat{\mathbf{r}} \tag{7.24}$$

and it causes the moon to have an acceleration

$$\mathbf{a} = \mathbf{F}/M_M. \tag{7.25}$$

Combining Eqs. (7.24) and (7.25), we obtain the moon's acceleration due to gravity,

$$\mathbf{a} = \frac{-GM_E}{r^2}\hat{\mathbf{r}}. \tag{7.26}$$

The acceleration is centripetal – radially inward toward the center of the earth – and the orbit of the moon is very nearly a circle. Comparing Eqs. (7.26) and (7.23), and canceling a common factor of r, we find that the requirement that gravity supply the centripetal acceleration for uniform circular motion of speed v is satisfied if

$$v^2 = \frac{GM_E}{r}. \tag{7.27}$$

At precisely this speed, the force of the earth's gravity makes the moon fall just the right amount to stay in its circular orbit. This is true not only of the moon but also for any satellite of any planet, including artificial ones. It is the basic mechanism of the solar system.

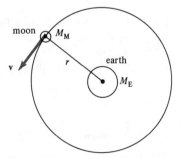

Figure 7.11 Quantities pertaining to the moon's orbital motion.

Example 4

Assuming no atmosphere, what horizontal speed would have to be imparted to a golf ball in order for it to orbit the earth at its surface?

Taking the radius of the orbit as the radius of the earth and using Eq. (7.27), we have

$$v^2 = \frac{GM_E}{R_E} = \frac{(6.67 \times 10^{-11} \text{ N m}^2/\text{kg}^2)(6.0 \times 10^{24} \text{ kg})}{6.4 \times 10^6 \text{ m}},$$

$$v = 7900 \text{ m/s},$$

which is over 20 times the speed of sound!

We can also obtain a relationship between the period T of a circular orbit and the radius of the orbit. Using Eqs. (5.65) and (5.58), we can express the centripetal acceleration as

$$a = \omega^2 r = \frac{4\pi^2 r}{T^2}. \tag{7.28}$$

Equating this to the acceleration caused by gravity from a body of mass M,

$$a = \frac{GM}{r^2}, \tag{7.29}$$

we obtain

$$T^2 = \frac{4\pi^2 r^3}{GM}. \tag{7.30}$$

This relationship, a shining success of the Copernican revolution, is a special case of Kepler's third law, which we will encounter later in Chapter 16.

7.6 OTHER EXAMPLES OF UNIFORM CIRCULAR MOTION

Many physical systems besides orbiting satellites exhibit uniform circular motion, and many kinds of force can supply the necessary centripetal acceleration.

In a simple conical pendulum (Fig. 7.12) a mass, supported by a rope, swings in a horizontal circle of radius r. The weight mg and the vertical components of tension $T \cos \theta$ balance,

$$T \cos \theta - mg = 0. \tag{7.31}$$

The horizontal component of tension supplies the necessary centripetal force

$$T \sin \theta = m\omega^2 r. \tag{7.32}$$

For a rope of length L, we can replace r by $L \sin \theta$ in Eq. (7.32). Eliminating T from Eqs. (7.31) and (7.32) we then obtain

$$\cos \theta = \frac{g}{\omega^2 L}. \tag{7.33}$$

As ω is increased, θ increases.

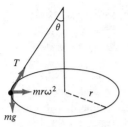

Figure 7.12 Simple conical pendulum.

It is important to note that there is no outward, or *centrifugal*, force acting on the conical pendulum mass in the fixed frame of reference in which we analyzed the problem.* The inward pull of the rope is needed not to counterbalance any outward force, but to turn the mass continuously away from straight-line motion and toward the center of the circle.

As another example, an airplane making a horizontal circular turn has a force diagram (Fig. 7.13) closely resembling that for the conical pendulum. The plane must be banked through an angle such that **F**, the reaction of the air on the plane, not only balances the

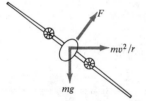

Figure 7.13 An airplane banks to make a turn.

weight of the plane but also produces a resultant force of magnitude mv^2/r toward the center of the turn.

Centrifuges separate milk from cream, or uranium 238 from uranium 235, by spinning a mixture of particles suspended in a liquid solution. When the liquid is spun, it tends to flow outward initially. As a result its density increases in the outer region of the centrifuge and decreases in the inner region, creating an inward-directed pressure gradient. The pressure gradient quickly halts the outward flow by supplying the centripetal force required to maintain a nearly circular flow of the liquid. But the centripetal force required

---

*The absence of centrifugal force in this frame is surprising because of our childhood experience that if *we* are the mass, holding on to the end of the rope and swinging around with it, we seem to be in the grip of a powerful outward tendency – we have to pull on the rope to keep from flying outward. In Chapter 9 we will analyze this situation from the child's point of view (i.e., in a frame of reference fixed with respect to the child and circling uniformly with respect to the earth). Since this frame is constantly accelerated with respect to the earth, it is not an inertial frame, and the analysis of Chapter 9 will show that an extra outward pull, called an *inertial* or *pseudo* force, indeed occurs in this frame in accordance with what the child feels. But there is no such centrifugal force in our original Earth-fixed, inertial frame of reference.

to maintain a *particle* in circular motion as the centrifuge spins, $m\omega^2 r$, is greater for heavy particles, whereas the pressure gradient experienced by heavy and light particles in a given region of the solution is the same. The result is that the heavy particles migrate outward.

Uniform circular motion in a *vertical* circle, such as that of a rider on a Ferris wheel, introduces a new feature. In this case, although a constant centripetal force is needed for uniform circular motion, the component of gravity impelling the rider toward the center of the circle varies as he moves around the circle. Hence the other forces acting on him (supplied by the seat, seatbelt, etc.) must also vary to compensate. Compare, for example, the situation at the top and bottom of the ride in Fig. 7.14a. At the bottom the nongrav-

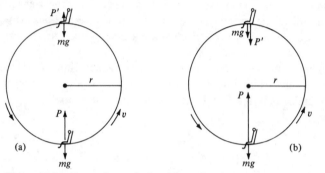

Figure 7.14 Forces in vertical uniform circular motion at the top and bottom of the cycle, for (a) $v^2/r < g$, (b) $v^2/r > g$.

itational force (labeled P) must be opposite to and larger than gravity to supply the necessary centripetal force mv^2/r:

$$-mg + P = \frac{mv^2}{r}. \tag{7.34}$$

Here P is supplied by the upward push of the seat; a seatbelt or bar would not be needed no matter how fast the wheel turned. At the top, on the other hand, gravity acts toward the center and (for low velocities) the other forces on the rider (labeled P') act outward and reduce the centripetal force:

$$mg - P' = \frac{mv^2}{r}. \tag{7.35}$$

Here again P' is supplied by the upward push of the seat, for normal operation at low velocities, but now the push is weaker than it was at the bottom: $P' = m(g - v^2/r)$ versus $P = m(g + v^2/r)$. If the Ferris wheel were speeded up to dangerous (and presumably illegal) velocities $v^2/r > g$, P' would need to reverse sign and point inward to help supply the needed centripetal force (Fig. 7.14b). The push of the seat under the rider can only give $P' \geq 0$; to provide a downward P' ($P < 0$), seatbelts would be essential. If seat belts were not used, the centripetal force would be insufficient to maintain uniform circular motion and the rider would fly out of his seat.

The motion of a car on the loop-the-loop at an amusement park, or of a pail of water swung in a vertical circle, is more complicated because the angular speed need not be constant in these cases. We shall later derive a formula for acceleration for nonuniform circular motion (in Example 1 of Chapter 17), two consequences of which can easily be stated here:

(i)   When the angular speed is not constant there is a tangential component of acceleration which is not present in uniform circular motion.

(ii) The centripetal component of acceleration is still v^2/r, just as it is for uniform circular motion.

In view of (ii), Eq. (7.35) for the force in the centripetal direction is still applicable. We find the surprising result that the riders and water will not fall out if the motion is fast enough. The loop-the-loop car, for example, is above the riders at the top of the swing, so the seat can only push downward at this point [$P' \leq 0$ in the sign convention of Eq. (7.35)]. Nor surprisingly, then, any unsecured object falls out [i.e., the downward force exceeds the required centripetal force and Eq. (7.35) cannot be balanced] at low speeds. But when the speed at the top is such that $v^2/r > g$, a downward-directed P' is just what is needed to supply sufficient centripetal force to keep the rider on track. In this case fast motion is safer than slow!

7.7 A FINAL WORD

That 1/20th of an inch would have been enough of an accomplishment for any ordinary lifetime. But for Newton, it was barely the beginning. The list of his scientific and mathematical discoveries leaves us breathless.

But not everything he did was scientifically respectable. He spent years of his life immersed in alchemy, Biblical chronology, and other arcane pursuits. In his view, these studies were part and parcel of his search for a system of the world. He was also more than an amateur politician. He was twice elected to Parliament, and, in the year 1705, Queen Anne knighted him, making him Sir Isaac Newton. He was also given a sinecure – a safe lifetime position as Warden of the Mint. In that capacity, he was responsible for the coin of the realm – and for capturing and interrogating forgers. In 1693, he suffered a nervous breakdown. Some people today think he was suffering from mercury poisoning, possibly contracted during his experiments in alchemy. Some evidence for that has been found by chemical analysis of hairs, but not all historians agree. In any case, he recovered from his illness and went on to become President of the Royal Society, a position he held from 1703 to his death in 1727. But compared to his accomplishments, the personal details of Newton's life hardly matter.

Newton gave us not only a series of scientific discoveries, but more important, a coherent view of how and why the universe works. That view has dominated Western thought from his time right down to our very own. Isaac Newton was a human being with faults and flaws – maybe even more than his share of them. But he was also a giant, almost unparalleled in our history.

Problems

The Universal Law of Gravity

1. Three point masses, each of mass m, are fixed at the corners of an equilateral triangle of side a, as shown in the figure below. Calculate the gravitational force on any one of the masses.

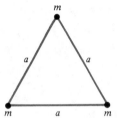

2. Using the astronomical data of Appendix D, calculate the gravitational force between the moon and (a) the earth, (b) the sun. Obtain the ratio of these forces. Does your result make sense – why doesn't the sun's pull tear the moon away from the earth?

3. At what distances from the center of a uniformly dense planet is the acceleration of gravity half of the value at the surface?

4. Suppose you lived in Flatland, a two-dimensional plane universe. What dependence of force on distance would allow you to make "lines of force" arguments?

Gravitational Acceleration on the Surface of a Planet or Moon

5. Using the values of g, G, and the radius of the earth (6.4×10^6 m), calculate the mass of the earth.

6. Knowing that the mass of the moon is about 1% of the earth's mass, and its radius about one-fourth of the earth's radius, calculate the acceleration of gravity on the surface of the moon.

7. Suppose that an apple is 100 km above the surface of the earth and that the earth somehow uniformly expanded its radius by 100 km while the mass remained constant. Would the force of gravity on the apple be more, less, or the same after the expansion as before? Why?

8. Because a spherically symmetric dense object can be considered to have all its mass concentrated at its center when calculating the effects of gravity, a weight hung on the end of a string (forming a plumb line) normally points toward the center of the earth. But a plumb line 2 km from the center of a hemispherical mountain that is 1 km in radius and as dense as the earth exhibits a small deviation. State the direction and find the approximate magnitude of the angle of deviation.

Falling in Orbital Motion

9. Which planet falls farther toward the sun in one second, Mercury or the earth? Explain your reasoning. Using the astronomical data of Appendix D, calculate how far the earth falls in one second in its orbit around the sun.

Circular Orbits

10. The planet Mars has a satellite, Phobos, which has an orbital radius of 9.4×10^6 m and a period of 7 h 39 min. Calculate the mass of Mars from these data.

11. Can a spy satellite hover constantly over, say, New York? Explain.

12. What is the radius of the orbit of a geosynchronous communications satellite which at all times is directly above a point on the equator? Estimate the approximate time delay in a telephone conversation between Europe and America via the satellite link (the speed of the signal is 3×10^{10} cm/s).

13. Taking the radius of the orbit of Mars about the sun as 1.52 times that of the earth, determine the number of years it takes for Mars to make one revolution about the sun.

14. Suppose a satellite could be put into orbit in an evacuated circular tunnel inside the earth. Would the speed of the satellite in such an orbit be greater or less than that of the golf ball in Example 4? How would the angular speed of the satellite differ from that of the golf ball?

Other Examples of Uniform Circular Motion

15. Consider a simple conical pendulum with a string of length L that can only sustain tensions up to a maximum T_c. In terms of T_c and L, what is the maximum sustainable angular speed ω_c beyond which the string breaks?

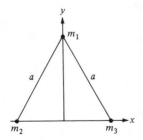

16. Three masses are arranged at the corners of an equilateral triangle.

(a) What gravitational force does each mass feel because of the other two?

Suppose we want to spin the triangle in its own plane so that the gravitational forces found in (a) exactly supply the centripetal forces.

(b) About what axis should the triangle be spun? Refer to the diagram and give your answer as the coordinates of the point through which the axis should pass.

(c) How fast should the triangle be spun?

17. A 0.5-kg mass, mounted on the end of a motor-driven light rod, moves with constant speed in a vertical circle of radius 1 m.

(a) If the speed of the mass is 3.0 m/s, determine the tension in the rod at the lowest point.

(b) Is the rod more likely to break when the mass is at the top or at the bottom of the circle? Why?

(c) Determine the minimum speed the mass must have so that the rod will remain under tension (will not go into compression) at the highest point.

CHAPTER 8

FORCES

I don't know what I may seem to the world, but, as to myself, I seem to have been only a boy playing on the sea shore, and diverting myself in now and then finding a smoother pebble or a prettier shell than ordinary, whilst the great ocean of truth lay all undiscovered before me.

Isaac Newton

8.1 THE FUNDAMENTAL FORCES: CLASSIFICATION AND UNIFICATION

Using crude water clocks to time balls rolling down inclined planes, Galileo searched for and found a description of how bodies fall. His law of falling bodies, however, wasn't a fundamental law of nature. Within half a century it was superseded by a deeper insight into nature – Newton's universal law of gravity. Through the genius of Newton, the force of gravity,

$$\boxed{\mathbf{F} = -G\,\frac{M_1 M_2}{r^2}\,\hat{\mathbf{r}}.}$$
(8.1)

was revealed as a fundamental force of nature.

Inspired by Newton, physicists in the eighteenth century sought to identify, classify, and mathematically describe the numerous forces observed in nature. Knowledge of these forces provided physics with a certain predictive power, because according to Newton's second law, $\mathbf{F} = m\mathbf{a}$, forces shape the motion of all things. Through painstaking experiments, these physicists reached empirical descriptions of forces in the world about them: tensions, spring forces, friction, viscosity, electricity, magnetism, chemical action. As the number of forces grew, so did the applications in an increasingly industrialized world. Yet a question which confronted these physicists was: Are all these forces fundamental, or can they be reduced to more basic forces?

Not until late in the eighteenth century did a second force emerge as fundamental – the electric force. The French engineer Charles Augustin Coulomb assumed that, analogous to the gravitational force between two masses, the electric force between two charges is proportional to the product of the charges. Experimentally, he found that the electric force is similar to gravity in another way: the force between two charges decreases as the square of the distance between them. Summarized mathematically, the electric force $\mathbf{F}$ between two charges q_1 and q_2 which are separated by a distance r is known as Coulomb's law and written as

$$\boxed{\mathbf{F} = K_e\,\frac{q_1 q_2}{r^2}\,\hat{\mathbf{r}}.}$$
(8.2)

Just as G is a universal constant for gravity, K_e is a universal constant for electricity.

Magnetism was also identified as a fundamental force of nature. The attraction or repulsion between two magnets could be described by a force between pairs of magnetic poles. The progress of physics appeared to be a triumph of Newtonian mechanics: the forces of nature were successively reduced to attractions and repulsions between particles.

The first 40 years of the nineteenth century, however, saw a growing reaction against such a division of phenomena in favor of some kind of correlation of forces. The turn inward to unification of forces was spearheaded by Oersted, Ampère, and Faraday. By the middle of the century they had succeeded in unifying two hitherto disparate forces, electricity and magnetism, into one – electromagnetism. The process was crowned in the theory of James Clerk Maxwell, who expressed the unification by a set of equations which interrelate electric and magnetic phenomena. Soon tensions, spring forces, friction, viscosity, chemical actions, and even light were recognized as arising fundamentally from the electromagnetic force. Based on Maxwell's success the search for a common mathematical description, or unification, of forces had begun.

With the twentieth century came the discovery of radioactivity, the probing of atoms, and the subsequent realization that more than just gravity and electromagnetism would be needed to explain this new world. Both a new dynamics – the laws of quantum mechanics, which supersede Newton's laws on atomic and subatomic distance scales –

and new forces were needed. Because Newton's laws do not apply on the distance scales at which the new forces act, we shall not pursue the study of these forces in this book. Nevertheless many themes, such as the quest for the fundamental constituents of matter and the fundamental forces of nature, carry over from classical mechanics, and it is interesting to survey briefly what has been found.

Experiments probing atoms revealed that inside an atom there is a compact center – the nucleus – composed of positively charged protons and neutral neutrons. Negatively charged electrons orbit the nucleus, held by the electric force from the protons. The natural question which arose is: What holds the nucleus together? Physicists realized that neither gravity nor electromagnetism held the compact nucleus together, but that a new force was at work. Aptly named the *strong force*, it overcomes the electric repulsion between protons and holds the nucleus together. Unlike gravity and electricity, the strong force does not extend to great distances; it has a limited range – the size of a nucleus, 10^{-13} cm. Outside this range, the strong force has virtually no effect. If it did, we wouldn't be here; matter would collapse into dense lumps of subatomic particles.

Natural radioactivity could not be explained in all instances by any of the known forces – strong, electromagnetic, or gravitational. Another force was implicated in the decay of nuclei – the *weak force*. This force is intrinsically weaker than the strong force, and has an even more limited range, about 10^{-16} cm or 1/1000th the size of the nucleus. Because of its severely limited range its most common manifestations in nuclei are feeble indeed, about 10^6 times weaker than the strong force. Nevertheless, the weak force plays an essential role in the release of nuclear energy in stars, and in causing some stars ultimately to explode.

The behavior of the four fundamental forces of nature – strong, electromagnetic, weak, and gravitational – is reasonably well understood, but nobody knows why there should be four of them. Albert Einstein spent the last 20 years of his life unsuccessfully searching for a way to unify two of the forces, gravity and electromagnetism. Twentieth-century physics has become a story of attempting to explain all the complexities of physics as aspects of similar systems, a search for unification of the fundamental forces. The water clocks and inclined planes of Galileo have been replaced by increasingly larger, more energetic particle accelerators. Unified theories are emerging that bring together the weak and electromagnetic forces, as well as more comprehensive theories that attempt to give a coherent account of how all these forces may have evolved from simpler laws in the infancy of the universe. The early universe ultimately may be the only experimental test for such theories. It may be the great ocean of truth that still lies undiscovered before us.

8.2 THE STRENGTH OF GRAVITATIONAL AND ELECTRIC FORCES

One of the great and deep mysteries of physics is that the laws describing the gravitational force and the electric force have essentially the same mathematical character. They are

$$\mathbf{F} = -G\,\frac{M_1 M_2}{r^2}\,\hat{\mathbf{r}}, \tag{8.1}$$

and

$$\mathbf{F} = K_e\,\frac{q_1 q_2}{r^2}\,\hat{\mathbf{r}}. \tag{8.2}$$

It seems almost a minor point that each law contains an unspecified universal constant. Yet for applications to the real world, it is essential to know what those constants are.

For gravity, we already know that G is related to the acceleration g of a falling body near the surface of the earth and the mass and radius of the earth:

$$g = GM_{E}/R_{E}^{2}. \tag{7.11}$$

The radius of the earth has been known since antiquity, and we also know g, so measuring G immediately tells us the mass of the earth M_{E}. The determination of G was one of the great classic experiments in physics.

Henry Cavendish, a British physicist, performed the historic measurement of G in 1798. Deeply inspired by Newton, Cavendish regarded the *Principia* as the model for exact sciences, and the search for the forces between particles guided his scientific explorations. But Cavendish had fitful habits of publication; he left unpublished whatever did not fully satisfy him. Luckily the determination of G was an experiment in which he took pride.

Figure 8.1, which is adapted from Cavendish's 1798 article, shows the apparatus he invented for his delicate experiment which measured forces equal to one-billionth of the weights of the bodies involved. The two small lead balls are attached to a rigid rod forming a dumbbell which is suspended by a thin fiber that allows the dumbbell to rotate freely. When the two larger lead balls are placed near the ends of the dumbbell, the smaller masses are attracted to the larger ones by the gravitational force **F**. This force, although extremely small, nevertheless exerts a torque which rotates the dumbbell and twists the fiber. The fiber opposes the twisting with a torque that equals the angle of rotation times a known coefficient. By measuring the deflection of a beam of light reflected off the small mirror attached to the fiber, the angle and thus the force **F** can be determined. Because the balls are spherical, their masses act as if concentrated at their centers. Knowing the masses m and M, and the distance r between their centers when the twisting stops, we can calculate the value of G through Eq. (8.1). As stated in Chapter 7, that value is found to be $G = 6.67 \times 10^{-11}$ N m^2/kg^2.

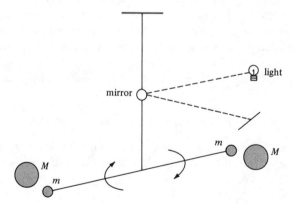

Figure 8.1 Schematic of Cavendish's apparatus for measuring G.

By this experiment, Cavendish rendered the universal law of gravitation complete. The law was no longer a proportionality relation as Newton had stated it, but an exact law through which quantitative analysis could be made. It was the most important contribution to the study of gravitation since Newton.

Through a similar experiment in 1787, Coulomb showed that the electric force between two charges is analogous to gravity: it decreases as the inverse square of the distance between the charges. Likewise the value of the electric constant K_e could be measured; its value is 9.0×10^9 N m²/C², where 1 C, one coulomb, is the unit of charge. But what is charge?

The early Greeks had discovered that amber attracts bits of straw and identified that property of amber with charge. Charge is that which creates electric force; even today we can reduce its essence no further. We don't know exactly what charge is any more than we know what mass is. Unlike mass, charge comes in two different varieties – positive and negative – terms introduced by Benjamin Franklin. Like charges repel whereas opposite charges attract. On the other hand, gravity is always attractive. There is also a fundamental unit of charge – the charge of the proton (or, exactly equal in magnitude and opposite in sign, the charge of the electron). All charges come in multiples of this unit of electricity; the charge of a proton is 1.6×10^{-19} coulombs.

Atoms consist of positively charged nuclei and negatively charged electrons. An electron is about 2000 times lighter than a proton. The simplest atom, hydrogen, consists of one proton and one electron separated by 10^{-8} cm. This separation distance is much larger than the nucleus, so we picture a negative charge outside the positive nucleus. The force that holds the electron to the proton to make a hydrogen atom is the electric force.

To construct heavier atoms in our model, we first construct nuclei with more protons and neutrons in them. Since the overall charge of atoms is zero, there are as many electrons around the nucleus as there are protons in the nucleus. Although the electrons repel each other, they are held to the nucleus by the electric force.

Atoms in turn can attract other atoms to make larger composites called molecules. The force that holds the atoms together to form molecules is again the electric force, in this case a residue of it which extends outside the basically neutral atom when its electron orbits are distorted by the presence of neighboring atoms. Atoms and molecules can form larger agglomerations which we see as liquids and solids. These too are held together by electric forces.

Given the strengths and nature of the various fundamental forces, we can ascertain the phenomena that each of them controls. Gravity acts on all matter, as reflected by its dependence on the masses of objects. Although its strength diminishes universally with distance, the effects of gravity are nevertheless felt across the far reaches of the universe. Gravity holds together planets and stars, organizes solar systems and galaxies; it orders the universe.

Electricity, in giving rise to tensions, spring forces, friction, viscosity, and chemical actions, governs the everyday world around us. It is intrinsically stronger than gravity and is the dominant force on our size scale, for objects as large as mountains and as small as atoms. On larger scales, gravity dominates because the gravitational attractions of all individual masses add whereas the electrical attractions and repulsions of individual electrons and protons tend to cancel. The cancelation is almost perfect in a large body, essentially because of the very strength of the electric force, which causes charge to flow rather easily and neutralize most charge excesses.

On distance scales of nuclear size or smaller, the strong and weak forces organize matter. The strong force, intrinsically stronger than gravity and electromagnetism, dominates within this distance. Physicists speculate that the lack of influence of the strong force outside nuclei is analogous to the relative unimportance of the electrical force on scales larger than the earth: the strong force is *so* strong that it neutralizes its sources within the nucleus essentially perfectly.

Table 8.1 summarizes the four fundamental forces of nature, their respective ranges, and their respective strengths (as estimated for the forces acting between two protons at short range). Note the utter unimportance of gravity on the atomic scale of 10^{-8} cm. Only on the scale of a mountain, 1 km or 10^{13} atoms on a side, 1 km^3 or 10^{39} atoms in volume, does the mass become large enough to compensate the relative strength factor of 10^{-39}.

Table 8.1 Characteristics of the Four Fundamental Forces

Force	Relative strength	Range	Importance
Strong	1	10^{-13} cm	Holds nucleus together
Electromagnetic	10^{-2}	Infinite	Controls everyday phenomena – friction, tensions, etc.
Weak	10^{-2}	10^{-16} cm	Nuclear transmutation
Gravitational	10^{-39}	Infinite	Organizes large-scale phenomena and universe

8.3 CONTACT FORCES

The fundamental forces of nature are *action-at-a-distance* forces: their effects can be experienced when the particles are not in contact. A second category of forces is *contact forces*. These are forces which two objects exert on each other when they are physically in contact with each other, as for example, when a book rests on a table.* Contact forces are not fundamental forces; instead, they arise from electric forces acting in complicated ways among enormous numbers of atoms.

The contact forces are described by *empirical rules*, which are experimental summaries of the net result of all the complications. Most of these empirical descriptions were deduced by eighteenth-century scientists who were unaware of the underlying fundamental forces. Even today, when the fundamental forces are better understood, quantitative deduction of the contact forces from fundamental laws remains difficult and the empirical rules are still commonly used.

---

*The implication that contact forces drop abruptly to zero when two bodies separate, though valid on macroscopic distance scales, is an idealization. On the atomic scale the contact force falls off continuously though rapidly when the distance of separation increases, as one would expect from the electrical origin of the force.

The tension in a rod, wire, rope, or string is an example of a contact force. The atoms in any of these objects are bound together by electric forces which tend to keep the atoms a certain distance apart, called the equilibrium distance. When you pull on one end of the wire or rope each atom electrically tugs on its neighbors, and the pull is transmitted to the other end, much as in a chain link. The net result of all the electric forces acting on the atoms is the macroscopic force – the tension. The foregoing argument provides the atomic explanation of the empirical rules stated for tension in Chapter 6: tension acts along the wire or rope, and if the weight of the object is negligible or has no component along its length, the pull is transmitted to the other end undiminished in strength. If the weight cannot be neglected, the tension varies along the wire or rope in such a way as to hold it up against gravity.

Compression in a rigid rod is another example of contact force. When you push on one end, each atom pushes on its neighbors, attempting to maintain the equilibrium separation, and the push is transmitted to the other end of the rod.

In addition to their role in transmitting forces, tension and compression also change the body sustaining them, elongating or shortening it. We can obtain a qualitative picture of what is happening by carrying our previous reasoning about electric forces a step further. Consider, for example, a wire under tension. As stated before, the atoms in the wire have an equilibrium position at which electric forces tend to keep each atom. The new point we now wish to emphasize is that when the wire is put under tension, each atom is pulled a tiny bit away from the equilibrium distance. The electric forces try to put the atoms back into the equilibrium position, but some elongation remains while the wire is under tension.

The empirical law for the change in length of a wire is simple: the force exerted by a wire (on an object) is proportional to the change in length of the wire. This is known as Hooke's law (after Robert Hooke). Hooke's law also applies to compressions.

Springs obey Hooke's law especially well. Here the elongation is greatly increased by a geometrical effect, the straightening out of the coiled wire.

Hooke's law is expressed mathematically as follows:

Hooke's law:

$$F = -kx,$$

(8.3)

where x is the change in length of the wire or spring from its equilibrium value. In discussing Eq. (8.3) one often uses the spring as a model, and adopts a terminology associated with springs to describe the stretching of straight wires and many other objects as well as springs. The constant k is called the spring constant and is a measure of the stiffness of the spring; the stiffer the spring, the larger the value of k. The direction of the force always opposes the displacement of the end of the spring from its unstretched position. When $x > 0$, the spring is stretched and F is negative; when $x < 0$, the spring is compressed and F is positive. The spring force always acts to restore the spring to its unstretched length, as Fig. 8.2 illustrates.

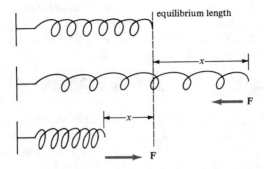

Figure 8.2 The force exerted by a spring described by Hooke's law.

In addition to tension and compression, which we already made use of in Chapter 6, and Hooke's law relating tension and compression to change of length, there are further varieties of contact force which will play a role in the applications of Newton's laws in the present chapter. For example, whenever any object is pressed against another there is a contact force between the two objects, known as the *normal force*. This force is a result of repulsion between the atoms of the two objects. The empirical rules for a normal force are that its magnitude depends on how hard the object is pressed, but the direction of the normal force acting on an object is always perpendicular to the surface.* Figure 8.3 indicates normal forces between different objects.

Other important examples of contact forces include friction and viscosity. We shall discuss these forces in subsequent sections.

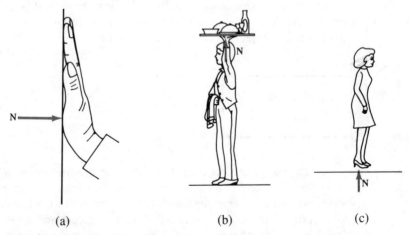

Figure 8.3 Illustration of the normal force exerted by (a) wall against hand, (b) hand against tray, (c) floor against person.

*The concept of normal force can also be applied when one object is pulled away from another in a direction normal to the surface and the pull is resisted by adhesion. In this case the force is the result of attraction between atoms.

To bring out in more detail how contact forces arise microscopically, a typical electric force between two neighboring atoms is plotted in Fig. 8.4. One sees immediately that the simplicity of the fundamental $1/r^2$ force between two charges has gotten submerged in complications. The attraction at long range is the resultant of canceling attractions and repulsions among the electrons and nuclei of the two atoms; it falls off with distance as r^{-7} or so. The repulsion at short range is governed by quantum effects which tend to keep the electrons of the two atoms apart.

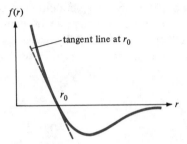

Figure 8.4 The dependence of the force $f(r)$ between two atoms on their separation distance r, and linear approximation $f'(r_0)(r - r_0)$ to $f(r)$ near r_0.

In the absence of external disturbance the equilibrium distance in Fig. 8.4 is r_0, where the electric force vanishes. Greater separation distances, such as occur when a wire is pulled from one end, are resisted by the attractive long-range force, giving rise to tension. Closer spacing, such as occurs under compression, is resisted by the short-range repulsion giving rise to the normal force.

The microscopic basis for Hooke's law can be found by drawing a straight line in Fig. 8.4 tangent to the force curve at r_0. This linear relation between force and displacement corresponds to Hooke's law, and it is a good approximation to the force near the equilibrium point.

What factors determine the spring constant k? If we consider the stretching of a straight wire of length L by an attached weight, as pictured in Fig. 8.5, the force on one atom in the wire due to its neighbor is

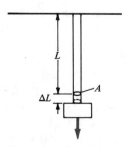

Figure 8.5 Stretching of a loaded wire.

$$f = f'(r_0)(r - r_0)$$

$$= -|f'(r_0)|(r - r_0) \qquad (8.4)$$

where $f'(r_0)$ is the (negative) slope of the force curve at r_0. The total force of tension on all the atoms in a cross section of area A is of order*

$$F = -\frac{A}{A_0}|f'(r_0)|(r - r_0), \qquad (8.5)$$

where A_0 is the area of cross section occupied by a single atom. The overall stretching of the wire, ΔL, is the stretching between a neighboring pair of atoms, $r - r_0$, times the total number of such stretchings in length L, namely L/L_0, where L_0 is the separation between neighboring atoms. Thus we have $\Delta L = (r - r_0)L/L_0$ or $r - r_0 = L_0\,\Delta L/L$ and

$$F = -\frac{A}{A_0}|f'(r_0)|\frac{L_0}{L}\,\Delta L$$

or

$$F = -\frac{|f'(r_0)|L_0}{A_0}A\,\frac{\Delta L}{L} = -YA\,\frac{\Delta L}{L}. \qquad (8.6)$$

The quantity Y is called Young's modulus and contains the dependence on microscopic properties $[L_0/A_0 \approx r_0/r_0^2 = r_0^{-1}$ and $f'(r_0)]$ of the particular atoms in the wire. The dependence on geometrical properties has been separated out in the factor $A\,\Delta L/L$. Hooke's constant k for the straight wire is YA/L; note that it depends on both geometrical and microscopic factors. To produce an ordinary spring, one coils the wire. This makes it much less stiff (smaller k, larger $\Delta L = x$ for a given pull) because of an extra geometrical effect: most of ΔL comes from straightening the coil.

The dependence of k on $f'(r_0)$ (which depends on the material used) and on the geometry of the wire is a reminder that the spring force is not fundamental. But not until a wire is stretched by a *large* amount does the full complexity of the contact force become evident. Beyond some point, the force in Fig. 8.4 is no longer well represented by the linear approximation in Hooke's law. Moreover, when stretched too far, the wire does not return to its original length when the applied force is removed; the stretching has caused permanent deformation by moving or creating imperfections in the wire which reduce the number of atomic bonds in a cross-sectional area. The nature of such imperfections, and the eventual breaking point of the wire, depend not only on the interatomic force law and the overall geometry of the wire, but also on the history of how the wire has been treated.

8.4 APPLICATION OF NEWTON'S LAWS

Now that we have increased our repertoire of empirical rules for forces, we return to their use in Newton's laws. The success of Newtonian mechanics was the identification

---

*This relation is only approximate because each atom is pulled by several other neighboring atoms at various angles with respect to the direction along the wire.

of forces and the subsequent dynamical explanation of the motion of objects influenced by these forces. Generations of students have learned Newton's laws by solving all sorts of dull as well as interesting problems in mechanics. Even before Newton the great sixteenth-century humanist Erasmus, when he was a student, wrote a letter to a friend saying how dull mechanics lectures were. Nobody will ever know how many minds, eager to learn the secrets of the universe, found themselves studying inclined planes and pulleys instead, and decided to switch to some more interesting profession. Nevertheless, through solving problems, you do understand physics.

Our task is to apply Newton's laws and analyze the motion of objects. Let's list a few extremely useful steps which, once mastered, will allow you to solve a huge class of problems in mechanics. These duplicate in part the steps suggested in Section 6.5 for treating the special case of equilibrium, but the procedures are so important that they bear some repetition:

1. Draw a *free-body diagram* for every object whose motion is to be analyzed. This entails drawing each object separated from all others and clearly indicating the external forces acting on each by arrows which start or end on the object.
2. Label all external forces acting on the objects. In labeling, use the fact that action–reaction pairs of forces are equal and opposite.
3. Choose a coordinate system for each object under consideration. The choice should simplify the equations as much as possible; for example, it is often useful to place one axis along the direction of the acceleration.
4. Apply Newton's second law in component form to each object:

$$\sum F_x = ma_x, \qquad \sum F_y = ma_y, \qquad \sum F_z = ma_z.$$

This requires resolving forces into components.
5. Write down, or use directly in the second law, any equations of constraint that restrict the motion of the objects under analysis. For example, the vertical component of acceleration for an object constrained to slide along a floor vanishes. Similarly, if two weights are connected by an extensionless cord running over a pulley, their accelerations are equal in magnitude.
6. By now you should have as many equations as unknown quantities. Whenever possible first solve for the unknowns algebraically, then substitute numbers and units to obtain quantitative answers.

The following examples illustrate the method used to apply Newton's laws. Each body in these examples is treated as a point mass so that the forces are assumed to act at one point and there is no torque (we shall deal with some problems involving torque in later sections). In addition, the masses of pulleys and strings are considered negligible. Although these assumptions may appear artificial (where can you buy a massless string?), it is understanding the method which is important now.

Example 1
A 60-kg passenger is riding in an elevator. Find the force exerted by the floor on the passenger when the elevator is

(a)   accelerating upward at 3.0 m/s^2,
(b)   accelerating downward at 3.0 m/s^2, and
(c)   moving downward with a constant speed of 4.0 m/s.

In drawing a free-body diagram of the passenger (unflatteringly represented by a block) we have only two forces to consider, gravity mg and the normal force N.

If we choose the positive z axis to point vertically upward, the second law $\Sigma F_z = ma$ implies

$$N - mg = ma,$$

which tells us that the normal force is $N = m(g + a)$. For case (a) we use $a = +3.0$ m/s^2 and get $N_a = 770$ N. For case (b), the acceleration is downward and therefore negative, $a = -3.0$ m/s^2. Substituting this into our expression for N, we get $N_b = 410$ N. For case (c) the acceleration is zero, and the normal force is simply equal to the weight: $N_c = 590$ N.

Example 2

A parcel slides down a chute so smooth that friction is negligible. Calculate the acceleration of the parcel as well as the normal force from the chute.

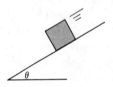

The free-body diagram for the parcel is shown below. Note that the normal force N is perpendicular to the inclined chute.

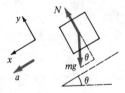

Choosing the x axis to point down the plane, in the direction of the acceleration, we have the following:

$$\sum F_x = ma \quad \text{implies} \quad mg \sin \theta = ma,$$

$$\sum F_y = 0 \quad \text{implies} \quad N - mg \cos \theta = 0.$$

The second equation tells us that the normal force is $N = mg \cos \theta$. The first equation tells us that the acceleration of the parcel is $a = g \sin \theta$. This is the relation that enabled Galileo to determine g. Motion on an inclined plane is uniformly accelerated, but can be made much slower than free fall, and thus more easily measured, by taking θ small.

Example 3

Two blocks, one of mass $m_1 = 1.0$ kg, the other with mass $m_2 = 2.0$ kg, are pushed along a frictionless surface by a force of 2.0 N. Find the acceleration of the blocks and the force of block 1 on block 2.

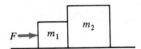

Following the steps outlined in the text, we first draw a free-body diagram for each block. The forces acting on block 1 are gravity m_1g; normal force N_1; outside push F; and, because 1 pushes on 2, 2 pushes back on 1 (by the third law) with a force we'll call P. A similar set of forces act on block 2, as shown in the free-body diagram. Note that F does not act directly on block 2, so it is not shown acting on it. The effect of F is felt indirectly, through the force P arising from contact with block 1.

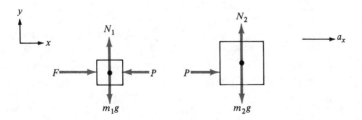

The blocks accelerate to the right and are constrained to move along the horizontal surface, so we'll choose that to be the direction of the x axis. Then Newton's second law $\sum F_y = ma_y$ for block 1 tells us that $N_1 - m_1g = 0$, whereas applying $\sum F_x = ma_x$ to block 1 gives

$$F - P = m_1a_x.$$

In this equation there are two unknowns, P and a_x, so we need one more equation. Applying the second law to block 2 (and using the constraint that the acceleration has the same value for both blocks) we get $N - m_2g = 0$ and

$$P = m_2 a_x.$$

Substituting this value for P into our previous equation, we have

$$F - m_2 a_x = m_1 a_x,$$

which allows us to solve for a_x:

$$a_x = F/(m_1 + m_2).$$

(We could also have obtained this result by thinking of the two blocks as one block of mass $m_1 + m_2$ acted upon only by the force F in the x direction.) Substituting numbers, we find

$$a_x = 0.67 \text{ m/s}^2.$$

To find the force P that block 1 exerts on 2, we simply substitute the value of a_x into the equation for P,

$$P = m_2 a_x = (2.0 \text{ kg})(0.67 \text{ m/s}^2) = 1.33 \text{ N}.$$

The direction of P is shown in the free-body diagram.

Example 4

If the mechanism for transmitting push P from block 1 to block 2 in Example 3 is a massless spring rather than direct contact between the blocks, what is the compression of the spring? (Assume constant compression, and use spring constant $k = 10^3$ N/m.)

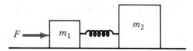

Let us focus our attention on the forces and motion in the x direction. Since the spring is massless, the acceleration of the whole system is still $a_x = F/(m_1 + m_2)$ as in Example 3. And because the separation between blocks is assumed constant, each block still accelerates at this same rate.

The forces on block 1, the spring, and block 2 are as indicated in the free-body diagram:

Applying Newton's law $\Sigma F_x = ma_x$ to the spring we obtain

$$P_1 - P_2 = m(\text{spring})a_x.$$

But because the mass of the spring is assumed negligible, we have $P_1 = P_2$. Newton's law for each block then takes the same form as in Example 3. We find from Example 3

$$P_2 = m_2 a_x = \frac{m_2 F}{m_1 + m_2}$$

for the force on the spring. The compression of the spring is

$$|x| = \frac{P_2}{k} = \frac{m_2 F}{(m_1 + m_2)k}.$$

Substituting numbers, we find $|x| = 2/15$ cm.

Example 5

A 3.0-kg monkey holds on to a light rope which passes over a frictionless pulley and is attached to a 4.0-kg bunch of bananas. What is the acceleration of the monkey?

Since we have two objects, we draw a free-body diagram for each. Because tensions in ropes always pull, the tension T acting on the monkey is vertically upward. The same tension T pulls vertically upward on the bananas as well.

Let's label the monkey's mass, position, and acceleration by m_1, z_1, and a_1 measured vertically upward, and the bananas' mass, position, and acceleration by m_2, z_2, and a_2 measured vertically upward.

Applying $\Sigma F_{z_1} = m_1 a_1$ to the monkey, we have

$$T - m_1 g = m_1 a_1. \tag{a}$$

In addition, the fixed length of the rope imposes the constraint that the bananas accelerate downward at the same rate as the monkey accelerates upward, $a_2 = -a_1$. Using this result to eliminate a_2 from the second application of Newton's law, we obtain

$$T - m_2g = -m_2a_1.$$ (b)

We are left with two equations (a) and (b) and two unknowns. Subtracting (b) from (a) to eliminate T we find $-m_1g + m_2g = m_1a_1 + m_2a_1$; solving for the acceleration a_1 we obtain

$$a_1 = (m_2 - m_1)g/(m_2 + m_1).$$

Since the bananas are heavier than the monkey ($m_2 > m_1$), the monkey accelerates upward, while the bananas accelerate downward at the same rate. Substituting values, we find $a_1 = 1.4 \text{ m/s}^2$. The acceleration is substantially less than g because the combined monkey–rope–banana system behaves like a linear chain, with the pulls at either end partially canceling whereas the inertia of the monkey combines with that of the bananas.

If the monkey and bananas are replaced by simple weights, this arrangement is called an *Atwood's machine*. If m_1 and m_2 are nearly the same, the acceleration can be made quite small, permitting an accurate measurement which determines g rather well.

Example 6
A puck of mass m on a frictionless table is attached to a mass M by a light string which passes through a hole in the table. What must be the speed of the puck if it is to move in a circle of radius r while M remains at rest?

We know that the circular motion of the puck must be sustained by a centripetal force. Here the tension in the string (created by the weight Mg) provides that force. The free-body diagrams are shown below:

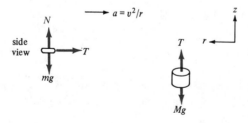

Since the puck is moving in a circle, it has a centripetal acceleration $a = v^2/r$. That's how the speed of the puck enters the problem.

For M let's choose the positive z axis to be vertically upward, and for the puck, we choose the radial direction outward to be the positive r direction (the direction opposite to the centripetal acceleration). Applying the second law $\Sigma\, F_z = 0$ to the mass M (which is not accelerating), we find

$$T - Mg = 0.$$

For the puck, recalling that both the tension and acceleration act inward, $\Sigma\, F_r = ma_r$ implies $-T = -mv^2/r$, so

$$T = mv^2/r.$$

Substituting for T and solving for the speed, we find

$$v = \sqrt{Mgr/m}.$$

8.5 FRICTION

In 1699 the French scientist Guillaume Amontons investigated the losses caused by friction in machines. From his studies he found the empirical relationship that frictional forces from a surface do not exceed an amount proportional to the normal force exerted by the surface on the object,

$$f \leq \mu N,$$

where μ is the *coefficient of friction*. Later Coulomb noted that μ varies with the two materials that are in contact.

Friction is an inescapable example of a force produced by electricity. At times we wish that we could do away with it, so as, for example, to improve engine performance, yet without it (a condition approached on ice or a waxed floor) we couldn't walk. Even though a highly polished object may appear smooth, when examined through a microscope it appears very rough, having countless surface irregularities, as shown in Fig. 8.6. When two objects are placed in contact, the many contact points resulting from the (microscopic) rough edges tend to interlock or even become welded together by electric forces. When one object moves across another, these tiny welds continually rupture and reform. In addition the interlocking obstacles must be overcome by lifting or deformation or abrasion. The net result of these complex causes is friction – a force parallel to the surface which opposes the motion of the object. Since the number of welds is proportional to pressure from the object on the surface, the force of friction is proportional to the normal force on the object, as Amontons found.

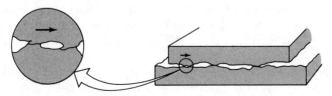

Figure 8.6 Microscopic examination of a highly polished surface reveals irregularities.

The frictional forces acting between surfaces at rest with respect to each other are called forces of *static friction*. Suppose that you have a block at rest on a horizontal surface. By Newton's second law, the force of friction is zero, as Fig. 8.7a illustrates.

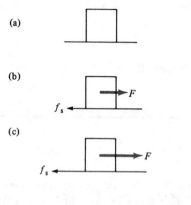

(a)

(b)

(c)

(d)

$F = \mu_s N$ slipping occurs

Figure 8.7 The force of static friction increases up to a maximum value equal to $\mu_s N$.

Now suppose you apply a small measurable force **F** to it as in Fig. 8.7b, and observe that the block doesn't move. By Newton's second law, the force of static friction is equal in magnitude and opposite in direction to **F**. Now suppose that you increase **F** and note that the block still doesn't move. The force of static friction increases as well, always being equal to **F** in magnitude, as Fig. 8.7c shows. As **F** is further increased, there will be a definite value of **F** for which the block slips, as Fig. 8.7d illustrates. The smallest force necessary to start motion is the maximum force of static friction. These observations can be summarized by the relation for the magnitude of force of static friction f_s,

Static friction:

$$f_s \leq \mu_s N$$

(8.7)

where μ_s is the coefficient of static friction which depends on the two surfaces in contact and N is the normal force. Static friction always acts parallel to the surface and opposes the intended motion of an object in its rest frame. Table 8.2 lists values of μ_s for various materials.

Once the block begins to move, *kinetic friction* acts on the block. This frictional force is usually less than static friction (e.g., it is harder to form welds on a moving contact). The magnitude of the kinetic friction f_k approximately obeys the empirical relationship

Kinetic friction:

$$f_k = \mu_k N \qquad (8.8)$$

where N is the normal force and μ_k is the coefficient of kinetic friction which depends on the two surfaces in contact. Table 8.2 lists a few values of μ_k. The force of kinetic friction is always opposite to the velocity of the object.

Table 8.2 Coefficients of Static and Kinetic Friction

Material	μ_s	μ_k
Steel on steel	0.78	0.42
Nickel on nickel	1.10	0.53
Teflon on Teflon	0.04	0.04
Oak on oak (parallel to grain)	0.62	0.48
Oak on oak (perpendicular)	0.54	0.32
Ice on ice	0.05	0.04

Example 7

A spring with pointer attached is used as a force scale. First the spring is calibrated by hanging a known mass $M = 1$ kg from it vertically; the mass stretches the spring (moves the pointer) 5 cm. Then the spring is used to pull a wooden block of mass $m = 0.5$ kg horizontally across sandpaper. The spring is stretched by 3 cm just before the block starts to move, but only 2 cm when it moves at constant velocity. What is the spring constant k, and what are the coefficients of friction μ_s and μ_k?

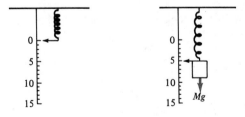

The equilibrium condition for the calibration is $k|x| = Mg$, where $|x|$ is the distance the spring is stretched, so the spring constant is

$$k = \frac{Mg}{|x|} = \frac{(1 \text{ kg})(9.8 \text{ m/s}^2)}{0.05 \text{ m}} = 196 \frac{\text{N}}{\text{m}}.$$

For the block on the sandpaper, the equations of equilibrium are

$$N = mg, \qquad k|x| = f.$$

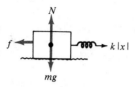

Just at the point of slipping, the static friction becomes $f_s = \mu_s N = \mu_s mg$. Therefore

$$\mu_s = \frac{k|x|}{mg} = \frac{(196 \text{ N/m})(0.03 \text{ m})}{(0.5 \text{ kg})(9.8 \text{ m/s}^2)} = 1.2.$$

When the block moves at constant velocity, the spring force is balanced by kinetic friction, $f_k = \mu_k N = \mu_k mg$. Since the spring displacement is now only two-thirds as great, the force is only two-thirds as great and we have

$$\mu_k = \frac{k|x|}{mg} = 0.8.$$

Example 8

A block rests on an inclined plane which has a variable angle θ. The angle θ is increased from zero, and at 40° the block slips. What is the coefficient of static friction?

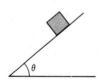

In a free-body diagram for the block, static friction is directed up the plane (because the block tends to slip downward).

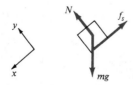

Choosing the positive x axis along the plane and the positive y axis perpendicular to the plane, we have the following:

$$\sum F_x = 0 \quad \text{implies} \quad mg \sin \theta - f_s = 0,$$

$$\sum F_y = 0 \quad \text{implies} \quad N - mg \cos \theta = 0.$$

When $\theta = 40°$, the force of static friction is maximum and we can substitute $f_s = \mu_s N$. But from our second equation, the normal force $N = mg \cos 40°$. Substituting all this into our first equation, we get

$$mg \sin 40° - \mu_s mg \cos 40° = 0,$$

which tells us that

$$\mu_s = \tan 40° = 0.84.$$

Note that only when *maximum* static friction is acting on an object can you use $f_s = \mu_s N$.

When θ is greater than $40°$, the block slides down the plane, with $\sum F_x = mg \sin \theta - \mu_k mg \cos \theta = ma$, giving for the acceleration

$$a = g(\sin \theta - \mu_k \cos \theta).$$

This equation is also sometimes relevant at $\theta < 40°$, but here it must be treated with care. At small θ ($\tan \theta < \mu_k$) it seems to predict an acceleration up the plane. This correctly describes the case in which an initially downward sliding block decelerates until it comes to rest, but at that point we must switch over to the equilibrium ($a = 0$) condition. Remember that friction opposes motion; it cannot of itself produce motion up the plane. In the usual situation $\mu_s > \mu_k$ there is also an intermediate range of angles ($\mu_s > \tan \theta > \mu_k$) where static equilibrium can be satisfied, but $a > 0$ once the block is set in motion. Here equilibrium is only metastable; a sufficiently strong blow will start the block accelerating downwards.

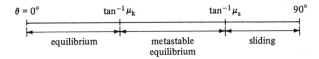

Example 9
A ladder of weight w and length L leans against a wall. The wall is frictionless, but the ground has coefficient of friction μ_s. At what angle θ does the ladder start to slip?

If we assume the ladder is uniform, the weight acts through its center as shown in the free-body diagram. The force exerted on the ladder by the wall is normal to the wall, but the force exerted by the ground has components both normal and parallel to the ground owing to the presence of friction.

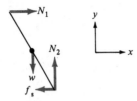

Newton's second law gives

$$\sum F_x = 0 \quad \text{implies} \quad N_1 - f_s = 0,$$

$$\sum F_y = 0 \quad \text{implies} \quad N_2 - w = 0.$$

To determine f_s we must also balance torques. Taking torques around the point of contact with the ground and following the convention that counterclockwise torques are positive, we have

$$w \frac{L}{2} \sin \theta - N_1 L \cos \theta = 0,$$

so

$$N_1 = f_s = \frac{w}{2} \tan \theta.$$

At the point of slipping, $f_s = \mu_s w$, so

$$\tan \theta = 2\mu_s.$$

Note that the force exerted by the ground is not generally directed along the ladder, but must pass through the intersection point of N_1 and w to avoid giving a net torque about that point.

If the wall is rough, its force on the ladder includes a parallel component f_{s1}. There are now four unknown forces, N_1, N_2, f_s, and f_{s1}, but still only three independent equations! This simply means that the solution is not unique*; the ladder can be kept in equilibrium for a given angle θ by many possible sets of forces.

---

*To be sure, more equations can be obtained by setting $\sum \tau_{iz} = 0$ about different points. But this does not provide *independent* conditions because, as we proved in Section 6.6, having $\sum F_i = 0$ and $\sum \tau_i = 0$ about one point *ensures* that $\sum \tau_i = 0$ about all points.

Friction is messy, and the rules given above are idealizations. To give an example of the complications that arise in practice, consider the coefficient of friction for a copper block on a copper plate. The value of μ_s given in handbooks is about 1.6. But if we took great care to prepare clean smooth surfaces, working in a vacuum to avoid oxidation and prevent even a thin layer of air from getting between the blocks and the plate, the atoms on the surface would forget which piece they belonged to and would fuse together. The coefficient of friction would be huge! The value of μ given in the handbook refers to a typically oxidized, dirty surface as found in normal use. Furthermore, to tell the whole truth, μ is not generally independent of velocity and N. It is not fundamental at all. The use of Eqs. (8.7) and (8.8) with constant handbook values of μ simply provides a rough and ready first estimate, which is often all one needs in practice.

8.6   DRIVING ON CURVED ROADWAYS

The motion of an automobile around a curve is an important example bringing into play many of the principles in this chapter. Consider an automobile rounding a circular curve whose center is a distance R to the right. The force acting on the car must supply a net centripetal acceleration v^2/R to the right.

If the curve is unbanked, the centripetal acceleration must be supplied entirely by friction. Figure 8.8a shows the car, with the reaction $\mathbf{F}$ of the road as well as the force due to gravity drawn (for simplicity) as acting through the center of gravity. In Fig. 8.8b we exhibit the vectorial relation that must exist among $\mathbf{F}$, gravity, and centripetal force. Figure 8.8c shows the free-body diagram with $\mathbf{F}$ split into normal and frictional components.

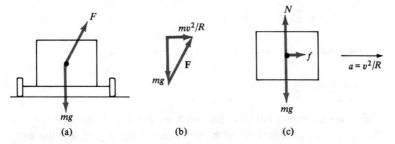

(a)   (b)   (c)

Figure 8.8 Forces on a vehicle rounding an unbanked curve. In (c), N and f are net forces not drawn at their points of application; the individual normal and frictional forces applied at each tire will be shown in Fig. 8.10.

Newton's second law gives

$$N - mg = 0, \qquad\qquad\qquad\qquad\qquad (8.9)$$

$$f = \frac{mv^2}{R} . \qquad\qquad\qquad\qquad\qquad (8.10)$$

The maximum strength of friction* is $f = \mu_s N = \mu_s mg$, so the car goes off course (skids) if

$$\frac{v^2}{R} > \mu_s g. \tag{8.11}$$

At high speed, a small R and a slippery road (low μ_s) are dangerous.

If the curve is banked, less reliance is placed on friction. The reaction force **F** required to keep the car on course is exactly the same as before (Fig. 8.8b), but now more of it can be supplied by the normal force. Figure 8.9 shows the optimally banked case in which the road surface is perpendicular to the required reaction force **F** and friction does not come into play at all. In this case Newton's second law gives

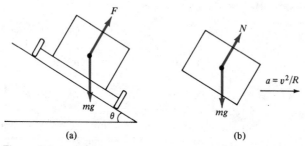

(a)   (b)

Figure 8.9 Vehicle rounding a banked curve.

$$N \cos \theta - mg = 0, \tag{8.12}$$

$$N \sin \theta = \frac{mv^2}{R}. \tag{8.13}$$

Solving for the optimal banking angle, one obtains

$$\tan \theta = \frac{v^2}{gR}. \tag{8.14}$$

Engineers construct banked turns with an average speed in mind. Vehicles making the turn at greater speeds than the road is planned for require a component of friction to stay on the road, but less than an unbanked road would require.

In reality the road acts on the car not at the center of gravity, but at the points of contact with the tires as pictured in Fig. 8.10. So torques must be considered, and it is

---

*Here we use the fact that static friction applies in the radial direction when the car has no velocity component in that direction.

In the direction of the car's rolling motion a new category of friction applies: *rolling friction*. Kinetic friction does not apply because (as will be discussed in Sec. 14.11) a rolling tire is at rest just at the instantaneous point of contact with the road. Otherwise it would be skidding! Rolling friction works by a somewhat different mechanism, involving peeling off of the tire surface and also continual deformation of both the tire and the road surface underneath, which has the effect that the tire is always climbing out of a slight depression it has made. The use of wheels is advantageous because rolling friction is much less than the ordinary kinetic friction associated with sliding. The motion of a car is opposed by kinetic friction only when the tires skid.

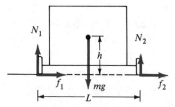

Figure 8.10 Torques on a car making an unbanked turn.

here that the virtues of a low center of gravity and broad wheelbase in preventing overturn become evident. Newton's second law now gives

$$N_1 + N_2 - mg = 0, \tag{8.15}$$

$$f_1 + f_2 = \frac{mv^2}{R}. \tag{8.16}$$

It is convenient to consider the torques about the center of gravity. They must satisfy

$$f_1 h + f_2 h + N_2 \frac{L}{2} - N_1 \frac{L}{2} = 0, \tag{8.17}$$

where h is the height of the center of gravity above the road and L is the length of the axle. This condition applies, in spite of the car's turning motion, because the car does not rotate in the plane of the diagram. Solving these three equations, we find

$$N_1 = \frac{mg}{2} + \frac{mv^2 h}{RL}, \tag{8.18}$$

$$N_2 = \frac{mg}{2} - \frac{mv^2 h}{RL}. \tag{8.19}$$

The normal forces on the two sides are not equal. If equilibrium requires negative N_2, the road cannot supply it and the car will overturn. This occurs when

$$\frac{v^2}{R} > \frac{gL}{2h}, \tag{8.20}$$

so large L and small h are desirable for stability. Vehicles are normally engineered to skid before they overturn on an unbanked curve. Comparing Eqs. (8.11) and (8.20), we see that this requires

$$\frac{L}{2h} > \mu_s. \tag{8.21}$$

8.7 MOTION IN A RESISTIVE MEDIUM

In the *Principia* Newton considered the motion of objects in resistive media – for example, dust particles falling through air or marbles falling in water. The resistive force in these cases is *viscosity*. It is another contact force, extremely complicated to work out in detail.

But when an object moves at low velocity through a fluid, such as a gas or liquid, the viscosity is approximately proportional to the velocity. Mathematically we write this as

$$\mathbf{F}_{vis} = -K\eta\mathbf{v}. \tag{8.22}$$

The minus sign indicates this force is always opposite to the velocity of the object. The proportionality constant K depends on the size and shape of the object, and η, the coefficient of viscosity, depends on the internal friction between different layers of the fluid.

In general, calculating K is laborious, but a century ago George Stokes found the result for a sphere of radius R (known as Stokes's law):

$$K = 6\pi R. \tag{8.23}$$

Thus the viscous force acting on a sphere takes the simpler form*

$$\mathbf{F} = -6\pi R\eta\mathbf{v}. \tag{8.24}$$

From Stokes's law we see that K has units of meters, so that by Eq. (8.24) the coefficient of viscosity η has units of newton-seconds per square meter. This coefficient depends on temperature: for liquids it decreases with increasing temperature, but for gases it increases with temperature. Table 8.3 lists the coefficient of viscosity for several fluids.

A simple laboratory demonstration of the viscous drag force is provided by dropping a marble into a beaker of a very viscous liquid, like glycerin. The marble appears to fall with a constant speed. The forces acting on the marble are gravity, mg downward, and viscous force $-6\pi R\eta\mathbf{v}$ upward. If the marble is falling with a constant speed v_L, its acceleration is zero, so by Newton's second law,

$$mg - 6\pi R\eta v_L = ma = 0. \tag{8.25}$$

Table 8.3 Coefficients of Viscosity at 20°C (Unless Noted)

Liquid	η (10^{-3} N s/m^2)	Gas	η (10^{-5} N s/m^2)
Water (0°C)	1.792	Air (0°C)	1.71
Water	1.005	Air	1.81
Castor oil	9.86	Ammonia	0.97
Glycerine	833	Hydrogen	0.93

*This force law is fairly accurate for the small, relatively slow raindrops and oil drops treated in this chapter. It is not accurate for cannonballs or skydivers. The resistive force for such large objects at their normal speeds is approximately proportional to their cross-sectional area times the square of the speed. For more detailed discussion of this subject see, for example, D. Roller and R. Blum, *Physics: Volume I, Mechanics, Waves, and Thermodynamics* (Holden-Day, San Francisco, 1981), pp. 395–404.

In other words, the viscous force is equal to the weight. Solving for the speed v_L, called the *limiting* or *terminal velocity*, we find

$$v_L = \frac{mg}{6\pi R \eta}. \qquad (8.26)$$

Example 10

Find the terminal velocity of a small raindrop assumed to have a radius of 40 μm, in still air.

According to Eq. (8.26) we need to know η, R, and the mass of the raindrop. The η required is that of the medium the drop falls through – air, not water; from Table 8.3 we have $\eta = 1.8 \times 10^{-5}$ N s/m² for air. The radius $R = 4 \times 10^{-5}$ m is given. The mass is simply the density of water, $\rho = 10^3$ kg/m³, times the volume, which is assumed to be spherical:

$$m = \rho \tfrac{4}{3}\pi R^3.$$

Substituting, we have

$$v_L = \rho \tfrac{4}{3}\pi R^3 g/(6\pi R\eta) = \tfrac{2}{9}\rho g R^2/\eta.$$

Inserting values, we find $v = 0.2$ m/s.

Note that this result is strongly dependent on radius; the sort of fine droplets found in a mist fall very slowly but large raindrops fall much more rapidly. However, for raindrops larger than 40 μm, Stokes's law in not accurate because turbulence (not considered here) sets in. Furthermore, drops of radius >500 μm become nonspherical in falling. Thus for large drops the detailed relation between v_L and R differs from that given above.

Equation (8.26) tells us that terminal velocity is proportional to the weight of an object. In other words, that heavier bodies fall faster! Can this be true? Is the world really Aristotelian? The answer is that we are *including* air resistance – precisely what Aristotle thought must always be present, and Galileo preferred to ignore. But now we have Newton's laws, which provide the framework to treat the question systematically and to appreciate the circumstances in which Aristotle's ideas apply and those in which Galileo's approximation to the full dynamics is accurate.

For any spherical object falling at low velocity in a viscous medium, not necessarily at terminal velocity, Newton's second law implies

$$m\frac{dv}{dt} = mg - 6\pi R\eta v. \qquad (8.27)$$

This is a differential equation. Presently we will solve it and see how the velocity depends on time. But first, let's see what we can learn from the differential equation without solving it.

Let's look at what Eq. (8.27) describes. At the instant we drop the object, it is at rest, which means $v = 0$; therefore the viscous force $-6\pi R\eta v$ is also *momentarily* zero. Consequently, we have $m\, dv/dt = mg$ at that instant. In other words, the object starts with acceleration $dv/dt = g$, just as Galileo said it should. Because the ball accelerates, the speed and viscous force increase. As a result, the right-hand side of Eq. (8.27), $mg - 6\pi R\eta v$, becomes smaller than mg, and the acceleration decreases. As the velocity increases (of course, the velocity increases ever more slowly as time goes on because the acceleration is getting smaller) the right-hand side approaches zero. If it were zero, gravity and the viscous force would balance each other, so the object would have zero acceleration and would fall with the terminal velocity.

The object starts out being Galilean and ends up being Aristotelian. A key question is, how long does this take? If it takes hours to reach terminal velocity, we can forget about Aristotle; the effects of viscosity can be ignored. But if it takes only a fraction of a second, then the object spends most of its time falling at terminal velocity.

We can figure out whether the velocity v approaches the terminal velocity slowly or quickly by a dimensional analysis of the differential equation (8.27). This means that we take a look at the units of each term. If we divide Eq. (8.27) by the mass m we get

$$\frac{dv}{dt} = g - \frac{6\pi R\eta}{m}\,v.$$

Since each term must have the same units, distance/time^2, the factor multiplying v on the right must have units 1/time, so its reciprocal has units of time. Let's denote this reciprocal by t_0. Thus, by definition,

$$t_0 = \frac{m}{6\pi R\eta}, \tag{8.28}$$

and t_0 has units of time. The differential equation now becomes

$$\frac{dv}{dt} = g - \frac{v}{t_0}.$$

What "time" does t_0 represent physically? To find out we let $t \to \infty$ in the differential equation. The velocity approaches the terminal velocity v_L, and dv/dt approaches 0, so in the limit we get

$$0 = g - \frac{v_L}{t_0}, \qquad \text{or} \qquad v_L = gt_0.$$

In other words, t_0 is the time it would have taken to reach the terminal velocity v_L if the acceleration were always equal to g. The number t_0 is called the "characteristic time." We will now solve the differential equation and see that the value of t_0 will tell us whether the velocity v approaches terminal velocity v_L quickly or slowly, depending on whether t_0 is small or large.

The differential equation (8.27) may appear more complicated than any we have seen so far, but actually it is a familiar one in disguise. Since $g = v_L/t_0$ we can rewrite the differential equation as follows:

$$\frac{dv}{dt} = -\frac{v - v_L}{t_0}.$$

Now let $w = v - v_L$, the difference between the actual velocity v and the terminal velocity v_L. Then $dw/dt = dv/dt$ since v_L is constant, so w satisfies the simpler differential equation

$$\frac{dw}{dt} = -\frac{1}{t_0} w.$$

This is the differential equation for the exponential function which we've seen earlier in Chapter 3 (Problem 3.9). Its solution is

$$w(t) = w(0)e^{-t/t_0}$$

where $w(0) = v(0) - v_L$. But $v(0) = 0$ so $w(0) = -v_L$. Replacing $w(t)$ by $v(t) - v_L$ we see that

$$v(t) - v_L = -v_L e^{-t/t_0}$$

so the solution is

$$v(t) = v_L(1 - e^{-t/t_0}). \tag{8.29}$$

The characteristic time t_0 appears in the denominator of the exponential term, so it governs the rate at which this exponential tends to zero. If t_0 is small, the exponential decays very rapidly and v quickly approaches the terminal velocity v_L. If t_0 is large the exponential decreases more slowly and it takes a longer time for v to approach v_L.

We can now systematically predict which motions will follow Galileo's prescription and which will appear Aristotelian. A heavy ball (large m) falling in air (very small η), according to Eq. (8.28), takes a very long time to reach terminal velocity; it tends to behave in the way Galileo described, unless it falls very far. On the other hand, a marble (let's say with $m = 0.01$ kg, $R = 0.01$ m) in glycerin ($\eta = 0.8$ N $\cdot$ s/m^2) takes a time $t_0 = 0.06$ s to approach terminal velocity; in other words it is close to terminal almost all of the time we are watching it.

Example 11

If $v_L = gt_0$ and $v(t) = v_L(1 - e^{-t/t_0})$, check that $v(t)$ has the physically expected limiting behaviors: (a) $v(t) \to v_L$ as $t \to \infty$ and (b) $v(t) \approx gt$ if t/t_0 is small. Find $v(t_0)$.

(a) As $t \to \infty$ the exponential term goes to 0 and $v(t) \to v_L$ as expected.

(b) To find a linear approximation for $1 - e^{-t/t_0}$ we use the second fundamental theorem of calculus,

$$e^u - 1 = \int_0^u e^x \, dx.$$

If u is small the integrand is nearly 1 so $e^u - 1 \approx u$, or

$$1 - e^u \approx -u.$$

Taking $u = -t/t_0$ we get

$$v(t) = v_L(1 - e^{-t/t_0}) \approx v_L t/t_0 = gt$$

if t/t_0 is small. In other words, the object essentially undergoes free fall until the velocity becomes large enough to make air resistance appreciable. The curve below is the graph of $v(t)$ and the dashed lines indicate the limiting behaviors.

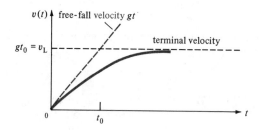

As for $v(t_0)$, we find by substituting $t = t_0$ in Eq. (8.29)

$$v(t_0) = v_L(1 - e^{-1}).$$

Evaluating the exponential, we find $v(t_0) = 0.63v_L$. Thus we have another quantitative interpretation of the time t_0: it is the time required for the object to reach 63% of its terminal velocity.

8.8 THE OIL-DROP EXPERIMENT

To conclude this chapter, all of its manifold threads – fundamental forces, their precise strength, contact forces, and the application of Newton's laws of motion – are now going to be drawn together in recounting one of the most famous experiments of modern physics.

The fundamental electric force on a charge q_1 due to a second charge q_2,

$$\mathbf{F} = K_e \frac{q_1 q_2}{r^2} \hat{\mathbf{r}} \tag{8.2}$$

can be broken into the product of q_1 and an *electric field* $(K_e q_2/r^2)\hat{\mathbf{r}}$ due to the second charge. The total electric force on q_1 is likewise the product of q_1 and the total electric field due to the summed effects of all other charges. Because the force has this form, a study of the motion of a charged particle in an electric field can determine its charge. The electron was discovered by J. J. Thomson in 1896. Nevertheless, because of the difficulty of observing the motion of a single electron accurately, some time passed before its charge was determined. Instead of observing a single electron, a more feasible approach was to study the motion of small droplets that carry only a small net excess of electronic charges. In 1906 Robert A. Millikan, then an assistant professor at the University of Chicago, devised an ingenious experiment which for the first time made it possible to measure the charge on an individual droplet. Through Millikan's experiment it became possible to verify that electricity in gases and chemical solutions is built out of discrete units of electric charge, and to determine what that unit is.

Figure 8.11 Robert A. Millikan's original apparatus to measure the electron charge. (Courtesy of the Archives, California Institute of Technology.)

Millikan's original apparatus is shown in Fig. 8.11. Millikan used oil for the very same reason mankind spent three hundred years improving clock oils: oil droplets scarcely evaporate. Therefore the viscous force would not change during an experiment. Sprayed from an atomizer, the droplets would acquire a charge q due to friction (remember, friction is a result of electrical forces between atoms) as they pass through the nozzle of the atomizer. The charged droplets then fall through a hole in the uppermost of two metal plates, which Millikan connected to a 10,000-volt battery. While between the plates, the droplets experience an electric force in addition to gravity, as shown in Fig. 8.12. The electric field **E** is downward but because of the negative charge of the droplets (and electrons), the force q**E** is upward.

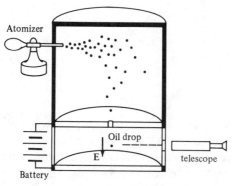

Figure 8.12 Schematic of Millikan's oil-drop apparatus.

Through a viewing device placed a couple of feet from the chamber, Millikan could watch individual droplets illuminated by light which passed through water so as not to heat the air in the chamber. Because of their extremely small size, the droplets appear as stars on a black background. By increasing the voltage, he could make droplets rise; those droplets with greater charge rose more quickly. By reversing the voltage, he could make them fall faster. In addition, he could change the charge on a droplet by sending a stream of ions into the chamber. Millikan's fascination with the acrobatic motion of droplets lightened the long, solitary hours he spent in the lab squinting through the eyepiece and recording hundreds of measurements.

By adjusting the voltage on the plates (and hence the electric field) certain droplets could be suspended when the upward electric force equaled the weight of the droplet:

$$qE = mg. \tag{8.30}$$

One might think that Millikan could determine the charge q directly from Eq. (8.30). But the mass of the droplet also appears in Eq. (8.30), and although Millikan knew the relation

$$m = \tfrac{4}{3}\pi R^3\rho \tag{8.31}$$

and the density ρ, the radius R varies from drop to drop. If a drop is small enough to be balanced by a not overly large field, its radius is too small to measure directly by light. So Millikan had to determine the radius by a second measurement on the *same* droplet, using the rate of fall when the field was reduced. Or more generally, since in practice the perfect balance of Eq. (8.30) is hard to achieve, he measured the rate of rise and fall of a given drop at two different voltages.

Precisely what did Millikan do to determine the charge of the electron? First, he studied the motion of a droplet drifting upward between the plates of Fig. 8.12. According to Newton's second law this motion can be described by

$$m\frac{dv}{dt} = qE - mg - 6\pi R\eta v. \tag{8.32}$$

In this case the electric force pushes the negatively charged droplets upward but the viscous force acts downward (opposite v) in the same direction as gravity. By setting $dv/dt = 0$ in Eq. (8.32), we find the terminal velocity to be

$$v_1 = \frac{qE - mg}{6\pi R\eta}. \tag{8.33}$$

The characteristic time to reach this terminal velocity turns out to be the same as when there is no electric field and is given by Eq. (8.28):

$$t_0 = \frac{m}{6\pi R\eta}. \tag{8.28}$$

One finds that a typical droplet of size such that qE and mg nearly balance reaches terminal velocity very quickly, and that the terminal velocity is quite slow. Using a stopwatch to time a droplet moving between fiducial marks engraved on the viewing device, Millikan could measure the terminal velocity. When he rewrote Eq. (8.33) in the form

$$v_1 = \frac{qE - \frac{4}{3}\pi R^3 \rho g}{6\pi R \eta} \tag{8.34}$$

by using Eq. (8.31) to relate the mass of the droplet to its radius, only q and R remained unknown.

A second measurement was provided by turning the electric field off and watching the free fall of the same droplet. When the droplet is simply falling under the force of gravity, the terminal velocity is given by Eq. (8.26),

$$v_2 = \frac{mg}{6\pi R \eta}, \tag{8.26}$$

which we again rewrite as

$$v_2 = \frac{2R^2 \rho g}{9\eta} \tag{8.35}$$

with the aid of Eq. (8.31). In effect, this second measurement is used to find the radius of the droplet, R.

Using Eq. (8.35) we can eliminate R in Eq. (8.34), and with some algebra solve for q. The result is

$$q = \frac{18\pi\eta^{3/2}}{E\sqrt{2g\rho}} v_2^{1/2} (v_1 + v_2). \tag{8.36}$$

By measuring v_1 and v_2 Millikan determined the charge of a droplet.

To be very precise, Millikan actually used $\rho - \sigma$ in place of ρ, where σ is the density of air. The reason is that air provides an additional upward buoyant force on a droplet, which is equal to the weight of the air displaced by the droplet (this is known as Archimedes' law). The weight of air displaced is just the density of air times the volume of the droplet. Accounting for this force is equivalent to saying that the weight of the droplet is reduced to $mg - m_a g = (m - m_a)g$, where m_a is the mass of air displaced: $m_a = \frac{4}{3}\pi R^3 \sigma$. Since the density of air is about one-thousandth that of oil, the correction is barely necessary. With this correction taken into consideration, Eq. (8.36) becomes

$$q = \frac{18\pi\eta^{3/2}}{E\sqrt{2g(\rho - \sigma)}} v_2^{1/2} (v_1 + v_2). \tag{8.36}$$

A correction to Stokes's law which we shall not discuss here was also necessary to obtain precise results.

Through hundreds of delicate measurements, Millikan, the patient experimentalist, discovered that the charge on a droplet always comes out an integral multiple (like 1, 2, 3, etc.) of the smallest charge he found. Here was confirmation that charges come in integral multiples of a fundamental charge – the charge of the electron.

By reevaluating the coefficient of viscosity for air, and reducing errors caused by temperature and air currents, Millikan succeeded in determining the charge e of the electron with an error of 0.1%. The value he published in 1913 was $e = -(1.603 \pm 0.002) \times 10^{-19}$ coulomb, which served physics for a generation and is within experimental error of the most recent value. For his momentous efforts, Millikan received the Nobel Prize in 1923.

The quest for the fundamental constituents of matter goes on. Today some physicists are searching for fractionally charged particles called quarks. Based upon a symmetry classification for elementary particles, quarks are the building blocks of particles which exist inside nuclei and carry charges of $+\frac{2}{3}e$ and $-\frac{1}{3}e$. Modifications of Millikan's historic experiment are used by some of these quark hunters.

8.9 A FINAL WORD

When Millikan made his measurements, alone in his laboratory, like any scientist, he recorded what he had done in a notebook. Afterward, he would gather his results, write a scientific paper, and publish it for all the world to see. But his notebooks, the raw data of his experiments, were for his own eyes only. Figure 8.13 shows a page from Millikan's notebook. Before we criticize what we see, let's remember what Millikan was doing. He was measuring, for the first time ever, one of the fundamental constants of nature. His task was to make his measurements in the most careful, dispassionate way possible, then publish all of his results so that other scientists could judge whether he'd done it properly. The page in Figure 8.13 is dated March 15, 1912. He writes down the temperature and barometric pressure, then he starts recording data: the times for a droplet to move between fiducial marks under pure gravity (G), and with the field (F). Then he calculates the velocities, uses logarithms to multiply them together (he didn't have a hand calculator), and finally he gets his result.

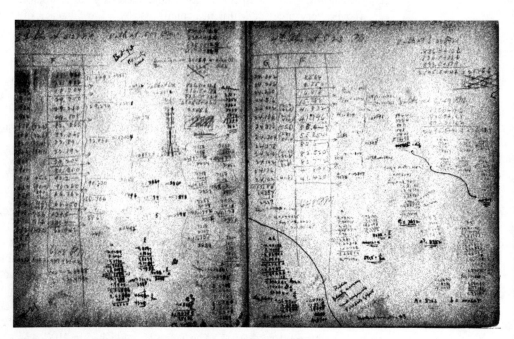

Figure 8.13 Page from Robert A. Millikan's research notebook.
(Courtesy of the Archives, California Institute of Technology.)

On one page he writes: "One of the best ever . . . almost exactly *right*" – What's going on here? How can it be right if he's supposed to be measuring something he doesn't *know*? On another page he writes: "Beauty. Publish!" One might expect him to publish everything! On another page, the usual stuff, then: "4% too low – something wrong." Not 4% too low but publish anyway, like a good scientist. Then something very revealing: ". . . distance wrong." He's found an excuse for not publishing it. More pages . . . "Beauty, one of the best," and so on for pages and pages.

Now, you shouldn't conclude that Robert Millikan was a bad scientist. He wasn't – he was a great scientist, one of the best. What we see instead is something about how real science is done in the real world. What Millikan was doing was not cheating. He was applying scientific judgment. He had a pretty clear idea of what the result ought to be – scientists almost always think they do when they set out to measure something. So, when he got a result he didn't like, he wouldn't just ignore it – that *would* be cheating. Instead, he would examine the experiments to see what went wrong. Now that seems reasonable, but it's actually a powerful bias to get the result he wants, because you can be sure that when he got one he liked, he didn't search as hard to see what went right.

Experiments must be done that way. Without that kind of judgment, the journals would be full of mistakes. So, then, what protects us from being misled by somebody whose "judgment" leads to a wrong result? Mainly, it's the fact that sooner or later, someone else with a different prejudice will make another measurement. You see there *is* a real answer; it's part of nature. And, so long as that's true, sooner or later the truth will come out. Much is written in textbooks about the scientific method, especially that picking the results you like is a cardinal sin. Don't believe everything you read. Science is a difficult and subtle business, and there is no method that assures success.

Problems

The Strength of Gravitational and Electric Forces

1. Using the values of G, g, and the radius of the earth, calculate the mass of the earth. From your value for the mass determine the density of the earth by treating the earth as a solid sphere. The average density of rocks on the earth's surface is about 2.7 g/cm^3; comparing this value to your calculated value, what can you conclude about the interior of the earth?

2. Compare the electric force between the proton and electron in a hydrogen atom with the gravitational force between them ($m_p = 1.67 \times 10^{-27}$ kg, $m_e = 9.11 \times 10^{-31}$ kg).

Contact Forces

3. A book rests on a table which stands on the floor.

   (a) Is the normal force acting on the book the reaction force to its weight? Explain.
   (b) List the vertical forces acting on the table. How strong is the normal force exerted by the floor on the table?

Application of Newton's Laws

4. A 1.5-kg block on a frictionless table is attached to a spring of spring constant k = 0.5 N/m. If the spring is stretched 3.0 cm and released, what is the acceleration of the block at the instant it is released?

5. A mover pulls on a 20.0-kg crate resting on a floor with a force of 80.0 N at an angle of 37°, but the crate does not move. What is the normal force of the floor on the crate?

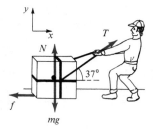

6. Rework Example 3 of this chapter for the case of F applied to the larger block from the other side. Explain why the force of one block on the other is different in this case.

7. Three identical boxes of mass m = 2.0 kg are pulled along a horizontal frictionless table by a force F = 18 N. Find

(a) the acceleration of the blocks,
(b) the tension in each string.

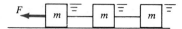

8. A lamp hangs vertically from a cord in a posh elevator which is accelerating upward at 2.5 m/s². If the tension in the cord is 35 N, what is the mass of the lamp?

9. Blocks A and B, of masses 3.0 kg and 5.0 kg, respectively, are attached by a light cord which passes over a frictionless pulley as shown. The horizontal surface below A is also frictionless. Find

(a) the acceleration of the system,
(b) the tension in the cord.

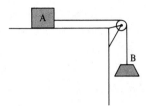

10. Inside a revolving space station, a container appears to rest on the outside wall as shown. If the mass of the container is 5.0 kg and the radius of the outer wall is 12.0 m, with what speed must the station be revolving so that the normal force on the container is 49 N?

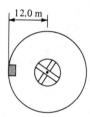

11. A marble is placed in a hemispherical bowl of radius R. If the bowl is spun around its vertical axis with constant angular speed ω, the marble eventually settles at a distance r from the axis. Find r as a function of ω.

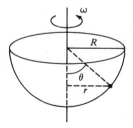

Friction

12. A 150-N cart rests in the aisle of a jet which is cruising horizontally with constant speed. The coefficient of static friction between the cart and floor is 0.4; the coefficient of kinetic friction is 0.2. The frictional force acting on the cart is (a) 0 N, (b) 60 N, (c) 30 N, (d) 150 N.

13. A delivery truck is loaded with crates whose coefficient of static friction with the floor is 0.3. If the truck is moving at 50 km/h, what is the shortest distance within which the truck can stop without the crates sliding?

14. A horizontal force of 20 N is applied to a 1-kg book resting against a wall. The book is on the verge of slipping. Determine the coefficient of static friction between the book and the wall.

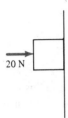

15. A hockey puck slides 20.0 m on a frozen pond in 8.0 s before coming to rest. Find the coefficient of kinetic friction between the ice and puck.

16. A 2.0-kg block of ice slides down a chute which is inclined at 53°. If the coefficient of friction between the ice and chute is 0.1, calculate the acceleration of the block.

17. A child rides in a rotor ride at an amusement park. The rotor consists of a hollow cylinder of radius R which rotates about a vertical axis with an angular speed ω. With the child against the spinning wall of the rotor, the floor drops out, but the child remains "pinned" to the wall.

 (a) Draw a free-body diagram for the child "pinned" to the wall.
 (b) For a given coefficient of static friction between the child and the wall, determine the minimum ω that will keep the child pinned to the wall. Express the centripetal acceleration for this ω in terms of g.

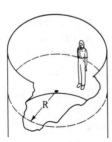

18. Two blocks are connected over a massless, frictionless pulley as shown. The mass of A is 8.0 kg and the coefficient of kinetic friction is 0.20. If block A slides down the plane with constant speed, what is the mass of B?

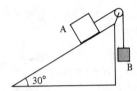

19. A weightless ladder of length L leans against a wall at an angle θ. The wall is frictionless, but the ground has coefficient of friction μ_s. A man of weight W climbs a distance d up the ladder. At what value of d does the ladder start to slip?

20. A bug walks radially outward on a record which is rotating at an angular speed $\omega = 45$ rpm (revolutions per minute). If the coefficient of static friction between the bug and the record is 0.08, how far can it walk before it slips?

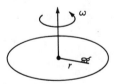

Motion in a Resistive Medium

21. If a stone is thrown vertically upward in the air, does it take more time to go up to its highest point, or to come down? Explain.

22. Verify by differentiation that $v(t) = gt_0(1 - e^{-t/t_0})$, where t_0 is specified by Eq. (8.28), is a solution to Eq. (8.27). Using this solution for $v(t) = dx/dt$, find $x(t)$. What is $x(t_0)$?

The Oil-Drop Experiment

23. Show that $v(t) = t_0(qE/m - g)(1 - e^{-t/t_0})$ is a solution of Eq. (8.32), where t_0 is given in Eq. (8.28). Prove that it is consistent with the terminal velocity given in Eq. (8.33).

24. Estimate t_0 and v_2 in Millikan's experiment. The density of the oil is $\rho \approx 0.85$ g/cm^3, and a typical drop has a radius $R \approx 10^{-6}$ m.

25. In a search for quarks, modern printing technology is used to provide highly uniform oil droplets of mass 10^{-7} g. Starting from rest, these droplets fall down a high tower in a *vacuum*. Over a vertical distance d near the top the droplets are subjected to a *transverse* electric force NeE where E is the electric field strength, e is the charge on one electron, and N is the net number of electron charges on the droplet. Subsequently the droplets fall through a farther distance D purely under the influence of gravity. The object is to check whether N is always an integer.

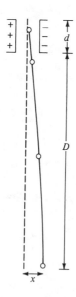

Find an expression for the transverse displacement $x(N)$ of the drop when it reaches the bottom of the tower, due to the electric force (you may take $D \gg d$). To form a preliminary judgment as to whether the measurement is sensitive to a quark charge, calculate the differential displacement $\Delta x = x(N + \frac{1}{3}) - x(N)$ for $d = 2$ m, $D = 20$ m, $eE = 1.6 \times 10^{-9}$ dyn (corresponding to $E = 10^3$ V/cm) and compare it to the droplet radius (computed using $\rho = 0.85$ g/cm^3 for the oil). How might you improve the sensitivity of the measurement?

CHAPTER

9

FORCES IN ACCELERATING REFERENCE FRAMES

From the beginning it appeared to me intuitively clear that, judged from the standpoint of such an observer [moving relative to the earth], everything would have to happen according to the same laws as for an observer who, relative to the earth, was at rest.

Albert Einstein, *Autobiographical Notes* (1949)

9.1 INERTIAL AND NONINERTIAL REFERENCE FRAMES

We have already introduced Galileo's ideas on relative motion in Chapter 4. We defined inertial frames – frames in which the law of inertia holds – and remarked that an observer in any inertial frame deduces the same laws of motion, and has no way of determining whether he is at rest or moving in an absolute sense. Galileo was able to provide striking examples of these ideas, such as a stone dropped from the mast of a moving boat, and to deduce a vitally important application – the earth need not be considered the stationary hub around which the heavens revolve.

However, Galileo did not have a clear-cut dynamical framework within which to derive his ideas. And exactly how to treat motion in a rotating frame, or indeed in any noninertial frame – one that is accelerated relative to an inertial frame – remained obscure.

It was only after Newton's second law was discovered that Galileo's ideas could be derived in a clear-cut way. Moreover, Newton's laws could be used in accelerated as well as inertial frames. This allowed Newton to supplement his description of circular motion as viewed from an inertial frame (where, as we have seen, some physical force must supply a centripetal acceleration) with a treatment of circular motion as felt by an observer riding along with the circling object. This treatment clears up the apparent discrepancy between the absence of any outward force in the description we have given of circular motion in inertial frames, and the undeniable sensation of outward force felt by anyone who rides a merry-go-round or circling vehicle.

9.2 GALILEAN RELATIVITY

Within Newton's laws of motion, things that do not change form valuable landmarks. In the next two chapters we shall elucidate the quantities called energy and momentum which remain constant in the course of the most complex motions. And in the present chapter we begin with the form of the equations themselves, which remains unchanged under certain transformations of the coordinate system. This technique of studying the behavior of Newton's laws under coordinate transformations (changes in the frame of reference) is the mathematical way of proving Galileo's conjectures that different inertial frames are equivalent, and, in later sections, of ascertaining the behavior of forces in noninertial frames.

To recount briefly, Newton's three laws are

(1) A body acted on by no forces moves with constant velocity.
(2) $\mathbf{F} = m\mathbf{a}$.
(3) $\mathbf{F}_{12} = -\mathbf{F}_{21}$.

Recognizing that the first law is but a special case of the second (as will be discussed in Chapter 11), we see that the explicit dependence on kinematics enters via the acceleration in the second law.

Let us consider Newton's laws in two different reference frames called S and S′. First suppose they are related by a simple translation of the origin, as in Fig. 9.1a, so the position $\mathbf{r}$ of a body in S is related to the position $\mathbf{r}'$ of the same body in S′ by the equation

$$\mathbf{r} = \mathbf{r}' + \mathbf{r}_0. \tag{9.1}$$

If the shift of origin is time independent ($\mathbf{r}_0$ independent of t), velocities and accelerations are the same in both reference frames:

$$\frac{d\mathbf{r}}{dt} = \frac{d\mathbf{r}'}{dt}, \tag{9.2}$$

$$\frac{d^2\mathbf{r}}{dt^2} = \frac{d^2\mathbf{r}'}{dt^2}. \tag{9.3}$$

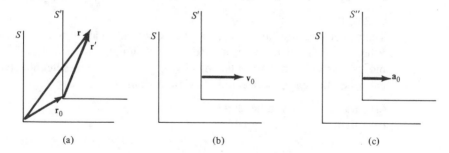

Figure 9.1 Two reference frames with (a) shifted origin, (b) constant relative velocity, (c) constant relative acceleration.

Thus the two frames are equivalent for discussing Newton's laws of motion, although the precise location of an object will be different in the two frames. The laws of motion are the same in London and in Leipzig.

Next, suppose S′ moves at a constant velocity with respect to S as in Fig. 9.1b, so the position of a body in S is related to its position in S′ by the equation

$$\mathbf{r} = \mathbf{r}' + \mathbf{r}_0 + \mathbf{v}_0 t. \tag{9.4}$$

In this case the velocities measured with respect to the two frames will differ because

$$\frac{d\mathbf{r}}{dt} = \frac{d\mathbf{r}'}{dt} + \mathbf{v}_0, \tag{9.5}$$

but only by a constant amount, so if Newton's first law holds in one frame it will also hold in the other. The acceleration will be the same, because

$$\frac{d^2\mathbf{r}}{dt^2} = \frac{d^2\mathbf{r}'}{dt^2}. \tag{9.6}$$

Thus once again the two frames are equivalent for discussing Newton's laws of motion, though the initial location and velocity of an object will be different in the two frames. This is described by saying that the laws of motion are *invariant* under transformations (9.1) and (9.4). This invariance of the laws of motion is called *Galilean relativity*.

Note that Galilean relativity is built into the form of the fundamental forces of gravity and electricity. Equations (8.1) and (8.2) for the gravitational and electrical forces between objects A and B depend not on the absolute positions $\mathbf{r}_A$ and $\mathbf{r}_B$, which would be different in a displaced or moving frame, but on the distance of separation $|\mathbf{r}_A - \mathbf{r}_B|$, which is unchanged by coordinate shifts such as (9.1) or (9.4).

It can also be shown that Newton's laws are invariant under a change in the orientation of frame S′ relative to S (that is, if S′ is rotated as well as translated relative to S). This, too, is part of Galilean relativity: the laws of motion in a tilted frame are the same as in a horizontal one.

Example 1

The center of a toy cart contains a hole in which a ball rests on a compressed spring. When the spring is released, it kicks the ball straight up relative to the cart. If the cart moves with constant horizontal velocity v_{x0} on a table top, determine the motion of the ball when the spring is released, as viewed from:

(a)   frame S' fixed in the cart.
(b)   frame S fixed in the table top.

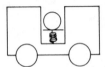

By Galilean relativity, the same laws of motion apply to the ball in both frames. As discussed in Chapter 4, the horizontal motion is governed by the law of inertia (since $F_x = a_x = 0$), and the vertical motion is governed by the law of falling bodies ($a_y = -g$).

However, the trajectory does not look the same in the two frames. In frame S', in which the cart is at rest, the motion is straight up and back down into the hole, whereas in frame S the trajectory is parabolic because the ball shares in the inertia of the moving cart:

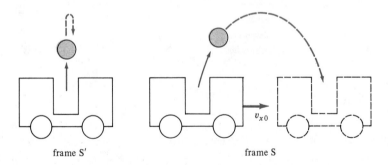

frame S'   frame S

In terms of the differential equation, the solutions differ because the initial condition on v_x is different in frame S. In any case, the ball is always directly above the hole, and eventually falls back into the hole in either frame.

Analytically, in frame S' where the cart is at rest, the purely vertical motion is that found in Section 3.8:

$$x' = 0,$$

$$y' = v_{y0}t - \tfrac{1}{2}gt^2.$$

The parabolic trajectory in frame S can be found directly by the methods of Chapter 4. But it is also easily obtained by taking horizontal and vertical components in Eq. (9.4),

$$x = x' + v_{x0}t,$$

$$y = y',$$

where we have taken S' and S to overlap at $t = 0$. Substituting the coordinate transformations into the previously obtained expressions for x' and y', we find

$$x = v_{x0}t,$$

$$y = v_{y0}t - \tfrac{1}{2}gt^2,$$

from which the parabolic trajectory is found by eliminating t.

Example 2

While the laws of physics are the same in all inertial frames, human laws are often specific to frames of reference fixed on the earth's surface. This is obviously the case with speed limits. A more subtle example occurs in American football. Once the ball is advanced beyond its starting point in a play, the ball cannot be thrown forward, but one can "lateral" it (throw it sideways or backwards). We now show that this rule is inconsistent unless it refers to a specific frame of reference (in practice, one fixed on the earth's surface).

Consider a runner moving straight downfield with constant velocity $\mathbf{v}_0 = v_{y0}\hat{\mathbf{j}}$ relative to the earth. Suppose he throws the ball sideways, with velocity $\mathbf{v}' = v_x'\,\hat{\mathbf{i}}$ in a frame

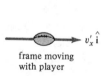

frame moving
with player

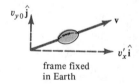

frame fixed
in Earth

moving along with him. The runner perceives his throw as a legal lateral, but a spectator standing still perceives that the ball's motion relative to a frame fixed in the earth is, by Eq. (9.5),

$$\mathbf{v} = \mathbf{v}' + \mathbf{v}_0 = v_x'\,\hat{\mathbf{i}} + v_{y0}\,\hat{\mathbf{j}},$$

which has an illegal forward component. So the rule must refer to a specific frame to avoid ambiguity.

Not all football officials are aware of this subtlety. In the 1982 Stanford–California football game, a remarkable five-lateral play occurred. Officials running along with the play called it the way they saw it, *legal*, although films made by a camera fixed in the stands showed the final pass traveling two yards forward.

9.3   INERTIAL FORCES

We turn now to the study of noninertial frames. When a frame of reference S'' has acceleration $\mathbf{a}_0$ relative to an inertial frame S, as in Fig. 9.1c, the acceleration $\mathbf{a}$ of a body with respect to S will differ from its acceleration $\mathbf{a}''$ with respect to S'' by $\mathbf{a}_0$,

$$\mathbf{a} = \mathbf{a}'' + \mathbf{a}_0. \tag{9.7}$$

According to Newton's second law, the force experienced by the body in frame S is

$$\mathbf{F} = m\mathbf{a}, \tag{9.8}$$

and the force $\mathbf{F}''$ experienced in frame S'' must satisfy

$$\mathbf{F}'' = m\mathbf{a}''. \tag{9.9}$$

We deduce from Eq. (9.7) that these forces must differ by $m\mathbf{a}_0$:

$$\boxed{\mathbf{F}'' = \mathbf{F} - m\mathbf{a}_0.} \tag{9.10}$$

The extra force

$$\mathbf{f} = -m\mathbf{a}_0 \tag{9.11}$$

experienced in the accelerated frame is called an *inertial force*. It is often given the alternative names of *pseudo* force or *fictitious* force, though there is nothing fictitious about the effects of such a force in an accelerated frame.

It is evident from Eq. (9.11) that an inertial force is always proportional to the mass. Another special property is that there are frames (the inertial frames) in which the inertial force vanishes.

Although the treatment here may seem completely different from the case of unaccelerated frames, there is a certain parallel in the underlying idea. When we merely displace the origin, the shapes of trajectories as well as the forces in frames S and S' are the same. When S' moves at uniform velocity with respect to S, trajectories look different in the two frames, as we saw in Example 1. When S'' accelerates relative to S, the forces look different as well. But in all cases the *form* of the fundamental law $\mathbf{F} = m\mathbf{a}$ remains unchanged.

9.4   INERTIAL FORCES IN A LINEARLY ACCELERATING FRAME

The simplest case of inertial force occurs in frames that accelerate at a constant rate $\mathbf{a}_0$ in a straight line. In this case the inertial force $\mathbf{f} = -m\mathbf{a}_0$ is constant and uniform.

An example is provided by an elevator with acceleration a_0. According to Eq. (9.10), an object of mass m experiences gravitational and inertial forces

$$F'' = m(g - a_0) \tag{9.12}$$

in the frame of the elevator. If the object is placed on a scale on the elevator floor, it presses down with force F'' on the scale, which thus registers an apparent weight

$$W'' = m(g - a_0).$$ (9.13)

The object is held at rest with respect to the elevator by a reactive normal force of the same magnitude, exerted upward by the scale.

Example 3

A freight elevator initially accelerates downward at 2.0 m/s^2, then descends at constant speed, then decelerates at 2.0 m/s^2 to a stop. A 100-kg load is on a scale that rests on the floor. What weight does the scale register during each portion of the trip?

The scale registers its normal weight $mg = 980$ N during the constant-speed portion of the trip. During the downward acceleration the scale reads

$$W'' = (100 \text{ kg}) (9.8 - 2.0) \frac{\text{m}}{\text{s}^2} = 780 \text{ N},$$

and during the downward deceleration (that is, upward acceleration) it reads

$$W'' = (100 \text{ kg}) [9.8 - (-2.0)] \frac{\text{m}}{\text{s}^2} = 1180 \text{ N}.$$

In setting the maximum load for the elevator, the construction engineers should take into account the extra effective weight exerted on the floor during an upward acceleration.

A body in true free fall (one acted on only by gravity in an inertial frame) provides an extreme example of these relationships. Such a body has acceleration $\mathbf{a}_0 = \mathbf{g}$ relative to the inertial frame, so the net force in a frame fixed in the *body* is $\mathbf{F}'' = m(\mathbf{g} - \mathbf{a}_0)$ $= \mathbf{0}$. That is, objects are *weightless* in this frame.

Weightlessness occurs not only in a body falling straight down, but also in orbiting spaceships. In a frame fixed with respect to a spaceship, the ship, an astronaut inside it, and an apple he releases are all weightless. This leads to dramatic and visible effects: the astronaut and the apple can float in the spaceship. One can consider the same phenomenon in a frame fixed with respect to the earth. Here the spaceship, astronaut and apple all have weight; nevertheless, the astronaut and apple can float because they orbit the earth at the same rate as the ship. If the astronaut stands on a spring scale in the spaceship, the scale is falling toward Earth at the same rate he is, so he exerts no pressure on the spring and the scale reads zero.

If the acceleration $\mathbf{a}_0$ is in a different direction than the gravitational force felt in an inertial frame, the apparent direction as well as magnitude of gravity is shifted in the accelerated frame. Consider, for example, a glass of water resting on a horizontal table top in an airplane accelerating horizontally in the x direction for takeoff. The sum of gravitational and inertial forces on the water in its rest frame is

$$\mathbf{F}'' = \mathbf{F} - m\mathbf{a}_0$$ (9.14)

$$= -mg\hat{\mathbf{k}} - ma_0\hat{\mathbf{i}}.$$

The surface of the water takes up a position normal to the force $\mathbf{F}''$, as shown in Fig. 9.2. The water can be described as experiencing an "effective g" of magnitude F''/m directed along $\mathbf{F}''$.

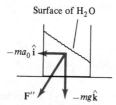

Figure 9.2 Gravitational and inertial forces on a glass of water in a horizontally accelerating airplane.

9.5 CENTRIFUGAL FORCE

An important and common case of inertial force occurs in frames that rotate in uniform circular motion with respect to an inertial frame. Examples include the earth in its daily rotation, a merry-go-round, and many other familiar phenomena.

Relative to the inertial frame, each point in the uniformly rotating frame has centripetal acceleration

$$\mathbf{a}_0 = -\frac{v^2}{r}\hat{\mathbf{r}} = -\omega^2 r\hat{\mathbf{r}}, \tag{9.15}$$

where ω is the angular speed of the frame, r is the distance of the point from the axis of rotation, and $v = \omega r$ is the speed of the point. The inertial force

$$\mathbf{f}_c = -m\mathbf{a}_0 = \frac{mv^2}{r}\hat{\mathbf{r}} = m\omega^2 r\hat{\mathbf{r}} \tag{9.16}$$

points radially outward and is therefore called the *centrifugal force*. The centrifugal force grows with distance from the axis of rotation.

The forces on a circling body can always be described either in an inertial frame or in the rotating frame, and it is important to distinguish the two descriptions clearly. Consider, for example, the horizontal motion of an object circling a post to which it is tethered by a rope, as in Fig. 9.3a. In Fig. 9.3b we show the force diagram in the inertial frame. Here there is no centrifugal force; the uniform circular motion requires a centripetal acceleration $v^2 r$, and the tension in the rope supplies it. Figure 9.3c shows the force diagram in the frame rotating with the object. In this frame the object does not accelerate; indeed, it does not move at all. The tension of the rope is exactly balanced by the centrifugal force.

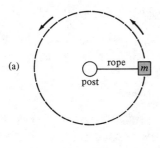

Figure 9.3 (a) Uniform circular motion of an object roped to a post. (b) Force diagram in inertial frame. (c) Force diagram in rotating frame.

9.6   EFFECT OF THE EARTH'S ROTATION ON *g*

If the earth were exactly spherical and did not rotate, the weight of an object would point directly toward the center of the earth. But this is not exactly true when the earth's rotation is taken into account.

Let us analyze the weight of an object of mass m, measured on a scale situated at latitude θ on the earth's surface (see Fig. 9.4). If the earth did *not* rotate, the only forces on the object would be the gravitational attraction $m\mathbf{g}_0$ of the earth toward its center, and the push $\mathbf{N}_0$ of the scale, which would be equal and opposite to $m\mathbf{g}_0$.

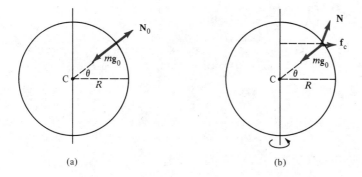

Figure 9.4 Forces felt by a mass on scale on (a) nonrotating Earth and (b) rotating Earth.

On the rotating earth, the true gravitational force $m\mathbf{g}_0$ still points directly toward the center C. But now the centrifugal force $\mathbf{f}_c = m\omega^2 r\hat{\mathbf{r}}$ also acts on the object. This force is directed outward perpendicular to the earth's axis of rotation, as shown in Fig. 9.4b, not outward from its center. In terms of the earth's radius R its magnitude is therefore $m\omega^2 R \cos\theta$. The equilibrium condition for the object is

$$\sum \mathbf{F}_i = \mathbf{0} = m\mathbf{g}_0 + (m\omega^2 R \cos\theta)\hat{\mathbf{r}} + \mathbf{N} \qquad (9.17)$$

where $\mathbf{N}$, the push of the scale, is slightly modified from $\mathbf{N}_0$ as a result of the centrifugal force (the shift indicated in Fig. 9.4 is greatly exaggerated).

The *negative* of $\mathbf{N}$ is what we call the weight of the object and label $m\mathbf{g}$. In terms of $m\mathbf{g}$, Eq. (9.17) becomes

$$m\mathbf{g} = m\mathbf{g}_0 + (m\omega^2 R \cos\theta)\hat{\mathbf{r}}, \qquad (9.18)$$

as depicted in Fig. 9.5.

Applying the law of cosines to the triangle diagrammed in Fig. 9.5, we obtain

$$g^2 = g_0^2 + (\omega^2 R \cos\theta)^2 - 2g_0\omega^2 R \cos^2\theta. \qquad (9.19)$$

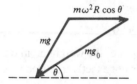

Figure 9.5 Relation between g and g_0.

Numerically $\omega^2 R \cos\theta$ is much smaller than g or g_0, so the second term on the right can be dropped, leaving

$$g^2 = g_0^2 \left(1 - \frac{2\omega^2}{g_0} R \cos^2\theta\right). \qquad (9.20)$$

Taking the positive square root of both sides and using the fact that $\sqrt{1-\varepsilon}$ is approximately $1 - \frac{1}{2}\varepsilon$ for small ε,* we find [taking $\varepsilon = (2\omega^2 R \cos^2\theta)/g_0$]

$$g \approx g_0 - R\omega^2 \cos^2\theta. \qquad (9.21)$$

*Because $(\sqrt{1-\varepsilon} - 1)(\sqrt{1-\varepsilon} + 1) = (1 - \varepsilon) - 1 = -\varepsilon$, we have

$$\sqrt{1-\varepsilon} - 1 = \frac{-\varepsilon}{\sqrt{1-\varepsilon} + 1}.$$

If ε is small, the right-hand term is nearly $-\frac{1}{2}\varepsilon$, so $\sqrt{1-\varepsilon}$ is nearly $1 - \frac{1}{2}\varepsilon$.

The reduction in g due to centrifugal force is greatest at the equator ($\theta = 0$) and zero at the poles.

Empirically, the variation of g with latitude follows the form of Eq. (9.21) but the measured coefficient of the $\cos^2 \theta$ term is about 50% greater than our prediction. This is due to the extra effect of the equatorial bulge (another consequence of the rotation). The measured values of g range from 983.1 cm/s^2 at the North Pole to 978.1 cm/s^2 at the equator.

The observed dependence of g on latitude establishes conclusively that the earth is not an inertial frame. Which frames *are* inertial?

Empirically the answer is simple. Inertial frames are at rest, or move at constant velocity, with respect to the "fixed stars," or more precisely the distant galaxies. On large scales the distant galaxies are rather uniformly distributed throughout space and provide a definite frame of reference.

Why the distribution of distant galaxies should determine which frames are inertial, and thus influence centrifugal forces on the earth, is a deeper question, which requires Einstein's general theory of relativity for its answer.

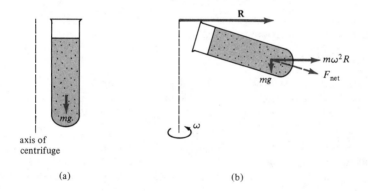

Figure 9.6 (a) A test tube in a centrifuge. (b) When the centrifuge rotates, centrifugal force makes the free end of the test tube swing out.

9.7 CENTRIFUGES

Suppose we have a test tube containing small particles suspended in a liquid. If the particles are heavier than the liquid, they will settle to the bottom, but if the particles are extremely small, this will take a long time. To speed up the process, we attach the test tube to a centrifuge, a mechanical device whose operation depends on centrifugal force.

Initially the tube hangs vertically, as in Fig. 9.6a. The centrifuge is carefully balanced with other tubes (not shown in the figure). When the centrifuge is spun about its central vertical axis, the tubes feel a centrifugal force (in the frame rotating with the centrifuge) pointing in the horizontal direction. The resultant of gravity and centrifugal force acts like an *effective gravity*, which, at high rates of spin, is much stronger than gravity alone and points almost horizontally (Fig. 9.6b). The tubes, being suspended on smooth pivots, rise until they are oriented along the direction of the net force they feel (Fig. 9.6b). The liquid surface in the tube likewise orients itself normal to the net force it feels, in qualitatively the same manner as our earlier example of the glass of water in an accelerating airplane (Fig. 9.2). And a particle suspended in the liquid moves in the direction of the net force *it* feels, that is, essentially toward the bottom of the tube (but not quite, since the net force varies in magnitude and, slightly, in direction with the distance of the particle from the axis of rotation). The useful outcome is that because $\mathbf{F}_{net}$ is much greater than $m\mathbf{g}$ for high rates of spin, the suspended particles settle to the bottom much more rapidly than they would otherwise.

Some typical numbers for the operation of ordinary centrifuges, and for the much faster devices called ultracentrifuges, are given in Table 9.1. Here f denotes the frequency of rotation (the number of cycles per second). Since 2π radians correspond to one cycle, the number of radians per second is $\omega = 2\pi f$. The last column of the table gives the acceleration $\omega^2 R$ in units of g, which is a measure of the increase in F_{net} when $\omega^2 R$ is much greater than g (Fig. 9.6b).

To estimate the drift speed of a particle of mass m in suspension, let us list the forces felt by the particle in the frame rotating with the centrifuge. First, there is the gravitational force $m\mathbf{g}$. Second, the centrifugal force is $m\omega^2\mathbf{R}$, where R is the distance from the axis of rotation to the particle. Third, by Archimedes' principle of buoyancy, there is a correction that reduces the sum of the first two forces to

$$(m - m_0)(\mathbf{g} + \omega^2\mathbf{R}),$$

where m_0 is the mass of the liquid displaced by the particle. This correction takes into account pressure differences in the liquid which tend to keep any small volume of the liquid itself in equilibrium; a body immersed in a liquid feels a net force only to the extent that its mass differs from the mass of an equal volume of liquid. Finally, any motion resulting from the net force is opposed by a resistive force which, for small slowly moving particles, has the form $\mathbf{F}_{vis} = -K\eta\mathbf{v}$ as discussed in Section 8.7, where η is the viscosity of the fluid. Thus the overall force on the particle is

$$\mathbf{F} = (m - m_0)(\mathbf{g} + \omega^2\mathbf{R}) - K\eta\mathbf{v}. \tag{9.22}$$

As discussed in Section 8.7, the velocity of the particle reaches a limiting value when $\mathbf{F}_{vis}$ just balances the other forces, leaving the total force in (9.22) equal to zero. In the case of a centrifuge, $\omega^2 R$ is so much larger than g that we can ignore $\mathbf{g}$, so the

Table 9.1 Properties of a Typical Centrifuge and Ultracentrifuge

	f (s^{-1})	$\omega = 2\pi f$	R (cm)	$\omega^2 R$ (cm/s^2)	$\omega^2 R/g$
Centrifuge	30	188	10	3.6×10^5	360
Ultracentrifuge	10^3	6.28×10^3	10	4×10^8	4×10^5

limiting drift velocity of the particle will be approximately in the direction of **R**. The corresponding speed is then equal to

$$v = \frac{(m - m_0)\omega^2 R}{K\eta}.$$ (9.23)

Thus the faster the rotation, the faster the particle will settle to the bottom.

For the special case of a small spherical particle at low speed, we found in Chapter 8 that K has the value

$$K = 6\pi r,$$ (8.23)

where r is the radius of the particle, and we also have

$$m - m_0 = \tfrac{4\pi}{3} r^3 (\rho - \rho_0),$$ (9.24)

where ρ and ρ_0 are the densities of the particle and liquid, respectively. Putting these relations into Eq. (9.23) we find for the drift speed

$$v = \frac{2r^2 (\rho - \rho_0)\omega^2 R}{9\eta}.$$ (9.25)

Note that the drift speed increases as the square of the radius of the particle, so large particles indeed move to the bottom of the tube much faster than small ones. To give a numerical example, for bacteria with radius $r = 10^{-4}$ cm and $\rho = 1.1\rho_0$ suspended in a tube of water ($\rho_0 = 1$ g/cm^3 and, from Table 8.3, $\eta = 10^{-3}$ N s/m^2) and spun 30 rps in the centrifuge of Table 9.1, one finds $v \approx 10^{-2}$ cm/s. This is 400 times the natural rate of settling in an unrotated test tube, and provides a convenient settling of about 10 cm in 15 min. On the other hand, a virus of radius $r = 3 \times 10^{-6}$ cm settles a thousand times more slowly; convenient separation of such small particles requires the higher rotation rate of an ultracentrifuge.

9.8   A FINAL WORD

After Newton's great work, the principles of mechanics remained undisturbed for 200 years. And then, in the late nineteenth century, discoveries in electromagnetism created a crisis. The Scottish physicist James Clerk Maxwell developed equations that unified electricity and magnetism and predicted electromagnetic waves whose speed, as derived from these equations, exactly equaled the known speed of light. But Maxwell's equations also implied that in a vacuum electromagnetic waves would travel at the same speed in all inertial frames. Contrary to Eq. (9.5), $\mathbf{v} = \mathbf{v}' + \mathbf{v}_0$, which applies to mechanical objects, the velocity of light would not differ by the relative velocity $\mathbf{v}_0$ between two frames. Attempts to avoid this contradiction were foreclosed by a variety of observations and experiments, culminating in the experiment of two Americans, Michelson and Morley, who proved conclusively that light travels no more and no less rapidly in the direction the earth moves about the sun than in other directions.

One reaction to the discovery that Galileo's ideas on moving reference frames might not be exactly valid at high velocities was to restudy those ideas and attempt to define their essence. The French physicist and mathematician Henri Poincaré was the first to focus explicitly on a ''principle of relativity'' in Galileo's work, and call it by that name.

In 1905 Albert Einstein, then 26 years old, published his theory of special relativity, which showed, in a deep and satisfying way, how to resolve the problem raised by Maxwell's equations and the Michelson–Morley experiment. In the spirit of Galileo, Einstein based his work on the relativity principle, that the laws of physics should be the same in all inertial frames. To preserve this principle in the light of the new discoveries in electromagnetism, however, he had to replace Eq. (9.4), $\mathbf{r} = \mathbf{r}_0 + \mathbf{r}' + \mathbf{v}_0 t$, by a more general law in which time as well as space is transformed in passing from one frame to another. Einstein's law differs from the old one at high velocity, but closely approximates Eq. (9.4) for the low-velocity situations that concerned Galileo and Newton. Thus the most profound aspect of Galileo's ideas has turned out to be the relativity principle, not the specific transformation law seventeenth-century physicists drew from it, though the latter is highly accurate for motions which we study in this text at speeds much lower than that of light.

Problems

Forces in a Linearly Accelerating Frame

1. A 10-kg mass slides on a table with coefficient of friction 0.5. Compare the force required to slide it if the table is

   (a) fixed on earth
   (b) in an elevator accelerating upward at 1 m/s^2
   (c) on the moon
   (d) in a spacecraft orbiting the earth.

2. Just as the plane is about to take off, a physicist looks out the window and sees a drop of water rolling down the inside. It leaves a wet track at a 10° angle to the vertical.

   (a) Does the drop roll toward the front of the plane or the back?
   (b) What is the plane's horizontal acceleration?

Centrifugal Force

3. A passenger jet airplane traveling 600 mph executes a circular turn at constant altitude. Find the minimum turn radius such that the passengers experience less than 10% change in the effective value of g during the turn (this is in the comfortable range). What is the minimum turn radius if the plane is traveling 300 mph near landing?

4. We have stated several times that for most purposes the earth is a good approximation to an inertial frame of reference. Make this statement quantitative by calculating the acceleration of a body on the earth's surface (at the equator) due to the earth's rotation. State your answer in terms of units of g.

5. **(a)** How fast would the earth have to rotate to make centrifugal force pull loose objects off the earth?
   (b) If the rotation was barely fast enough to make this happen, from which part of the earth would the objects be pulled off?
   (c) Compare your answer in part (a) to the minimal speed required for an object to orbit the earth, as deduced in Chapter 7 by arguments in an inertial frame.

6. A space station of radius 10 m spins to provide its inhabitants with a sense of artificial "gravity" when afloat in space. The rate of spin is chosen to simulate $g = 10$ m/s^2 at the floor.

   (a) Find the length of the "day" as seen in the spacecraft through a porthole in the floor.
   (b) What is the difference in the magnitude of the apparent gravitational acceleration between the head and feet of a 180-cm astronaut?

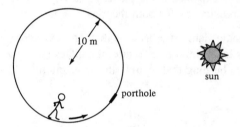

7. A student is undergoing a fraternity initiation. He finds himself in a windowless box with a 4 × 20 m floor. Three straps hang from the roof of the box, one at the center and one at either end. Suddenly the box lurches into motion; the strap in the center of the box inclines 10° from the vertical toward one of the long walls. At the same time the student finds the following note: "You won't be let out of the box until you can decide whether it's sliding sideways or riding on a circular track of radius 100 m. The wheels are under the two end straps."

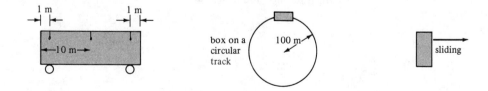

   (a) How can the student tell which type of motion the box is undergoing?
   (b) If the box is sliding sideways, how fast is it accelerating?
   (c) If the box is running in a circle, how fast is it going?

(d) It's always nice to have a check on one's work. In circular motion not all the straps are the same distance from the center of the circle. Describe an additional measurement which, combined with this fact, yields a check that we have the right velocity and radius of circular motion.

8. An imaginary planet called Zog is perpetually cloudy, and is spherical with a radius of 6000 km and mean density of 5.0 g/cm^3. Every 20 h the sky goes from being light to dark and back to light again. Physicists on Zog are divided into two schools regarding the brightening of their cloudy sky; one holds that the planet spins in front of an external light source, the other that the planet does not spin but that the brightening is like a wave traveling in the atmosphere. Design an experiment using accurate measurements of the apparent gravitational acceleration on the surface of Zog to settle the brightening question. For your experiment:

(a) Describe the results you would expect if the atmospheric wave hypothesis is correct.

(b) Describe the results expected if the planet rotates.

(c) Make numerical predictions for both theories.

9. The hot gases in a candle flame rise because they are less dense than the cooler air around them. This is an important feature of candles since the rising gas includes wastes from the burning that would otherwise smother the candle.

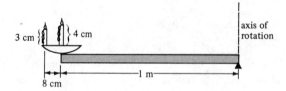

Consider two candles on a movable arm, as shown above. The two candles have height 3 and 4 cm, and are 8 cm apart. Both are enclosed in the same windscreen.

(a) If we spin the arm, which way do the candle flames point?

(b) Can we spin the arm so that the exhaust from one candle extinguishes the other? If so, how fast should we spin it?

CHAPTER 10

ENERGY: CONSERVATION AND CONVERSION

. . . You see, therefore, that living force [energy] may be converted into heat, and that heat may be converted into living force, or its equivalent attraction through space. All three, therefore – namely, heat, living force, and attraction through space (to which I might also add light, were it consistent with the scope of the present lecture) – are mutually convertible into one another. In these conversions nothing is ever lost. The same quantity of heat will always be converted into the same quantity of living force. We can therefore express the equivalency in definite language applicable at all times and under all circumstances.

James Prescott Joule, "On Matter, Living Force, and Heat" (1847)

10.1 TOWARD AN IDEA OF ENERGY

The law of conservation of energy is one of the most fundamental laws of physics. No matter what you do, energy is always conserved. So why do people tell us to conserve energy? Evidently the phrase "conserve energy" has one meaning to a scientist and quite a different meaning to other people, for example, to the president of a utility company or to a politician. What then, exactly, is energy?

The notion of energy is one of the few elements of mechanics not handed down to us from Isaac Newton. The idea was not clearly grasped until the middle of the nineteenth century. Nevertheless, we can find its germ even earlier than Newton. The essence of the idea of the conservation of energy can be seen in the incredibly fertile experiments that Galileo performed with balls rolling down inclined planes.

It is astonishing how many results Galileo squeezed out of his simple experiments. Bodies fall much too fast to be timed by the crude water clocks of the seventeenth century, but by slowing down the falling motion with his inclined planes, Galileo showed that uniformly accelerated motion was a part of nature. That alone was an achievement to crown him a genius.

Galileo did more with inclined planes. He arranged that once a ball had finished rolling down one inclined plane it would proceed to roll back up another, which could be more or less inclined than the first. Here he discovered a suggestive fact: no matter what the incline of the first plane and no matter what the incline of the second, the ball would finally come to rest on the second plane at the same height as that at which it had started on the first. He concluded that if the second plane were horizontal, the ball would continue rolling with the same speed forever. In other words, he discovered the law of inertia.

Once we have grasped the law of inertia, we can easily see why the ball starts up the second inclined plane after rolling down the first. But that does not tell us why it always reaches the same vertical height it started with. The ball almost seems to remember its origin. We prefer to say that something is conserved, rather than remembered. The name we give to the conserved quantity is *energy*.

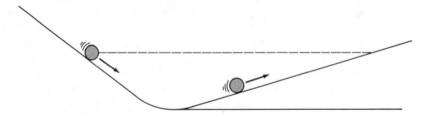

Figure 10.1 Galileo's experiment with inclined planes and rolling balls.

When, following Galileo's experiment as illustrated in Fig. 10.1, one lifts a ball from the table up to its starting point on the inclined plane, one endows it with a form of energy called *potential energy*. *Energy that a body has by virtue of its location is called potential energy.* If one then releases the ball it starts rolling, picking up speed. By the time it reaches the bottom of the incline, it is back at the level of the table. If it previously had potential energy due to its height above the table, that energy is now gone. In place of the potential energy the ball has the energy of motion. The energy that had been present in potential form has not been lost; rather it has been transformed into another form. *Energy of motion* is called *kinetic energy*.

As the ball continues, ascending the second inclined plane, it slows down. It is losing kinetic energy, but in return it is regaining energy of position, potential energy, as its height increases. When the ball finally comes to rest, all its kinetic energy has been transformed back into potential energy. This happens when the ball's height above the table is precisely what it was at the beginning of the experiment. That is the modern scientific view of what Galileo observed.

Earlier we would have described the same experiment in different terms. We would have said: To lift the ball, Galileo applies a force opposing gravity. When he releases the ball, the force of gravity makes it roll down the plane. At the bottom of the plane, its inertia makes the ball roll back up the second plane. We have a valid description without the concept of energy, so why talk about energy?

We need the concept of energy because it expresses one fact our old description didn't prepare us for: the ball ends up at the same height it started at, never any higher, and, if we ignore air resistance and friction, never any lower. Something is the same at the end of the ball's motion as it was at the beginning. That something is its energy. The recognition of energy conservation increases our understanding and simplifies analysis of the motion.

The concept of energy was invented precisely because something is conserved. Then why are politicians and gas company executives telling us to conserve energy? To answer this burning question, we need a precise and quantitative definition of energy.

10.2   WORK AND POTENTIAL ENERGY

In ordinary speech, work refers to any exertion maintained for some time. But in physics the word *work* is used more precisely, to describe an energy transfer from one thing to another carried out by a force acting over distance.

For the simple case of a constant force of magnitude F moving an object a distance h parallel to the force, the work W done by the force is defined as

$$W = Fh. \tag{10.1}$$

For example, if you lift an object of mass m in such a way that it doesn't accelerate, then the lifting force is equal to the weight mg of the object. Since mg is constant, the work done to raise the object up to height h is

$$W = mgh. \tag{10.2}$$

Note that since the work all goes into potential energy U in the example, the gravitational potential energy in the uniform field near the earth can be given the simple form

$$U = mgh. \tag{10.3}$$

Now that we have a definition of work, let's go back and follow energy conservation through the various stages of Galileo's experiment. First, Galileo lifts the ball a height h, performing work on it by applying a force to balance the preexisting force of gravity. We say that work is done *by the force that Galileo applies*, or alternatively that work is done *against the force of gravity*. In any case overall energy is conserved; the work represents a transfer from the world outside the ball (namely, from Galileo) to the ball.

Since there is no net force, the ball is displaced without accelerating; thus all the work goes into the potential rather than kinetic energy:

$$W = U = mgh.$$

Next Galileo releases the ball, and acting now solely under the preexisting force of gravity, it rolls down the incline. During this stage gravity does work on the ball. Again overall energy is conserved, the work representing a transfer from potential to kinetic energy. Finally, when the ball rolls back up an incline to the original height, energy is still conserved. Here work is done against gravity; it represents a transfer back from kinetic to potential energy.

What if the force is not constant? Suppose a force acts on a particle moving in a straight line and that the force has strength $F(z')$ directed along this line when the particle is at z'. How much work is done by this variable force in moving the particle from $z' = a$ to an arbitrary position $z' = z$? The work will depend on z and we denote it by $W(z)$. We *define* $W(z)$ by

$$W(z) = \int_a^z F(z') \, dz'. \tag{10.4}$$

This definition can be motivated as follows.

First, we want $W(a) = 0$ because no work is done if we don't move the particle at all. Now suppose we move the particle from the position z to a nearby position $z + h$, where h is a small positive number. The work done by the force from z to $z + h$ is the difference

$$W(z + h) - W(z).$$

We can express this same work in another way. Over the interval from z to $z + h$, the force $F(z')$ will change, but it has some average value which we can call $F(\bar{z})$. If we treat the force as though it had the constant value $F(\bar{z})$ over the internal from z to $z + h$, then the work it does is $F(\bar{z})h$, the product of the force times the distance. Therefore

$$W(z + h) - W(z) = F(\bar{z})h,$$

or, dividing by h,

$$\frac{W(z + h) - W(z)}{h} = F(\bar{z}). \tag{10.5}$$

Now we let h shrink to zero on both sides of this equation. The left-hand side becomes dW/dz and the right-hand side becomes $F(z)$, so in the limit we find

$$\frac{dW}{dz} = F(z). \tag{10.6}$$

In other words, the derivative of the work is $F(z)$. Therefore, by the second fundamental theorem of calculus we have

$$W(z) - W(a) = \int_a^z F(z') \, dz'. \tag{10.7}$$

Since $W(a) = 0$ this gives us Eq. (10.4).

Note that if the force is constant, the integral is the constant force times the length of the interval, which agrees with Eq. (10.1).

The definition of work given above in Eq. (10.4) deals with displacements parallel to the applied force, but this is not the most general possibility. For example, when gravity works on a ball on Galileo's inclined plane, the force of gravity is always vertically downward, but the ball is displaced not straight downward, but along the inclined path. How do we handle this?

The nonparallel vectors – downward force $m\mathbf{g}$ and displacement $\mathbf{r}$ along the incline – are shown in Fig. 10.2. The work done by gravity to bring the ball to the bottom of the incline must equal the work Galileo did to lift it up there, mgh. From the figure we see that $h = r \cos \theta$, so we can write the work done as

$$W = mgh = mgr \cos \theta. \tag{10.8}$$

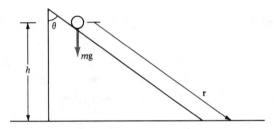

Figure 10.2 Work done by gravity is $m\mathbf{g} \cdot \mathbf{r} = mgh$.

Taking another look at Fig. 10.2, we realize that $mg \cos \theta$ is the component of the weight in the direction of the displacement $\mathbf{r}$. We already have a notation perfectly suited for this purpose: the dot product. As you recall from Chapter 5, the dot product is equal to the product of the component of one vector along a second vector times the magnitude of the second vector. Thus we can describe the work done here as

$$W = mgr \cos \theta = m\mathbf{g} \cdot \mathbf{r} = \mathbf{F} \cdot \mathbf{r}. \tag{10.9}$$

In general, the work done by a constant force $\mathbf{F}$ acting through a displacement $\mathbf{r}$ is the displacement multiplied by the component of the force in the same direction:

$$W = \mathbf{F} \cdot \mathbf{r}. \tag{10.10}$$

The problem of defining work is most complicated in the general case where the force $\mathbf{F}$ may vary both in magnitude and direction, and acts on a particle moving along a curve from A to B, as illustrated in Fig. 10.3. As in Eq. (10.4) for a variable-strength force directed along a straight path, we will use an integral to define work. This integral must take into account both the force $\mathbf{F}$ and the curve along which the particle is moved.

Figure 10.3 A variable force $\mathbf{F}$ acting along a curve.

Here's an intuitive procedure for arriving at such an integral. Describe the curve by its position vector, say

$$\mathbf{r} = x\hat{\mathbf{i}} + y\hat{\mathbf{j}} + z\hat{\mathbf{k}}. \tag{10.11}$$

It seems reasonable to say that for a force $\mathbf{F}$ which produces a small displacement $d\mathbf{r}$ the amount of work done is the dot product

$$dW = \mathbf{F} \cdot d\mathbf{r}. \tag{10.12}$$

The total work done in moving the particle from point A to point B is obtained by adding all these small amounts of work. We indicate the summation process by the integral symbol and write

$$W = \int_{A}^{B} \mathbf{F} \cdot d\mathbf{r}. \tag{10.13}$$

This notation resembles that in (10.4), but the symbol in (10.13) is a new kind of integral. It is called a *line integral* or a *contour integral* because the integration takes place along a curve joining A and B. Such an integral can be defined in terms of ordinary integrals of the type with which we are familiar. To do this, we consider the position vector as a function of time t and write

$$\mathbf{r} = \mathbf{r}(t).$$

The initial point A of the curve corresponds to some value of t, say $t = t_a$, so $A = \mathbf{r}(t_a)$. Similarly, $B = \mathbf{r}(t_b)$ for some time $t = t_b$.

We can interpret the dot product $\mathbf{F} \cdot d\mathbf{r}$ to mean $(\mathbf{F} \cdot d\mathbf{r}/dt)\, dt$. This suggests that we simply *define* the line integral $\int_{A}^{B} \mathbf{F} \cdot d\mathbf{r}$ by the equation

$$\int_{A}^{B} \mathbf{F} \cdot d\mathbf{r} = \int_{t_a}^{t_b} \mathbf{F} \cdot \frac{d\mathbf{r}}{dt}\, dt. \tag{10.14}$$

The integral on the right is our common garden variety of integral over an interval from $t = t_a$ to $t = t_b$. It incorporates the curve as well as the force because in the integrand on the right the force is to be evaluated at the position $\mathbf{r}$ along the curve, and the derivative $d\mathbf{r}/dt$ also depends on $\mathbf{r}$. Note that since $\mathbf{F}$ is a function of position and the position $\mathbf{r}(t)$ depends on time, the force as well as $d\mathbf{r}/dt$ depends on time. Thus the integrand is the following function of t:

$$\mathbf{F}(\mathbf{r}(t)) \cdot \frac{d\mathbf{r}}{dt}.$$

Equation (10.14) is taken to be the definition of the line integral $\int_{A}^{B} \mathbf{F} \cdot d\mathbf{r}$, and then Eq. (10.13) is used to define the work done by $\mathbf{F}$.

Example 1

A particle is moved from the origin $(0,0)$ to the point $(2,2)$ by a variable force $\mathbf{F}$ whose value at each point (x, y) is given by

$$\mathbf{F}(x, y) = \sqrt{y}\,\hat{\mathbf{i}} + (x^2 + y)\hat{\mathbf{j}}.$$

Calculate the line integral

$$\int_{(0,0)}^{(2,2)} \mathbf{F} \cdot d\mathbf{r}$$

along each of the following paths:

(a)   The line segment described by

$$\mathbf{r}(t) = t\hat{\mathbf{i}} + t\hat{\mathbf{j}}, \quad 0 \le t \le 2.$$

(b)   The parabolic arc described by

$$\mathbf{r}(t) = t\hat{\mathbf{i}} + \tfrac{1}{2}t^2\hat{\mathbf{j}}, \quad 0 \le t \le 2.$$

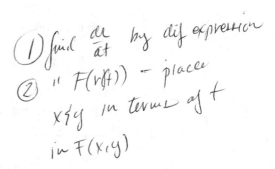

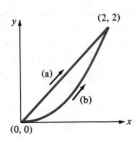

For the path in (a) we have

$$\frac{d\mathbf{r}}{dt} = \hat{\mathbf{i}} + \hat{\mathbf{j}}.$$

On this path, $\mathbf{r} = x\hat{\mathbf{i}} + y\hat{\mathbf{j}}$, where $x = t$ and $y = t$. Therefore, putting $x = t$ and $y = t$ in $\mathbf{F}(x, y)$ we find

$$\mathbf{F}(\mathbf{r}(t)) = \sqrt{t}\,\hat{\mathbf{i}} + (t^2 + t)\hat{\mathbf{j}},$$

so the dot product of $\mathbf{F}(\mathbf{r}(t))$ with $d\mathbf{r}/dt$ is

$$\mathbf{F}(\mathbf{r}(t)) \cdot \frac{d\mathbf{r}}{dt} = \sqrt{t} + t^2 + t.$$

Integrating, we have

$$\int_{(0,0)}^{(2,2)} \mathbf{F} \cdot d\mathbf{r} = \int_0^2 (\sqrt{t} + t^2 + t)\,dt = \tfrac{4}{3}\sqrt{2} + \tfrac{8}{3} + 2 = 6.55.$$

Along path (b) we have $d\mathbf{r}/dt = \hat{\mathbf{i}} + t\hat{\mathbf{j}}$. On this path, $\mathbf{r} = x\hat{\mathbf{i}} + y\hat{\mathbf{j}}$ with $x = t$ and $y = \tfrac{1}{2}t^2$. Therefore, putting $x = t$ and $y = \tfrac{1}{2}t^2$ in $\mathbf{F}(x, y)$, we find

$$\mathbf{F}(\mathbf{r}(t)) = \sqrt{\tfrac{1}{2}t^2}\,\hat{\mathbf{i}} + (t^2 + \tfrac{1}{2}t^2)\hat{\mathbf{j}} = \tfrac{1}{2}\sqrt{2}t\hat{\mathbf{i}} + \tfrac{3}{2}t^2\hat{\mathbf{j}}.$$

The dot product of $\mathbf{F}(\mathbf{r}(t))$ with $d\mathbf{r}/dt$ is

$$\mathbf{F}(\mathbf{r}(t)) \cdot \frac{d\mathbf{r}}{dt} = \frac{\sqrt{2}}{2}t + \frac{3t^3}{2} \; .$$

Integrating this we obtain

$$\int_{(0,0)}^{(2,2)} \mathbf{F} \cdot d\mathbf{r} = \int_0^2 \left(\frac{\sqrt{2}}{2}t + \frac{3t^3}{2} \right) dt = \sqrt{2} + 6 = 7.41.$$

This example shows that the line integral of a given force from one point to another may depend on the path joining the two points.

Let us examine several applications which illustrate different facets of our definition of work:

(i)   A librarian holding a large stack of books stationary in front of him does no work on the books because the books don't move. And yet, his belief that he is doing work on *something* is correct even by the precise physicist's definition. As his arms strain to counteract the torque exerted by the books his muscles move, sagging and then retensing repeatedly. He is doing work on his muscles, even though no net work gets done on the books.

(ii)   A pendulum swinging around steadily in a horizontal circle affords an example of a situation in which the full apparatus of the line integral in Eq. (10.13) is needed to define work. The result in this case is that the tension in the rope does no work, because the force is always applied perpendicular to the instantaneous displacement (Fig. 10.4). Though surprising at first sight, this result is reasonable when you consider that the pendulum is neither gaining nor losing energy.

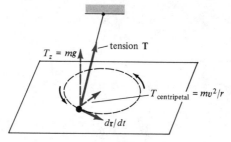

Figure 10.4 The tension acting on a pendulum moving in a horizontal circle has two components, each normal to the instantaneous displacement and therefore doing no work.

(iii)   Keeping track of the relative directions of $\mathbf{F}_{grav}$ and $d\mathbf{r}$, we see that the work $W_{grav} = \int \mathbf{F}_{grav} \cdot d\mathbf{r}$ done by gravity on a ball is $F_{grav}h = mgh$ when the ball falls from a height h, but $-mgh$ when the ball rises through a height h. Thus the work done by gravity is related to the gravitational potential-energy change by

$$W_{\text{grav}} = -\Delta U.$$

This is the opposite of the work done by an external force $\mathbf{F}_{\text{ext}} = -\mathbf{F}_{\text{grav}}$ in lifting a ball against gravity,

$$W_{\text{ext}} = \int \mathbf{F}_{\text{ext}} \cdot d\mathbf{r} = \Delta U.$$

In the following example we discuss another case in which work done on a system by an external force is converted into potential energy, the stretching of a spring.

Example 2

What amount of potential energy is stored in a spring which is stretched from equilibrium to a distance x?

We'll use Eq. (10.4) to calculate the work, which in turn is equal to the potential energy stored in the spring. When we stretch a spring a distance x, the spring exerts a force

$$F = -kx. \tag{8.3}$$

Therefore, to stretch it, we need to apply a force $-F = +kx$, which is, of course, in the same direction as the displacement of the spring. Thus the work we do is

$$W_{\text{ext}} = \int_0^d kx\,dx = \frac{1}{2}kd^2.$$

Since the potential energy stored in the spring is equal to the work done on it, we have

$$U = \tfrac{1}{2}kd^2.$$

The potential energy is the same whether the spring is stretched or compressed.

The dimensions of work are those of force times distance. The SI unit of work is one newton-meter, or one *joule* (after a British physicist whom we shall encounter again later), abbreviated 1 J. In the cgs system the unit of work is one dyne-centimeter, or one *erg*. In the British system, one foot-pound is the basic unit of work and has no other name. Using conversions between the fundamental units, we can verify that

$$1\ \text{J} = 10^7\ \text{erg} = 0.738\ \text{ft lb}. \tag{10.15}$$

For example, suppose a car is rolling slowly along a road and a man pushes backward on it with a constant force of 50 lb, bringing it to rest within a distance of 10 ft. The work done by the car on the man is 500 ft lb. (Note that the work done *by the man on the car* is -500 ft lb, a negative quantity because the force and displacement are in opposite directions.)

10.3 KINETIC ENERGY AND THE CONSERVATION OF ENERGY

When the ball reaches the bottom of Galileo's incline, where $h = 0$, it will have lost all its potential energy U, converting it to kinetic energy K. We know exactly how large K will be: it will be equal to U. But K is the energy of motion; it should depend, not on where the ball is, as U does, but on how fast it is going. What exactly is the connection between kinetic energy and speed?

Let's consider a different but related experiment which leads us easily to the answer. If we simply drop the ball in free fall from a height h, we recall from Example 1 of Chapter 2 that the ball acquires a speed satisfying

$$v^2 = 2gh.$$

Multiplying by $\frac{1}{2}m$ to get mgh on the right, we have

$$\tfrac{1}{2} mv^2 = mgh. \tag{10.16}$$

We know that gravity does work $W = mgh$ on the falling ball, converting it into kinetic energy K. So by Eq. (10.16), K must have the value

$$K = \tfrac{1}{2} mv^2. \tag{10.17}$$

Using this example as motivation, we now *define* the kinetic energy K of *any* body of mass m moving with speed v by the equation

$$\boxed{K = \tfrac{1}{2} mv^2.} \tag{10.17}$$

With the aid of this definition plus our expression

$$W = \int_A^B \mathbf{F} \cdot d\mathbf{r} = \int_{t_a}^{t_b} \mathbf{F} \cdot \frac{d\mathbf{r}}{dt}\, dt \tag{10.14}$$

for work, we shall prove that the relationship found in the above example holds generally: the work done by a net force $\mathbf{F}$ equals the change in kinetic energy. To begin the proof we use Newton's second law to replace $\mathbf{F}$ by $m\, d\mathbf{v}/dt$ in the integrand of Eq. (10.14), and $d\mathbf{r}/dt$ by $\mathbf{v}$, obtaining

$$\int_A^B \mathbf{F} \cdot d\mathbf{r} = \int_{t_a}^{t_b} m\frac{d\mathbf{v}}{dt} \cdot \mathbf{v}\, dt. \tag{10.18}$$

The dot product $d\mathbf{v}/dt \cdot \mathbf{v}$ is related to the speed v. Recalling that $v^2 = \mathbf{v} \cdot \mathbf{v}$ we know that

$$\frac{d}{dt}(v^2) = \frac{d}{dt}(\mathbf{v} \cdot \mathbf{v}) = \mathbf{v} \cdot \frac{d\mathbf{v}}{dt} + \frac{d\mathbf{v}}{dt} \cdot \mathbf{v} = 2\frac{d\mathbf{v}}{dt} \cdot \mathbf{v}, \tag{10.19}$$

and hence

$$\frac{d\mathbf{v}}{dt} \cdot \mathbf{v} = \frac{1}{2}\frac{d}{dt}(v^2). \tag{10.20}$$

Substituting this into the last integral and denoting the work done by W_{AB} we get

$$W_{AB} = \int_{t_a}^{t_b} \frac{d}{dt}\left(\frac{1}{2}mv^2\right)dt. \tag{10.21}$$

Since the integrand is the derivative of $\frac{1}{2}mv^2(t)$ we can evaluate the integral by the second fundamental theorem of calculus to obtain

$$W_{AB} = \frac{1}{2}mv^2(t_b) - \frac{1}{2}mv^2(t_a). \tag{10.22}$$

The quantity

$$K = \frac{1}{2}mv^2(t)$$

is the kinetic energy of the particle at time t, and we have just shown that the total work done by $\mathbf{F}$ is indeed the change in kinetic energy,

$$W_{AB} = \int_A^B \mathbf{F} \cdot d\mathbf{r} = K_B - K_A. \tag{10.23}$$

In the case of a body falling vertically a distance h in the earth's uniform gravitational field, the work $\int \mathbf{F} \cdot d\mathbf{r}$ done by the gravitational force is mgh. In the previous section we identified this work with the negative of the gravitational potential-energy difference:

$$-(U_B - U_A) = \int_A^B \mathbf{F} \cdot d\mathbf{r}.$$

The sign is such that the positive work done by gravity on a falling object lowers its potential energy; i.e., $U_B - U_A$ is negative. Again motivated by this case, we now *define* the negative of the change in potential energy in the general situation by the same relation

$$-(U_B - U_A) = \int_A^B \mathbf{F} \cdot d\mathbf{r}. \tag{10.24}$$

We now have two expressions relating the work done on an object, one to kinetic energy, the other to potential energy. Combining Eqs. (10.23) and (10.24) we see that

$$-(U_B - U_A) = K_B - K_A. \tag{10.25}$$

Rearranging terms, we find the conservation law

$$\boxed{U_A + K_A = U_B + K_B.} \tag{10.26}$$

If we do no work that transfers energy into our system, the total energy of the system, potential plus kinetic, is conserved.

The integral in Eq. (10.24) describes the *change* in potential energy, not the potential energy itself. The same type of integral can sometimes be used to define the potential energy itself (as will be done later), but there is a small technical difficulty that must be faced: the value of U_B is arbitrary. Suppose we see a stone of mass m on a table and ask for its potential energy. One may reply, as we did in Eq. (10.3), that it is mgh. But from what level is h measured? From the floor? From the street level outside? Or from sea level? In practice, potential energy is always calculated with respect to some arbitrary level of reference, often the lowest level a body attains during a given discussion, where U_B is set equal to zero. The actual choice of the original level is usually irrelevant because we are interested in *differences* or *changes* in the potential energy between two locations and the difference is independent of the choice of origin.

Let's see how we can use energy conservation in describing the motion of a frictionless sliding block (we replace Galileo's rolling ball with a sliding block here to avoid the complications associated with rotation). Let h denote the maximum or initial height above the table, and let z be an arbitrary height, as indicated in Fig. 10.5. Then mgz is the potential energy when the block is at height z, and $\frac{1}{2}mv^2$ is the kinetic energy. The conservation law says that

$$mgz + \tfrac{1}{2}mv^2 = \text{const.} \tag{10.27}$$

Since the left-hand side is the same anywhere along the block's path, we can determine the constant by evaluating the left-hand side at any point where we know all the quantities. For example, if we take the initial point where $z = h$, we know that $v = 0$, so the value of the left-hand side is simply mgh, which is the constant. Thus we can write

$$mgz + \tfrac{1}{2}mv^2 = mgh. \tag{10.28}$$

From this statement of conservation of energy, we can find the speed of the block at any height z.

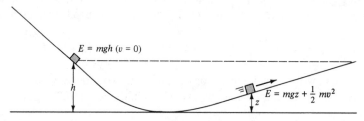

Figure 10.5 Conservation of energy applied to Galileo's experiment.

None of this depends in any way on how steeply the planes are inclined. In fact, the planes could be vertical; then we would have a freely falling body. Regardless of whether a body is slipping down an inclined plane or freely falling, if it starts at height h, its speed v can be found from Eq. (10.28), which when solved for v gives

$$v = \sqrt{2g(h - z)}. \tag{10.29}$$

The speed depends only on the vertical distance fallen. For example, Eq. (10.29) agrees at $z = 0$ with the expression $v = \sqrt{2gh}$ given earlier in this chapter for a freely falling body.

The law of conservation of energy is useful for analyzing motion not only on inclined planes and for falling bodies but also for many other phenomena such as pendulums and springs. The following is a generalization of the procedure we used to apply the law:

> (i)   Define your system.
> (ii) Pick one reference position for $U = 0$ and use it consistently.
> (iii) Write down the total energy of the system at the point, say A, where you want to determine some unknown quantity (like speed or height); $E_A = U_A + K_A$.
> (iv) Find another point, say point B, where you know everything about the object's motion and write down the total energy at that point; $E_B = U_B + K_B$.
> (v)   Conservation of energy implies that $E_A = E_B$; equate the two energies and solve for the unknown quantity.

So far, we know two types of potential energy which U can represent. One is the potential energy a mass m has owing to its height z above the earth's surface. If we take $U = 0$ at $z = 0$, this has the value

$$U_{grav} = mgz. \tag{10.3}$$

The other is the potential energy stored in a spring either stretched or compressed by an amount x. If we take $U = 0$ at $x = 0$, this has the value

$$U_{spr} = \tfrac{1}{2} kx^2. \tag{10.30}$$

The following examples apply the law of conservation of energy to various problems.

Example 3

A pendulum bob is pulled aside from the vertical through an angle θ and released. Find the speed of the bob and the tension in the string at the lowest point of the swing, assuming $L = 0.3$ m, $\theta = 30°$, and $m = 0.5$ kg.

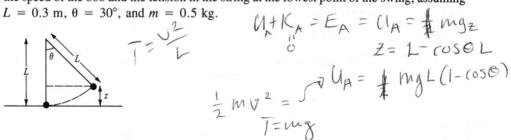

We anticipate using the law of conservation of energy, so let's measure all vertical heights (and therefore potential energy) from the lowest point of the swing, point B. Now the total energy of the bob at point A, just before being released, is

$$E_A = mgz,$$

where z is the vertical height above B. Using geometry, we see that $z = L - L \cos \theta = L(1 - \cos \theta)$, so the initial energy is

$$E_A = mgL(1 - \cos \theta).$$

The total energy at B is purely kinetic energy because we have chosen $z = 0$ there, so $E_B = \frac{1}{2} m v_B^2$. Therefore the law of conservation of energy, $E_A = E_B$, implies

$$mgL(1 - \cos \theta) = \frac{1}{2} m v_B^2.$$

Solving for the speed v_B we find

$$v_B = \sqrt{2gL(1 - \cos \theta)} \ .$$

Substituting numbers, this turns out to be 0.9 m/s in our case.

One might wonder why the tension in the string, which is pulling on the bob at all times, doesn't supply energy to it as gravity does, thereby introducing an extra term into the energy conservation equation. The answer is that the tension does no work on the bob because it acts at right angles to the motion, $\int \mathbf{F} \cdot d\mathbf{r} = 0$.

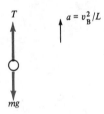

To find the tension in the string – a force – we need to use Newton's second law. Applying the second law with the free-body diagram for the bob, we see that a combination of weight and tension is causing the bob to move in its circle of radius L. This is nonuniform motion along a circle, but we recall from the end of Section 7.6 that the centripetal component of acceleration is v^2/L even for nonuniform motion. Therefore we have at the lowest point

$$T_B - mg = m v_B^2/L,$$

which implies

$$T_B = mg + m v_B^2/L.$$

Substituting for v_B from our law of conservation of energy we get for the tension in the string when the bob is at point B

$$T_B = mg + 2mg(1 - \cos \theta) = mg(3 - 2 \cos \theta).$$

This result has the reasonable property that the tension at the bottom of the swing is greater the higher the swing started, and reduces to $T_B \approx mg$ when the swing starts from $\theta \approx 0$. Numerically the tension turns out to be 6.2 N for a swing started from $\theta = 30°$ with $m = 0.5$ kg.

A combination of the law of conservation of energy and Newton's laws provides a powerful method for attacking a variety of problems in classical physics. Even though the law of energy conservation for a system such as ours was derived from Newton's laws, which means we could have solved the problem by employing Newton's laws alone, the use of energy conservation increases our understanding and simplifies analysis of the motion.

Example 4

A toy dart gun consists of a massless spring that, when compressed 0.5 m, can project a 20-g rubber dart vertically upward to a height of 3.0 m. Determine the spring constant.

Both the dart and spring can have energy. Initially the system (dart and spring) has potential energy stored in the spring as well as gravitational potential energy of the dart. At its highest point, the dart possesses an increased gravitational potential energy and no kinetic energy and the spring, no longer being compressed, has no potential energy in it. The law of conservation of energy implies that at these two points the energy must be the same. Let's formulate mathematically what we've just stated verbally. Initially the energy of the system is $E_A = U_s + U_d + K_d$, being the sum of the potential energy of the spring ($U_s = \frac{1}{2} kz_A^2$, which corresponds to taking $z = 0$ at the uncompressed position of the spring) the gravitational potential energy of the dart (in general $U_d = mgz$, but initially we take $z = z_A = -0.05$ m), and the kinetic energy of the dart (initially $K_d = 0$). At the highest point, we have $E_B = U_s + U_d + K_d = 0 + mgz_B + 0$. The relation $E_A = E_B$ implies

$$\tfrac{1}{2} kz_A^2 = mg(z_B - z_A) = mgh.$$

Solving for k, we find

$$k = 2mgh/z_A^2 = 2(0.02 \text{ kg}) (9.8 \text{ m/s}^2) (3.0 \text{ m})/(0.05 \text{ m})^2$$

$$= 4.7 \times 10^2 \text{ N/m}.$$

Example 5

An 80-kg rock climber is climbing a vertical wall. A rope attached to him passes through an aluminum ring (called a *carabiner*) 40 m below and is anchored (*belayed* in the jargon of rock climbers) a further 5 m below. The belay is 70 m above the ground. The climber slips and falls vertically. When the rope becomes taut, it acts as a spring. The carabiner is assumed to be frictionless. The rope is assumed to be massless and has a spring constant of 520 N/m.

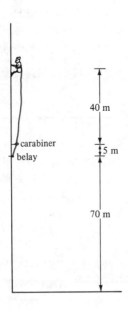

(a) If nothing breaks, how far does the rope stretch?

(b) Now assume that the breaking point of the rope is a tension of 19,000 N, and that the carabiner breaks if the force exerted on it exceeds 22,000 N. Does the rope or the carabiner break in the fall described above?

(c) If the carabiner breaks, does the climber hit the ground?

When the rope becomes taut the climber will be 40 m below the carabiner, corresponding to a height of 35 m above the ground. Choose the zero point of the z axis at 35 m above the ground and choose the zero level of the gravitational potential energy at the same point. Then the total energy of the climber plus the rope

$$E = K + U(\text{climber}) + U(\text{rope})$$

is initially $E = 0 + mgz_0 + 0$ where $z_0 = 80$ m. As the climber falls, energy conservation gives

$$E = mgz_0 = \tfrac{1}{2} mv^2 + mgz \qquad\qquad (z > 0),$$

$$E = mgz_0 = \tfrac{1}{2} mv^2 + mgz + \tfrac{1}{2} kz^2 \qquad (z < 0)$$

(the rope acts as a spring only at $z < 0$ where it is taut). To visualize the relationship between E, U, and K (which cannot be less than 0 physically) in problems such as this, it is very helpful to draw an energy diagram:

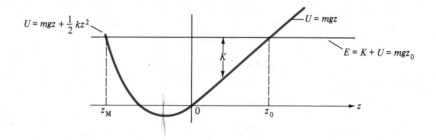

(a) The condition for maximum stretch is $K = 0$,

$$mgz_0 = 0 + mgz + \tfrac{1}{2} kz^2.$$

This quadratic equation has two roots,

$$z = -\frac{mg}{k}\left(1 \pm \sqrt{1 + \frac{2kz_0}{mg}}\right),$$

one positive and one negative. The positive root, which is obtained by using the minus sign in front of the radical, lies outside the region where the spring potential acts physically: the rope is not taut at $z > 0$. That is, the positive root is extraneous. We need the *negative* root. Numerically we find $z_M = -17.1$ m, so the maximum stretch is $-z_M = 17.1$ m.

If you look at force rather than energy to do this problem, note that $mg + kz = 0$ is *not* the condition for maximum stretch. Rather it is the minimum of the total potential-energy curve $U = mgz + \tfrac{1}{2} kz^2$ in the energy diagram, and the kinetic energy $K = E - U$ is actually *maximal* here as one sees from the diagram. Above this point gravity dominates and the man accelerates downward; below this point the spring force dominates and the man decelerates.

(b) The maximum tension T in the rope is

$$k(-z_M) = 520 \times 17.1 \text{ N} = 8892 \text{ N},$$

well below the breaking point of the rope. The force on the carabiner is twice this as the rope passes through it. So the carabiner will break before the rope does, a most desirable safety feature! In our case the force on the carabiner is 17784 N, below its breaking point.

$T \quad T$

(c) Now suppose the carabiner is defective and breaks at a lower tension. We'll confine our attention to the simplest case, in which the carabiner breaks as soon as the falling man reaches $z = 0$.* In this case the rope becomes taut only 25 m above the ground. We repeat the foregoing analysis with z_0 increased from 80 m to 90 m, and find that the rope stretches 18.0 m, leaving the climber $25 - 18 = 7$ m above the ground.

10.4 GRAVITATIONAL POTENTIAL ENERGY

A swinging pendulum, a plunging roller coaster, and a vibrating guitar string are all examples of potential energy changing into kinetic energy, and back into potential energy, and so on in an energy volley. In each of these examples, the system is endowed with potential energy by changing where it is – pulling aside a pendulum, raising a roller coaster, plucking a guitar string. And once the system is released, the volley begins as potential energy and is converted back and forth into kinetic energy. It is also possible to get a system started by giving it kinetic energy. A baseball thrown by an outfielder, an arrow fired from a hunter's bow, and a planetary probe launched by NASA are examples of giving a system kinetic energy and allowing part of that energy to change into gravitational potential energy.

For a rocket, however, we need to think more about its potential energy; it isn't mgh. For we know from Chapter 7 that the force of gravity on an object of mass m is mg only near the surface of the earth. Newton's universal law of gravity expresses the general case:

$$\mathbf{F} = -G\frac{mM_e}{r^2}\,\hat{\mathbf{r}} \tag{7.2}$$

where all distances are measured from the center of the earth. So we need to consider the change in the force of gravity as the rocket moves away from the earth.

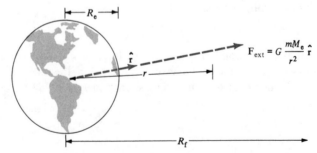

Figure 10.6 The work done to move a mass m from R_e to R_f.

Let's calculate the work done in sending a rocket from the surface of the earth, where $r = R_e$, to some distance R_f far away. First of all, to lift the rocket we must apply a force $\mathbf{F}_{ext} = -\mathbf{F}_{grav}$ equal and opposite to the force of gravity given by Eq. (7.2). For simplicity, let's assume that we send the rocket straight out, in a radial direction. The

*It can be shown that this case produces as large a stretching as any other.

force $\mathbf{F}_{ext}$ and displacement will be in the same direction as shown in Fig. 10.6. Since this is linear motion the work we do is

$$W_{ext} = \int_{R_e}^{R_f} F_{ext}\, dr.$$

r^{-2}

Substituting for F_{ext}, which is a function of r, we have*

$$W_{ext} = \int_{R_e}^{R_f} \left(G\,\frac{mM_e}{r^2} \right) dr = GmM_e \int_{R_e}^{R_f} \frac{dr}{r^2}. \tag{10.31}$$

Knowing that an antiderivative of $1/r^2$ is $-1/r$, we can evaluate the integral and obtain

$$W_{ext} = GmM_e \left(\frac{1}{R_e} - \frac{1}{R_f} \right). \tag{10.32}$$

Since $1/R_e$ is bigger than $1/R_f$, the external work done is a positive quantity, which it had to because we needed to supply energy to get the rocket to R_f.

This last result holds if the motion is along any path from the surface of the earth to any point in space a distance R_f away. We will prove this in Example 6, but to get an idea of why this is so, consider a special path from distance R_e to distance R_f which consists of steps alternating along radial lines and circular arcs (perpendicular to the radial lines) as shown in Fig. 10.7.

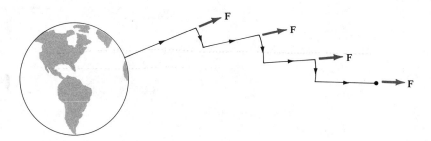

Figure 10.7 A path consisting of radial lines and circular arcs from the surface of the earth to any point in space.

Since the force $\mathbf{F}_{ext}$ is always radial, no work is done along the circular arcs because $\mathbf{F}_{ext}$ is perpendicular to the displacement there. So the total external work done is obtained by adding work done in the radial directions alone, and this is the same result we got before,

$$W_{ext} = GmM_e \left(\frac{1}{R_e} - \frac{1}{R_f} \right). \tag{10.32}$$

*Here we ignore the fact that in actual rocket propulsion, the mass of the rocket changes as it burns up fuel. The principles of rocket propulsion, which enable one to take this complication into account, will be described in Section 11.5.

According to our bookkeeping, the work we do on the rocket increases its potential energy. The potential energy is higher at R_f than it would be on the surface of the earth by an amount $\Delta U = W_{ext}$ or

$$\Delta U = U(R_f) - U(R_e) = G\frac{mM_e}{R_e} - G\frac{mM_e}{R_f}. \tag{10.33}$$

From this expression we identify the potential energy as

$$U(r) = \frac{-GmM_e}{r} \tag{10.34}$$

plus an arbitrary constant. For events taking place near the earth's surface one often takes the reference position where U vanishes at $r = R_e$ by setting the constant equal to GmM_e/R_e. But for objects escaping to large distances it is common to take the reference position at $r = \infty$ by setting the constant equal to zero.

By conservation of energy, if the rocket stopped at R_f and fell back to the earth, it would arrive with a kinetic energy $\frac{1}{2}mv^2$ equal to $U(R_f) - U(R_e)$, the value of which is given in (10.33).

Example 6

Prove that the work done on the rocket is independent of the path from the surface of the earth to any point in space at distance R_f.

Let's say the rocket moves along a path with position vector $\mathbf{r}(t)$ at time t. At time $t = 0$ it is at the surface of the earth, so $R_e = r(0) = |\mathbf{r}(0)|$. At some time t_1 it is at a distance R_f, so $R_f = r(t_1) = |\mathbf{r}(t_1)|$. The work we do to send the rocket from $\mathbf{r}(0)$ to $\mathbf{r}(t_1)$ is the line integral

$$W_{ext} = \int_{\mathbf{r}(0)}^{\mathbf{r}(t_1)} -\mathbf{F} \cdot d\mathbf{r} = \int_0^{t_1} -\mathbf{F} \cdot \frac{d\mathbf{r}}{dt}\, dt,$$

where $\mathbf{F}$ is the gravitational force given by (7.2). The position vector satisfies $\mathbf{r} = r\hat{\mathbf{r}}$ so its derivative is

$$\frac{d\mathbf{r}}{dt} = r\frac{d\hat{\mathbf{r}}}{dt} + \frac{dr}{dt}\hat{\mathbf{r}}.$$

Taking the dot product of this with $-\mathbf{F}$ we get

$$-\mathbf{F} \cdot \frac{d\mathbf{r}}{dt} = -r\mathbf{F} \cdot \frac{d\hat{\mathbf{r}}}{dt} - \frac{dr}{dt}\mathbf{F} \cdot \hat{\mathbf{r}}.$$

But $\hat{\mathbf{r}}$, being a unit vector, has constant length so it is always perpendicular to its derivative $d\hat{\mathbf{r}}/dt$. Since $-\mathbf{F}$ has the same direction as $\mathbf{r}$, it too is perpendicular to $d\hat{\mathbf{r}}/dt$, so $\mathbf{F} \cdot d\hat{\mathbf{r}}/dt = 0$ and the foregoing equation becomes

$$-\mathbf{F}\cdot\frac{d\mathbf{r}}{dt} = -\frac{dr}{dt}\,\mathbf{F}\cdot\hat{\mathbf{r}} = \frac{GmM_e}{r^2}\frac{dr}{dt},$$

where we have used (7.2) for $\mathbf{F}$. Therefore the integral for work becomes

$$W_{\text{ext}} = \int_0^{t_1} \frac{GmM_e}{r^2}\frac{dr}{dt}\,dt = \int_0^{t_1} \frac{d}{dt}\left(\frac{-GmM_e}{r}\right)dt.$$

The integrand is now a derivative, so we can evaluate the integral by the second fundamental theorem to obtain

$$W_{\text{ext}} = GmM_e\left(\frac{1}{r(0)} - \frac{1}{r(t_1)}\right) = GmM_e\left(\frac{1}{R_e} - \frac{1}{R_f}\right),$$

the same formula we obtained for linear motion.

The result that the work done on the rocket by gravity is path independent,

$$W_{AB}(1) = W_{AB}(2), \tag{10.35}$$

implies that the work done by gravity in going around a closed loop (Fig. 10.8) is zero:

$$W_{AB}(1) + W_{BA}(2) = W_{AB}(1) - W_{AB}(2) = 0. \tag{10.36}$$

Figure 10.8 Work done by a conservative force in going around a closed loop is zero.

A force obeying this property is called a *conservative force*. The gravitational and electrical forces are conservative; friction is not. The force considered in Example 1 of this chapter is nonconservative, since we found the work done by it to be path dependent. It is possible to associate a potential energy with a force, as we have done for gravity, if and only if the force is conservative.

In early science-fiction stories on extraterrestrial travel, like those of Jules Verne and H. G. Wells, human beings ventured into space, not in sleek rocket ships, but in shells fired out of colossal cannons. You might wonder, as these writers certainly did, whether it is really possible to release humanity from the shackles of the earth's gravity in such a way. How fast should a shell be fired to escape from the earth and never return?

We can figure out the speed necessary to escape from the earth by employing conservation of energy. If the shell starts out with a speed v, it has kinetic energy $\frac{1}{2}mv^2$. As it flies away from the earth, that energy gradually turns into potential energy. By the time the speed drops to zero, the shell is at a distance, say R_f, where the potential energy it gained is equal to its initial kinetic energy:

$$\tfrac{1}{2} m v^2 = G m M_e \left(\frac{1}{R_e} - \frac{1}{R_f} \right). \tag{10.37}$$

If the shell is to escape completely from the earth's gravitational field, it should be far away: R_f must be infinite. Then the term $1/R_f$ in the last equation is zero. Therefore, in order to escape, the minimum speed v_e with which the shell must start is given by setting the initial kinetic energy equal to the required increase in potential energy:

$$\tfrac{1}{2} m v_e^2 = G m M_e / R_e. \tag{10.38}$$

Solving for v_e, we find the *escape speed* (usually called the *escape velocity*),

$$v_e = \sqrt{2 G M_e / R_e}. \tag{10.39}$$

Note that this value doesn't depend on the mass of the shell, or on its initial direction. Whether it is a molecule of air escaping from the atmosphere, or an interplanetary probe on its way to Neptune, all objects need the same initial speed to escape from the earth.

Example 7

Estimate the escape velocity from the earth.

We know from Chapter 7 that

$$g = G M_e / R_e^2,$$

so we can write the escape velocity from the earth in the convenient form

$$v = \sqrt{2 g R_e}.$$

Knowing that g is about 10 m/s^2 and the radius of the earth is approximately 6.4×10^3 km, we easily find that the escape velocity is about 11 km/s, or 7 mi/s ($= 25{,}000$ mph), a large but not impossible speed to attain.

10.5 POTENTIAL ENERGY AND STABILITY

Now that we've mastered the ideas of work and potential energy, we shall use them to generate insights into the stability of mechanical systems. Suppose that you have an object of mass m, acted upon by a force $F(x)$ which vanishes at a certain point. For simplicity let the object move only in the x direction, and let $x = 0$ be the point where $F(x)$ vanishes. Then by Newton's second law the object will remain at $x = 0$ if placed at rest. We say the object is in *equilibrium* at the point $x = 0$ where the net force vanishes, and that the place where $F = 0$ is the *equilibrium position*.

What are the properties of the potential energy near the equilibrium point? It is helpful to begin by recalling that in one dimension the work done by a force $-F(x)$ opposing a preexisting force $F(x)$ is

$$W = -\int_0^x F(x') \, dx'. \tag{10.4}$$

If all the work goes into increasing potential energy we have

$$U(x) - U(0) = W = -\int_0^x F(x')\, dx';$$ (10.40)

the integral of force $-F$ gives the change in potential energy. Then by the first fundamental theorem of calculus we find the important relation that the derivative of potential energy gives the force $-F$:

$$F = -\frac{dU}{dx}.$$ (10.41)

For example, in a uniform gravitational field, $U = mgz$ and the gravitational force $-dU/dz$ has strength mg and acts in the $-z$ direction.

An immediate consequence of Eq. (10.41) is that $dU/dx = 0$ at our equilibrium position $x = 0$. In addition you may remember from Chapter 3 that $dU/dx = 0$ at a maximum or minimum of $U(x)$. Thus we arrive at the important conclusions that *at the equilibrium position of an object the net force on it vanishes; $dU/dx = 0$ if the force is associated with a potential energy, and $U(x)$ is at a maximum or minimum* (or more rarely, U may be constant or have a point of inflection).

When $U(x)$ is maximal or minimal at $x = 0$, it will most commonly have the form

$$U(x) - U(0) = \tfrac{1}{2} kx^2$$ (10.42)

near $x = 0$. Here the main feature is the quadratic dependence on x. We have chosen to call the coefficient $\tfrac{1}{2} k$ to make the next equation come out conveniently; k is positive at a minimum and negative at a maximum. These two possibilities are illustrated in Fig. 10.9, where $U(0)$ has been set equal to zero for convenience. The force corresponding to Eq. (10.42) is

$$F = -\frac{dU}{dx} = -kx.$$ (10.43)

We recognize (10.43) as having the form of Hooke's law, although many physical systems other than springs can give this behavior.*

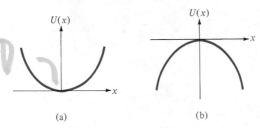

(a)   (b)

Figure 10.9 Graph of the potential-energy function $U(x) = \tfrac{1}{2} kx^2$ for (a) $k > 0$, (b) $k < 0$.

*Other forms that give a maximum or minimum, for example $U(x) - U(0) = \alpha x^4$, or αx^6, . . . , are also possible, but rarer. So are points of inflection such as $U(x) - U(0) = \beta x^3$, or βx^5, ...

We can gain a physically intuitive understanding of these results by considering a marble and a bowl. Admittedly the three-dimensional motion of the marble differs from one-dimensional motion in the potential $U(x)$, even if the bowl has the parabolic shape of the $U(x)$ curve, but the qualitative similarity of behavior is nonetheless instructive. Two positions of the marble and bowl are pictured in Fig. 10.10. If you place the marble at rest at the bottom of the bowl, it remains there; it is at the minimum of potential energy and there is no net force acting on it. Similarly, if you place the marble on top of an inverted bowl, as in Fig. 10.10b, the marble is again perfectly happy to remain at the maximum of its potential energy because there are no net forces acting on it. In each case, the marble is in equilibrium.

(a)   (b)

Figure 10.10 Bowl corresponding to a potential with (a) stable equilibrium, (b) unstable equilibrium.

Although the marble is at rest in both cases, the situations are not equivalent. The reason is this: If you push the marble when it's at the bottom of the bowl, it rolls around for a while, but eventually comes to rest at the equilibrium position, where it was before. If, however, you disturb in the slightest way the marble resting on top of the bowl, it rolls down the bowl and probably off the table; it will never return to its original position. The fact that an object has no force on it does not guarantee that there is stability.

Evidently, there are two types of equilibrium. When the marble is at the bottom of the bowl, it is said to be in *stable equilibrium*. When it's on top of the inverted bowl, it is said to be in *unstable equilibrium*.

Physically, the difference between the two types of equilibrium is as follows: When the equilibrium is stable, a disturbance produces a restoring force back toward the equilibrium position. This is a direct consequence of the fact that the potential energy is a minimum at that point. *If the disturbance increases the potential energy of the object, then the object will fall back to its original position, to a point of lower potential energy, when released.*

Just as the force resulting from the bowl and gravity causes the marble to move toward the bottom, we can imagine the force associated with the potential energy in Fig. 10.9 pushing an object back toward its equilibrium position. The force, given by Eq. (10.41),

$$F = -\frac{dU}{dx},$$

(10.41)

has sign opposite to the slope of the potential-energy curve. So, as depicted in Fig. 10.9a, if the object is moved in the positive x direction, where the slope of the potential is positive, the force is in the opposite direction, back toward the equilibrium position.

Likewise, if the object is moved in the negative x direction, the force is in the opposite direction, restoring the object to its equilibrium position.

When the equilibrium is unstable, a slight push causes a force on the object directed *away* from where it was. *If the disturbance lowers the potential energy, the object will continue to move farther from its original position.* As shown in Fig. 10.9b, if the particle is moved in the positive x direction, the slope of the potential-energy curve is negative and hence the force is positive. The force in the positive direction pushes the object farther from its original position.

There is also a third kind of equilibrium called *neutral equilibrium* which can be illustrated by a marble lying on a flat surface. If the marble is displaced slightly its potential energy doesn't change and it suffers neither a restoring force nor a repelling force. Neutral equilibrium corresponds to regions where the potential-energy curve is constant.

Mathematically, the different types of equilibrium can be distinguished by the value of the second derivative of U at the point of equilibrium. As illustrated in Figs. 10.9 and 10.10, at a point of stable equilibrium the second derivative is positive, while at a point of unstable equilibrium the second derivative is negative.

Pendulums, roller coasters, and guitar strings are examples of systems with points of stable equilibrium. These are cases in which something set into motion continues in motion for a while, as kinetic energy turns into potential energy and back into kinetic and so on. As time goes on, more and more of that energy turns into heat, until finally the system has the least energy it possibly can have, no kinetic energy (it comes to rest), and its lowest possible potential energy. That is why stable equilibrium always occurs at a minimum of potential energy.

Unstable systems include a pencil balanced on its point, a row of dominoes, and a house of cards. All these systems have excess potential energy, which can easily be turned into other forms, and eventually will be.

Example 8

Why is a tall, thin parfait glass less stable than a squat, cylindrical glass?

Consider a cylindrical glass of radius R with center of gravity a distance h above the table (a). It's potential energy is $U = mgh$. If the glass is tilted slightly, its center of gravity is raised (b). Therefore the glass is stable against small displacements. The potential energy reaches its maximum, $U = mg\sqrt{h^2 + R^2}$, when the center of gravity is directly above the corner of the glass (c). This is the limit of the stable region. If pushed beyond this point the glass can lower its center of gravity by tilting further, so the glass falls (d). The critical angle for stability,

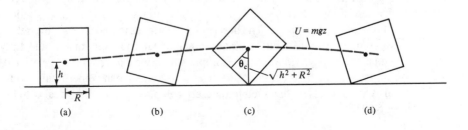

(a)   (b)   (c)   (d)

$$\theta_c = \tan^{-1} \frac{R}{h},$$

can be enlarged by increasing the radius or lowering the center of gravity. A parfait glass is less stable than a squat glass because it has a high center of gravity and a small base.

The same conclusions can be reached by studying torques. Gravity acts downward through the center of gravity and the normal force exerted by the table acts upward through the point of contact. In its upright position, the glass is in equilibrium (a). For slight tilting the torque acts to restore the glass to its stable upright position (b). When the center of gravity comes directly above the corner of the glass the restoring torque vanishes (c). Beyond this point the torque is destabilizing (d). Overall there are two equilibrium points: the stable equilibrium point at (a), and the unstable equilibrium point at (c), which also represents the upper bound of displacements that will return to (a).

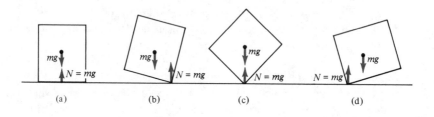

Ideas of equilibrium and stability are applied to areas besides physics as well. One example is an ecological system. Simple ecological systems, such as a small jungle, tend to be stable. When there are enough little animals around for big animals to eat, the system is in equilibrium. If the predators eat too many prey, however, then there aren't enough little animals and the big animals start dying. The presence of fewer predators allows the little animals to recover, and the big animals recover too.

Another example of a stable system is a well-designed government. The term "checks and balances" is used to express the idea that the system automatically responds in a way that opposes disturbances. On the other hand, it is easy to think of political and economic systems which are unstable.

This kind of example brings us very far from our starting point, the simple equation $F = -kx$. Indeed, the approach in this section has been different from our usual one. Up to now the study of physics has typically been a matter of taking some phenomenon (say, a falling body), separating it into its component parts (say, gravity and air resistance), then concentrating on what seems fundamental and describing it precisely and completely. By contrast, in this section we have used the simple equation $F = -kx$ as a metaphor for describing the behavior of increasingly complicated systems – a guitar string, a row of dominoes, or a jungle. The description continues to make sense, even though it becomes more and more difficult to write mathematical equations that would fully describe the situation. In physics, a simple example which we can analyze in detail and use to extend our insight into nature is called a *model*. Workers in many other fields, such as economics, also try to represent the often immensely complex systems they deal with by mathematical models.

Example 9

The potential energy between two gas molecules is often described by the empirical function

$$U(r) = -U_0[2(r_0/r)^6 - (r_0/r)^{12}]$$

where r is the distance between the two molecules and U_0 and r_0 are positive constants. What is the separation distance at which the molecules feel no force, and what is the value of the potential energy there?

The intermolecular force can be obtained from this potential by taking the derivative

$$F = -\frac{dU}{dr} = U_0\left(\frac{-12r_0^6}{r^7} + \frac{12r_0^{12}}{r^{13}}\right).$$

This force is positive (radially outward) at small r, negative at large r, and vanishes at $r = r_0$. The value of the potential at the point where $F = 0$ is

$$U(r_0) = -U_0.$$

The intermolecular potential and force are plotted below.

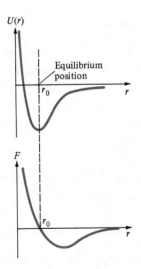

In the region near $r = r_0$, the graph of F can be approximated by the tangent line at r_0. Because

$$F'(r_0) = -72U_0/r_0^2$$

the linear approximation is

$$F \approx -72U_0 \frac{r - r_0}{r_0^2} .$$

By integrating $-F$ we obtain a corresponding quadratic approximation for U:

$$U \approx -U_0 + 36U_0 \frac{(r - r_0)^2}{r_0^2} .$$

In other words, the intermolecular force and potential behave like a spring at small displacements from equilibrium, as discussed in Chapter 8, with spring constant $k = 72U_0/r_0^2$.

10.6 HEAT AND ENERGY

Early attempts to formulate the idea of energy conservation ran into a rigid prejudice: Newton's laws embodied all truth about nature; if a law could not be based on Newton's mechanics, it was not physics but mysticism. Nevertheless, a dozen scientists in the first half of the nineteenth century proposed in some form that energy is conserved. Two of these were James Prescott Joule, a practical-minded brewery owner working in England, and Herman von Helmholtz, a German physiologist. Through their work, the law of conservation of energy became scientifically respectable. But to achieve this, they had to confront the apparent nonconservation of energy in the world.

There are numerous phenomena for which energy – at least in the form of kinetic and potential energy – appears not to be conserved. A box sliding across the floor eventually comes to rest; a rubber ball dropped from some height bounces less and less until it to eventually comes to rest. However, if we could look at a sliding box or bouncing ball closely enough to see atoms and molecules, we would better understand what is happening. For example, as a box slides along the floor, atoms and molecules of the box and floor interact and are pulled from their equilibrium positions. However, the atoms aren't free to move far; the electric forces pull them back to their equilibrium positions. When they spring back, they hit the adjacent atoms and set them moving, and so on. This motion goes on inside the box and inside the table. Energy is not really lost; it's converted into kinetic energy of atoms and molecules.

Thermal energy is the name we give to energy in the form of hidden motion of atoms and molecules. Technically, heat is transferred between a system and its environment as a result of temperature changes. But the term *heat* is often used generically to encompass thermal energy as well. There are many processes in nature which turn kinetic energy, the organized energy of the motion of an entire large body, into thermal energy, the unorganized motion of atoms. Friction is one example; viscosity is another. In an automobile engine, or a steam engine, the opposite process occurs: heat is turned into work.

If it weren't for the fact that energy can turn into heat, it would have been easier to discover the law of conservation of energy. But in the nineteenth century, the idea of atoms and molecules in constant motion had not yet developed. Instead, before the discovery of the law of conservation of energy, there was a different theory. Heat was thought to be a kind of fluid, called *caloric*. In this theory, heat itself was a conserved quantity.

The caloric theory was not idle speculation but a detailed mathematical theory. A caloric theorist would describe how an iron rod in a fire becomes hot by saying that caloric is flowing from the fire into the iron rod; he would know exactly how much heat, or caloric, was required to bring the rod to a given temperature. If the hot rod were immersed in water, a caloric theorist would say that some of the caloric leaks out of the metal into the water, thereby warming up the water; by applying the law of conservation of caloric, the precise rise in temperature of the water could be predicted. The missing element, the barrier to discovering the law of conservation of energy, was the fact that it is possible to change work or kinetic energy into heat and vice versa.

The credit for the law of conservation of energy goes not to the first of the many people who discovered it, but to the last, because that person pinned it down so well that it didn't need to be discovered again. James Prescott Joule is said to have become interested in heat by a desire to develop more efficient engines for the family brewery. In experiments conducted between 1837 and 1847 in the brewery and at his own expense, Joule established the law convincingly.

Joule made careful measurements of exactly how much work turned into a specific amount of heat. The invention of the steam engine had led the way to measuring energy changes. Almost from their beginning, steam engines were rated according to their *duty*, which referred to how heavy a load an engine could lift using a given supply of fuel. Joule used such a practical engineering approach to determine first if electric motors could be made economically competitive with steam engines (and improve the brewery), and later to quantify the relation between work and heat.

Joule's famous experiments involved an apparatus, as shown in Fig. 10.11, in which slowly descending weights turn paddle wheels in a container of water. As the weights fall and paddles turn, the water temperature rises. Through precise measurements of the work done by the weights and the rise in temperature of the water, Joule uncovered the relationship between heat and work. The unit of heat is the calorie, which is the amount of heat required to raise the temperature of one gram of water 1°C. Joule found that this unit of heat is related to units of energy according to*

$$1 \text{ calorie} = 4.18 \text{ joules.} \tag{10.44}$$

(Of course, he didn't call them joules – that came later.)

Being a practical man and possessing a limited mathematical background, Joule was content with experimenting in his lab. Despite his irrefutable experiments, many physicists of the time wanted a rigorous, mathematical theory of the law of conservation of energy. Such respectability was bestowed by the German physiologist and physicist, Herman von Helmholtz. In a paper published in 1847, Helmholtz showed, much as we did in Section 10.3, that the law of conservation of energy follows from Newton's laws. With his proof in hand, Helmholtz went on to apply the law to various cases, such as gravity, electricity, and friction.

In addition to thermal energy, there can also be hidden potential energy. For example, the potential energy of a carbon atom in a piece of coal and an oxygen molecule in the

---

*The "calorie" employed as the unit of food energy is actually a kilocalorie, equivalent to 4180 J.

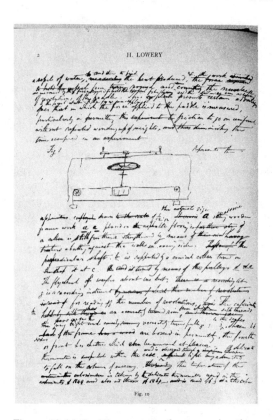

Figure 10.11 Joule's apparatus for measuring the conversion of work into heat.

air is far greater than the potential energy in a carbon dioxide molecule, produced when the carbon and oxygen combine as the coal burns. This is chemical energy, our name for the electrical potential energy built into the structure of molecules, crystals and so on. When coal burns, of course, the released potential energy turns into heat.

An even greater source of hidden potential energy is nuclear energy, the potential energy built into the structure of nuclei. Nuclear energy can be converted into thermal or electrical energy either by fission, the breakup of very heavy nuclei, or fusion, the coalescence of light nuclei.

Once the various forms of energy have been recognized, we can trace the evolution of energy transformations backward in time. For example, the energy you use to lift a book was, at an earlier stage, chemical energy stored in muscles. And it was whatever brand of hamburger you had for lunch that provided that chemical energy. For jet engines, the energy to fly you to a vacation resort comes from potential energy stored in the jet fuel. The fossil fuel, in turn, came from plants and animals living on the earth millions of years ago. Their energy ultimately came from sunlight. Most energy available on earth came originally from the sun, with the exception of that generated by nuclear reactions. And scientists are confident that the sun itself is a cosmic nuclear reactor.

Thus most of the energy that we use started out as nuclear energy, and the nuclear energy that exists in the universe is a legacy to us from the instant in which the universe originated – the Big Bang. Further back than that, we can't trace it. The primordial stock of energy from the Big Bang is still the same as it was at that first instant, 20 billion years ago, but that energy has been converted into various forms.

We began our discussion with the observation that in many cases it appears that energy is not conserved; such cases involve friction, viscous force, and other means of dissipating energy. Joule and Helmholtz showed us that energy is *always* conserved and that it can be *transformed* from one form into another. Recognition of this fact has been instrumental in providing us with our present understanding of the history of the cosmos.

Example 10

Starting at a height of 25 m, a sled of mass 20 kg slides down a hill with a 30° slope. If the sled starts from rest and has a speed of 15 m/s at the bottom of the hill, calculate the energy dissipated by friction along its path and calculate the coefficient of friction.

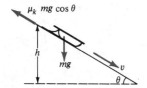

In sliding down the hill, the sled converts mgh of potential energy into $\frac{1}{2} mv^2$ of kinetic energy and an amount of work W done against friction. Therefore W is

$$W = mgh - \tfrac{1}{2} mv^2 = (20 \text{ kg}) (10 \text{ m/s}^2) (25 \text{ m}) - \tfrac{1}{2} (20 \text{ kg}) (15 \text{ m/s})^2$$

$$= 5000 \text{ J} - 2250 \text{ J}$$

$$= 2750 \text{ J}.$$

The sled slides a distance $h/\sin \theta$ down the hill, with friction exerting a backward force $\mu_k N = \mu_k mg \cos \theta$ during this time. So the work done against friction is

$$W = (\mu_k mg \cos \theta) (h/\sin \theta).$$

Combining the two equations for W we find

$$\mu_k mgh \cot \theta = mgh - \tfrac{1}{2} mv^2$$

or

$$\mu_k = (\tan \theta) (1 - v^2/2gh).$$

Using the given values for θ, v, and h, we find $\mu_k = 0.32$. As a check, note that for a frictionless incline ($\mu_k = 0$, $\tan \theta \neq 0$), our equation yields $\frac{1}{2} mv^2 = mgh$, the well-known result for conservative motion in a uniform gravitational field.

10.7 MECHANICAL ADVANTAGE AND EFFICIENCY OF MACHINES

A *machine* is a device to change the magnitude and/or direction of a force. An input force F_i applied to the machine is transformed into a more convenient or useful output force F_o applied by it. A measure of the amount by which the force is transformed is the *mechanical advantage*, defined as the ratio

$$M = F_o/F_i. \tag{10.45}$$

Machines can be analyzed either in terms of force or in terms of energy conservation.

For example, suppose one uses a wrench to tighten a bolt that is difficult to turn by hand (Fig. 10.12). In terms of torque about the center of the bolt, the input torque $F_i r_i$ applied is transmitted to the bolt, where it supplies an output torque $F_o r_o$ with force F_o sufficiently large to overcome the resistive force acting through the small bolt of radius r_o:

$$F_i r_i = F_o r_o. \tag{10.46}$$

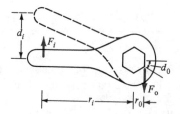

Figure 10.12 A wrench turns a bolt.

This is a simple example of a *lever*. In the alternative analysis based on energy conservation, the input work $F_i d_i$ must equal the output work $F_o d_o$ if we ignore friction:

$$F_i d_i = F_o d_o. \tag{10.47}$$

We have gained a reduction in input force by a corresponding sacrifice in the distance through which the output force acts. The mechanical advantage $M = F_o/F_i = r_i/r_o$ is the same in both analyses because $d_o/d_i = r_o/r_i$.

Another example is loading a heavy weight mg into a truck by means of an *inclined plane* of length r and height h (Fig. 10.13). Let us analyze the problem in terms of energy. The work to be done is mgh. By using the incline, the trucker can reduce the force he applies to only $mg \sin \theta$, but of course he must still do the same amount of work, so the distance he pushes the object increases from h to $r = h/\sin \theta$:

$$(mg)h = (mg \sin \theta)r. \tag{10.48}$$

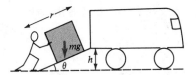

Figure 10.13 Use of an inclined plane.

The mechanical advantage in this case is

$$M = \frac{r}{h} = \frac{1}{\sin \theta}.$$ (10.49)

A third basic type of machine is the *pulley*. A single pulley (Fig. 10.14a) merely changes the direction of the force, if friction can be ignored so that tension is transmitted without loss from one side of the pulley to the other. Thus its mechanical advantage is $M = 1$. A block and tackle is a pulley system arranged as in Fig. 10.14b. It is used to lift a weight. The applied tension F_i is transmitted undiminished to each rope segment, yielding a mechanical advantage equal to the number of parallel ropes holding up the load (e.g., $M = 2$ in Fig. 10.14b). The reader is encouraged to verify that this result can also be obtained readily by use of energy conservation.

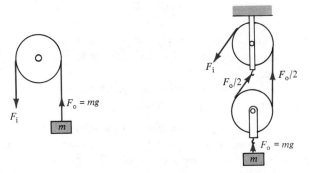

Figure 10.14 (a) Single pulley. (b) Block and tackle.

A fourth basic type of machine is the *wheel and axle* (Fig. 10.15). It is easy to show that the mechanical advantage is $M = R_2/R_1$, either by using energy conservation or by considering torques.

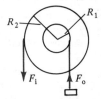

Figure 10.15 Wheel and axle.

All machines are subject to friction, which reduces the work output below the ideal results we have calculated. The *efficiency* of a machine is the ratio of its work output to its work input:

$$\text{Eff} = \frac{W_o}{W_i},$$
(10.50)

and is always less than 100% in practice.

10.8 POWER

Power is defined to be the rate at which work is done with respect to time:

$$P = \frac{dW}{dt}.$$
(10.51)

Looking at the definition of work given in Eq. (10.14) from a fixed time t_a to an arbitrary time t,

$$W = \int_{t_a}^{t} \mathbf{F} \cdot \frac{d\mathbf{r}}{d\tau} \, d\tau,$$
(10.14)

we see that we can identify the integrand at time t with dW/dt by the first fundamental theorem of calculus. So, noting that $d\mathbf{r}/dt = \mathbf{v}$, we also have

$$P = \mathbf{F} \cdot \mathbf{v}.$$
(10.52)

Power may be used to counteract friction, as when a car drives at constant speed on a level roadway; to increase kinetic energy, as when a car accelerates; or to increase potential energy, as when a car climbs a hill.

The obvious units of power are collected together with the units of work in Table 10.1. In addition, two commonly used units deserve mention:

(1) The *horsepower* (hp) was introduced by James Watt (1736–1819), who, for comparison with the steam engine he had developed, estimated the average power output of a horse on the basis of actual measurements. It has the value

$$1 \text{ hp} = 550 \text{ ft lb/s} = 746 \text{ W}.$$
(10.53)

Table 10.1 Units of Work and Power

	SI	cgs	U.S. customary
Work	joule (J) = newton-meter	erg = dyne-centimeter	foot-pound
Power	joule per second = watt (W)	erg per second	foot-pound per second

(2)   In honor of Watt, one joule per second is called one watt (W). The *kilowatt* is 1000 W. Standard for electrical usage, the kilowatt (kW) can also be used for other forms of power. Since it is so common a power unit, the *kilowatt-hour* (kW h) has been based on it. This is the work done when a steady power of 1 kW is supplied for one hour.

When devices are operating one pays for energy, not power, therefore the electric bill is in kilowatt-hours. On the other hand, when one buys a new device it is rated for power, since power determines the acceleration capability of a vehicle or the brightness of a lightbulb.

Example 11

A marble of radius 1 cm and mass 10 g falls through a tube of glycerin at terminal velocity. At what rate is gravity supplying power to overcome viscosity?

The limiting velocity is

$$v_L = \frac{mg}{6\pi R\eta} = 0.63 \; \frac{m}{s} \,. \tag{8.26}$$

Making use of Eq. (10.52), we find that the power supplied by gravity is

$$P = mgv_L = 0.061 \text{ W}.$$

10.9   A FINAL WORD

Once you understand something, it's almost impossible to put yourself in the position of not understanding it, and try to figure out how people thought about the problem before. Up to the time when Joule did his experiments, it was thought that heat itself was the conserved quantity and couldn't be created out of work. But how could people for thousands of years before Joule not realize that by rubbing their hands together they could warm them? Even more to the point, by the 1830s, long before Joule's experiments, railroads were strung across England. The burning of coal in locomotives eventually set the train in motion – converting heat into work. How could people living at this time, riding on railroads, not have believed that it was possible to turn heat into work?

It was not through lack of trying to understand, as we see from the thoughts of a young French military engineer, Sadi Carnot. Although Carnot died at the age of 32 from scarlet fever, he is one of the important figures of nineteenth-century science. Carnot discovered what we now call the second law of thermodynamics, which is one of the most profound laws in physics. He succeeded in doing this without knowing the law of conservation of energy, which is the first law of thermodynamics. Yet it may have been easier to discover the second law of thermodynamics without knowing the first law.

Carnot reasoned by analogy. His idea of how heat worked was by analogy to a water wheel. As water runs down and over a water wheel, it makes the wheel turn, yet the water is conserved. That is, to get work out of the water doesn't require using up water. Instead water falling from a large height to lower height turns the water wheel. Carnot thought that heat worked in exactly the same way. He thought that caloric, starting out at high temperature, was capable of doing work on the way down to low temperature, analogous to water flowing over the water wheel. His analogy also suggested the idea that came to be the second law of thermodynamics: once heat runs downhill from high to low temperature, it will not run uphill again, back up to the high temperature. That seemed obvious from the water wheel analogy.

A piece of coal has potential energy stored in it in the form of chemical bonds. The process of combustion, say in a steam locomotive, transforms that energy directly into heat at high temperature. Some of the heat is able to drive a piston, producing work and setting the locomotive into motion. Eventually all the heat, including the part turned into work, winds up as thermal energy once again, as the locomotive loses kinetic energy through frictional processes. But the thermal energy is no longer at the high temperature of the burning coal, but rather at the low temperature of the air outside. The net result of all this, aside from getting you from one place to another, has been to transform potential energy into high-temperature thermal energy and finally into low-temperature thermal energy. And once that's done, it is no longer possible to get that thermal energy back into the form of the original potential energy so it can be used again. That's what the second law of thermodynamics says.

Although Carnot did not know that, unlike water, some of the heat actually is transformed into work before changing back into heat again, this limitation did not prevent him from discovering the second law of thermodynamics.

Energy from the sun, originally at very high temperature, is stored temporarily in coal, oil, and other fossil fuels. We can release it into heat at high temperature (but lower than the temperature of the sun) and, whether we use it for useful work or not, it always winds up as low-temperature heat (heat at ambient temperature). All of that energy is perfectly conserved; not a bit is ever lost. But the value of the energy has been degraded – it has become less useful. The world has run down a little bit, and it will never be wound up again.

This is the real nature of our energy crisis. We are not using up energy, we are just transforming it into less useful forms.

Problems

Work

1. Suppose that two forces, $\mathbf{F}_1$ and $\mathbf{F}_2$, act on an object, with $F_1 = 10$ N, $F_2 = 16$ N, such that the two forces have opposite directions. What is the total work done on the object when it moves a distance of 2 m under the action of these forces?

2. Two springs A and B are identical except that A is stiffer than B (larger spring constant). On which spring is more work done if

   (a) they are stretched by the same force,
   (b) they are stretched by the same amount?

3. A gardener pushes a lawn mower on a horizontal lawn, with a force 20.0 N applied at an angle of 37° from the horizontal, for a distance of 15.0 m. Calculate the amount of work done by the gardener.

4. A force $\mathbf{F} = (3.0\text{ N})\hat{\mathbf{i}} - (7.0\text{ N})\hat{\mathbf{j}}$ moves an object through a displacement $\mathbf{r} = (4.0\text{ m})\hat{\mathbf{i}} + (3.0\text{ m})\hat{\mathbf{j}} + (2.0\text{ m})\hat{\mathbf{k}}$. Find the amount of work done on the object by this force.

5. The force exerted on an object is described by $F = F_0(x/b - 1)$, where F_0 and b are constants. Find the work done by this force on the object as the object moves from $x = 0$ to $x = 3b$ by

   (a) plotting $F(x)$ and graphically determining the area under the curve,
   (b) evaluating the integral analytically.

6. One way to evaluate a line integral such as Eq. (10.13) is to write $d\mathbf{r} = dx\,\hat{\mathbf{i}} + dy\,\hat{\mathbf{j}}$ and $\mathbf{F} \cdot d\mathbf{r} = F_x\,dx + F_y\,dy$ (we confine ourselves to a plane for simplicity). The work done in moving a particle from A and B is then written

$$W = \int_A^B (F_x\,dx + F_y\,dy).$$

Suppose the force is $\mathbf{F} = (2x^2 + 3y^2)\hat{\mathbf{i}} + 5xy\,\hat{\mathbf{j}}$, A is the origin, and B has coordinates (2,2). Find the work done in moving from A and B along the following paths:

   (a) Straight along the x axis to $x = 2$, then straight in the y direction to point B. (For example, $\mathbf{r}(t) = t\hat{\mathbf{i}}$ for $0 \le t \le 2$ and then $\mathbf{r}(t) = 2\hat{\mathbf{i}} + (t-2)\hat{\mathbf{j}}$ for $2 \le t \le 4$.)
   (b) Straight along the line $x = y$ from A to B.

Work and Potential Energy

7. The work done to stretch a certain spring a distance of 1.0 cm from equilibrium is 0.2 J. If 0.2 J work were done in additional stretching of the stretched spring, how far from equilibrium would the spring be stretched?

8. Show that the work done against gravity in the uniform field near the earth's surface depends only on the vertical distance an object is moved.

Conservation of Combined Potential and Kinetic Energy

9. A pendulum bob is pulled aside and released. A peg is located as shown in the sketch. How high above the peg will the bob rise?

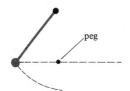

10. A truck at rest on the top of a hill is allowed to roll down, and at the bottom it has a speed of 5 mph. If it is allowed to roll down the hill again, but this time starting with an initial speed of 3 mph, what will be its speed at the bottom of the hill? (Ignore any work done against frictional forces and drag.)

11. A good athlete can sprint at a rate of 10 m/s. If the athlete can convert all this kinetic energy into potential energy with the aid of a fiberglass pole, how high can he pole vault?

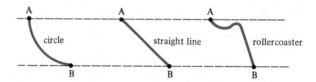

12. Three identical blocks slide down the frictionless surfaces shown below after being released from the same height A:

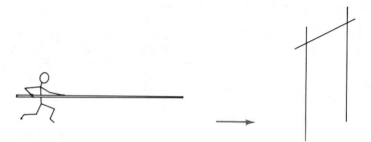

    (a) How do the speeds of the blocks at point B compare?
    (b) Which block do you think reaches point B first? Why?

13. A mass attached to a vertical spring is gently lowered to its equilibrium position, at which point the spring is stretched a distance d. If the same object is attached to the same vertical spring but allowed to fall instead, to what maximum distance is the spring stretched during its subsequent motion down and back up?

14. A block of ice slides down the frictionless incline shown below and compresses the spring. Taking the mass of the block to be 1.5 kg, the spring constant to be 3.0×10^2 N/m, and the distance the block slides down the incline before striking the spring as 1.2 m, find the maximum distance the spring is compressed.

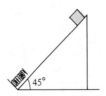

15. A spring of spring constant 3.0×10^3 N/m propels a small (0.5-kg) block up a frictionless plane inclined at 37°. If the spring was initially compressed 3.0 cm, how far up the plane from the point of release will the block travel before coming momentarily to rest? When the block returns and hits the spring, how much will it compress the spring?

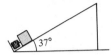

16. For the pendulum of Example 3, find the tension at an arbitrary angle α from the vertical (of course $\alpha \leq \theta$, the angle from which the pendulum is released).

17. A 2.0-kg block on a horizontal, frictionless surface is pressed against a spring of spring constant 1.5×10^3 N/m. The spring is compressed a distance of 8.0 cm. When released, what speed will the block acquire?

18. A pendulum bob of mass m hanging at the end of a string of length L is struck so that it has a speed v. What must v be in order for the string to go slack (tension becomes zero) at the top of the swing (180° from where the bob began)?

19. A small block of mass m slides down a frictionless incline, starting from rest at height h, and around a circular track of radius b. Find the speed of the block at point C.

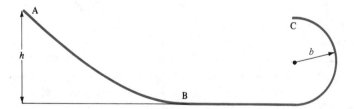

For the same track, determine the minimum height h (in terms of the radius b) such that the block barely makes it to point C without falling off. (*Hint.* Use the condition that if the block barely makes it to point C, the normal force exerted on the block by the track at that point will be zero. Also recall that the centripetal component of acceleration is v^2/b even if the speed varies on the circular track.)

20. A Near Eastern potentate, wearing a heavy 15-lb gold crown, rides the loop-the-loop at an amusement park. He enjoys the experience, especially at the very top of the loop, where (on this particular ride) the push of the crown on his head falls to zero momentarily. But the crown feels almost unbearably heavy as he travels the lower part of the loop. How strong is the push of the crown on his head at the very bottom of the loop? [*Hint.* To get started, refer to Eq. (7.34), but remember that the speed increases on the lower parts of our frictionless loop.]

21. A child sits on a hemispherical mound of ice. Assuming the mound to be frictionless, if the child is given a slight nudge and slides off, find the angle from the horizontal at which the child leaves the mound.

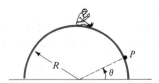

22. Consider the rock climber in Example 5.

    (a) For the stated parameters, how far above the carabiner can he safely climb?

    (b) At first sight one might think that a stiffer rope would be safer, because it wouldn't stretch as far. But a stiffer rope also means that the maximum tension in the rope is greater. Find, in the approximation $z_0 \gg |z_M|$, how much the spring constant of the rope can be increased before the carabiner is broken by the fall described in Example 5.

Gravitational Potential Energy

23. A ballistic missile is fired vertically from the earth with a speed of 9.0 km/s. If atmospheric friction is neglected, how far above the earth's surface will it rise? (Use $R_e = 6400$ km, $M_e = 6.0 \times 10^{24}$ kg.)

24. Many people back in 1958 were surprised when the first artificial satellite, Sputnik I, increased in speed as it spiraled back to Earth. As with all satellites, it fell to the earth because friction with the outer atmosphere caused it to lose energy, but as it spiraled closer and closer to the earth, its speed increased. Explain why it speeded up as it lost energy.

25. Calculate the ratio of the escape velocity to the speed necessary for an object to orbit the earth at its surface.

26. Using the data in Appendix D, determine the escape velocity from
    (a) Mercury,   **(b)** Mars.

27. Find the increase in speed needed for a satellite orbiting the earth in a high circular orbit at a speed of 4.5 km/s to escape completely from the earth.

Potential Energy and Stability

28. Determine the point(s) of stable equilibrium for a particle whose potential energy is described by

$$U(x) = 4x^3 - 6x,$$

where x is in meters and U in joules.

Heat and Frictional Dissipation of Energy

29. How much work does a 60-kg hiker do in climbing 1 km vertically? How many kilocalories (i.e., "calories of food energy") does this represent? Taking into account that muscle contraction generates about 4 J of thermal energy in the hiker's body for every joule of work done, by how many degrees would the hiker (considered as approximately 60 kg of water, the body's main constituent) heat up if he couldn't perspire or radiate energy?

30. Trace the following forms of energy back to their sources insofar as you possibly can:

(a) the calories in an apple,
(b) the light from a lightbulb,
(c) the motion of a wind-up wristwatch.

31. Suppose you are driving along a fog-shrouded road at speed v, and suddenly a brick wall appears a distance R in front of you. Would it be better to slam on the brakes (and hope that you stop before reaching the wall) or swerve in a circular arc of radius R to avoid the wall? To help you decide, compare the force necessary to change all the car's kinetic energy entirely into work done against friction with that necessary to turn it in a circle.

32. A 0.03-kg marble is dropped into a cylinder containing glycerin. Starting from a height of 0.15 m, the marble reaches the bottom of the cylinder with a speed of 1.4 m/s. What fraction of the marble's initial energy is dissipated by the viscous force of the glycerin?

33. A driver traveling at 15 mph slams on the brakes and skids 35 ft before stopping. Find the coefficient of friction between the car and the road. If the car had been moving at 60 mph, how far would it have skidded?

34. A toy car of mass 0.02 kg travels along the frictionless track shown below. The spring ($k = 3.0 \times 10^2$ N/m) is initially compressed 0.05 m.

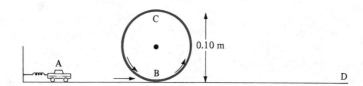

(a) How much work is done by the normal force as the car moves from B to C?
(b) What is the speed of the car at C?
(c) If at D a small parachute is released, bringing the car to rest on the flat, how much energy is dissipated by the air resistance?

35. A 25-g bullet fired at 400 m/s penetrates 10 cm into a block of wood, which remains stationary.

(a) What is the energy dissipated by friction between the block and bullet?
(b) If the frictional force were constant during the deceleration, what must its value have been?

36. A spring of spring constant 5.0×10^2 N/m, initially compressed 0.07 m, fires a 0.4-kg block across a rough horizontal surface. If the coefficient of friction between the block and the surface is 0.5, how far from its starting point will the block travel before coming to rest?

37. A 0.4-kg block is given a speed of 1.5 m/s as it is projected up a plane inclined at 30°. If the coefficient of friction between the block and plane is 0.6, how far up the plane will the block slide?

Mechanical Advantage

38. Use energy conservation to find the mechanical advantage of the block and tackle pictured below (assume the pulleys are weightless and frictionless). Check your answer by balancing forces.

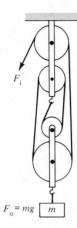

39. A fancy wooden nutcracker has an arm of length L which turns a screw of radius R and pitch P (pitch is the distance between successive threads). As the arm is turned, the screw slowly presses down on the nut, cracking it. Which two of the four basic types of machine (lever, inclined plane, pulley, wheel and axle) are combined in this device? Draw F_i and F_o. Find the mechanical advantage of the nutcracker.

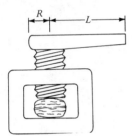

Power

40. If the lawn mower of Problem 3 moves at $\frac{1}{2}$ m/s, how much power does the gardener expend in moving it, measured in watts? In horsepower?

41. A human being, in normal (nonstrenuous) daily activities, expends about 100 W of power.

    (a) How many kilocalories (calories of food energy) does he need per day to maintain this activity?
    (b) How many additional watts does a 60-kg mountain climber expend in raising his gravitational potential energy if he climbs 1 km in three hours at a steady rate?

42. For the block of Problem 36, what is the power dissipated by friction as a function of time as the block moves across the surface (starting at the moment when the block parts company with the spring)?

CHAPTER  11

THE CONSERVATION OF MOMENTUM

And with respect to the general cause, it seems manifest to me that it is none other than God himself, who, in the beginning, created matter along with motion and rest, and now by his ordinary concourse alone preserves in the whole the same amount of motion and rest that he placed in it. For although motion is nothing in the matter moved but its mode, it has yet a certain and determinate quantity, which we easily see may remain always the same in the whole universe, although it changes in each of the parts of it.

René Descartes, *Principles of Philosophy* (1644)

11.1 THE UNIVERSE AS A MACHINE

The seventeenth century was the century when science assumed its modern form and the scientific spirit infected Europe. It was the time when Aristotle's view of nature was rejected and Galileo's great book of the universe was adopted. The new science was nourished by an optimism that mankind could discover the laws of nature.

One of the most significant and influential figures in seventeenth-century natural philosophy was René Descartes. Early in his life, Descartes rebelled against the traditions in which he had been thoroughly educated. He sought new foundations for knowledge, foundations which could underpin certainty in our knowledge of nature. Convinced of the indubitable logic of mathematics, Descartes came to identify mathematics with physics.

Descartes is credited with having been the first person to state the law of inertia correctly. Unlike Galileo, who claimed that inertial motion was horizontal (always parallel to the surface of the earth), Descartes realized that in the absence of force a body would continue to move in a straight line in any direction.

Descartes admired and at the same time criticized Galileo's efforts at resolving what he considered to be mundane questions, such as free fall and the motion of the pendulum. Descartes had greater designs; he wanted to create a system of the world as a whole. He saw the universe as a great machine, the mechanical universe of our title, set into motion by the Creator who created matter and the laws of motion. Descartes's mechanical universe is no different from artifacts produced by skillful artisans; it is only a matter of scale. Like clockwork, the universe inexorably follows purely mechanical laws.

His idea of a mechanical universe, however, presented a problem. What keeps the machine going? Like clocks, machines run down and need to be wound up from time to time. Descartes had to confront the question of whether the universe needs periodic intervention from the Creator to keep it going.

To resolve this dilemma, Descartes invented a new law. Writing hypothetically, so as not to arouse the ire of the Church, which had silenced Galileo only years earlier, Descartes proposed that the total quantity of motion in the universe is constant, preserved by God who created it in the beginning and constantly sustains it. The "quantity of motion" is something we now call *momentum* **p**, a vector quantity which is the product of the mass of an object and its velocity:

$$\mathbf{p} = m\mathbf{v}. \tag{11.1}$$

Mathematically, we write Descartes's law as

$$\sum_{i=1}^{n} \mathbf{p}_i = \mathbf{p}_1 + \mathbf{p}_2 + \mathbf{p}_3 + \cdots + \mathbf{p}_n = \text{const} \tag{11.2}$$

where $\mathbf{p}_i$ is the momentum of the ith body, and the summation of the momenta of all the bodies (here there are n of them) is the total momentum. Since the total momentum is constant, if one object in the mechanical universe slows down, another must speed up.

In Chapter 10 we encountered our first example of a conservation law: conservation of energy. Descartes's law on the conservation of "quantity of motion" is another such important law distinct from conservation of energy. Because energy changes into heat, causing machines to run down, Descartes had to postulate a new law to explain the mechanics of the universe and to prevent his clockwork from eventually stopping. To find the origin of Descartes's ad hoc law, we turn once again to the laws that irrevocably altered the way mankind viewed the universe – the laws of Sir Isaac Newton.

11.2 NEWTON'S LAWS IN RETROSPECT

In Chapter 6 we introduced Newton's three laws, upon which mechanics is based, yet in subsequent chapters we have made very unequal use of the three laws. Our mainstay has been the second law. It is $\mathbf{F} = m\mathbf{a}$ that governs the motion of objects in Newton's clockwork universe. In fact, we've mainly used this one law because for many purposes the other two laws are contained in it.

To recount briefly, Newton's three laws are:

1. The law of inertia.

2. $\Sigma\mathbf{F} = \dfrac{d(m\mathbf{v})}{dt}$.

3. The law of action and reaction.

Suppose we have a single particle of mass m and momentum $m\mathbf{v}$. According to the second law,

$$\mathbf{F} = \frac{d(m\mathbf{v})}{dt} = m\frac{d\mathbf{v}}{dt}.$$

Now if there are *no forces* acting on this particle, then this expression is equal to zero:

$$m\frac{d\mathbf{v}}{dt} = \mathbf{0},$$

We know how to solve this equation. If the derivative of something is zero, then that something must be constant. That something here is velocity. The velocity $\mathbf{v}$ must be constant if no force acts on the particle, and constant velocity means constant speed and constant direction. But this is the law of inertia: a body acted upon by no force will keep moving along a straight line with constant speed. Therefore the first law is contained in the second law.

What about the third law? Let's imagine an object either at rest or moving at constant velocity. In this case, no external force acts on the object. Imagine further that you take a closer look and notice that the object is in fact a compound body, composed of two particles which aren't necessarily touching each other, as illustrated in Fig. 11.1a. Now we have two particles, which we'll unimaginatively label particles 1 and 2.

(a)   (b)

Figure 11.1 (a) A compound body composed of two separate particles. (b) Internal action–reaction forces on individual particles.

By the second law, zero acceleration implies that no net force acts on the two particles. Whatever forces do act between particles 1 and 2, they must add up to zero, as shown in Fig. 11.1b:

$$\mathbf{F}_{12} + \mathbf{F}_{21} = \mathbf{0},$$

$$\mathbf{F}_{12} = -\mathbf{F}_{21}.$$

If the internal forces did not add up to zero then (blurring our vision again) the compound body would have a net acceleration. But by our assumption, it does not. Therefore, if particle 1 applies a force on particle 2, particle 2 must apply *an equal and opposite force* on particle 1. This is the precisely what the third law states. At least in the case of two bodies moving without external force, Newton's third law is contained within the second.

If our compound body is composed of three or more particles, it is no longer necessarily true that the second law implies the third. Overall, the scope of the laws is as follows. The second law states how motion is determined by forces, but does not specify them. The third law provides important partial information about forces. On the basis of information available in his time, Newton was able to specify the properties of one of the basic forces: gravity. To achieve the more detailed description of other forces that we surveyed in Chapter 8, further experimentation has been necessary and continues to this day.

11.3 THE CENTER OF MASS

Newton's laws as stated apply to a point particle described by its position, velocity, and acceleration. But every object we see in nature is really a compound body made up of smaller parts, ultimately of atoms, and we have often applied Newton's laws to such extended objects. In the present section we show why such applications to compound bodies are justified, and specify what we mean by the velocity and acceleration of a compound body.

The key is to find the point of a body where we can think of all its mass as being concentrated. This point is called the *center of mass*. In symmetric bodies it is identified with the geometric center of an object.

The mathematical definition of center of mass for two mass points m_1 and m_2 located at x_1 and x_2 on the x axis, as in Fig. 11.2, is

$$\bar{x} = \frac{m_1 x_1 + m_2 x_2}{M},$$

(11.3)

where

$$M = m_1 + m_2.$$

(11.4)

One can think of $\bar{x}$ as an average, in which each position x_i is weighted by the fraction $m_i/(m_1 + m_2)$ of the total mass at that point. If x_2 and x_1 are spaced a distance D apart, we find

$$\bar{x} = \frac{x_1 m_1 + (x_1 + D)m_2}{M} = x_1 \frac{m_1 + m_2}{M} + \frac{Dm_2}{M}$$

$$= x_1 + \frac{Dm_2}{M}.$$

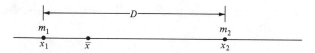

Figure 11.2 Center of mass of two points.

If $m_1 = m_2$, the center of mass is halfway between x_1 and x_2, as one would expect from symmetry.

Extending the same definition to y and z coordinates, we have

$$\bar{y} = \frac{m_1 y_1 + m_2 y_2}{M}, \tag{11.5a}$$

$$\bar{z} = \frac{m_1 z_1 + m_2 z_2}{M}. \tag{11.5b}$$

The three scalar equations (11.3), (11.5a), and (11.5b) can be expressed as one vector equation

$$\bar{r} = \frac{m_1 \mathbf{r}_1 + m_2 \mathbf{r}_2}{M}, \tag{11.6}$$

where $\bar{\mathbf{r}} = \bar{x}\hat{\mathbf{i}} + \bar{y}\hat{\mathbf{j}} + \bar{z}\hat{\mathbf{k}}$. These definitions are readily generalized to n different mass points of total mass M by using the weights m_i/M at each point. For example, (11.3) becomes

$$\bar{x} = \frac{\sum\limits_{i=1}^{n} m_i x_i}{M} \tag{11.7}$$

whereas (11.6) becomes

$$\boxed{\bar{\mathbf{r}} = \frac{\sum\limits_{i=1}^{n} m_i \mathbf{r}_i}{M},} \tag{11.8}$$

where

$$M = \sum\limits_{i=1}^{n} m_i. \tag{11.9}$$

The preferred role played by the center of mass in dynamics can be seen by differentiating $\bar{\mathbf{r}}$. If the individual masses m_i do not change, differentiation of (11.8) yields

$$M\frac{d\bar{\mathbf{r}}}{dt} = \sum\limits_{i=1}^{n} m_i \frac{d\mathbf{r}_i}{dt}. \tag{11.10}$$

Recognizing that the right-hand side of this relation is the total momentum of the system, we can write Eq. (11.10) as

$$M\bar{\mathbf{v}} = \mathbf{p}_{\text{tot}}. \tag{11.11}$$

Differentiating once again and employing Newton's second law for each mass point, we find

$$M\frac{d^2\bar{\mathbf{r}}}{dt^2} = \sum_{i=1}^{n} m_i \frac{d^2\mathbf{r}_i}{dt^2} = \sum_{i=1}^{n} \mathbf{F}_i \tag{11.12}$$

or, in other words,

$$M\bar{\mathbf{a}} = \frac{d\mathbf{p}_{\text{tot}}}{dt} = \mathbf{F}_{\text{tot}}. \tag{11.13}$$

Equation (11.13) looks very similar to Newton's second law for the motion of a single particle. The differences are:

(i) M is the total mass, not the mass of a single particle.

(ii)   The acceleration $\bar{\mathbf{a}}$ is the acceleration of a conceptual point, the center of mass, not the acceleration of any particular mass point. As Fig. 11.2 shows, there may be no particle at the center of mass.

(iii) $\mathbf{F}_{\text{tot}}$ is the total force on the system, not the force on a particular mass point.

Moreover, when we express the force on each particle as a sum of external forces plus the "internal forces" exerted by all the other particles, and then add all the forces to obtain the total force, we find that the internal forces cancel in pairs by Newton's third law. So $\mathbf{F}_{\text{tot}}$ is really the sum of all the *external* forces – a great simplification, for example, in the common case in which the external force is gravity and the internal forces include all the complications of atoms interacting in a solid. In words, Eq. (11.13) says:

The center of mass $\bar{\mathbf{r}}$ of a system of particles moves under external forces as if all the mass were concentrated at $\bar{\mathbf{r}}$ and all forces acted at $\bar{\mathbf{r}}$.

When we talk about the velocity and acceleration of a compound body, we are referring to that of the center of mass. If, for example, no net force acts on a compound system initially at rest, the center of mass of the system remains at rest even though individual parts of the system may move.

Example 1

A 180-lb man tries to step out of a 90-lb canoe, initially at rest, onto a lakeside pier. What happens if he tries to step 2 ft laterally without holding on to the pier?

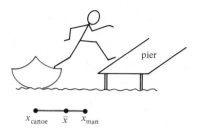

$$x_{canoe} \qquad \bar{x} \qquad x_{man}$$

The canoe has no keel, so it is not a bad approximation to neglect the lateral reaction of the water on the canoe during the brief time in which the action is taking place. Thus there is no external lateral force on the system man-plus-canoe, and the $\bar{x}$ of the system remains at rest. But for the man, initially positioned in the center of the canoe at $\bar{x}$, to displace himself 2 ft to one side while keeping

$$\bar{x} = \frac{180x_{man} + 90x_{canoe}}{180 + 90}$$

constant, the canoe must be displaced 4 ft the other way and the man is in great danger of falling into the lake!

In terms of forces internal to the system, the man pushes off the boat, moving it one way, in order to get the boat to apply an equal and opposite reaction force moving him the other way. This is a safe maneuver in a heavy power boat, but risky in a light canoe.

In the following chapters, we shall often find the center of mass useful. It is important to realize that center of mass is a different concept from center of gravity. The center of gravity of an extended body is the point at which the resultant of gravitational forces on the body acts. The center of mass of the body is the point at which Newton's law acts in an especially simple form – a dynamical statement valid for any type of force. In a region of uniform gravitational acceleration, the center of mass $\bar{r}$ and the center of gravity r_{CG} coincide, because the gravitational force $m_i g$ acting on each particle is just proportional to the mass m_i in this case. But for a body sufficiently large that the nonuniformity of the gravitational acceleration must be taken into account, r_{CG} is shifted from $\bar{r}$ toward the side of the body where gravity is strongest. For example, the moon's center of gravity lies somewhat closer to the earth than the moon's center of mass, because the gravitational force $GM_E m_i / r^2$ on a particle of mass m_i is stronger nearer the earth.

11.4 THE LAW OF CONSERVATION OF MOMENTUM

Now that we have derived the consequence of Newton's second law for a system of n particles,

$$M\bar{a} = \frac{d}{dt}(m_1 v_1 + m_2 v_2 + \cdots + m_n v_n) = F_{ext}, \qquad (11.14)$$

let's use this result to deduce the law of conservation of momentum. Suppose that when viewed from afar, the center of mass of the system is observed not to be accelerating. We can conclude that no net external forces act on the system. This allows us to simplify Eq. (11.14) to

$$\frac{d}{dt}(m_1\mathbf{v}_1 + m_2\mathbf{v}_2 + m_3\mathbf{v}_3 + \cdots + m_n\mathbf{v}_n) = \mathbf{0}. \tag{11.15}$$

We know how to solve this equation: if the derivative of something is zero, that something must be constant. Thus the momentum of the system, identified as the sum of the individual momenta of each of the bodies, is equal to a constant:

$$\boxed{m_1\mathbf{v}_1 + m_2\mathbf{v}_2 + m_3\mathbf{v}_3 + \cdots + m_n\mathbf{v}_n = \mathbf{const.}} \tag{11.16}$$

Although the individual momenta of the bodies may change, the *total* momentum of the system remains constant. Since momentum is a vector quantity, Eq. (11.16) implies three equations when written out in components:

$$m_1v_{1x} + m_2v_{2x} + m_3v_{3x} + \cdots + m_nv_{nx} = \text{const}, \tag{11.17a}$$

$$m_1v_{1y} + m_2v_{2y} + m_3v_{3y} + \cdots + m_nv_{ny} = \text{const}, \tag{11.17b}$$

$$m_1v_{1z} + m_2v_{2z} + m_3v_{3z} + \cdots + m_nv_{nz} = \text{const}. \tag{11.17c}$$

From the way we obtained the law of conservation of momentum, it appears as if this law is limited in its applications because we assumed that no external forces acted on the system of particles. However, the scope of the law is much broader than our derivation suggests, for two reasons.

First, there are many cases in which an external force, usually gravitational or electric, acts on a system, but is negligibly weak compared to internal forces for the brief duration of an explosion or collision. Examples include the explosion of a rocket falling in the earth's gravitational field and the collision of a proton with a nucleus in the electrical environment of a solid target. In examples of this type, momentum is conserved to a very good approximation.

Second, if something applies a force to a system, then the system, no matter how intricate, exerts an equal but opposite force on that something. And if we include that external thing as part of a new, larger system, then the whole system conserves momentum. That's why the total momentum of the universe is constant – there's no larger system containing it.

If we concentrate on only one part of the universe, then there may be external forces acting on that system for some time. These forces are called *impulsive forces*. Their actions over time are related to momentum in much the same way that work is connected to energy; impulsive forces are a bookkeeping device which allows us to concentrate on only one part of the universe. In Section 11.8 we will discuss impulsive forces.

Example 2

A firecracker at rest explodes into three fragments. Two fragments, each having the same mass, fly off perpendicular to each other with the same speed $V = 8$ m/s. The third fragment has half the mass of either of the other fragments. Find the velocity $\mathbf{v}$ of the third fragment.

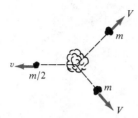

Before the explosion, the firecracker was at rest and had zero momentum. So, after the explosion, the momenta of the fragments must add up to zero. Let's introduce the coordinate system shown below in which the momenta of the fragments are indicated. (We've indicated that the unknown momentum of the third fragment must be in the opposite direction to the vector sum of the momenta of the other two fragments.)

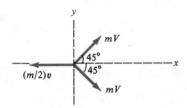

With m the mass of one of the identical fragments, conservation of momentum in the x direction implies that

$$0 = mV_x + mV_x - \tfrac{1}{2} mv_x,$$

or

$$0 = mV \cos 45° + mV \cos 45° - \tfrac{1}{2} mv_x.$$

Solving for v_x, we get

$$v_x = 4V \cos 45° = 4(8 \text{ m/s})(\tfrac{1}{2} \sqrt{2}) = 16 \sqrt{2} \text{ m/s}.$$

Conservation of momentum in the y direction implies that

$$0 = mV \sin 45° - mV \sin 45° + \tfrac{1}{2} mv_y,$$

which implies $v_y = 0$, as we've already indicated in the figure.

Example 3

A cannon of mass M fires a shell of mass m. Give a method of determining the speed of the shell by purely mechanical means.

This problem was solved by Benjamin Thompson (later Count Rumford), an American-born adventurer and scientist, while working for the British during the Revolutionary War. Thompson mounted the cannon horizontally as a pendulum. Before the explosion the momentum of the system is zero, because both the cannon and the shell are at rest. After the explosion, the motion is initially horizontal. The shell moves at speed v whereas the cannon recoils with a speed V in the opposite direction. The sum of the two momenta must remain zero. Taking the direction of the velocity of the shell as the positive direction, we obtain from conservation of momentum [Eq. (11.17a)]

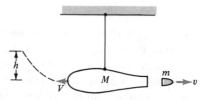

$$mv - MV = 0.$$

Solving for v, we find

$$v = (M/m)V.$$

Thompson determined V by letting the pendulum motion convert the initial kinetic energy of the cannon into potential energy,

$$\tfrac{1}{2}MV^2 = Mgh,$$

and measuring the height h. Substituting, one obtains the shell speed in terms of known quantities:

$$v = \frac{M}{m}\sqrt{2gh}.$$

[Note that the momentum of the system (cannon plus shell) is conserved only during the explosion; over the longer time span of the subsequent motion net external forces act on both the cannon and the shell.]

One can also use Thompson's method to find how much chemical energy is transformed into kinetic energy of the shell and cannon during the explosion. The amount is

$$K = \tfrac{1}{2}MV^2 + \tfrac{1}{2}mv^2.$$

In terms of measured quantities this is

$$K = \tfrac{1}{2}MV^2 + \tfrac{1}{2}\frac{M^2}{m}V^2 = Mgh\left(1 + \frac{M}{m}\right).$$

11.5 ROCKET PROPULSION

Among the most interesting applications of momentum conservation is rocket propulsion. The rocket ejects gas at high speed in one direction, and is accelerated by recoil in the opposite direction.

In normal operation gas is emitted continuously over a period of time, but to grasp the essential features, let's begin by considering a rocket that fires its exhaust matter intermittently, like bullets from a gun. Specifically, we consider a small time interval from time t to a later time $t + h$. We assume that some exhaust matter is expelled at time t and that no further matter is expelled during this small time interval.

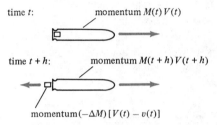

time t: momentum $M(t)V(t)$

time $t + h$:   momentum $M(t + h)V(t + h)$

momentum $(-\Delta M)[V(t) - v(t)]$

Figure 11.3 Momentum (measured relative to the earth) before and after a rocket expels matter of mass $-\Delta M = -[M(t + h) - M(t)]$.

At time t the rocket has mass $M(t)$ and velocity $V(t)$. During the small time interval the expelled matter has mass $M(t) - M(t + h)$ and is expelled backward with velocity $-v(t)$ relative to the rocket [or $V(t) - v(t)$ relative to the earth]; it maintains this velocity throughout the interval. The rocket recoils forward, changing its velocity to $V(t + h)$ at time $t + h$.

At time $t + h$ the system consists of two parts, a rocket with momentum equal to $M(t + h)V(t + h)$ and exhaust matter with momentum

$$[M(t) - M(t + h)][V(t) - v(t)].$$

The law of conservation of momentum states that the momentum of the new system is equal to that of the old, so

$$M(t)V(t) = M(t + h)V(t + h) + [M(t) - M(t + h)][V(t) - v(t)].$$

This can be rearranged to give

$$M(t + h)[V(t + h) - V(t)] = -v(t)[M(t + h) - M(t)]. \qquad (11.18)$$

Dividing by h and letting $h \to 0$ we find

$$M(t)\frac{dV}{dt} = -v(t)\frac{dM}{dt}. \qquad (11.19)$$

This is the basic differential equation satisfied by a rocket with variable mass. It is easy to understand physically. The left-hand term is the product of the instantaneous mass of the rocket and its acceleration. The right-hand term is the accelerating force on the rocket caused by the thrust developed by the rocket engine. To obtain a large thrust we want to expel the mass rapidly (make $-dM/dt$ large) at a large velocity [make $v(t)$ large].

Equation (11.19) can also be written as

$$\frac{dV}{dt} = -\frac{v(t)}{M(t)}\frac{dM}{dt} = -v(t)\frac{d}{dt}[\ln M(t)].$$

If the exhaust matter is expelled with constant velocity v, this equation can be integrated at once to give

$$V(t) - V(0) = -v[\ln M(t) - \ln M(0)] = v \ln \frac{M(0)}{M(t)}. \qquad (11.20)$$

For example, if we start from rest and expel half the mass, the rocket attains the speed $V(t) = v \ln 2$. In view of the slow rate of change of the logarithm, attempts to increase the rocket speed much beyond the exhaust speed v from a standing start achieve diminishing returns. The physical reason is that when $V > v$ the exhaust is emitted *forward* relative to the earth, so the rocket, while still gaining some speed, actually loses momentum in a reference frame fixed with respect to the earth.

Example 4

The Saturn V launch vehicle put the first man on the moon in 1969. The first stage of this vehicle had the following characteristics: total thrust 3.4×10^7 N, rate of fuel consumption 1.4×10^4 kg/s, duration of first stage burn 150 s, initial mass 2.8×10^6 kg. Estimate the speed of the vehicle at the end of the first stage, neglecting gravity and air resistance.

The thrust $-v\, dM/dt$ has magnitude 3.4×10^7 N and $dM/dt = -1.4 \times 10^4$ kg/s, so the velocity at which exhaust is expelled relative to the rocket is

$$v = \frac{\text{thrust}}{-dM/dt} = \frac{3.4 \times 10^7 \text{ N}}{1.4 \times 10^{11} \text{ kg/s}} = 2.4 \times 10^3 \text{ m/s}.$$

The mass of the vehicle after 150 s is

$$\begin{aligned} M(t) &= M(0) - t(1.4 \times 10^4 \text{ kg/s}) \\ &= 2.8 \times 10^6 - 150(1.4 \times 10^4) \\ &= 0.7 \times 10^6 \text{ kg}, \end{aligned}$$

only a quarter of the initial value. Therefore we estimate the velocity at the end of the first stage to be

$$V(t) = v \ln \frac{M(0)}{M(t)} = 2.4 \times 10^3 \ln 4 = 3.3 \times 10^3 \frac{\text{m}}{\text{s}}.$$

The actual velocity was only about 2.8×10^3 m/s, the difference being attributable to gravity and air resistance.

Since the first stage only achieves 2.8 km/s, whereas escape velocity is 11 km/s (Example 7 of Chapter 10), further stages are needed after the stage-one engines and empty fuel containers are ejected to lighten the load.

The motion of an actual rocket is complicated by the presence of external forces such as the earth's gravity. If external forces are present, then, instead of (11.19), the basic differential equation of motion is

$$M(t) \, \frac{dV}{dt} \, = \, -v(t) \, \frac{dM}{dt} \, + \, F(t) \qquad\qquad (11.21)$$

where $F(t)$ represents the sum of all external forces acting on the rocket at time t. We shall not discuss the solution of this equation here.

Unlike an airplane, a rocket needs no air to react against or supply lift. Air is simply a source of resistance for a rocket. But the related process of jet propulsion utilizes air, taking it in at the front of the missile or engine, employing it to oxidize the chemical fuel, and then expelling the exhaust out the rear. Thus a jet undergoes less mass change than a rocket. In the idealized case in which a jet-propelled missile undergoes no mass change, merely taking in air that is originally at rest and expelling the same mass of gas at velocity $V - v$, the missile speed is limited to $V \le v$. For if V exceeded v, the gas would gain forward momentum by being sucked through the jet engine; therefore the missile would *lose* momentum and velocity.

11.6   ENERGY AND MOMENTUM CONSERVATION IN COLLISIONS

In deriving the law of momentum conservation, we did not need to specify the nature of the forces acting between the bodies. These forces between the bodies could have been electrical, gravitational, or other, and it would make no difference whatsoever; momentum is conserved regardless of the nature of participating forces. The law of conservation of momentum, like that of conservation of energy, is a vast and powerful principle. It makes an overall statement about nature without fussing over the details of the force involved.

We now have two conservation laws which may be applied together as a potent aid to understanding certain kinds of problems. These are problems in which two objects initially moving in some direction briefly interact, as in a collision, and move off in different directions. The word *interact* in physics means that some force is applied between the objects. We may not know the details of the force, and we may not know whether the objects have touched or interacted at a distance without direct contact, but we can conclude that for a brief time the particles have exerted forces on each other from the fact that the motion of each object is changed. By using the laws of conservation of energy and momentum, we can predict much about the subsequent motion of particles in this type of encounter.

Although it might not appear that the world is simply a collection of bodies colliding with each other, this type of problem is important in modern physics. Twentieth-century physics is an exploration of quantum mechanics – the physics of atoms and elementary particles. What physicists want to know in quantum mechanics is the hidden, invisible internal structure of atoms, nuclei, and even protons and neutrons themselves. But there are very few ways to find what goes on inside a nucleus.

One way to study the nucleus is to accelerate it to a large momentum, smash it into another nucleus, and observe what happens. Such experiments are undertaken in colossal particle accelerators, such as the one at Fermilab, near Chicago, where protons collide with other protons. The debris of these collisions can be seen as tracks of tiny bubbles produced inside chambers of liquid hydrogen. Figure 11.4 is a photograph of tracks made in a bubble chamber. From such tracks, momenta can be measured and particles identified. Although this procedure lacks a certain delicacy, no one has found another kind of experiment to probe the inside of a nucleus. Someone once described it as similar to attempting to learn music by listening to a piano falling down a flight of stairs.

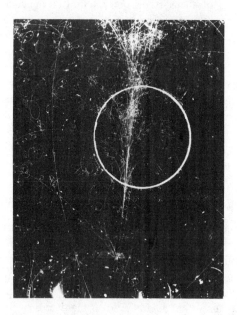

Figure 11.4 Particle tracks produced in a bubble chamber as a result of a high-energy collision. (California Institute of Technology.)

In preparation for applying the equations of conservation of energy and momentum, let's first write energy and momentum in terms which will allow us to compare them easily. We know that the momentum of a body is given by

$$\mathbf{p} = m\mathbf{v}, \tag{11.1}$$

and the kinetic energy is given by

$$K = \tfrac{1}{2}mv^2.$$

Recalling that $v^2 = \mathbf{v} \cdot \mathbf{v}$, we can use the result $\mathbf{v} = \mathbf{p}/m$ from Eq. (11.1) and substitute into the kinetic energy expression:

$$K = \tfrac{1}{2}m(p/m)^2,$$

$$K = \frac{p^2}{2m}. \tag{11.22}$$

This result clearly shows the connection between kinetic energy and momentum and allows us to simplify many equations.

Suppose we have two objects, of mass m_1 and m_2, on a horizontal plane. We consider the case in which one object, which we'll call the target, is at rest; the second object hits the target, as shown in Fig. 11.5. We'll study the result of this collision. Although we don't know anything about the nature of the forces that operate between the bodies when they're close together, we nonetheless will discover properties of the encounter by applying the laws of conservation of energy and momentum.

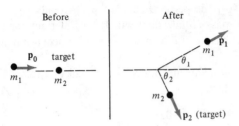

Figure 11.5 Collision of two objects.

Initially the target has zero momentum; the momentum of the other object we'll call $\mathbf{p}_0$. After the encounter, the target moves off with momentum $\mathbf{p}_2$ and the other body with momentum $\mathbf{p}_1$. What can we say about $\mathbf{p}_1$ and $\mathbf{p}_2$? Conservation of momentum implies that the total momentum of the system before the encounter is equal to the total momentum of the system after the encounter:

$$\mathbf{p}_0 = \mathbf{p}_1 + \mathbf{p}_2. \tag{11.23}$$

Since the motion takes place on a horizontal plane, the potential energy of the bodies in the earth's gravitational field doesn't change and we need not worry about it (in any case, it would not change appreciably during a sufficiently brief collision). Furthermore, if no energy is dissipated as heat in a collision, we say the collision is *elastic*. (If some kinetic energy is transformed into heat, we call the collision *inelastic*.) We'll assume that this particular collision is elastic. In an elastic collision, conservation of energy implies that the kinetic energy before the encounter is equal to the kinetic energy of the objects after the encounter. Using Eq. (11.22), we equate kinetic energies and get

$$\frac{p_0^2}{2m_1} = \frac{p_1^2}{2m_1} + \frac{p_2^2}{2m_2}. \tag{11.24}$$

Equations (11.23) and (11.24) contain all the information provided by momentum and energy conservation in elastic scattering. To avoid lengthy calculations, we confine ourselves here and in the problems to special cases which lead to algebraic simplifications.

THE CONSERVATION OF MOMENTUM

If the mass is the same for both objects, $m_1 = m_2 = m$, Eq. (11.24) simplifies to

$$\frac{p_0^2}{2m} = \frac{p_1^2}{2m} + \frac{p_2^2}{2m}.$$

Multiplying by $2m$ we get

$$p_0^2 = p_1^2 + p_2^2. \tag{11.25}$$

Equation (11.23), being a vector equation, represents three scalar equations which we could proceed to solve analytically. But before solving the problem in that way, let's look again at the vector equation. It says that the vector $\mathbf{p}_0$, whatever it is, is the sum of two vectors $\mathbf{p}_1$ and $\mathbf{p}_2$. In other words, the vectors form a triangle if both $\mathbf{p}_1$ and $\mathbf{p}_2$ are nonzero.

Also, Eq. (11.25) tells us that in the equal-mass case the square of the length of $\mathbf{p}_0$ is equal to the sum of the squares of the lengths of $\mathbf{p}_1$ and $\mathbf{p}_2$. The vector triangle obeys the Pythagorean theorem, so $\mathbf{p}_1$ is perpendicular to $\mathbf{p}_2$. Already we have a prediction: when two objects of equal mass collide they will fly off at right angles to each other. The final state in Fig. 11.5 has been drawn to show this special relationship occurring in equal-mass scattering.

We can also solve the problem algebraically, instead of geometrically. Let's take p_0^2 and work out the dot product:

$$p_0^2 = \mathbf{p}_0 \cdot \mathbf{p}_0 = (\mathbf{p}_1 + \mathbf{p}_2) \cdot (\mathbf{p}_1 + \mathbf{p}_2),$$

$$p_0^2 = p_1^2 + p_2^2 + 2\mathbf{p}_1 \cdot \mathbf{p}_2.$$

Comparing with Eq. (11.25) we conclude that

$$\mathbf{p}_1 \cdot \mathbf{p}_2 = 0.$$

There are two ways in which this last result can hold true. One solution is to say that the vectors are perpendicular to each other, the conclusion reached earlier. The other solution, however, wasn't analyzed when we solved the problem geometrically: one of $\mathbf{p}_1$ or $\mathbf{p}_2$ can be zero.

If $\mathbf{p}_2 = \mathbf{0}$, then the target remains fixed, indicating that the other object missed the target completely, an unexciting prospect. If, however, $\mathbf{p}_1 = \mathbf{0}$, then we have a head-on collision in which particle 2 (the target) flies off and particle 1 is left standing still. You can easily demonstrate this case by making two pennies collide head-on.

The equal-mass collision treated above is an important special case. It applies to billiard balls as well as to protons of speed much less than the speed of light.* Repeating the argument for the more general case of unequal-mass particles, one obtains similar but more complicated relations; the first object no longer comes to rest in a head-on collision, and the angle between the two objects is no longer 90° after a non-head-on collision.

For any masses, the two laws of conservation of momentum and energy working together give us a powerful means of analyzing collisions. These laws transcend the

---

*At velocities approaching the speed of light there are relativistic corrections which modify the kinematics presented here.

details of the forces involved; they are a bookkeeping device that tells us what is the same before and after the collision. The following examples illustrate this approach to understanding collisions.

Example 5

A glider of mass $M = 0.60$ kg has a piece of clay at one end. Initially it is traveling with a speed $V_0 = 0.20$ m/s along an air trough[†] when it collides with a glider of mass $m = 0.40$ kg which is moving at $v_0 = 0.10$ m/s. It is observed that the two gliders move off together at a speed V.

(a) Determine the speed V.
(b) How much kinetic energy is dissipated in the collision?

(a) Since the two gliders stick together, we expect that the collision is inelastic; some amount of kinetic energy must be dissipated in order to have the two gliders move off as one. Therefore we first apply conservation of momentum:

$$MV_0 + mv_0 = (M + m)V$$

where after the collision both gliders have the same speed V. Solving for V we find

$$V = (MV_0 + mv_0)/(M + m),$$

which numerically is 0.16 m/s.

(b) The energy dissipated is equal to the difference between the initial and final kinetic energy of the system:

$$\text{energy dissipated} = (\tfrac{1}{2}MV_0^2 + \tfrac{1}{2}mv_0^2) - \tfrac{1}{2}(M + m)V^2,$$

which turns out to be 1.2×10^{-3} J.

Example 6

A proton traveling with a speed $v_0 = 600$ m/s interacts with another proton initially at rest. From particle tracks, it is seen that one proton moves off at 60° from the initial direction. Assume the collision is elastic.

(a) What is the direction of the other proton after the interaction?

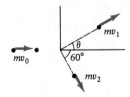

[†]An air trough is a nearly frictionless device used for demonstrations of linear momentum conservation. The glider slides on a cushion of air, created by blowing compressed air through small holes in the trough.

(b)   What are the speeds of both protons after the interaction?

(a)   Since we have an elastic collision between two identical particles, the discussion in the text tells us that the particles will move off at right angles to each other. Therefore the angle θ in the diagram above is equal to 30°.

(b)   To find the speeds of the protons, let's use the conservation of momentum equation in component form. The initial momentum in the x direction is mv_0, so we have

$$mv_0 = mv_1 \cos 30° + mv_2 \cos 60°$$

where v_1 and v_2 are our unknowns. In the y direction, the initial momentum is zero (by the way we chose our coordinate axes), so conservation of momentum for that component implies

$$0 = mv_1 \sin 30° - mv_2 \sin 60°.$$

From this last equation we can solve for v_1,

$$v_1 = v_2 \sin 60°/\sin 30° = \sqrt{3}v_2,$$

and substitute this value into our momentum equation for the x direction:

$$v_0 = \sqrt{3}v_2 \sin 60° + v_2 \cos 60°.$$

Solving for v_2, we find

$$v_2 = v_0/(\sqrt{3} \sin 60° + \cos 60°) = \tfrac{1}{2}v_0,$$

which tells us that $v_2 = 300$ m/s. Once we have this value, it is an easy matter to substitute back into the equation $v_1 = \sqrt{3}v_2$ and find that $v_1 = 520$ m/s.

Example 7

A compact car with mass $m_A = 1300$ kg, and a sports car with mass $m_B = 1000$ kg approach an intersection, each traveling at 14 m/s. They collide and move off together at an angle θ as indicated in the diagram. Find
(a)   the angle θ,
(b)   the speed of the entangled cars after the collision,
(c)   the amount of energy dissipated in the collision.

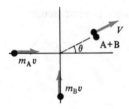

We know that the total momentum before the collision is equal to the total amount after the collision. Let's write this out in components. For the x direction, we have

$$m_A v = (m_A + m_B)V \cos\theta$$

where both cars have the same speed after the collision. Because we have two unknowns, V and θ, we need another equation before we can solve for them. In the y direction, conservation of momentum implies

$$m_B v = (m_A + m_B)V \sin\theta.$$

Dividing our second equation by the first, we can solve for θ:

$$\frac{m_B}{m_A} = \frac{\sin\theta}{\cos\theta} = \tan\theta,$$

$$\tan\theta = 1000/1300,$$

which leads to $\theta = 37.6°$. As a check, note that $\theta \to 0$ in the limit $m_B/m_A \to 0$, and $\theta \to \pi/2$ in the limit $m_A/m_B \to 0$, consistent with the expectation that if one object has nearly all the mass, the combined system will continue to move in the same direction as the initial heavy object after the collision.

To find the speed of the cars after impact, we can use either momentum equation and solve for V; the result is $V = 10$ m/s. The difference between the initial kinetic energy and the final kinetic energy is the energy dissipated:

$$\text{energy dissipated} = (\tfrac{1}{2}m_A v^2 + \tfrac{1}{2}m_B v^2) - \tfrac{1}{2}(m_A + m_B)V^2,$$

which turns out to be 1.1×10^5 J.

If momentum and energy conservation determine so much about collisions independently of details of the particles and the forces, why do physicists scatter protons to learn about their structure and the forces between them?

To find the answer, note that in two dimensions the final state of an elastic scattering such as the one shown in Fig. 11.5 can be described by four parameters, say, p_1, p_2, θ_1, and θ_2. But there are only three conservation laws in two dimensions: conservation of p_x, p_y, and energy. So momentum and energy conservation do not determine *everything* about the collision for a given initial momentum; one extra parameter – either an angle or a magnitude of momentum – must be specified to determine a unique collision. In Example 6, for instance, the final direction of one proton was specified. The property of the initial state that was left out of our description, thereby leaving one parameter of the final state unspecified, is the distance of nearest approach the moving particle would make with the target particle if it continued on a straight line (Fig. 11.6). This distance is measured between the center of mass of the two particles and is called the *impact parameter*. An expert billiard player is able to control the impact parameter and thereby determine the final direction of both billiard balls. But a nuclear physicist scattering subatomic particles such as protons is in the same position as a novice billiard player attempting a difficult shot; he cannot control the impact parameter. In fact it varies randomly from one collision to another, resulting in a distribution of final states.

The relation between final states and impact parameters depends on the nature of the force between the bodies. In the case of a contact force such as that between two billiard

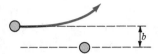

Figure 11.6 Impact parameter b of a collision.

balls, only impact parameters less than twice the radius of a ball result in scattering. On the other hand, two protons scatter slightly even at large impact parameters because of the long-range Coulomb repulsion. Careful study of the distribution of final angles and momenta yields information on the nature of the force and the structure of the scattering objects, and that is why physicists scatter protons and other small particles.

11.7 CENTER-OF-MASS COORDINATES

In a two-body collision, the kinematics is quite complicated if both of the incoming bodies are moving before the collision and both of the outgoing bodies are moving after the collision. Substantial simplifications result when one of the bodies is initially at rest, as we showed in detail in Section 11.6. Here we found for the equal-mass case that the initially moving body stops dead in its tracks for a head-on impact, or the bodies emerge at right angles to one another for an oblique impact.

The frame of reference in which the second body is at rest is called the *rest frame*. Often, as in a typical nuclear physics scattering experiment, the second body is at rest with respect to the laboratory, so the simplicities of the rest frame also apply in the *laboratory frame*.

Another frame of reference in which the kinematics simplifies is the *center-of-mass frame*, i.e., a coordinate system attached to the center of mass. If no external forces act on the system, the center of mass is not accelerated. So even though individual particles within the system may collide and be accelerated by internal forces, the center of mass travels with constant velocity $\bar{\mathbf{v}}$ relative to the laboratory frame S_L. Thus the center-of-mass frame S_C can be reached from S_L by a Galilean transformation of constant velocity $\bar{\mathbf{v}}$ and is itself an inertial frame. A body that had velocity $\mathbf{v}_i$ with respect to the laboratory will have velocity $\mathbf{v}'_i = \mathbf{v}_i - \bar{\mathbf{v}}$ with respect to the center of mass.

To see why the kinematics simplifies in the center-of-mass frame, let us begin by recalling the definition of the center-of-mass coordinate in an arbitrary frame:

$$\bar{\mathbf{r}} = \frac{1}{M} \sum m_i \mathbf{r}_i. \tag{11.8}$$

Taking the time derivative of both sides, we obtain

$$\frac{d\bar{\mathbf{r}}}{dt} = \bar{\mathbf{v}} = \frac{1}{M} \sum m_i \mathbf{v}_i. \tag{11.26}$$

But in the special case of the center-of-mass frame, $\bar{\mathbf{r}}$ is fixed so its time derivative $d\bar{\mathbf{r}}/dt$ vanishes on the left-hand side of Eq. (11.26). Recalling that we have labeled the particle velocities in this particular frame $\mathbf{v}'_i$, then, we can write (11.26) as

$$\sum m_i \mathbf{v}'_i = \mathbf{0} \qquad (11.27)$$

for this case. In other words, recognizing the sum in (11.27) as the total momentum $\mathbf{p}_{\text{tot}}$ of the system, we have

$$\mathbf{p}_{\text{tot}} = \mathbf{0} \qquad (11.28)$$

in the center-of-mass frame.

The first major simplification in a two-body collision occurring in the center-of-mass frame is that the individual momenta before impact are equal in magnitude and opposite in direction:

$$m_1 \mathbf{v}'_1 + m_2 \mathbf{v}'_2 = \mathbf{0}. \qquad (11.29)$$

After impact the direction of each particle's motion will have changed, and (in an inelastic collision) its speed may have changed. The final velocities require new labels; let's call them $\mathbf{u}_i$. But since $\mathbf{p}_{\text{tot}}$ is conserved, the individual momenta are still equal in magnitude and opposite in direction as shown in Fig. 11.7:

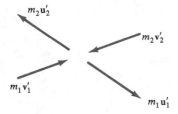

Figure 11.7 Kinematics of a two-body collision in the center-of-mass frame.

$$m_1 \mathbf{u}'_1 + m_2 \mathbf{u}'_2 = \mathbf{0}. \qquad (11.30)$$

To relate the final momenta to the initial momenta we need one more condition, usually obtained from energy considerations. If the collision is elastic, we can equate the initial and final kinetic energies in the center-of-mass frame:

$$K' = \frac{m_1 v_1'^2}{2} + \frac{m_2 v_2'^2}{2} = \frac{m_1 u_1'^2}{2} + \frac{m_2 u_2'^2}{2}$$

$$= \frac{p_{1\,\text{in}}^2}{2m_1} + \frac{p_{2\,\text{in}}^2}{2m_2} = \frac{p_{1\,\text{out}}^2}{2m_1} + \frac{p_{2\,\text{out}}^2}{2m_2}.$$

Taking momentum conservation into account, $\mathbf{p}_{1\,\text{in}} = -\mathbf{p}_{2\,\text{in}}$ (Eq. 11.29) and $\mathbf{p}_{1\,\text{out}} = -\mathbf{p}_{2\,\text{out}}$ (Eq. 11.30), we find

$$K' = \frac{p_{1\,\text{in}}^2}{2}\left(\frac{1}{m_1} + \frac{1}{m_2}\right) = \frac{p_{1\,\text{out}}^2}{2}\left(\frac{1}{m_1} + \frac{1}{m_2}\right).$$

We conclude that the magnitude of the momentum and thus the speed of each particle in the center-of-mass frame is unchanged by the collision:

$$p_{1 \text{ in}} = p_{1 \text{ out}}, \qquad v_1' = u_1', \tag{11.31a}$$

$$p_{2 \text{ in}} = p_{2 \text{ out}}, \qquad v_2' = u_2'. \tag{11.31b}$$

The only effect of an elastic two-body collision in the center-of-mass frame is to change the direction of the velocities. This is another advantage of working in center-of-mass coordinates.

Example 8

A particle of mass m_1 and initial velocity $\mathbf{v}_0 = v_0 \hat{\mathbf{i}}$ scatters elastically off a particle of mass m_2 that is initially at rest. The particles emerge at angles θ_1 and θ_2. At what θ_2 does the target particle have maximum transverse velocity v_{2y}?

The answer is rather surprising, but is easily obtained by considering the center of mass frame. By Eq. (11.26) the center of mass moves with velocity $\bar{\mathbf{v}} = \mathbf{v}_0 m_1/(m_1 + m_2)$ relative to the laboratory. In the center of mass frame, the incoming particles have velocities $\mathbf{v}_0 - \bar{\mathbf{v}}$ and $-\bar{\mathbf{v}}$, respectively. They emerge back-to-back with speeds unchanged. Because the target particle has speed $\bar{v}$ in this frame no matter what angle it is scattered into, its transverse velocity reaches a maximum value $\bar{v}$ when $\theta_{\text{cm}} = 90°$.

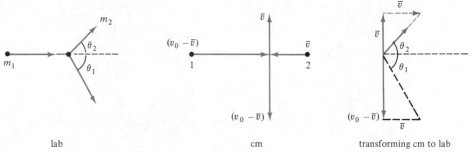

lab   cm   transforming cm to lab

To transfer the velocity back to the laboratory frame one adds $\bar{\mathbf{v}}$ which, being along the initial direction, leaves the transverse velocity $v_{2y} = \bar{v}$ unchanged but makes the angle of maximal v_{2y} in the laboratory $\theta_2 = 45°$.

Note that the corresponding θ_1 is

$$\theta_1 = \tan^{-1}[(v_0 - \bar{v})/\bar{v}] = \tan^{-1}(m_2/m_1),$$

which equals $45°$ only if the masses are equal. This is an example of the general statement that the particles emerge at right angles to one another only in the equal mass case.

The velocities of scattering particles are usually given in the laboratory frame whereas analysis is simpler in the center-of-mass frame, so it is helpful to relate the laboratory and center-of-mass energies. The kinetic energy of a two-body system is

$$K = \tfrac{1}{2}m_1 v_1^2 + \tfrac{1}{2}m_2 v_2^2$$

in the laboratory frame and

$$K' = \tfrac{1}{2}m_1 v_1'^2 + \tfrac{1}{2}m_2 v_2'^2$$

in the center-of-mass frame. To relate K to K' consider their difference

$$K - K' = \tfrac{1}{2}m_1(v_1^2 - v_1'^2) + \tfrac{1}{2}m_2(v_2^2 - v_2'^2).$$

Using $\mathbf{v}_1 = \mathbf{v}_1' + \bar{\mathbf{v}}$ and $\bar{v}^2 = \bar{\mathbf{v}} \cdot \bar{\mathbf{v}}$ we find

$$v_1^2 - v_1'^2 = (\mathbf{v}_1' + \bar{\mathbf{v}}) \cdot (\mathbf{v}_1' + \bar{\mathbf{v}}) - \mathbf{v}_1' \cdot \mathbf{v}_1' = 2\mathbf{v}_1' \cdot \bar{\mathbf{v}} + \bar{v}^2.$$

Similarly,

$$v_2^2 - v_2'^2 = 2\mathbf{v}_2' \cdot \bar{\mathbf{v}} + \bar{v}^2,$$

so the formula for $K - K'$ becomes

$$K - K' = m_1\mathbf{v}_1' \cdot \bar{\mathbf{v}} + m_2\mathbf{v}_2' \cdot \bar{\mathbf{v}} + \tfrac{1}{2}(m_1 + m_2)\bar{v}^2$$

$$= (m_1\mathbf{v}_1' + m_2\mathbf{v}_2') \cdot \bar{\mathbf{v}} + \tfrac{1}{2}M\bar{v}^2,$$

where $M = m_1 + m_2$. But $m_1\mathbf{v}_1' + m_2\mathbf{v}_2' = \mathbf{0}$ by Eq. (11.29), and the relation between K and K' can be written

$$K = K' + \tfrac{1}{2}M\bar{v}^2. \tag{11.32}$$

This result is strikingly simple: the kinetic energy as seen in the laboratory frame equals the kinetic energy in the center-of-mass frame *plus* the kinetic energy associated with a single particle containing the total mass M and moving with velocity $\bar{\mathbf{v}}$.

Equation (11.32) is especially useful for interpreting inelastic collisions. In an inelastic collision, some or all of the kinetic energy K' as seen in the center of mass frame may be converted into other forms of energy. For the same collision viewed in the laboratory, not all of the kinetic energy is available for conversion because some of it is tied up in the quantity $\tfrac{1}{2}M\bar{v}^2 = \tfrac{1}{2}\bar{p}^2/M$ associated with motion of the center of mass, which is conserved if total mass is constant. Therefore if one wishes to find how much kinetic energy is available for conversion into other forms, it is convenient to go to center of mass coordinates.

Example 9

A hydrogen atom of mass m_H and initial velocity v_0 in the laboratory collides with an electron that has mass m_e and is initially at rest. What fraction of the initial laboratory kinetic energy is available to increase the internal energy of the atom by an amount ΔE?

In the center-of-mass frame, all of the energy is available. A conversion

$$K' = \Delta E$$

is possible. The kinetic energy in the laboratory frame is initially $\tfrac{1}{2}m_H v_0^2$ and is related to K' by Eq. (11.32):

$$\tfrac{1}{2}m_H v_0^2 = K' + \tfrac{1}{2}(m_H + m_e)\bar{v}^2.$$

According to Eq. (11.26), the center-of-mass velocity is

$$\overline{v} = \frac{m_{\mathrm{H}} v_0}{m_{\mathrm{H}} + m_{\mathrm{e}}} \, .$$

Eliminating $\overline{v}$ from the previous two equations, we find

$$\frac{m_{\mathrm{H}} v_0^2}{2} = K' + \frac{m_{\mathrm{H}}^2 v_0^2}{2(m_{\mathrm{H}} + m_{\mathrm{e}})} \, ,$$

which, by rearrangement, becomes

$$K' = \frac{m_{\mathrm{H}} v_0^2}{2} \left(1 - \frac{m_{\mathrm{H}}}{m_{\mathrm{H}} + m_{\mathrm{e}}} \right)$$

$$= \frac{m_{\mathrm{H}} m_{\mathrm{e}} v_0^2}{2(m_{\mathrm{H}} + m_{\mathrm{e}})} \, .$$

Thus only a fraction

$$\frac{K'}{K} = \frac{m_{\mathrm{e}}}{m_{\mathrm{H}} + m_{\mathrm{e}}} \approx \frac{1}{1837}$$

of the laboratory-frame kinetic energy is available for conversion. Almost all of the initial kinetic energy is tied up in center-of-mass motion; a light electron cannot slow down a heavy proton very much.

If it is the hydrogen atom that is initially at rest, and the electron that has the initial momentum, the center of mass moves much more slowly than the electron, being dominated by the heavier proton. In this case a repeat of the above argument shows that $K'/K = m_{\mathrm{H}}/(m_{\mathrm{H}} + m_{\mathrm{e}})$; almost all of the laboratory-frame energy is available for conversion.

11.8 IMPULSE: COLLISION FORCES AND TIMES

Whenever we have a group of objects on which no net external forces act, we can apply the law of conservation of momentum to analyze the motion. Application of this law permits us to derive conditions on the motion of objects comprising the system without knowing the details of the forces involved.

Let's now examine what happens to individual objects participating in a collision, by focusing not on the entire system, but on a single object. When we narrow our vision down to only one body, that body *does* have a net external force acting on it, called an *impulsive force*. Since collisions normally are very short in duration, lasting about 10^{-6} to 10^{-1} s for macroscopic objects, the force acts on the object for a short time. Can we characterize the behavior of this fleeting force quantitatively?

By Newton's third law, when two objects collide, they experience equal but oppositely directed impulsive forces. But often, one of the objects has additional forces acting on it. For example, when a car crashes into a rail, the car only experiences an impulsive force, whereas the rail experiences not only an impulsive force of the same magnitude but forces from supports buried in the ground. Through the bookkeeping device of impulsive forces, though, we can focus on the motion of only one of the bodies, usually the object whose motion is more interesting.

We know that the force acting on an object during a collision changes the momentum of that object according to Newton's second law,

$$\mathbf{F} = \frac{d\mathbf{p}}{dt} .$$ (6.1)

If both sides are integrated over the time interval from t_1 to t_2, we obtain

$$\int_{t_1}^{t_2} \mathbf{F} \, dt = \int_{t_1}^{t_2} \frac{d\mathbf{p}}{dt} \, dt = \mathbf{p}(t_2) - \mathbf{p}(t_1),$$

by the second fundamental theorem of calculus. In other words we have shown that

$$\int_{t_1}^{t_2} \mathbf{F} \, dt = m\mathbf{v}_2 - m\mathbf{v}_1 .$$ (11.33)

To understand the significance of Eq. (11.33), consider the integral

$$\mathbf{I} = \int_{t_1}^{t_2} \mathbf{F} \, dt.$$ (11.34)

If $\mathbf{F}$ were constant, this integral would be the impulsive force multiplied by the time $\Delta t = t_2 - t_1$ during which it acts. It can be the same for a small force acting over a long time as for a large force applied briefly. It is a measure of the effect of the force acting over time and is called the *impulse* of $\mathbf{F}$. Equation (11.33) states that *impulse equals change in momentum*.

If $\mathbf{F}$ acts during a time interval Δt but is *variable*, one would have to employ the precise time dependence, which is usually not known, to calculate the impulse integral. But even in the absence of precise knowledge this integral provides physical insight. If we define the *average* force $\overline{\mathbf{F}}$ by the equation

$$\overline{\mathbf{F}} = \frac{1}{\Delta t} \int_{t_1}^{t_2} \mathbf{F} \, dt,$$ (11.35)

then Eq. (11.34) becomes

$$\mathbf{I} = \overline{\mathbf{F}} \, \Delta t,$$ (11.36)

whereas (11.33) can be written as

$$\overline{\mathbf{F}} \, \Delta t = \Delta \mathbf{p}.$$ (11.37)

This simple result, the connection between the average force acting on an object, its change of momentum, and the duration, provides us with additional insight into collisions.

The connection between change of momentum, average force, and time is illustrated by numerous examples in everyday life. A tennis player, for example, hits the ball with a great force to impart momentum to the ball on a serve. To impart the maximum momentum, the player "follows through" with the serve. This action extends the time of contact between the ball and racquet, and, according to Eq. (11.37), produces a greater change in momentum. In order to achieve a maximum change in momentum for an object, you should apply as large a force as possible over as long a time interval as possible.

On the other hand, the time of contact may be very small, yet the force very large. A golf ball crashing into a window is in contact with the window for a short time before coming to rest, yet a large force acts for that short time, causing considerable damage. The idea of large forces and short contact times also explains why a karate expert can break a concrete brick with a chop of the hand. The karate expert bears down on the brick with great speed, aiming for the hand to stop not on the surface of the brick, but somewhere inside. The large change in momentum of the hand creates, by the third law, a large change in momentum of the brick. And if the time of contact is short, the corresponding force is large enough to crack the brick. To maximize the force exerted on an object during a collision, you need, according to Eq. (11.37), to create the maximum change in momentum for the object in as short a time as possible.

Sometimes, though, we wish to change the momentum of an object with as small a force as possible. For example, when you jump from a table down to the floor, you instinctively bend your knees upon contact with the floor. This simple act extends the time of collision between you and the floor, thereby reducing the force exerted on you by the floor. Similarly, boxers learn how to "roll with the punch" by moving the head along with the punch. This movement extends the time of contact with the opponent's glove, reduces the force of impact needed to stop the punch, and usually avoids a broken jaw. The boxer's gloves themselves, along with car bumpers, are equipment built to spread collisions out over longer times, reducing the maximum force and deceleration.

11.9 A FINAL WORD

We've now mastered two of the three great conservation laws of mechanics. In a world in which everything seems to change, energy and momentum are the same now as they have always been and always will be. There's something strange, however, about those laws: Isaac Newton, the man who for the most part gave birth to mechanics, never thought of the conservation laws, even though they are direct consequences of $F = ma$. It is almost as if he purposely left something for others to discover.

There is another small mystery in what we've learned. Newton based his mechanics on three laws, yet we use only one law, $F = ma$, to describe how the world works. Newton may have had an inkling about the nature of conservation laws. That's what his third law is really about.

Take conservation of energy, for example. Energy is never created or destroyed, but it does flow from one place to another; that is what we've defined as work – a force acting through a distance. If a force acts on one body, an equal but opposite force acts on another body; that's the third law. So whenever work flows into one body, it always flows out of another. That is precisely the conservation of energy.

In addition, the rate of change of a body's momentum is equal to the force applied to it. But by the third law, another body has a reaction force acting on it. Therefore, as momentum flows into one body, the same amount must flow out of another. That's why momentum is conserved.

Newton's second law is a profound statement about how the world works. But as all-encompassing as it is, this law may not be quite so profound as the conservation laws it spawned.

Problems

The Center of Mass

1. Three masses $m_1 = 5$ g, $m_2 = 3$ g, and $m_3 = 2$ g are located at $5\hat{\mathbf{i}} + 7\hat{\mathbf{j}}$, $-2\hat{\mathbf{i}} + 6\hat{\mathbf{j}}$, and $-4\hat{\mathbf{i}} + 3\hat{\mathbf{j}}$, respectively. Locate the center of mass.

2. A projectile explodes while in flight. Fragments are blown in all directions as shown in the sketch. What can you say about the motion of the center of mass of the system after the explosion?

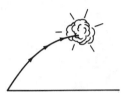

3. Suppose in a nightmare you find yourself locked in a light cage on rollers on the edge of a rapidly eroding cliff. Assuming that no external forces act on the system consisting of you and the cage, what could you do to move the cage away from the edge? What must you avoid doing? If you weigh 140 lb and the cage weighs 210 lb and is 10 ft long, how far can you move the cage?

Conservation of Momentum

4. Any good saloon brawler knows that to avoid getting your jaw broken with a punch, keep your mouth closed (literally and figuratively). How does conservation of momentum shed light on this advice?

5. A kernel of unpopped popcorn when heated explodes into a popped popcorn which shoots off in some direction. Must anything else be emitted in the "explosion"? Why? If something is emitted, what is it? If its mass is 10^{-3} that of the popcorn, at approximately what speed is it emitted?

6. Two particles move in a region free of external forces. At one instant they have momenta described by

$$\mathbf{p}_1 = (3\hat{\mathbf{i}} - 2\hat{\mathbf{j}} + \hat{\mathbf{k}}) \text{ kg m/s},$$

$$\mathbf{p}_2 = (-\hat{\mathbf{i}} + 5\hat{\mathbf{j}}) \text{ kg m/s}.$$

Some time later the momentum of one particle is

$$\mathbf{p}_1' = (8\hat{\mathbf{i}} - 6\hat{\mathbf{j}} - 5\hat{\mathbf{k}}) \text{ kg m/s}.$$

What is the momentum of the other particle?

7. Benjamin Thompson (Count Rumford) also used a method for determining the speed of a bullet or shell when it reaches the target. The bullet is fired horizontally into a block of wood mounted as a pendulum. The bullet stops in the wood, and the subsequent swing of the pendulum is measured. If the bullet has mass m and initial velocity v and the block has mass M,

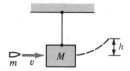

   (a) what is the horizontal velocity V of the pendulum just after impact?
   (b) Show that kinetic energy is not conserved during the impact, and use this result to find the heat generated during the impact.
   (c) Find v in terms of m, M, and the height h of the pendulum swing.

Motion with Changing Mass

8. A 20-kg child is riding on a 10-kg sled which is moving horizontally at 5 m/s. The child pushes a 2-kg block off the sled with a velocity of 1 m/s relative to the ground, in the direction opposite to the sled's motion. By how much does the sled's speed change?

9. A compressed massless spring is mounted on a toy train wagon, which is on a horizontal frictionless track. Initially everything is at rest. When the spring is released it fires a marble of mass m horizontally, and the train of mass M recoils in the opposite direction. If all the potential energy $\frac{1}{2}kx^2$ of the spring is converted into kinetic energy, calculate

   (a) the speed of the wagon and of the marble,
   (b) the kinetic energy of each of these bodies.

10. An empty box car of mass M and initial velocity V rolls under a coal hopper, which dumps coal of mass $2M$ vertically into the box car.

   (a) What is the velocity of the box car after receiving the coal?
   (b) Things look different in the frame of reference moving at velocity V with respect to the earth. Account for the initial and final momentum of the coal and the box car in this frame.

11. Consider once more the initially empty box car of Problem 10. Suppose now that the coal is fed continuously into the box car, falling vertically from above at the rate dM/dt. Using momentum arguments

(a)   write a differential equation relating the mass and velocity of the box car;

(b)   after eliminating time from the differential equation, solve it for V as a function of M.

12. Suppose that instead of keeping the exhaust speed v constant relative to the rocket as in Eq. (11.20), v is kept equal to the rocket velocity V. Solve Eq. (11.19) for this case [take $V(0) \neq 0$]. In particular, find the increase in V if 90% of the mass is ejected.

Energy and Momentum Conservation in Collisions

13. A pool player is about to make the shot indicated in the sketch. Is there any chance that the cue ball will end up in a pocket as well? Explain.

14. A 0.03-kg mass traveling at 0.08 m/s collides head-on with a 0.05-kg mass which is initially at rest. If the collision is elastic, find the speed of each mass after the collision.

15. A 3.0-kg object collides elastically with another object initially at rest and continues to move in the original direction at one-half of its initial speed. What is the mass of the second object?

16. An electron collides elastically with a hydrogen atom initially at rest. All motion occurs along a straight line. What fraction of the electron's initial energy is transferred to the hydrogen atom? (Take the mass of a hydrogen atom to be 1840 times that of an electron.)

17. A 5.0-kg wooden block is moving along a frictionless horizontal surface with a speed of 2 m/s when it collides elastically with a 1.0-kg ball on the end of a string as shown.

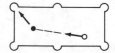

(a)   Find the velocity of each object immediately after the collision.

(b)   Determine how high the ball will rise.

18. A billiard ball moving at 2.5 m/s strikes an identical ball initially at rest and moves off with speed 1.5 m/s at an angle of 53° from its original direction. Find the speed and direction of the other ball.

19. A certain nucleus, at rest, decays into three particles, two of which are detected as shown in the figure. If $m_1 = 15 \times 10^{-27}$ kg, $m_2 = 8.0 \times 10^{-27}$ kg, $v_1 = 5.0 \times 10^6$ m/s, and $v_2 = 6.0 \times 10^6$ m/s,

   (a) find the momentum of the third particle of mass 10×10^{-27} kg,
   (b) calculate the energy involved in the decay.

20. Two objects of equal mass and initial speed collide inelastically and stick together. If they are observed to move off at one-third their initial speed, find the angle between the initial velocities of the objects. What fraction of the initial energy goes into heat?

21. A particle of mass m_1 and initial momentum p_0 collides elastically with a particle of mass m_2 that is initially at rest. If the second particle recoils at angle θ_2, find the magnitude of its momentum p_2 in terms of p_0, θ_2, m_1, and m_2.

22. Consider the collision of two protons, one initially at rest, in three dimensions. How many parameters are needed to specify the final state? How many parameters are determined by energy and momentum conservation? What properties of the initial state specify the remaining parameters?

Center-of-Mass Coordinates

23. An object moving with speed v in the laboratory collides with one of equal mass initially at rest. In the center of mass frame, half the initial kinetic energy is lost to heat on impact. What is the speed of each object in the center of mass frame after impact? What is the fractional kinetic energy loss in the laboratory?

24. A 10-kg dog is sitting in a 40-kg canoe when he notices a tasty lunch sitting unattended in another boat, which drifts past at 2m/s. The dog runs along his canoe until he keeps pace with the lunch, then steps over to the second boat. How fast is the lunch boat now moving relative to the canoe?

25. In a nuclear fission reactor, neutrons emitted at high speed in the fission process must be slowed down by collisions with inert nuclei such as ^{12}C so that they may induce further fission events.
   A fast neutron of initial velocity $v_0\hat{\mathbf{i}}$ collides elastically with a stationary ^{12}C nucleus.

   (a) What is the initial speed of each particle in the center-of-mass frame?
   (b) If the ^{12}C nucleus scatters into an angle θ in the center of mass, show that its final velocity in the laboratory frame is one-thirteenth of the vector

$$v_0(1 - \cos\theta)\hat{\mathbf{i}} + v_0 \sin\theta\hat{\mathbf{j}}.$$

(c) In such an elastic collision, what is the maximum fraction of its laboratory kinetic energy that the neutron can lose?

(d) If the average energy lost in such a collision is one-half of the maximum possible loss, what is the average number of collisions a neutron must undergo in order to reduce its kinetic energy from 1,000,000 eV to 1000 eV?

26. Three perfectly resilient balls of masses $M_1 >> M_2 >> M_3$ are held nearly in mutual contact with their centers in a vertical line, the largest being at the bottom and the smallest at the top. They are released simultaneously and fall together through a height H, whereupon the lowermost ball strikes the floor and rebounds perfectly elastically. Each of the remaining balls, in turn, collides perfectly elastically with the one beneath it. Ignoring the finite radii and small initial separations of the balls, show that the second and third balls should rebound to the heights $H_2 = 9H$ and $H_3 = 49H$. (*Hint*: It is easiest to analyze each collision in its center of mass, which is to a good approximation the rest frame of the heavier of the two balls participating in that collision.)

Collision Forces and Times

27. If a 1-kg lump of Jell-O and a 1-kg block of lead were dropped from three feet above your head, which would hurt more? Why? Give a crude estimate of Δt and $\overline{F}$ for the lump of Jell-O.

28. A 700-kg car traveling at 15 m/s (i.e., about 33 mph) runs into a wall and stops in 0.1 s. What average force is exerted on the car during this time? Compare this to the force of gravity on the car.

CHAPTER 12

OSCILLATORY MOTION

Another question concerns the oscillations of pendulums, and it falls into two parts. One is whether all oscillations, large, medium, and small, are truly and precisely made in equal times. The other concerns the ratio of times for bodies hung from unequal threads; the times of their vibrations, I mean. . . . As to the prior question, whether the same pendulum makes all its oscillations – the largest, the average, and the smallest – in truly and exactly equal times, I submit myself to that which I once heard from our Academician [Galileo]. He demonstrated that the moveable which falls along chords subtended by every arc [of a given circle] necessarily passes over them all in equal times. . . ,

As to the ratio of times of oscillations of bodies hanging from strings of different lengths, those times are as the square roots of the string lengths; or should we say that the lengths are as the doubled ratios, or squares, of the times.

Galileo Galilei, *Two New Sciences* (1638)

12.1 FINDING A CLOCK THAT WOULDN'T GET SEASICK

Navigation has provided one of the most persistent motives for measuring time accurately. All navigators depend on continuous time information in order to find out where they are and to chart their course. But until about two centuries ago, no one was able to make a clock that could keep time accurately at sea.

Early travelers noticed that the North Star, unlike other stars, does not change its position with respect to the earth; it appears to be suspended in the northern sky. The farther northward they traveled, the higher in the sky the North Star appeared; at the North Pole it would be directly overhead. By measuring the elevation of the North Star above the horizon with a sextant, a navigator can determine the distance from the North Pole and the latitude, as Fig. 12.1a illustrates.

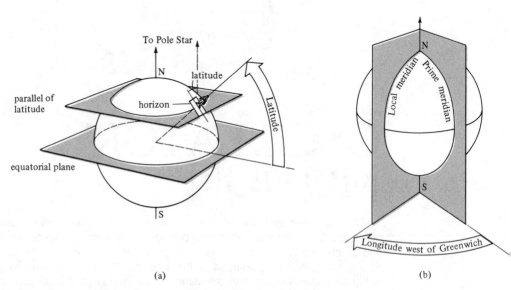

(a)   (b)

Figure 12.1 (a) Latitude determined by position of the North Star. (b) Longitude requires knowing a reference time.

On account of the earth's rotation, however, charting a course east or west presented a more complex problem. Only by knowing the time very accurately could a navigator calculate longitude. Because of the earth's rotation, the sun appears to travel across the sky from east to west at the rate of 1° in 4 min. (It takes the sun 24 h to move once around the earth, so it must move through 360° in 24 h, which is 360/24, or 15° in 1 h, or 1° in 4 min.) Now if a navigator determines the local time from the position of the sun and has a clock that very accurately tells the time in Greenwich, England (through which, by international agreement, the zero longitude line runs), he can easily determine his longitude. For every 4 min his clock (showing Greenwich time) differs from the local time, he is 1° of longitude away from Greenwich. Even at night, a navigator can determine the longitude by using star charts to determine the local time and by comparing that time to Greenwich time. What is essential is an accurate timepiece.

The earliest timekeeping devices, built by the ancient Egyptians, consisted of an alabaster bowl, wide at the top and narrow on the bottom, which had horizontal markings on the inside to tell the time. As water dripped out through a hole in the bottom of the bowl, successive lines were exposed. For centuries the basic design of the water clock remained unchanged; Galileo used such a water clock in his fertile experiments with balls rolling down inclined planes.

Sometime between the eighth and eleventh centuries, Chinese artisans constructed a water clock that had the characteristics of a mechanical clock. Falling water powered a wheel that contained small cups around its rim. As a cup filled with water, it became heavy enough to trip a lever, which allowed the next cup to move into place and advance the wheel by a step. In thirteenth-century Europe, variations of the Chinese water clock became popular. But aside from the fact that these clocks did not keep very good time, they also tended to freeze in the European winters.

Sand clocks (hourglasses) introduced in the fourteenth century avoided the freezing problem, but because of the weight of sand, they were limited to measuring short intervals of time. One of the chief uses of the hourglass was to determine a ship's position by *dead reckoning*. Sailors would throw a log overboard with a long rope attached to it, and then count knots, which were tied in the rope at equal intervals, as the rope played out for a specific amount of time. In this way sailors could crudely estimate the speed, or knots, at which the ship was moving. By knowing their speed and how long they had traveled in a certain direction, they could track their position.

The first truly mechanical clocks were built in the fourteenth century and consisted of pulleys and weights with escapements, similar to present-day cuckoo clocks. The accuracy of these early mechanical clocks depended on the friction between parts, the driving weights, and the skill of the craftsman constructing it. No two clocks would show the same time, let alone keep accurate time. What was needed was some sort of periodic, repeating device whose frequency was essentially a property of the device itself.

12.2 SIMPLE HARMONIC MOTION

The event that led to accurate timepieces was the analysis of *periodic* or *harmonic motion* – motion that repeats itself in equal intervals of time. When an object moves back and forth over the same path in harmonic motion, we say that it is *oscillating*. We now explore this type of motion.

In Section 10.5, we discussed the stability of an object subject to a linear restoring force

$$F = -kx,$$

which acts on it whenever it moves away from the equilibrium point $x = 0$. This force, for example, describes how a spring behaves (Section 8.3). Associated with this force is the potential energy

$$U = \tfrac{1}{2} kx^2.$$

We used this potential energy as a model to study stability and discovered that it typified many things: a marble in the bottom of a bowl, a mass at the end of a spring, a pendulum, a guitar string, and even a complex political or ecological system. All these systems have the property that if they are disturbed from equilibrium, the restoring force that acts on them tends to move them back into equilibrium.

These systems have another property in common: when disturbed from equilibrium, they tend to overshoot that point when they return, on account of inertia. Then the restoring force acts in the opposite direction, trying to return the system to equilibrium. The result is that the system winds up oscillating back and forth, like a marble in a bowl, a mass on a spring, and a guitar string. Figure 12.2 shows successive "snapshots" of a mass oscillating on the end of a spring.

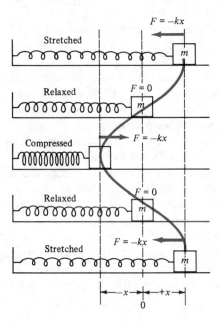

Figure 12.2 Snapshots of a mass oscillating on the end of a spring.

The horizontal displacement of the mass, plotted as a function of time, traces out a path resembling a sine or cosine curve, as suggested by Figure 12.2.

To determine this curve mathematically, we apply Newton's second law, $F = ma$, with $F = -kx$, the spring force, and $a = d^2x/dt^2$, the second derivative of the displacement $x = x(t)$, considered as a function of time t. Newton's law becomes

$$m \frac{d^2x}{dt^2} = -kx$$

or, dividing by m,

$$\frac{d^2x}{dt^2} = -\frac{k}{m} x. \tag{12.1}$$

This is a differential equation satisfied by the displacement $x(t)$.

We have already seen a differential equation of this type in connection with our study of centripetal acceleration in Chapter 5. In Section 5.8 we showed that a particle moving in uniform circular motion with position vector

$$\mathbf{r} = x\hat{\mathbf{i}} + y\hat{\mathbf{j}},$$

where

$$x = r \cos \omega t \tag{12.2}$$

and

$$y = r \sin \omega t \tag{12.3}$$

with r and ω constant, has an acceleration vector

$$\mathbf{a}(t) = \frac{d^2 x}{dt^2} \,\hat{\mathbf{i}} + \frac{d^2 y}{dt^2} \,\hat{\mathbf{j}},$$

which satisfies the relation

$$\mathbf{a}(t) = -\omega^2 \mathbf{r}. \tag{5.64}$$

Taking horizontal and vertical components of (5.64) we see that the cosine function $x(t)$ in (12.2) and the sine function $y(t)$ in (12.3) separately satisfy the same type of differential equation:

$$\frac{d^2 x}{dt^2} = -\omega^2 x \quad \text{and} \quad \frac{d^2 y}{dt^2} = -\omega^2 y.$$

Moreover, each of these is of type (12.1), with the constant ω^2 replacing k/m.

A differential equation of the form

$$f''(t) = -\omega_0^2 f(t), \tag{12.4}$$

where $f(t)$ is a function of t and ω_0 is a nonzero constant, is called the differential equation of *simple harmonic motion*. We have just seen that the horizontal and vertical components of uniform circular motion undergo simple harmonic motion, as does the displacement of a mass oscillating on the end of a spring. Figure 12.3 illustrates how uniform circular motion can be related to an oscillating mass-and-spring system.

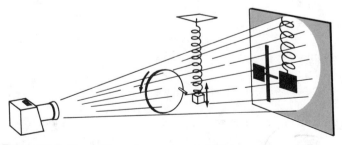

Figure 12.3 The shadow of a peg on an object executing uniform circular motion and an oscillating mass-and-spring system both exhibit simple harmonic motion.

In Section 12.3 we will learn how to find all functions $f(t)$ satisfying the differential equation of simple harmonic motion (12.4), but first we describe another physical problem which leads to the same type of differential equation.

Example 1

A U-shaped tube contains a liquid of density ρ. The total length of the column of liquid is L and the (uniform) cross-sectional area of the inside of the tube is A. At time $t = 0$ the liquid level on the left side is a height $x(0)$ above its equilibrium level (indicated by the dotted line in the figure), and the liquid level on the right is below equilibrium by $x(0)$. Show that the height $x(t)$ executes simple harmonic motion.

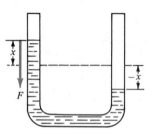

When x is not zero, there is a restoring force due to the weight of the displaced liquid. The mass of the displaced liquid (the mass excess on the left side relative to the right) is $m_d = \rho A(2x)$, so

$$F = -gm_d = -2g\rho Ax.$$

The sign is negative because this is a restoring force, acting in the opposite direction from the displacement x. The force has the form $-kx$ with $k = 2g\rho A$, so the water sloshes back and forth like a mass on a spring even though the system looks nothing like a spring!

In Newton's second law, $F = ma$, the mass on the right-hand side is the total mass of the liquid ρAL. Thus the equation of motion is

$$-2g\rho Ax = \rho AL \frac{d^2x}{dt^2},$$

which simplifies to

$$\frac{d^2x}{dt^2} = -\frac{2g}{L}x.$$

This is the differential equation (12.4) of simple harmonic motion with

$$\omega_0^2 = \frac{2g}{L}.$$

Note that the motion does not depend on the density ρ or the cross-sectional area A of the tube.

12.3 THE GENERAL SOLUTION OF THE DIFFERENTIAL EQUATION OF SIMPLE HARMONIC MOTION

Since many different physical problems lead to the differential equation of simple harmonic motion it is important to know what all the solutions look like. In courses on differential equations it is shown that if $\omega_0 \neq 0$ all solutions of the differential equation

$$f''(t) = -\omega_0^2 f(t) \tag{12.4}$$

must necessarily have the form

$$\boxed{f(t) = A \cos \omega_0 t + B \sin \omega_0 t,} \tag{12.5}$$

where A and B are constants.

The right-hand side of (12.5) is called a *linear combination* of $\cos \omega_0 t$ and $\sin \omega_0 t$ with arbitrary constant multipliers A and B, and is said to be the *general solution* of (12.4).

Two arbitrary constants A and B appear in the general solution because the differential equation is of second order. Somewhere in the process of solving a second-order differential equation two integrations are required, each producing an arbitrary constant of integration. In physical problems where $f(t)$ represents position, the constants A and B are determined by the initial position $f(0)$ and the initial velocity $f'(0)$. In fact, if we put $t = 0$ in (12.4) and (12.5) we find

$$A = f(0), \qquad B = f'(0)/\omega_0. \tag{12.6}$$

Therefore the general solution in (12.5) can be expressed as follows:

$$f(t) = f(0) \cos \omega_0 t + \frac{f'(0)}{\omega_0} \sin \omega_0 t. \tag{12.7}$$

By rearranging the constants in the general solution (12.5) we can express the result in terms of the cosine (or sine) alone. For example, if we introduce new constants C and θ_0, where

$$C = \sqrt{A^2 + B^2} \qquad \text{and} \qquad \theta_0 = \tan^{-1}\left(\frac{-B}{A}\right),$$

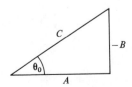

then we have (see accompanying triangle)

$$A = C \cos \theta_0 \quad \text{and} \quad -B = C \sin \theta_0,$$

and Eq. (12.5) becomes

$$f(t) = C \cos \theta_0 \cos \omega_0 t - C \sin \theta_0 \sin \omega_0 t.$$

This can be written more compactly as

$$f(t) = C \cos(\omega_0 t + \theta_0). \tag{12.8}$$

In other words, the general solution $A \cos \omega_0 t + B \sin \omega_0 t$ is merely a disguised form of a cosine curve. It can also be expressed as a sine curve (see Problem 8).

Example 2

A 0.5-kg mass is attached to a spring with spring constant $k = 112$ N/m and set into motion. If a clock is started when the oscillator has a displacement $x(0) = 0.3$ m and a velocity $v(0) = 6.0$ m/s, what is the displacement of the oscillator at time t?

According to Eq. (12.5) the displacement is given by

$$x(t) = A \cos \omega_0 t + B \sin \omega_0 t$$

where $\omega_0 = \sqrt{k/m} = \sqrt{224} = 15$ rad/s. By Eq. (12.6) we have

$$A = x(0) = 0.3 \text{ m} \quad \text{and} \quad B = x'(0)/\omega_0 = 0.4 \text{ m}.$$

Therefore the displacement of the oscillator at any instant is

$$x(t) = (0.3 \text{ m}) \cos(15\ t) + (0.4 \text{ m}) \sin(15t),$$

where t is in seconds.

When the answer is put in the alternative form (12.8), we have $C = \sqrt{A^2 + B^2} = 0.5$ m, $\theta_0 = \tan^{-1}(-B/A) = -53° = -0.9$ rad, and $x(t) = 0.5$ m $\cos(15t - 0.9)$.

When the solution is written in the form (12.8) the constants C and θ_0 have a simple geometric interpretation, illustrated in Figure 12.4. The extreme values of $f(t)$, which

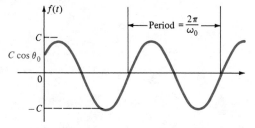

Figure 12.4 Amplitude and period of simple harmonic motion.

occur when $\cos(\omega_0 t + \theta_0) = \pm 1$, are $\pm C$. When $t = 0$ the initial displacement is $f(0)$ $= C \cos \theta_0$. As t increases, the position $f(t)$ oscillates between the maximum value $+C$ and the minimum $-C$. The number C is called the *amplitude* of the motion. The angle $\omega_0 t + \theta_0$ is called the *phase angle* and θ_0 itself is called the initial value of the phase angle.

The time T required for one complete cycle is called the *period*. Since the phase angle moves through 2π radians in one cycle we have

$$T = 2\pi/\omega_0.$$

The reciprocal of the period is called the *frequency f*:

$$f = 1/T = \omega_0/(2\pi).$$

This is the number of cycles an object in simple harmonic motion completes in one second. In SI units frequency is measured in *hertz*, abbreviated Hz; 1 Hz $=$ 1 cycle/s. The number ω_0 is called the *angular frequency*; it describes the change in phase angle (measured in radians) in unit time. (Sometimes for brevity we call ω_0 the frequency rather than the angular frequency. This should not cause any confusion.)

Any object which executes simple harmonic motion is called a *simple harmonic oscillator*, abbreviated SHO. The word *harmonic* comes from music. Musical instruments generally vibrate harmonically, and from the time of the Pythagoreans in the sixth century B.C. the study of musical harmony was an important stimulus to the understanding of oscillation. The term *simple refers to the linearity of the restoring force*, to the absence of external forces such as friction or viscosity, and to the presence of only a single frequency (whereas, in general, periodic motion is a composite with more than one frequency acting simultaneously). The world is full of simple harmonic oscillators. As diverse in appearance as springs, vibrating atoms, and electrical circuits, they all have in common the linear restoring force. In other words it is the differential equation of motion, not size, shape, substance, or appearance, that describes a SHO. Once we understand completely the SHO, we can put it in our pocket and explore the world of problems it solves.

12.4 EXAMPLES OF SIMPLE HARMONIC OSCILLATORS

To interpret the foregoing results physically we consider first the mass oscillating on the end of a spring subject to Hooke's law $F = -kx$, as described in Section 12.2. There we found that the displacement x satisfies the differential equation

$$\frac{d^2x}{dt^2} = -\frac{k}{m} x, \qquad (12.1)$$

so in this case we have a SHO with

$$\omega_0 = \sqrt{\frac{k}{m}}. \qquad (12.9)$$

If the initial displacement is A, and the initial velocity is 0, the solution of (12.1) is

$$x(t) = A \cos \omega_0 t. \tag{12.10}$$

The angular frequency ω_0 depends on the physical characteristics of the system, namely, the spring constant k and the mass m. The stiffer the spring, the larger the value of the spring constant, and, by Eq. (12.9), the larger the number of oscillations in one second. In other words, stiffer springs make the system oscillate more rapidly. This makes sense because a stiffer spring exerts a greater force and tends to accelerate the mass more. Equation (12.9) also tells us that the greater the mass, the slower the oscillations. We expect larger values of m to produce slower oscillations because of inertia. Since the angular frequency depends only on the characteristics of a particular mass and spring, we refer to ω_0 as the *natural angular frequency*; it is the angular frequency at which the system will naturally oscillate. The frequency of the oscillations is independent of the amplitude A.

Once we have $x(t)$ for a particular SHO, we can find everything there is to know about the oscillator. For example, to find the velocity of the mass we differentiate (12.10) with the result

$$\frac{dx}{dt} = -A\omega_0 \sin \omega_0 t. \tag{12.11}$$

We could differentiate once more to find the acceleration, or simply use (12.1) to obtain

$$\frac{d^2x}{dt^2} = -\omega_0^2 x = -A\omega_0^2 \cos \omega_0 t. \tag{12.12}$$

Example 3

A particle of mass 0.25 kg attached to a spring undergoes simple harmonic motion with an amplitude of 0.15 m and a frequency of 100 Hz. Determine

  (a) its angular frequency,
  (b) the spring constant,
  (c) its maximum velocity,
  (d) its maximum acceleration.

  (a) Using $\omega_0 = 2\pi f$, we have $\omega_0 = 630$ rad/s.
  (b) Since $\omega_0^2 = k/m$, we have

$$k = m\omega_0^2 = (0.25 \text{ kg})(630 \text{ rad/s})^2 = 9.9 \times 10^4 \text{ N/m}.$$

(To end up with units of N/m, we drop the radians because radians are dimensionless.)
  (c) From Eq. (12.11),

$$v(t) = -A\omega_0 \sin \omega_0 t, \tag{12.11'}$$

we see that the maximum velocity occurs when the sine is -1,

$$v_{max} = A\omega_0 = (630 \text{ rad/s})(0.15 \text{ m}) = 94 \text{ m/s}.$$

  (d) Arguing as in (c), we find that Eq. (12.12) implies that the maximum acceleration is

$$a_{max} = A\omega_0^2 = (630 \text{ rad/s})^2(0.15 \text{ m}) = 6.0 \times 10^4 \text{ m/s}^2.$$

Example 4

A spring of spring constant k hangs unstretched. A mass m is attached to the free end and released. Show that even though the mass is acted on by gravity as well as the spring force, it executes simple harmonic motion about its equilibrium point, and that the frequency is independent of g.

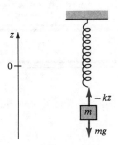

Let's call the initial height of the unstretched spring $z = 0$. When the mass is attached, the equilibrium point is where $-kz = mg$, that is, at $z_0 = -mg/k$. The force on the mass at an arbitrary height z can be written

$$F_z = -kz - mg = -k(z - z_0).$$

In other words, the mass feels a linear restoring force about the equilibrium point $z = z_0$. Newton's law $F = ma$ gives the differential equation

$$m \frac{d^2z}{dt^2} = -k(z - z_0),$$

which can also be written as

$$\frac{d^2(z - z_0)}{dt^2} = -\frac{k}{m}(z - z_0).$$

This tells us that $z - z_0$ executes simple harmonic motion,

$$z - z_0 = A \cos(\omega_0 t + \theta_0)$$

with angular frequency $\omega_0 = \sqrt{k/m}$, a number independent of g. Of course, g does affect the location of the equilibrium point since $z_0 = -mg/k$.

The displacement z itself is given by

$$z = z_0 + A \cos(\omega_0 t + \theta_0)$$

for some constants A and θ_0 which we will now determine from the given initial conditions. The velocity at time t is *(handwritten: ① if t=0 then $\frac{dz}{dt}=0$ ② ③ so $\theta_0=0$)*

$$\frac{dz}{dt} = -A\omega_0 \sin(\omega_0 t + \theta_0).$$

Since the mass is at rest at time $t = 0$ the velocity is 0 at $t = 0$ and we find $\theta_0 = 0$, so the formula for z becomes

$$z = z_0 + A \cos \omega_0 t.$$

But when $t = 0$ we also have $z = 0$. Hence $A = -z_0$ and we get

$$z = z_0 - z_0 \cos \omega_0 t.$$

The largest value of z is $z = 0$ (when $\cos \omega_0 t = 1$) and the smallest is $z = 2z_0 = -2mg/k$ (when $\cos \omega_0 t = -1$). This makes sense physically because the stiffer the spring, the less it will stretch, whereas increasing the mass stretches the spring farther.

Example 5

Determine the displacement $x(t)$ and the period of oscillation of the liquid in Example 1.

In Example 1 we showed that the displacement $x(t)$ satisfies the differential equation of simple harmonic motion with angular frequency $\omega_0 = \sqrt{2g/L}$. Therefore the period is

$$T = \frac{2\pi}{\omega_0} = 2\pi \sqrt{\frac{L}{2g}}.$$

The displacement is given by

$$x(t) = A \cos(\omega_0 t + \theta_0)$$

for some constants A and θ_0. If the liquid is at rest when $t = 0$ we find (as in Example 4) that $\theta_0 = 0$. Also, $x(0) = A$, so

$$x(t) = x(0) \cos \omega_0 t.$$

12.5 ENERGY CONSERVATION AND SIMPLE HARMONIC MOTION

In Chapter 10, we argued that an object near a potential-energy minimum would oscillate because potential energy would turn into kinetic energy, which in order to conserve energy would turn back into potential energy. That was how we explained why the object doesn't stop at the equilibrium point where there's no force. Now we've obtained the same result from a solution of Newton's second law without mentioning energy at all. To connect the two approaches, let's examine the energy of a simple harmonic oscillator.

Specifically, let's consider a mass oscillating at the end of a spring as described in Section 12.4. The displacement is given by

$$x(t) = A \cos \omega_0 t \tag{12.10}$$

with $\omega_0 = \sqrt{k/m}$. We already know that the potential energy is given by

$$U(x) = \tfrac{1}{2} kx^2. \tag{10.42}$$

Using Eq. (12.10), we can express the potential energy as a function of time:

$$U(t) = \tfrac{1}{2} kA^2 \cos^2 \omega_0 t.$$

The kinetic energy of the system is $\tfrac{1}{2}mv^2$ where the velocity $v = dx/dt$ is given by

$$\frac{dx}{dt} = -A\omega_0 \sin \omega_0 t, \tag{12.11}$$

so the kinetic energy as a function of time is

$$K(t) = \tfrac{1}{2} mv^2 = \tfrac{1}{2} m\omega_0^2 A^2 \sin^2 \omega_0 t.$$

Therefore the total energy of the system is

$$E = K + U,$$

$$E = \tfrac{1}{2} m\omega_0^2 A^2 \sin^2 \omega_0 t + \tfrac{1}{2}kA^2 \cos^2 \omega_0 t.$$

At first glance it appears as if the total energy is not constant because it seems to depend on t. To show that the total energy is indeed constant, we recall that $\omega_0^2 = k/m$, and substitute this into the expression for E:

$$E = \tfrac{1}{2} m(k/m)A^2 \sin^2 \omega_0 t + \tfrac{1}{2} kA^2 \cos^2 \omega_0 t$$

$$= \tfrac{1}{2} kA^2 (\sin^2 \omega_0 t + \cos^2 \omega_0 t).$$

The term in parentheses is equal to one, so we have the result

$$E = \tfrac{1}{2} kA^2, \tag{12.13}$$

which is a constant. Figure 12.5 shows graphs of $U(t)$, $K(t)$, and the total energy E.

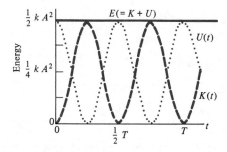

Figure 12.5 Graphs of potential, kinetic, and total energy for a simple harmonic oscillator.

Example 6

In Example 4, we attached a mass m to an initially unstretched spring with spring constant k and let it fall. We showed that from its starting point at $z = 0$ the mass executes simple harmonic motion about the equilibrium point at $z_0 = -mg/k$. Consider the same problem from the point of view of energy, and find the maximum kinetic energy.

The energy relationships can be seen especially clearly in an energy diagram. The potential energy of the mass [making the convention $U(0) = 0$] is

$$U = \tfrac{1}{2} kz^2 + mgz.$$

We can rewrite this potential as

$$U(z) = \frac{k}{2}\left(z^2 + \frac{2mgz}{k}\right)$$

$$= \frac{k}{2}\left(z + \frac{mg}{k}\right)^2 - \frac{(mg)^2}{2k}$$

$$= \frac{k}{2}(z - z_0)^2 - \frac{(mg)^2}{2k},$$

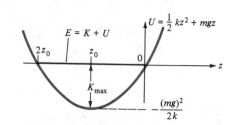

which is clearly a harmonic oscillator potential about $z = z_0$, plus a constant. The constant occurs because we set $U(z) = 0$ at $z = 0$ rather than at its minimum. It is evident from the energy diagram that $K_{max} = \tfrac{1}{2}(mg)^2/k$, and that the mass terminates its fall and turns around at $z = 2z_0$. The mathematical conditions which are satisfied at the turning point are $U(2z_0) = U(0)$, $K = 0$.

Our result for K_{max} can be checked directly from $K_{max} = \tfrac{1}{2}mv_{max}^2$ by using $v_{max} = A\omega_0 = A\sqrt{k/m}$ from Eq. (12.11) and recalling that the amplitude in Example 4 was $A = mg/k$. Putting all this together, we obtain

$$K_{max} = \tfrac{1}{2} mv_{max}^2 = \tfrac{1}{2} kA^2 = \tfrac{1}{2} k \left(\frac{mg}{k}\right)^2 = \frac{(mg)^2}{2k}.$$

When the treatment is made more realistic by adding friction, the ordered motion of the mass slowly loses energy and the mass settles down to the equilibrium point as indicated in the modified energy diagram below.

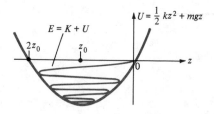

Equation (12.13) represents the initial amount of energy put into the oscillator by an outside agent performing work, as when you stretch a spring or wind up a watch. Because of friction, that energy is gradually transformed to heat; the energy of a real harmonic oscillator decreases with time and the amplitude of successive oscillations diminishes, rather than remaining constant as in Fig. 12.5. But the frequency of the oscillations remains constant; only the amplitude of successive oscillations decreases.

By understanding general principles and working out specific examples, we gain not only additional insights into the way things work, but often obtain technological improvements as well. Robert Hooke, after whom the spring force law is named, understood the essential feature of spring oscillations – that even in the presence of friction, the frequency remains constant. In the 1650s he experimented with the idea of using a metal spring to regulate the frequency of a clock. However, the first spring-controlled clock was built by Christian Huygens, a Dutch physicist. His idea was to use a spiral spring – the type still used today in mechanical watches.

In 1713 the British government offered a prize of £20,000 to anyone who could build a clock accurate enough to enable a seafarer to determine longitude to within one-half of a degree (about 35 mi). Among the many dexterous craftsmen who sought to win the ample award was an English clockmaker, John Harrison. For 40 years, he struggled to construct a spring-driven clock that could cope with rolling seas, temperature-induced expansion and contraction, and corrosive salt spray. Finally in 1761 he sent his son on a voyage to Jamaica to test his clock, but only after the government forced him to build an identical model lest the original be lost at sea. His masterpiece was a technological triumph – it allowed the navigator to determine longitude to within one-third of a degree.

12.6 THE SIMPLE PENDULUM

Because the angular frequency ω_0 does not depend on the amplitude of the motion of a harmonic oscillator, the time for each complete cycle, the period T, also does not depend on the amplitude; if the oscillator makes large oscillations, it moves rapidly, and if it makes smaller oscillations it moves more slowly. Even in the real world, where friction makes oscillations die down, the oscillator always takes the *same* amount of time for each cycle. As we have seen, the significance of this characteristic is that a simple harmonic oscillator can be used as a timing device. The discovery of this fact led immediately to the invention of the first accurate clocks. Even today, wristwatches which are accurate to within a few seconds per month use as timekeeping devices a kind of harmonic oscillator – a quartz crystal.

But the earliest clocks used a different oscillator – the pendulum. Galileo made the crucial discovery that a pendulum takes the same time per swing, even as its motion dies down, and he thereby laid the groundwork for improved timekeeping. In his *Two New Sciences* he eloquently summarized his observations.

Folklore places Galileo's discovery as having occurred in the Duomo, or Cathedral, in Pisa. The famous Leaning Tower of Pisa is actually the bell tower of the magnificent, high-ceiling cathedral. Hanging from the ceiling on a long cable is a lamp, which one day Galileo supposedly noticed swinging back and forth, probably just after it had been lit. Timing the swings by comparing them to his own pulse, Galileo realized that they always took the same time even as the swings became smaller and smaller.

This famous lamp, called Galileo's lamp, still hangs in the Cathedral at Pisa. However, there's one thing wrong with the tale: the church's records show that the lamp was installed in the 1650s, ten years after Galileo's death.

Setting fables aside, let's analyze the motion of a simple pendulum to find out precisely what factors determine the period. The idea is to use Newton's second law to find a differential equation which describes the motion of a pendulum and cast it into the form of Eq. (12.4),

$$\frac{d^2x}{dt^2} = -\omega_0^2 x. \tag{12.4}$$

In other words, we need to find an equation which says that the second derivative of the displacement is proportional to the negative of the displacement. Then, by comparison with Eq. (12.4), we'll know that the quantity multiplying $-x$ on the right-hand side of the equation is the square of the frequency with which a pendulum swings. Or if we prefer, we can describe the motion in terms of the angle $\theta(t)$ through which the pendulum swings, and $\theta(t)$ will play a role analogous to $x(t)$ in Eq. (12.4).

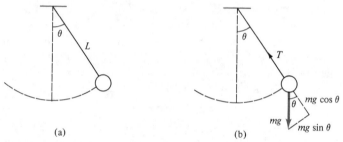

(a)   (b)

Figure 12.6 (a) The simple pendulum. (b) Force diagram for pendulum.

Let's consider the simple pendulum, like the one shown in Fig. 12.6a. We call it simple because we are idealizing the pendulum as a point mass of mass m at the end of a massless string of length L. An analysis based on Newton's second law, however, is not limited to this case; it's only easier. Suppose we start the pendulum in motion by pulling it aside and releasing it. When the pendulum moves through an angle θ from the vertical, it goes through a distance θL along a circular arc from its equilibrium position. The force responsible for restoring the pendulum to its equilibrium position (hanging straight down) is its weight mg. We can resolve this force into components parallel and perpendicular to the string, as illustrated in Fig. 12.6b. The perpendicular component, the one which is always tangent to the circular arc, causes the pendulum to accelerate back to its equilibrium position; from Fig. 12.6b we see that this component is $-mg \sin \theta$. The acceleration along this path is the second derivative of the displacement along the circular arc:

$$\frac{d^2}{dt^2} (L\theta) = L \frac{d^2\theta}{dt^2}.$$

Therefore along the arc Newton's second law implies

$$mL \frac{d^2\theta}{dt^2} = -mg \sin \theta.$$

Canceling the mass and dividing through by L, we get

$$\frac{d^2\theta}{dt^2} = -\frac{g}{L}\sin\theta.$$

This last result is not the equation for a simple harmonic oscillator; the second derivative of the displacement (here θ) is not proportional to $-\theta$ but to $-\sin\theta$. Furthermore, it can be shown that no elementary function will satisfy this differential equation.

The intrepid physicist, however, is not daunted by such minor obstacles. If physicists only worked on problems they knew how to solve exactly, they would accomplish very little. The essence of practical physics is to ignore what is unimportant and to approximate. As Table 12.1 indicates, when expressed in radians, θ is approximately equal to $\sin\theta$; the smaller the angle, the closer the agreement. Even for an angle of $\pi/4$ radians ($45°$) the difference between θ and $\sin\theta$ is only about 10%. Therefore, as long as we only consider small oscillations, we can safely replace $\sin\theta$ by the linear approximation θ in our equation:

$$\frac{d^2\theta}{dt^2} = -\frac{g}{L}\theta. \tag{12.14}$$

Now this equation is exactly like the simple harmonic oscillator equation; the variable now is θ, but that doesn't matter. We know that the solution is

$$\theta(t) = \theta_0 \cos\omega_0 t, \tag{12.15}$$

where θ_0 is the amplitude determined from the initial conditions. (This θ_0 should not be confused with the initial value of the phase angle introduced in Sec. 12.3.) This solution is reasonably good if the amplitude of the swings is small.

By comparing Eq. (12.14) with the simple harmonic oscillator equation, we see that the frequency of the oscillations is

$$\omega_0 = \sqrt{g/L}, \tag{12.16}$$

Table 12.1 Comparison of θ with $\sin\theta$ for Small Angles.

θ (deg)	θ (rad)	$\sin\theta$
0	0	0
0.5730	0.0100	0.0100
5.730	0.1000	0.0998
11.459	0.2000	0.1987
17.189	0.3000	0.2955
45.000	$\pi/4 = 0.7854$	0.7071

which means that the period is

$$T = 2\pi/\omega_0 = 2\pi\sqrt{L/g} \ .$$

Consequently, on any given planet, the frequency of a simple pendulum depends only on its length. Unlike a mass on a spring, for which the frequency $\sqrt{k/m}$ does depend on the mass, the frequency of a pendulum is independent of its mass. [When one removes the idealization of a point mass, as we shall do in Section 14.8, the equation for the frequency is different, but still independent of the mass.]

The reason the natural frequency is independent of the mass is exactly the same reason that the acceleration of a falling body on the surface of the earth doesn't depend on its mass: through Newton's second law, $F = ma$, and the universal law of gravity, $F = GmM_e/R_e^2$, the mass m cancels. The ingenious Isaac Newton used pendulums of different masses to test this cancelation with a precision of one part in a thousand. He realized that since pendulums of identical length but different mass have equal frequencies this proves exactly the same law as dropping a penny and a feather in a vacuum, but that the pendulum experiment works without being in a vacuum and is easier to observe.

Example 7

A pendulum has an amplitude of 20° and a length of 2.0 m. Find
    (a)   its natural frequency,
    (b)   its maximum velocity.
Using Eq. (12.15), we find

$$\omega_0 = \sqrt{g/L} = \sqrt{(9.8 \text{ m/s}^2)/(2.0 \text{ m})} = 2.2 \text{ rad/s} = 126°/\text{s}.$$

Differentiating Eq. (12.15), we obtain

$$\frac{d\theta}{dt} = -\theta_0\omega_0 \sin \omega_0 t.$$

The velocity of the pendulum along the circular arc is given by $v = L\, d\theta/dt$, so the maximum speed is

$$v_{\max} = \theta_0\omega_0 L = (20°)(\pi \text{ rad}/180°)(2.2 \text{ rad/s})(2.0 \text{ m}),$$

which turns out to be 1.5 m/s. Note that one must express θ_0 and ω_0 in radians, not degrees, to get the right answer.

12.7 GAINING INSIGHT THROUGH APPROXIMATIONS

We started out to study harmonic oscillations, motion executed by various things like pendulums and guitar strings. Understanding such periodic motion was crucial to the development of accurate timepieces. In our analysis we had to ignore air resistance and friction – idealizations we have often made. Yet for the pendulum, we approximated even further when we found out that it is not quite a harmonic oscillator, because its motion is along a circular arc rather than a straight line. Is physics imprecise?

Many people believe that physicists seek the most fundamental and precise equations that govern the behavior of the universe. But in fact, physicists don't have completely universal equations at their disposal. Newton's laws aren't such principles; his laws don't accurately describe objects as small as atoms or as large as galaxies. Although we understand atoms (quantum theory) and galaxies (general theory of relativity), we don't have one fundamental set of laws that explains both at the same time. Many physicists, though, search for such a law and believe that it soon will be within their grasp.

Suppose, however, that we already knew the fundamental laws that govern the universe. What would we do then? The obvious approach would be to write down those equations and find all the solutions. This would be excruciatingly difficult, since the laws presumably would be expressed as differential equations. But in principle it would seem it could be done; a differential equation can be solved numerically by a sufficiently powerful computer even if it is impossible to express the solution analytically by formulas. So, if we found all the solutions, we would unlock all the secrets of the universe. Or would we?

Solving all the equations of the universe numerically is something we would *not* want to do even if we could. The reason is very simple. The computer printout would be as complicated as the universe itself – and we already have the universe! What we want from physics is not the precise numerical results that describe exactly how everything behaves. Instead, what we seek is something much more subtle. We want understanding, insight and at best a kind of trained and dependable intuition about why things work the way they do.

By studying the differential equation of the simple harmonic oscillator we gained an understanding of how some things work, even though we don't know of any physical system that precisely satisfies this equation. Yet, as we look at the world around us, mentally armed with this equation and its solutions, we begin to see everywhere examples of things that we know have this equation buried somewhere deep in their behavior. Our understanding of how things work has been inexpressibly enriched once we grasp the idea of extracting from complicated phenomena simple, underlying elements. Harmonic motion is often one of those elements. The road to insight often does not go through meticulous, complete, and precise description. It usually starts out in quite a different direction, passing first through crude but clever estimations and approximations.

12.8 DAMPED OSCILLATIONS

In the real world, simple harmonic motion does not continue forever. The swings of a pendulum become smaller and smaller; a spring vibrates with steadily decreasing amplitude. This damping of the amplitude is caused by frictional and resistive forces, acting both internally within the oscillating system and externally in the surrounding medium (air or liquid) which retards the motion.

To study the damped motion we must consider Newton's second law with two force terms: the spring force and the damping force. We shall model the damping force after Stokes's law, in which resistance is assumed to be proportional to velocity. This is often valid for oscillations of sufficiently small amplitude. The damping force opposes the motion and thus can be written as

$$F_\text{d} = -\gamma \frac{dx}{dt} \tag{12.17}$$

where γ is a constant called the *damping coefficient*. Adding the damping force to the spring force, we find that Newton's second law takes the form

$$m\frac{d^2x}{dt^2} = -kx - \gamma \frac{dx}{dt}. \tag{12.18}$$

After rearranging terms and dividing through by m, this can be written as

$$\frac{d^2x}{dt^2} + \beta \frac{dx}{dt} + \omega_0^2 x = 0, \tag{12.19}$$

where $\beta = \gamma/m$ and, as before, $\omega_0^2 = k/m$.

If there is a spring force but no damping [second term zero in (12.19)], the general solution to (12.19) can be written $C \cos(\omega_0 t + \theta_0)$, as we found in Section 12.3. If there is damping but no spring force [third term zero in (12.19)], Eq. (12.19) again takes a simple form and the general solution turns out to be $x = Ce^{-\beta t} + D$ with C and D constant. It can readily be verified that this is a solution when $\omega_0 = 0$ by substituting into (12.19). If *both* the spring force and damping force are present, the solution of Eq. (12.19) is more complicated. As is shown in courses on differential equations, the general solution (for β not too large compared to ω_0) now has the form

$$x = Ce^{-bt} \cos(\omega_1 t + \theta_0). \tag{12.20}$$

where b and ω_1 are positive constants. This result, oscillations whose amplitude decreases with time, is plausible on physical grounds, and combines the oscillatory behavior and exponential decrease of the special cases described above. By substituting (12.20) into the differential equation (12.19) we will show that the equation is satisfied when b and ω_1 take on certain values depending on β and ω_0.

Taking the necessary derivatives of (12.20), we find

$$\frac{dx}{dt} = Ce^{-bt} [-b \cos(\omega_1 t + \theta_0) - \omega_1 \sin(\omega_1 t + \theta_0)],$$

$$\frac{d^2x}{dt^2} = Ce^{-bt}[(b^2 - \omega_1^2) \cos(\omega_1 t + \theta_0) + 2b\omega_1 \sin(\omega_1 t + \theta_0)].$$

Putting these results, as well as Eq. (12.20) for x, into Eq. (12.19) we find, after canceling Ce^{-bt},

$$(b^2 - \omega_1^2 - b\beta + \omega_0^2) \cos(\omega_1 t + \theta_0) + (2b\omega_1 - \beta\omega_1) \sin(\omega_1 t + \theta_0) = 0.$$

For this equation to hold for all t, the coefficients of $\sin(\omega_1 t + \theta_0)$ and $\cos(\omega_1 t + \theta_0)$ must vanish separately. We therefore require

$$b = \tfrac{1}{2} \beta \tag{12.21}$$

and

$$\omega_1^2 = \omega_0^2 + b^2 - b\beta.$$

Putting Eq. (12.21) into this second equation we find

$$\omega_1 = \sqrt{\omega_0^2 - \frac{\beta^2}{4}} = \sqrt{\frac{k}{m} - \frac{\gamma^2}{4m^2}}. \tag{12.22}$$

Thus the general solution to Eq. (12.19) for damped oscillations is, for $\beta < 2\omega_0$,

$$x = Ce^{-\beta t/2} \cos(\omega_1 t + \theta_0) \tag{12.23}$$

with ω_1 given by (12.22). Since we have found that Eq. (12.19) is satisfied for any choice of C and θ_0, they are arbitrary constants to be determined by the initial conditions. The damped oscillatory motion given by (12.23) is plotted in Fig. 12.7 for the particular case $\theta_0 = 0$.

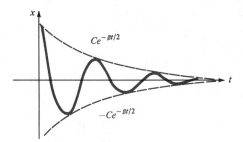

Figure 12.7 Damped oscillations.

Equation (12.23) for $x(t)$ confirms the statement made in Section 12.5 that the frequency of a damped oscillator remains the same throughout the motion, even though its amplitude decreases. This property is crucial for the use of oscillators in accurate timepieces.

The interpretation of Eqs. (12.21) and (12.22) for small β is quite natural and confirms our physical intuition. If β is zero, the motion is simple harmonic and the frequency ω_0 is just $\sqrt{k/m}$ as derived in Section 12.2. The greater β (the greater the resistive force), the more rapidly the envelope curves $Ce^{-\beta t/2}$ and $-Ce^{-\beta t/2}$ of Fig. 12.7 approach zero. Also, if β is increased, the frequency becomes smaller and the period becomes longer. In other words, friction slows down the motion.

If the resistive force is too great, the motion will not oscillate at all. This happens when $\omega_1 = 0$, that is, when $\beta = 2\omega_0$ or $\gamma = 2\sqrt{mk}$. In this case the motion is said to be *critically damped*. The amplitude x decreases rapidly to zero without any oscillation for this and all higher values of β.

In many practical devices such as the shock absorbers of an automobile, the designer attempts to build in approximately enough resistance to achieve critical damping. If the resistance is too small, the springs of the car oscillate through many periods each time the car hits an isolated bump, making the occupants uncomfortable, and might oscillate with increasing amplitude when the car hits a series of bumps, causing damage. This condition occurs when the shock absorbers have worn out; it is most simply tested by putting your weight on the trunk of the car, releasing it suddenly, and seeing whether the trunk oscillates more than once or twice in returning to equilibrium. On the other hand, if the resistance is too great, the springs are prevented from doing their job, which is to reduce the maximum force transmitted to the car body by spreading the impulse from the bump out over time. Because of this practical interest in achieving critical damping, the low-resistance case $\beta < 2\omega_0$ is often called *underdamped* and the high-resistance case $\beta > 2\omega_0$ is called *overdamped*.

12.9 FORCED OSCILLATIONS

Galileo was not the only famous member of the Galilei family; his father, Vincenzo, was an accomplished and articulate musician. Understandably, Vincenzo was interested in how sound was produced. In 1589 he published his work on the relationship of the lengths and tensions of strings to the tones they produced. This study may have been the first experimentally derived law ever to have been discovered to replace a rival law. In the sixteenth century, music was considered a branch of mathematics. The Pythagorean idea that harmonious tones are produced by strings whose lengths are in definite ratios dominated music theory. Vincenzo argued that the complex sounds of musical instruments had to be determined by ear, rather than by mathematics alone. His ideas instilled a keen ear and insatiable curiosity into the eldest of his seven children, Galileo.

Not only did Galileo uncover the factors that determine the frequency of a pendulum, but he also understood the phenomenon of resonance. He noted that the swings of a pendulum can be made increasingly large by repeated, timed applications of a small force, like a puff of air. This method of making a pendulum swing is an example of *forced oscillations* – vibrations induced by an external driving force. Galileo further realized that if the frequency of the external driving force exactly matches the natural frequency ω_0 of the system, a spectacular effect happens: the amplitude of the vibrations becomes exceedingly large. When a vibrating system is driven by a periodic force at the natural frequency of the system, we say that *resonance* occurs. Galileo knew that the phenomenon of resonance lay at the heart of the sounding boards of his father's clavichords and violins, and even of the power of a singer's voice.

Sometimes resonance can cause an oscillating system literally to break apart. Television commercials and movies have capitalized on this dramatic effect by showing wine glasses shattering when a singer hits a certain note. Is this impressive effect possible, or is it just Hollywood trickery?

Before we can answer, we need first to understand how and why resonance occurs. Perhaps the most familiar example of forced oscillations is a child's swing. Everyone knows how to push on a swing to make it oscillate with a large amplitude. If you want a child to swing high, you push in step with the motion of the swing, as Fig. 12.8a illustrates. By exerting a small force at the same point of each swing, you are timing your pushes to the natural frequency of the swing, which is a type of pendulum. The oscillations become larger and larger because you are adding energy to the system with each push. If, however, the pushes are not in step with the motion, as in Fig. 12.8b, the driving force opposes the motion and can cause the amplitude to diminish.

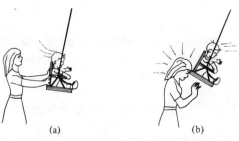

Figure 12.8 Forced oscillations of a swing by a force (a) in phase
and (b) out of phase with the natural frequency.

   The repeated application of a small force can create large-amplitude vibrations if
(a) the force is in step (or in phase) with the oscillating system and (b) the driving force
repeats with the same frequency as the natural frequency of the system. Under these
conditions, resonance occurs.

   Although the driving force may be small, the results of resonance may be spectacularly
large. In Fig. 12.9 a tuning fork is shown attached to a sounding box, which amplifies
the sound of the tuning fork. When an identical tuning fork is placed on a box nearby
and struck, the other begins to ring with what are called sympathetic vibrations. Here's
why: When a series of sound waves from the first tuning fork impinges on the second,
each compression of the air gives the fork a tiny push. Since these pushes occur at the
natural frequency of the tuning fork (remember that the forks are identical), they successively increase the amplitude of the vibration. The result is striking when you consider
how weak a disturbance sound is: a soft sound like the tuning fork is a change in air
pressure of about one part in 10^8, yet that is enough to cause the second tuning fork to
vibrate.

Figure 12.9 An example of resonance: a vibrating tuning fork causes
an identical tuning fork to vibrate.

12.10 DESCRIBING RESONANCE

Now that we know the conditions for resonance, let's describe the phenomenon mathematically. For a start, remember the simple harmonic oscillator equation,

$$m \frac{d^2x}{dt^2} = -kx,$$

which describes the motion of a mass m under the influence of a spring force $F = -kx$.
The natural frequency of the oscillations is specified by

$$\omega_0 = \sqrt{k/m}.$$

If the oscillator is disturbed, it vibrates at its natural frequency. In addition, remember that the harmonic oscillator is a potent model for all sorts of complex oscillatory systems. Each system will have a set of natural frequencies.

Example 8

As you are emptying a jug of water, does the frequency of the gurgles increase, decrease, or remain the same? In other words, does the sound change from a deep bass to a higher pitch or vice versa? Why?

First we note that as the liquid runs out, the air space inside the jug becomes larger. The air in the jug will have a resonant frequency at which it will oscillate. For a mass-and-spring system, the square of the natural frequency is inversely proportional to the mass; the more massive the system, the greater its inertia, and the more sluggishly it vibrates. Similarly, the natural frequency of the air inside the jug depends on the mass of the air. Therefore, as the space becomes larger and increases in mass, the natural frequency decreases, because it is more difficult to accelerate the larger mass. All this means that you hear the pitch becoming deeper. Try observing this effect. Can you predict what you'll hear when a jug is being filled?

Instead of allowing a system simply to oscillate at its natural frequency, suppose we push it back and forth with a force that oscillates with a frequency ω. We can describe the driving force by an oscillating mathematical function,

$$F(t) = F_0 \sin \omega t,$$

where F_0 is just a constant. With this new additional force, Newton's second law for the system is

$$m \frac{d^2x}{dt^2} = -kx + F_0 \sin \omega t.$$

Divide by m and rewrite this as

$$\frac{d^2x}{dt^2} + \omega_0^2 x = a_0 \sin \omega t, \tag{12.24}$$

where $a_0 = F_0/m$, a measure of the size of the driving force. This equation applies not only to a mass on a spring, but to *any* harmonic oscillator whose natural frequency is ω_0 and which is subject to a forcing function proportional to $\sin \omega t$.

In courses on differential equations it is shown that if $\omega \neq \omega_0$ the general solution of Eq. (12.24) can be written as

$$x(t) = x_1 + x_2 = C \cos(\omega_0 t + \theta_0) + A \sin \omega t \tag{12.25}$$

where C and θ_0 are arbitrary constants and where A depends on a_0, ω_0, and ω. Here the cosine term, which we have called x_1, is the familiar solution of the SHO obtained earlier without a driving force. Substituting $x = x_1 + x_2$ into the differential equation (12.24) we immediately note that x_1 makes the left-hand term vanish since $C \cos(\omega_0 t + \theta_0)$ is the general solution of $x'' + \omega_0^2 x = 0$. Thus x_1 drops out, and the second term x_2 satisfies exactly the same equation as $x(t)$, Eq. (12.24). Moreover, if $\theta_0 = 0$ the coefficient C depends only on $x(0)$ and not on the driving force $a_0 \sin \omega t$. This suggests that x_2 is the part of the solution that represents the response to the driving force.

Common sense suggests that if we continue to push an oscillator with some given frequency it will eventually try to oscillate with that frequency, so it is reasonable to expect that a function of the form $x_2 = A \sin \omega t$ will be a solution of (12.24). We now show this is so by substituting $x_2 = A \sin \omega t$ into (12.24) and determining A to satisfy the equation.

Differentiating x_2 twice and substituting into (12.24) we find

$$-\omega^2 A \sin \omega t + \omega_0^2 A \sin \omega t = a_0 \sin \omega t,$$

or

$$(-\omega^2 A + \omega_0^2 A - a_0) \sin \omega t = 0.$$

The only way this can be true for all t is for the coefficient of $\sin \omega t$ to vanish:

$$-\omega^2 A + \omega_0^2 A - a_0 = 0$$

or

$$A = \frac{a_0}{\omega_0^2 - \omega^2} \tag{12.26}$$

since $\omega \neq \omega_0$. Thus the part of the solution to Eq. (12.24) that represents the response to the driving force is

$$x_2(t) = \frac{a_0}{\omega_0^2 - \omega^2} \sin \omega t. \tag{12.27}$$

Let's pause for a moment and recall what we are looking for. We want to understand how and when resonance occurs. More dramatically, we want to know if it is possible to break a wine glass by singing the right note. The wine glass is represented by our basic equation, Eq. (12.24), with ω_0 its natural frequency (which you hear if you tap the wine glass), and $x(t)$ the distortion of the shape of the wine glass. The singer's voice (live or taped) causes the air to push the glass with a driving force $F_0 \sin \omega t$, leading to a disturbance of the glass $A \sin \omega t$ [plus $C \cos(\omega_0 t + \theta_0)$, which we're less interested in now]. The size of the resulting disturbance depends on A; if A becomes too large, the glass will shatter.

Let's interpret the equation for A. The wine glass rings with a definite frequency ω_0. If the sound waves striking the glass are from a bass note, that is, a low frequency ω which is much less than ω_0, $\omega \ll \omega_0$, then we can ignore ω^2 compared to ω_0^2 and Eq. (12.26) gives us

$$A \approx a_0/\omega_0^2.$$

For sound waves a_0 is very small; consequently A is also very small. This means that the glass vibrates only slightly when excited by the bass note frequency ω. In other words, nothing spectacular happens.

On the other hand, if the sound comes from a soprano, it has a high frequency, $\omega \gg \omega_0$. This time we ignore ω_0^2 compared to ω^2 in (12.26) and the resulting amplitude of the glass is

$$A \approx -a_0/\omega^2.$$

This effect is even smaller than that of a bass note because ω is bigger than ω_0. The minus sign tells us that the glass vibrates exactly opposite to the way it is being pushed by the sound waves.

But if the impinging sound waves have almost precisely the frequency ω_0, then something startling happens. According to Eq. (12.26) no matter how small a_0 is, as ω approaches ω_0, the resulting amplitude of the system becomes arbitrarily large, blows up, and so does the system. We have resonance.

In real life, a glass is seldom shattered by the voice of a singer. Instead an audio generator, a device that produces pure tones, tuned precisely to the natural frequency of the glass, is needed. And the natural frequency of the glass must be determined by a microphone held close to the glass which detects the frequency of the audio generator at which the glass vibrates the most. The difficulty in breaking glasses is fortunate, because sound of every possible frequency, although at low intensities, is always in the air. There wouldn't be a glass left in the world if they really broke easily.

The reason sound seldom shatters actual resonating glasses is resistive damping, which we'll take into account in the next section. But even in the idealized situation where friction and viscosity are ignored, the solution to (12.24) takes a different form than (12.25) when $\omega^2 = \omega_0^2$. In this case it can be shown that

$$x(t) = C \cos(\omega_0 t + \theta_0) - \frac{a_0}{2\omega_0} t \cos \omega_0 t. \tag{12.28}$$

The first term is a pure cosine as above, but the second term oscillates with increasing amplitude as t increases because of the presence of the factor t multiplying the cosine. In this undamped resonance the amplitude of the oscillation becomes arbitrarily large as t increases. An example of a solution $x(t)$ with $C = 0$ is shown in Fig. 12.10.

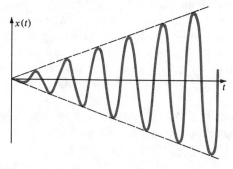

Figure 12.10 Resonant displacement of an oscillator in the absence of friction.

Breaking wine glasses is just one minor, albeit dramatic, example of the phenomenon of resonance. Many other things in everyday life exhibit resonance. Most cars have something that starts to rattle at certain motor speeds. This means that there is a natural harmonic oscillator somewhere (usually you can't quite pinpoint it) whose oscillation has the frequency of that motor speed. When the motor speed reaches that particular value, vibrations set in and the object begins to rattle. Another example is the rattling of windows when a jet airplane flies overhead. The windows have a natural frequency that is excited by the sound of the jet engines.

Similarly, the sounding board of a piano, which is a piece of wood with many natural frequencies, resonates when a vibrating string is attached to it. Resonance also occurs in the cavity of a violin; the air inside can have large oscillations for certain frequencies.

Sometimes the effects of resonance can be ominous. In an earthquake, seismic waves are sent out from the epicenter in a range of frequencies, mostly at low frequencies compared to audible sound, known as infrasound. What happens if a structure has a resonance at one of those frequencies? Buildings between 5 and 40 stories high are typically resonant at earthquake frequencies. In an earthquake, these structures can literally come apart because of resonance.

12.11 DAMPED FORCED OSCILLATIONS

The resonant amplitude derived in the foregoing section has at least two peculiar features. In the first place, the amplitude A in Eq. (12.26) is predicted to become infinite as $\omega \to \omega_0$, contrary to experience. In the second place, the amplitude abruptly changes sign from positive to negative at $\omega = \omega_0$. One can accept that the sign might be negative at $\omega > \omega_0$; after all, in this case the system is being driven faster than its natural frequency allows it to respond, so the oscillator is expected to lag behind the driving force. But the prediction of a discontinuous change from a large positive amplitude to a large negative one at $\omega = \omega_0$ seems peculiar.

A convenient demonstration of what actually happens is provided by the apparatus depicted at the upper left of Fig. 12.11 on p. 322. A horizontal wire or rod is pulled sinusoidally left and right with frequency ω by a motor (not shown). A pendulum suspended from the wire or rod is driven by this periodic motion (indicated by the dashed line in the figure) of its point of suspension. (To make the phase relations easier to follow visually it is helpful to attach orange or red spots to the horizontal wire as well as the pendulum bob.)

If $\omega \ll \omega_0$ (left-hand column of the figure), the pendulum swings with only modest amplitude, in the same direction as the point of suspension is moving. This is in accord with Eq. (12.26), which predicts that the displacement $x(t) = A \sin \omega t$ with which the pendulum bob swings away from the vertical along its arc has the same sign as the impressed acceleration $a_0 \sin \omega t$ when $\omega^2 < \omega_0^2$. The pendulum bob is said to be responding *in phase* with the driving force, as one might expect when the driving force varies more slowly than the natural frequency of the pendulum.

If the pendulum is driven much faster than its natural frequency ($\omega \gg \omega_0$, center column) it again swings with a modest amplitude. In this case the pendulum bob swings right as the point of suspension moves left, and vice versa, as predicted by the minus sign that Eq. (12.26) gives for A/a_0 when $\omega^2 > \omega_0^2$. We see that the pendulum bob moves

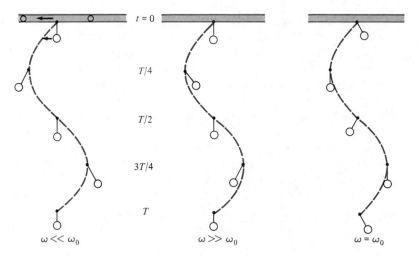

Figure 12.11 Motion of a pendulum driven at frequency ω, followed through one complete period from $t = 0$ to $t = T$. The dashed line traces the motion of the horizontally driven point of suspension.

through its cycle half a period behind the impressed force; it is said to be responding *180° out of phase*.

When the driving frequency is changed to $\omega = \omega_0$, the amplitude of the pendulum bob increases steadily over several periods, in a manner similar to Fig. 12.10. But eventually the swings stabilize at a large but finite amplitude. And, surprisingly, the pendulum bob moves through its cycle a quarter period behind the impressed force. It is said to be responding *90° out of phase* with the impressed force. Thus the peculiar features predicted by Eq. (12.26) do not in fact occur – the resonance amplitude is finite, and the phase lag is intermediate between the low-frequency and high-frequency limits, rather than jumping discontinuously from one to the other.

To describe accurately the phenomena observed at resonance, we must add resistive forces (which are always present for any oscillator in the real world) to the analysis of forced oscillations in the previous section. As in Section 12.8, we shall assume the resistive force has the form $-\gamma \, dx/dt$. There are now three forces acting on the oscillator: the spring force, a resistive force, and a driving force. Newton's second law becomes

$$m \frac{d^2x}{dt^2} = -kx - \gamma \frac{dx}{dt} + F_0 \sin \omega t.$$

Rearranging terms and dividing through by m we obtain

$$\frac{d^2x}{dt^2} + \beta \frac{dx}{dt} + \omega_0^2 x = a_0 \sin \omega t \qquad (12.29)$$

where $\beta = \gamma/m$, $\omega_0^2 = k/m$, and $a_0 = F_0/m$ as before.

It can be shown that the general solution of (12.29) can be written

$$x(t) = x_1 + x_2 = Ce^{-\beta t/2} \cos(\omega_1 t + \theta_0) + A \sin(\omega t - \alpha). \qquad (12.30)$$

The first term, $x_1 = Ce^{-\beta t/2} \cos(\omega_1 t + \theta_0)$, is the familiar solution (12.23) obtained for damped oscillations without a driving force. As before, ω_1 is given by Eq. (12.22), whereas C and θ_0 are constants determined by the initial conditions. However, their actual values are not needed to predict the behavior of the solution for large t because the exponential damping factor $e^{-\beta t/2}$ makes x_1 a transient which dies away as t increases, leaving the purely sinusoidal motion $x_2 = A \sin(\omega t - \alpha)$.

Substituting $x = x_1 + x_2$ into the differential equation (12.29) we immediately note that x_1 makes the left-hand side vanish and that the new term x_2 satisfies exactly the same equation as $x(t)$, Eq. (12.29).

The term $x_2 = A \sin(\omega t - \alpha)$ is the steady state reached when the transient dies away. As expected for a steady state, it oscillates with constant amplitude at the same frequency as the driving force. The term $\sin(\omega t - \alpha)$ expresses the fact that the phase angle for x_2 generally lags behind that of the driving force.

We can verify that Eq. (12.29) indeed has the solution $A \sin(\omega t - \alpha)$ and determine the values of A and α by substituting $x_2 = A \sin(\omega t - \alpha)$ into Eq. (12.29). Expanding $A \sin(\omega t - \alpha)$ into terms involving $\sin \omega t$ and $\cos \omega t$ and taking derivatives, we obtain

$$x_2 = A(\sin \omega t \cos \alpha - \cos \omega t \sin \alpha),$$

$$\frac{dx_2}{dt} = A\omega(\cos \omega t \cos \alpha + \sin \omega t \sin \alpha),$$

$$\frac{d^2 x_2}{dt^2} = A\omega^2(-\sin \omega t \cos \alpha + \cos \omega t \sin \alpha).$$

Substituting these results into Eq. (12.29) and collecting terms, we find

$$\left((-\omega^2 + \omega_0^2) \cos \alpha + \beta\omega \sin \alpha - \frac{a_0}{A} \right) \sin \omega t$$

$$+ \left[(\omega^2 - \omega_0^2) \sin \alpha + \beta\omega \cos \alpha \right] \cos \omega t = 0.$$

For this equation to hold at all t, the coefficients of $\cos \omega t$ and $\sin \omega t$ must vanish separately. We therefore require

$$\frac{\sin \alpha}{\cos \alpha} = \tan \alpha = \frac{\beta\omega}{\omega_0^2 - \omega^2} \tag{12.31}$$

and

$$A = \frac{a_0}{(\omega_0^2 - \omega^2) \cos \alpha + \beta\omega \sin \alpha}. \tag{12.32}$$

The phase angle α is conveniently displayed in the right triangle of Fig. 12.12. Here (12.31) and the Pythagorean theorem determine the ratios of the lengths of the sides, allowing one to read off

$$\sin \alpha = \frac{\beta\omega}{\sqrt{(\omega_0^2 - \omega^2)^2 + (\beta\omega)^2}},$$

$$\cos \alpha = \frac{\omega_0^2 - \omega^2}{\sqrt{(\omega_0^2 - \omega^2)^2 + (\beta\omega)^2}}.$$

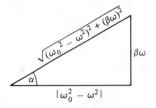

Figure 12.12 Geometric interpretation of α. The angle α is $<90°$ when $\omega_0^2 > \omega^2$, but $>90°$ when $\omega_0^2 < \omega^2$.

Inserting these results into Eq. (12.32) we find

$$A = \frac{a_0 \sqrt{(\omega_0^2 - \omega^2)^2 + (\beta\omega)^2}}{(\omega_0^2 - \omega^2)^2 + (\beta\omega)^2} = \frac{a_0}{\sqrt{(\omega_0^2 - \omega^2)^2 + (\beta\omega)^2}}.$$ (12.33)

To summarize, then, the general solution to (12.29) contains a transient term and a steady-state sinusoidal term:

$$x(t) = Ce^{-\beta t/2} \cos(\omega_1 t + \theta_0) + \frac{a_0}{\sqrt{(\omega_0^2 - \omega^2)^2 + (\beta\omega)^2}} \sin(\omega t - \alpha)$$ (12.34)

where C and θ_0 are determined by the initial conditions; α is given by Eq. (12.31), which we may write in the form

$$\alpha = \tan^{-1} \frac{\beta\omega}{\omega_0^2 - \omega^2};$$ (12.35)

and ω_1 is given by

$$\omega_1 = \sqrt{\omega_0^2 - \tfrac{1}{4}\beta^2}.$$ (12.22)

If there is no resistive force ($\beta = 0$), the amplitude in (12.33) takes on the magnitude $a_0/|\omega_0^2 - \omega^2|$ derived earlier in Eq. (12.26). In the presence of a small resistive force, A exhibits a resonance peak of finite height $\approx a_0/\beta\omega_0$ near $\omega = \omega_0$. The larger β becomes, the lower the peak height, as shown in Fig. 12.13.

Example 9

How do frictional forces prevent the amplitude of a driven oscillator from becoming infinite at resonance?

Let's first recall what we know about simple harmonic oscillators when friction is present, without any driving force. In Section 12.5 we argued that the effect of friction was to cause the amplitude (but not the frequency) of the oscillations to diminish with time. The reason for this is that the oscillator loses energy in the form of heat. Now if the oscillator is forced to oscillate, friction not only causes it to lose energy with each

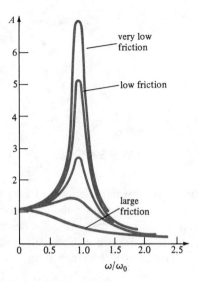

Figure 12.13 Resonance curves (A vs. ω/ω_0) for different amounts of frictional force acting on an oscillator.

oscillation but also to lag behind the driving force. Recall that we found that for a frequency much greater than the natural frequency, the amplitude is negative, indicating that the oscillator vibrates in a way opposite to the driving force. A similar effect is caused by friction and depends upon the amount of friction. As a result, the oscillator can never be completely in step with the driving force and moreover it keeps on losing energy. Consequently, the amplitude can never become infinite, only very large if friction is small.

In the absence of damping, the displacement

$$x = A \sin \omega t = \frac{a_0}{\omega_0^2 - \omega^2} \sin \omega t \qquad\qquad (12.27)$$

had the same sign as the driving term $a_0 \sin \omega t$ when $\omega < \omega_0$, and the opposite sign when $\omega > \omega_0$. This means that x can be written in the form

$$x = |A| \sin(\omega t - \alpha),$$

where $\alpha = 0$ for $\omega < \omega_0$, and then changes abruptly to $\alpha = 180°$ for $\omega > \omega_0$.

In the presence of damping, this abrupt phase change at the resonance frequency is replaced by the more gradual phase change of Eq. (12.35). The phase change is plotted in Fig. 12.14 for several values of β. The phase lag always passes through 90° at $\omega = \omega_0 = \sqrt{k/m}$. This behavior of the phase lag corresponds to what is actually seen in a physical system such as the driven pendulum of Fig. 12.11.

The solution (12.34) we have derived for damped forced oscillations applies not only to laboratory demonstrations of driven pendulums but to wine glasses, the cavities of

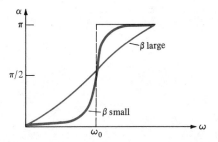

Figure 12.14 Variation of phase angle α with driving frequency.

various musical instruments, and earthquake-driven buildings as well. Whether the resonance is regarded as benign or troublesome depends on the application. One tries to make β small in the former case, large in the latter case. Thus the designer of an organ pipe tries to achieve a large resonant amplitude. But architects designing buildings in earthquake zones try to minimize the resonant response by increasing friction in the joints.

12.12 SWINGING AND SINGING WIRES IN THE WIND

One fascinating example of resonance is provided by telephone wires singing in the wind. Imagine a taut wire suspended in the wind. The air flow around a cross section of the wire is illustrated in Fig. 12.15a. This smooth flow of air around the wire becomes unstable if the wind speed is great enough. The wind tries to move around the wire and prevent a vacuum from forming. If the speed is too high, the wind can't achieve this in a smooth flow, and instead it forms an eddy on both sides, as in Fig. 12.15b. The eddies start to occur behind the wire, and after a short time these vortices begin peeling off on alternate sides and running downstream in the wake of the wire, as shown in Fig. 12.15c.

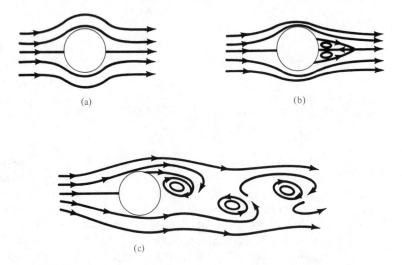

(a)   (b)

(c)

Figure 12.15 Formation of vortices which lead to singing wires.

This complicated yet very stable flow pattern was first explained by the aerodynamicist Theodore von Kármán. The full pattern with a line of vortices in opposite directions is called the von Kármán Vortex Street; Fig. 12.16 is an actual picture of the pattern.

Figure 12.16 Picture of a Von Kármán vortex street in a fluid. (Prandtl, L., and Tietjens, O. G. *Applied Hydro and Aerodynamics.* McGraw-Hill Book Co., New York (1934). By permission of publisher.)

Every time a vortex peels off the wire, it imparts a weak impulsive force. The reason is that the vortex has some momentum and the vortex and wire conserve momentum, so every one of the vortices peeling off gives a push to the wire. At some wind speed, the vortices will start peeling off at a resonant frequency of the wire and thereby set it into motion. This resonant effect is why wires in the wind at the right wind speed begin to sing. The ancient Greeks noticed the eerie sound produced by harps due to this effect; we call the effect an aeolian harp.

12.13 A FINAL WORD

On July 1, 1940, a new bridge was opened at the narrowest point in Puget Sound, connecting Tacoma to the Olympic peninsula. At the time it was the third-longest suspension bridge in the world. Right from the beginning, even before construction was completed, the bridge behaved in a peculiar way. Whenever there was a light breeze, ripples would run along the bridge. After a while local people began calling the bridge affectionately by the name of Galloping Gertie. To drive across the bridge on a windy day became a local sport because it was like riding a roller coaster, although it was disconcerting to people driving across the bridge to see the car in front of them disappear over the crest of a wave.

On November 7, 1940, four months after the bridge was opened, a new mode of oscillations showed up in the bridge in a prevailing southwesterly wind of about 42 mph. Instead of rippling motions down the bridge, twisting motions set in. The peculiarities of the bridge were being studied by an engineer from the University of Washington, Bert

Farquharson. He rushed down to take pictures of the new mode of oscillations. At 11 o'clock in the morning that day, the Tacoma Narrows bridge collapsed. An inquest into the collapse determined that the bridge had been built according to the best engineering standards of the day. No one was guilty of wrongdoing, but also no one could explain why the bridge collapsed.

A national commission investigating the collapse included aerodynamicist Theodore von Kármán of Caltech. He explained that vortices were pouring off the top and bottom of the bridge, driving the bridge at its resonant frequency, which eventually led to its collapse. His explanation was confirmed by experiments conducted in wind tunnels with structural models at both the University of Washington and Caltech. In spite of the confirmation, the bridge-building community was very reluctant to accept the explanation. Why? Bridge architects were concerned with static forces. They built in brute strength to confront maximum load, water flow, wind, etc. They didn't consider dynamic forces. Von Kármán said that the shape the roadway presented to the wind acted like an airplane wing. The displaced air formed vortices whose action induced vibrations in the deck. Since that disastrous event, all models for major bridges have been tested in wind tunnels, and bridge engineers have been forced to consider the aerodynamics of their designs.

Figure 12.17 depicts stages in the collapse of the Tacoma Narrows Bridge on that fateful day. In the twisting mode of Fig. 12.17b, the center line hardly moves at all – the vibrations go all around it. The car belongs to a man named Leonard Coatsworth who was a reporter on a local newspaper; he was the last person to try to cross the bridge. Farquharson himself tried to rescue a cocker spaniel out of Coatsworth's car. For his trouble, he was rewarded by the dog biting him on the hand, which was the only injury in the incident; the cocker spaniel, who never left the car, was the only fatality. A local college student named Winfield Brown decided to walk across the bridge that morning; it was a popular sport on a windy day. He came back crawling on his hands and knees, and reported one moment of sheer terror when the bridge had tilted so much under him that he looked straight down 200 ft into Puget Sound. Figure 12.17c shows the collapse of the bridge at 11:10 A.M.

At the same spot there is now a new suspension bridge built with modifications suggested by von Kármán. The principal changes were to make the bridge four lanes wide, to use open side trusses, and to place ventilating grills between lanes to equalize wind pressures above and below the deck. This bridge has never had the slightest difficulty. People still look at it nervously on windy days, but it never budges.

(a)

(b)

(c)

Figure 12.17 Collapse of the Tacoma Narrows Bridge. (Historical Photography Collection, University of Washington Libraries. Photo by Farquharson.)

SUMMARY OF EQUATIONS FOR OSCILLATORY MOTION

Differential Equation   *General Solution*

Simple Harmonic Motion:

$$\frac{d^2x}{dt^2} + \omega_0^2 x = 0 \quad (\omega_0 > 0)$$

$$x = x(0)\cos\omega_0 t + \frac{x'(0)}{\omega_0}\sin\omega_0 t$$

$$= C\cos(\omega_0 t + \theta_0)$$

Damped Oscillation:

$$\frac{d^2x}{dt^2} + \beta\frac{dx}{dt} + \omega_0^2 x = 0 \quad (0 < \beta < 2\omega_0)$$

$$x = Ce^{-\beta t/2}\cos(\omega_1 t + \theta_0),$$

$$\text{where} \quad \omega_1 = \sqrt{\omega_0^2 - \tfrac{1}{4}\beta^2}$$

Forced Oscillation without Damping:

$$\frac{d^2x}{dt^2} + \omega_0^2 x = a_0\sin\omega t \quad (\omega_0 > 0)$$

If $\omega \neq \omega_0$,

$$x = C\cos(\omega_0 t + \theta_0) + A\sin\omega t,$$

where $A = a_0/(\omega_0^2 - \omega^2)$.

If $\omega = \omega_0$,

$$x = C\cos(\omega_0 t + \theta_0) - \frac{a_0}{2\omega_0}t\cos\omega_0 t.$$

Forced Oscillation with Damping:

$$\frac{d^2x}{dt^2} + \beta\frac{dx}{dt} + \omega_0^2 x = a_0\sin\omega t$$
$$(0 < \beta < 2\omega_0)$$

$$x = Ce^{-\beta t/2}\cos(\omega_1 t + \theta_0)$$
$$+ A\sin(\omega t - \alpha),$$

where

$$\omega_1 = \sqrt{\omega_0^2 - \frac{1}{4}\beta^2},$$

$$A = a_0/\sqrt{(\omega_0^2 - \omega^2)^2 + (\beta\omega)^2},$$

$$\alpha = \tan^{-1}[\beta\omega/(\omega_0^2 - \omega^2)].$$

Problems

Simple Harmonic Motion

1. For a particle undergoing simple harmonic motion, determine whether the maximum value of each of the following occurs at the equilibrium point, or at one of the endpoints:

   (a) the force acting on the particle,
   (b) its speed,
   (c) its acceleration.

What changes could you make to a simple harmonic oscillator to double its

(d) maximum speed?
(e) maximum acceleration?

2. A 2.5-kg block hangs from a spring. If a 0.5-kg mass is attached to the block, the spring stretches an additional 0.05 m. Determine the frequency of oscillations if only the 2.5-kg block is attached to the spring.

3. A vertical spring is in equilibrium under its own weight. A small (0.10-kg) mass is gently attached to the bottom of the spring, without lowering it from its old to its new equilibrium point, and then let fall, whereupon it oscillates around its new equilibrium point. If the maximum speed of the mass is 0.20 m/s, find

(a) the spring constant,
(b) the amplitude of the oscillations,
(c) the frequency of the oscillations.

4. Two springs of spring constants k_1 and k_2 are attached to a block of mass m as shown. What will be the frequency of oscillations?

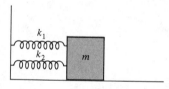

5. A mass is attached to a spring of spring constant k. If the spring is cut in half and the same mass suspended from one of the halves, how are the frequencies of oscillation related, before and after the spring is cut?

Initial Conditions and Simple Harmonic Motion

6. At time $t = 0$ an oscillator having a frequency of 35 Hz has a displacement $x(0) = 0$ and a velocity $v(0) = 20$ m/s.

(a) Find the displacement of the oscillator at any time.
(b) Find the maximum velocity of the oscillator.

7. Suppose that initially an oscillator has a displacement of 0.25 m, a velocity of -10 m/s, and a frequency of 10 Hz.

(a) Find its amplitude.
(b) Find the displacement at any instant.

8. Verify that $x(t) = C \sin(\omega_0 t + \phi_0)$, where C and ϕ_0 are constants, satisfies the differential equation (12.4).

9. In Example 2 of Chapter 7 we found that the gravitational force on a mass m inside a spherical earth of uniform density ρ is $F = kr$, directed toward the center, with $k = \frac{4}{3}\pi G\rho m$.

(a) For a mass dropped from the surface of the earth down a tunnel straight through the center of the earth, find the time τ_1 to pass entirely through the earth and back again.

(b) Compute the speed of the mass m as it passes through the center of the earth.

(c) Compare your result for τ_1 with the time τ_2 required to complete a circular orbit around the globe just above the earth's surface.

Energy Conservation and Simple Harmonic Motion

10. A simple harmonic oscillator of mass 0.40 kg has a frequency of 15 Hz and a total energy of 30 J. Find the amplitude of its oscillation.

11. From conservation of energy, show that the instantaneous speed of a simple harmonic oscillator is given by

$$v(t) = \omega_0 \sqrt{A^2 - x^2},$$

where $x = x(t)$ is the instantaneous displacement. When the oscillator is at one-third of its maximum displacement, what fraction of its total energy is in kinetic energy?

12. The springs of a car of mass 1000 kg give it a period when empty of 1.0 s for small vertical oscillations.

(a) Four persons, each of mass 60 kg, get into the car. How far down does the car body move?

(b) The car with its occupants is traveling along a horizontal road when it runs over a bump onto a new section of road, which is raised 10 cm above the old road. Assume that this suddenly raises the wheels and the bottom ends of the springs through 10 cm before the car body begins to move upward. In the ensuing rebound, what is the maximum "weight" (force exerted on his seat) of a 60-kg passenger? Ignore any effect of shock absorbers.

(c) What is the maximum upward velocity attained by the car and passengers? Again, ignore shock absorbers.

(d) Assuming the shock absorbers eventually stop the oscillations, how much energy must they absorb?

The Simple Pendulum

13. (a) What must be the length of a pendulum to have a period of 1.0 s on the surface of the earth?

(b) What will be the period of such a pendulum on the surface of the moon?

(c) How will the period of this pendulum on the earth change if the length is increased by 60%?

14. Qualitatively argue how by taking into account the mass of the supporting rod of a pendulum the frequency would change from that of a simple pendulum.

15. A simple pendulum of length 1.50 m makes 80 oscillations in 200 s. What is the local acceleration due to gravity?

Damped Oscillations

16. (a) Given that $x = Ce^{-\beta t/2} \cos(\omega_0 t + \theta_0)$, find the values of C and θ_0 that correspond to the initial conditions $v(0) = 0$, $x(0) = x_0$.

  (b) At what time is $|x(t)|$ largest for the initial conditions in (a)?

17. The amplitude $x = L\theta$ of a simple pendulum 1 m long with a 0.5-kg bob decreases from 5 cm to 2.5 cm during ten cycles.

  (a) Find γ (the coefficient in the damping term $\gamma \, dx/dt$ of Eq. (12.17)) for this pendulum.

  (b) Find the average power that a driving force would need to supply to maintain the amplitude at 5 cm. (You will need to make some modest approximations; state them.)

  (c) If a descending weight of 10 kg is used to supply this power in a grandfather clock, and a distance of 1 m is available for the weight to descend, find how often the clock needs to be rewound.

Forced Oscillations and Resonance

18. Suppose you have a marble which is free to roll inside a shallow bowl. Describe how by *horizontal* motion of the bowl alone, you can cause the marble to roll over the edge of the bowl.

19. Every child knows that by blowing across an empty soda bottle a sound can be produced; the frequency of the sound is the natural frequency of the air inside the bottle. Explain what happens to the frequency produced when the bottle is partially filled with liquid.

20. When the base of a vibrating tuning fork is touched to a table, the sound is amplified. Explain why.

21. In 1831 a bridge collapsed near Manchester, England, when soldiers marched across it in step. Since then, soldiers break step while marching across a bridge. Why?

22. (a) Order the pendulums A through E in the figure by their periods, beginning with the shortest.

  (b) Suppose the horizontal rod from which the pendulums are suspended is rather flexible. If pendulum E is given a slight push, which pendulum(s) will also begin to oscillate? Explain why.

23. Find the resonant frequencies for the following systems:

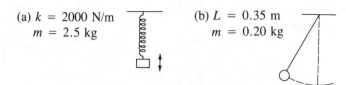

(a) $k = 2000$ N/m
$m = 2.5$ kg

(b) $L = 0.35$ m
$m = 0.20$ kg

24. A pendulum is forced to oscillate at a frequency $\omega = \frac{1}{4}\omega_0$, that is, at one-quarter of its natural frequency. Compare the amplitude and energy of the oscillations to those which occur at $\omega = \frac{1}{2}\omega_0$, assuming that damping is negligible and the term $C \cos(\omega_0 t + \theta_0)$ is absent.

25. Use the solution $x(t) = A \sin \omega t$, where A is given by Eq. (12.26), to find the total energy of a forced harmonic oscillator. What happens to the energy at resonance?

26. Verify that Eq. (12.28) is a solution of the differential equation (12.24) for $\omega = \omega_0$.

27. (a) Show that the amplitude A in Eq. (12.33) reaches its maximum value when $\omega^2 = \omega_0^2 - \beta^2/2$.

(b) Show that the maximum amplitude A is $a_0/(\beta\omega_1)$, where $\omega_1^2 = \omega_0^2 - \beta^2/4$.

(c) The difference $\omega_0^2 - \omega^2 = (\omega_0 + \omega)(\omega_0 - \omega)$ can be approximated by $2\omega(\omega_0 - \omega)$ when $\omega_0 - \omega$ is small. Use this to show that the expression under the radical sign in Eq. (12.33) is approximately

$$4\omega^2(\omega_0 - \omega)^2 + \beta^2\omega^2 = 4\omega^2[(\omega_0 - \omega)^2 + \beta^2/4]$$

in a narrow resonance ($\beta << \omega_0$), and deduce that $\omega = \omega_0 \pm \beta/2$ at the half maximum of A^2. This shows that the width of a narrow resonance is β.

(d) Suppose that at some frequency $\omega_a < \omega_0$ the oscillator is 45° out of phase with the driving force (i.e., $\alpha = 45°$), and at another frequency $\omega_b > \omega_0$ it is 135° out of phase. Derive the relation $\omega_b - \omega_a = \beta$. In conjunction with (c), this shows that the half maximum of A^2 occurs near $\alpha = 45°$ and 135°. (This relation becomes exact for a narrow resonance.)

Singing Wires

28. A cabinet next to a refrigerator contains pots and pans that make a vibrational noise when the refrigerator motor runs. What is the source of this sound? How can it be eliminated?

29. Suppose you were to simulate wind-whistling wires by waving a fork with long tines. Which way would you wave it, in the plane of the tines or perpendicular to it? After deciding which way will create a sound, check your prediction.

CHAPTER

13

ANGULAR MOMENTUM

Whereupon I computed what would be the Orb described by the Planets. . . . I found now that whatsoever was the law of the forces which kept the Planets in their Orbs, the areas described by a Radius drawn from them to the Sun would be proportional to the time in which they were described.

Isaac Newton

13.1 ROTARY MOTION

The world is full of things that exhibit rotary motion. They range in size from galaxies to electrons orbiting around atoms, and they include such familiar objects as orbiting planets, amusement park rides, flywheels, and bathtub vortices. What underlying principle explains the persistence of such motions? Can we describe them all in a unified way?

In Chapter 7 we studied two partial answers to such questions. One was that the moon circles the earth (or a planet circles the sun) by falling toward it continuously. Another was that an object can undergo uniform circular motion if a centripetal acceleration of magnitude v^2/r is supplied. But the first answer is restricted to gravitational attraction, the second is restricted to the special case of uniform circular motion, and each addresses the persistence of such motion rather obliquely.

The concept that unifies the description of all rotary motion is *angular momentum* and the persistence of such motions is most directly described in terms of the *conservation of angular momentum*. To be sure, angular momentum conservation, when applicable, can be derived from Newton's three laws. Any rotational problem can be solved directly by following Newton's three laws through time without introducing angular momentum. And yet, like energy conservation (to which the same comments applied), conservation of angular momentum is a most useful concept which both aids understanding and simplifies calculation.

13.2 TORQUE AND ANGULAR MOMENTUM

In Section 6.6 we gave the general definition of torque,

$$\boldsymbol{\tau} = \mathbf{r} \times \mathbf{F}. \tag{13.1}$$

As we see from this relation, torque is an unusual vector insofar as it depends on the distance r from a chosen reference point O to the point of application of the force $\mathbf{F}$ (Fig. 13.1). The torque tends to produce a rotation about point O, with the axis of rotation parallel to the vector $\boldsymbol{\tau}$.

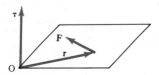

Figure 13.1 Vector relations for torque about a point O.

We now wish to bring out this relation between torque and rotation, with the aid of Newton's second law

$$\mathbf{F} = \frac{d\mathbf{p}}{dt}. \tag{6.1}$$

Substituting $d\mathbf{p}/dt$ for $\mathbf{F}$ in Eq. (13.1), we obtain

$$\boldsymbol{\tau} = \mathbf{r} \times \frac{d\mathbf{p}}{dt}.$$

Because

$$\frac{d}{dt}(\mathbf{r} \times \mathbf{p}) = \frac{d\mathbf{r}}{dt} \times \mathbf{p} + \mathbf{r} \times \frac{d\mathbf{p}}{dt}$$

by the product rule as applied to vectors (Sec. 5.6), the relation for $\boldsymbol{\tau}$ can be rewritten

$$\boldsymbol{\tau} = \frac{d}{dt}(\mathbf{r} \times \mathbf{p}) - \frac{d\mathbf{r}}{dt} \times \mathbf{p}.$$

The last term vanishes:

$$\frac{d\mathbf{r}}{dt} \times \mathbf{p} = \mathbf{v} \times (m\mathbf{v}) = \mathbf{0},$$

because the cross product of a vector with itself is zero, and we obtain the result

$$\boldsymbol{\tau} = \frac{d}{dt}(\mathbf{r} \times \mathbf{p}). \qquad (13.2)$$

Equation (13.2) bears a striking resemblance to Newton's second law,

$$\mathbf{F} = \frac{d}{dt}(m\mathbf{v}). \qquad (6.1)$$

On the left-hand side of (13.2) we have the torque, a "twisted" version of force. And on the right-hand side we have the time derivative of $\mathbf{r} \times \mathbf{p}$, a "twisted" version of momentum. This important quantity $\mathbf{r} \times \mathbf{p}$ is denoted by the symbol $\mathbf{L}$ and, in view of the analogy with momentum, is called *angular momentum*:

$$\boxed{\mathbf{L} = \mathbf{r} \times \mathbf{p}.} \qquad (13.3)$$

According to Eq. (13.2) the time rate of change of angular momentum equals the torque:

$$\boxed{\boldsymbol{\tau} = \frac{d\mathbf{L}}{dt}.} \qquad (13.4)$$

Like its linear analog $\mathbf{F} = d\mathbf{p}/dt$, Eq. (13.4) is a very important result. It has a number of significant consequences which we will explore in this chapter and the next.

When a system consists of two or more parts, the individual bodies composing it will generally be subject to both internal and external torques. For the system as a whole, however, the internal torques cancel, just as internal forces do, provided the internal forces between any pair of bodies are equal and opposite *and act along the line joining the two bodies*, as do gravitational and Coulomb forces. To see this, consider the pair of bodies in Fig. 13.2. The torque $\mathbf{r}_1 \times \mathbf{F}_{21}$ on body 1 about point O, due to the force $\mathbf{F}_{21}$ from body 2, is the same as $\mathbf{r}_{1\perp} \times \mathbf{F}_{21}$ in Fig. 13.2 because the component of $\mathbf{r}_1$ parallel to $\mathbf{F}_{21}$ does not contribute to the torque:

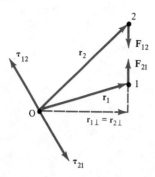

Figure 13.2 Torques exerted by a pair of bodies on each other.

$$\boldsymbol{\tau}_{21} = \mathbf{r}_1 \times \mathbf{F}_{21} = \mathbf{r}_{1\perp} \times \mathbf{F}_{21}. \tag{13.5}$$

Similarly the torque $\mathbf{r}_2 \times \mathbf{F}_{12}$ on body 2 about point O, due to the force $\mathbf{F}_{12}$ from body 1, is the same as $\mathbf{r}_{2\perp} \times \mathbf{F}_{12}$:

$$\boldsymbol{\tau}_{12} = \mathbf{r}_2 \times \mathbf{F}_{12} = \mathbf{r}_{2\perp} \times \mathbf{F}_{12}. \tag{13.6}$$

But $\mathbf{F}_{21} = -\mathbf{F}_{12}$ by Newton's third law, and one easily sees from Fig. 13.2 that $\mathbf{r}_{1\perp} = \mathbf{r}_{2\perp}$ if $\mathbf{F}_{12}$ and $\mathbf{F}_{21}$ have the same line of action. So the net internal torque of the pair vanishes,

$$\boldsymbol{\tau}_{21} + \boldsymbol{\tau}_{12} = \mathbf{r}_{1\perp} \times (\mathbf{F}_{21} + \mathbf{F}_{12}) = \mathbf{0},$$

about any point O. Proceeding as for internal forces, one can easily extend this result and show that the sum over all internal torques vanishes. Therefore only $\boldsymbol{\tau}_{\text{ext}}$ contributes to $d\mathbf{L}/dt$ in Eq. (13.4) for a system of particles:

$$\boldsymbol{\tau}_{\text{ext}} = \frac{d\mathbf{L}}{dt}. \tag{13.7}$$

When an external torque is present, integration of Eq. (13.7) over time yields

$$\int_{t_1}^{t_2} \boldsymbol{\tau}_{\text{ext}}\, dt = \mathbf{L}(t_2) - \mathbf{L}(t_1). \tag{13.8}$$

By analogy with the relation between impulse and momentum, the quantity on the left-hand side is sometimes called *angular impulse*, and the equation reads: "The angular impulse equals the change in angular momentum."

13.3 ANGULAR MOMENTUM CONSERVATION

As with its close relative, momentum, there are many circumstances in which the angular momentum of a system is conserved. First of all, the external torques applied to a body may vanish or cancel. In this case

$$\frac{d\mathbf{L}}{dt} = \mathbf{0},$$ (13.9)

so

$$\mathbf{L} = \mathbf{const}.$$ (13.10)

Because **L** is a vector, Eq. (13.10) is really three separate conservation laws for the components L_x, L_y, and L_z.

Second, even if external forces and torques are present, they may have a negligible effect during the brief duration of a collision or explosion. In this case the overall momentum and angular momentum of a system may be conserved in the collision or explosion:

$$\sum \mathbf{p}_i = \mathbf{const},$$
$$\sum \mathbf{L}_i = \mathbf{const},$$ (13.11)

even though strong internal impulses change the $\mathbf{p}_i$ and $\mathbf{L}_i$ of individual particles. For example, friction supplies both an external force and a torque to a rolling billiard ball, eventually bringing it to rest. But friction can be neglected in the collision of two billiard balls – both momentum and angular momentum are conserved.

Third, if something exerts an external torque on a system, then the system exerts an external but opposite torque on that something, as we saw in the previous section. If we include that external thing as part of a new, larger system, and that external thing is not subject to *other* torques, then the external torque on the whole system vanishes, so its angular momentum will be conserved. For example, in a classic lecture demonstration, a student stands on a turntable holding a bicycle wheel in the horizontal plane. When the student applies a torque to the bicycle wheel, it applies a countertorque to him. Neither the wheel's **L** nor the student's **L** is individually conserved, but the combined angular momentum of wheel, student, and turntable *is* conserved. So if the student spins the wheel clockwise (as viewed from above), starting from rest, the student must rotate counterclockwise to conserve the total angular momentum of the combined system (Fig. 13.3b). If the student now turns the wheel 180° over so it rotates counterclockwise, his rotation becomes clockwise to conserve total angular momentum (Fig. 13.3c). In this last

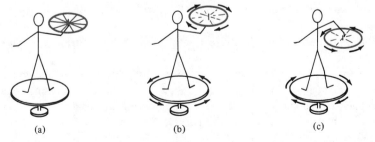

(a) (b) (c)

Figure 13.3 Angular momentum conservation for a student holding a bicycle wheel on a turntable.

case it is somewhat complicated to follow the torque in detail, but angular momentum conservation gives the final direction of rotation easily.

By continuing forever this process of enlarging the system we may include all objects within the universe in our system. In this case there is nothing left outside to exert a torque on it. So the total angular momentum of the universe is conserved.

Thus angular momentum joins energy and momentum as the third great conserved quantity of classical mechanics. These conservation laws all had their origin in Newton's laws. One can always solve a problem directly from Newton's laws, but use of the conservation principles leads to great simplifications and insight. We shall find that angular momentum conservation is just as helpful in simplifying rotational problems as momentum conservation was for problems of linear motion.

Example 1

A particle of mass m moves with constant velocity $\mathbf{v}$ along a straight line which is a distance b from the origin of a coordinate system.

(a)   Find the angular momentum of the particle at any instant.

(b)   Show explicitly that the angular momentum is conserved.

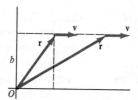

(a)   We know that the angular momentum is given by

$$\mathbf{L} = m\mathbf{r} \times \mathbf{v}. \tag{13.3}$$

The magnitude of the angular momentum is $L = mr_{\perp}v$, where $r_{\perp}$ is the component of the position vector $\mathbf{r}$ perpendicular to the velocity $\mathbf{v}$. From the diagram we see that this distance is b. Therefore

$$L = mr_{\perp}v = mbv.$$

By the right-hand rule, the direction of the angular momentum is into the plane of the page.

(b)   From our expression for L, we see that as the particle continues to move along the straight line, the component of $\mathbf{r}$ perpendicular to $\mathbf{v}$ remains equal to b. Thus the angular momentum is conserved. The reason for the conservation of angular momentum here is that no force acts on the particle (by Newton's first law), and therefore it feels no torque.

In Example 1 we could have chosen a different origin for the coordinate system, located at distance b' from the line of motion of the particle. Angular momentum would still be conserved in this case, but its value would be

$$L' = mb'v.$$

The conservation law always holds, but the choice of reference point is arbitrary. This situation reminds us of the similar arbitrariness in the choice of reference point about which torque is taken. In problems in which the general relation $\boldsymbol{\tau} = d\mathbf{L}/dt$ is used one can choose any reference point, but both $\boldsymbol{\tau}$ and $\mathbf{L}$ must be taken around the *same* point.

Example 1 also shows that a body need not be rotating to have angular momentum. Angular momentum is a very general property.

In the foregoing example energy, momentum, and angular momentum were all conserved. But it happened they were not needed to solve for the motion, which followed trivially from Newton's first law. In the next example the angular momentum of the system is again conserved (along with linear momentum) and it aids greatly in finding an easy solution for the motion.

Example 2

Two masses, 8.0 kg each, are connected by an extensible rod of negligible mass. Originally the rod is rotating about its center at an angular speed of 2 rad/s with the masses 0.4 m from the axis of rotation. What will be the rod's angular speed if the masses are pulled in to a distance of 0.3 m from the center?

This example is a prototype for rotating ice skaters, ballerinas, and students holding dumbbells on turntables, all of whom gain angular speed by pulling in their initially outstretched arms. In each case the friction of the ice on skates, the floor on dance slippers, the ball bearings on the turntable, or the pivot on the turning rod can be neglected on the time scale it takes to pull the arms in, so angular momentum is effectively conserved. We shall learn how to treat the distributed mass of a rotating person quantitatively in the next chapter. In the present example we analyze the system of the (massless) rod connecting (point) masses, which exhibits the same qualitative behavior while being simple enough to analyze completely with our present methods.

Each mass moves in a circle of radius r_1 about the axis of rotation, with constant speed v. At any instant, the velocity $\mathbf{v}$ of the mass is perpendicular to the radius vector $\mathbf{r}_1$. Therefore the magnitude of its angular momentum is

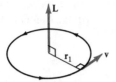

$$L = mvr_1 \sin 90° = mvr_1.$$

The direction of the angular momentum is perpendicular to the plane of the circle, as shown in the figure. Since the rod has been assumed massless, its angular momentum can be neglected.

The total angular momentum is the sum of the individual angular momenta. The angular momentum of both masses is in the same direction, so the magnitude of the total angular momentum is

$$L_1 = mvr_1 + mvr_1 = 2mvr_1$$

where r_1 is the distance from the axis and v is the speed of each mass. We can write this in terms of angular speed by recalling that $v = r_1\omega_1$:

$$L_1 = 2mr_1^2\omega_1.$$

When the masses are pulled in to a new distance r_2, the angular momentum will be

$$L_2 = 2mr_2^2\omega_2.$$

The force applied to pull in the masses acts radially inward along the rod, so $\mathbf{F}$ is parallel to $\mathbf{r}$, where $\mathbf{r}$ is the position vector of the mass relative to the center of the rod. Thus the torque $\boldsymbol{\tau} = \mathbf{r} \times \mathbf{F}$ acting on the mass about the center of the rod vanishes and angular momentum is conserved.

Conservation of angular momentum implies that $L_1 = L_2$. Equating the two angular momenta and solving for ω_2, we get

$$\omega_2 = \omega_1(r_1^2/r_2^2).$$

Substituting $\omega_1 = 2$ rad/s, $r_1 = 0.4$ m, and $r_2 = 0.3$ m, we find $\omega_2 = 3.6$ rad/s.

To pull the masses in, a centripetal force had to be supplied. It did work which went into rotational energy. We shall discuss this aspect of the problem in Chapter 14.

13.4 FORCE AND TORQUE

In equilibrium, as discussed in Sections 6.5 and 6.6, both the net force and the net torque on the system must vanish:

$$\sum \mathbf{F}_i = \mathbf{0}, \qquad\qquad\qquad\qquad\qquad (13.12a)$$

$$\sum \boldsymbol{\tau}_i = \mathbf{0}, \qquad\qquad\qquad\qquad\qquad (13.12b)$$

yielding a total of six scalar equations. For simplicity, in this book we usually consider only forces in a plane and torques about an axis normal to this plane. When both $\mathbf{F}_i$ and $\mathbf{r}_i$ are confined to the xy plane, the torque lies in the z direction. So we make frequent use only of the three conditions

$$\sum F_{ix} = 0, \tag{13.13a}$$

$$\sum F_{iy} = 0, \tag{13.13b}$$

$$\sum \tau_{iz} = 0. \tag{13.13c}$$

A system in equilibrium need not be at rest. Its center of mass may move with uniform velocity in the absence of any net force. And, for example, a rigid body may undergo uniform rotation about its center of mass in the absence of any net torque. Momentum and angular momentum are both conserved for a system in equilibrium.

It is important to realize that the equilibrium conditions for force and torque are quite distinct. A body can accelerate under the force of gravity, while experiencing no torque about its center of mass.* Conversely, the forces on a body may cancel, but be applied at different points so as to produce a nonzero torque.

As an example of this last point, Fig. 13.4 shows a *couple*. The forces balance, but the torque is nonzero. The torque about point 1 has the value $\mathbf{r}_{12} \times \mathbf{F}_2$.

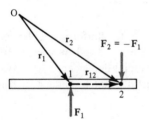

Figure 13.4 A couple.

A curious property of the couple is that its torque about an arbitrary reference point O, relative to which point 1 is situated at $\mathbf{r}_1$ and point 2 at $\mathbf{r}_1 + \mathbf{r}_{12}$, is independent of the reference point:

$$\mathbf{r}_1 \times \mathbf{F}_1 + \mathbf{r}_2 \times \mathbf{F}_2 = \mathbf{r}_1 \times (-\mathbf{F}_2) + (\mathbf{r}_1 + \mathbf{r}_{12}) \times \mathbf{F}_2$$

$$= \mathbf{r}_{12} \times \mathbf{F}_2. \tag{13.14}$$

The torque due to an unbalanced force, by contrast, depends on the reference point.

A couple is an exception to the rule, mentioned in Section 6.6, that for any set of forces a single *resultant* force can be found which gives the same total torque as well as

---

*Note, however, that the body does experience a torque about any other point. In general, if a rigid body is in rotational equilibrium but not translational equilibrium, the sum of the external torques is zero only about the center of mass.

force, and that the resultant can be balanced by a single *equilibrant* force. A couple has no resultant and no equilibrant. To balance a couple it is necessary to introduce *two* more forces, in other words, another couple.

An important special case in which a force does not produce a torque occurs for bodies orbiting around the origin under a *central force*. By central force we mean one of the form

$$\mathbf{F} = f(r)\hat{\mathbf{r}} \tag{13.15}$$

which has a radial direction and a magnitude depending only on r. The gravitational and Coulomb forces exerted by a point mass or charge are examples of central forces, but $f(r)$ need not vary as r^{-2}. A central force exerts a torque about almost all reference points. But if the reference point is placed at the force center itself, then $\mathbf{r}$ is always parallel to the force, so

$$\boldsymbol{\tau} = \mathbf{r} \times \mathbf{F} = \mathbf{r} \times f(r)\hat{\mathbf{r}} = \mathbf{0}. \tag{13.16}$$

The torque vanishes and continues to vanish as $\mathbf{r}$ changes direction in sweeping around the orbit. Therefore *angular momentum about the origin of a central force field is conserved*:

$$\boxed{\mathbf{L} = \mathbf{r} \times m\mathbf{v} = \mathbf{const},} \tag{13.17}$$

if no other forces act.

Note the generality of this result. It holds for uniform circular motion, with $L = rmv$. But it also holds for orbits that swing inward and outward, with v alternatively speeding and slowing. And it holds for any radial dependence $f(r)$.

In the next section we'll use this powerful result to understand Kepler's empirical law of equal areas for planetary motion.

13.5 KEPLER'S LAW OF EQUAL AREAS

As a young man, Johannes Kepler began an immensely fertile search for mathematical relations describing the orbits of planets.

One question he seized upon was the relation between a planet's distance from the sun and the time needed to complete one revolution (the orbital period). The orbital periods of the five planets visible to the naked eye had been known since antiquity. The ratio of their distances from the sun had been determined by Copernicus on the basis of his heliocentric model. [For a clear description of how this was done see A. P. French, *Newtonian Mechanics* (Norton, New York, 1971), pp. 246–9.] The greater the planet's distance from the sun, the longer its period. However, a precise mathematical ratio was lacking. Saturn, for example, is twice as far as Jupiter from the sun, but its 30-year period is not twice the 12-year period of Jupiter. No simple ratio relates the distances and periods of the other planets either. The orbital period of a planet increases with the planet's distance from the sun, but not in the same proportion as the distance. Nobody before Kepler is on record as having asked why this should be so.

Kepler theorized that there must be a force emanating from the sun which drives the planets in their orbits around the sun. The outer planets move more slowly because this driving force diminishes with distance. Kepler's proposal had revolutionary significance. For the first time an attempt was made not only to describe heavenly motion in geometrical terms, but to assign a physical cause to it. After a divorce of 2000 years, astronomy and physics met again. This reunion led to Kepler's three laws, the pillars on which Newton built his universe.

After years of false starts and drudgery searching for a more accurate description in the Copernican system of the motion of the planets, Kepler realized that the ancient idea of circular orbits had to be abandoned to conform to observed data. Through obstinate perseverance, he discovered on the basis of Tycho Brahe's new observations that the actual orbits correspond more to elliptical paths rather than the circular paths of the Copernican system. The sun is at one focus of the ellipse, and the earth, for example, moves around it as shown in Fig. 13.5. This is *Kepler's first law: the planets orbit the sun along elliptical paths.* We'll say more about this law in Chapter 16.

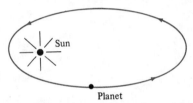

Figure 13.5 Illustration of Kepler's first law: the orbits of planets are ellipses with the sun at one focus.

Since antiquity, astronomers had observed that the planets do not move along their orbits at constant speed. Each planet moves faster when it is close to the sun than when it is far away. This means that if we draw the angles swept out by the earth moving around the sun in equal intervals of time at two different parts of the orbit, as illustrated in Fig. 13.6, one angle is smaller than the other. In searching for some regularity in this

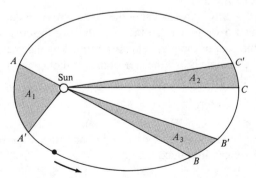

Figure 13.6 An illustration of Kepler's second law: the radius vector from the sun to a planet sweeps out equal areas in equal times.

orbit, Kepler discovered a law of amazing simplicity. As a planet moves in its orbit, *the radius vector from the sun to the planet sweeps out equal areas in equal times*; this is *Kepler's second law*. Kepler also succeeded in finding a precise relation between a planet's orbital period and its mean distance from the sun. This came to be known as his third law and we'll discuss it in Chapter 17.

Nearly 100 years after Kepler, Isaac Newton formulated his laws of mechanics and demonstrated that Kepler's laws can be deduced from his mechanics. Thus Newton realized Kepler's dream of assigning a physical cause to the geometrical motions of the planets. We'll now demonstrate precisely how Kepler's second law of equal areas arises from Newton's dynamics. But first we need a mathematical description of the law of equal areas.

Imagine a radius vector at some time t, $\mathbf{r}(t)$, drawn from the sun to a planet as in Fig. 13.7. If the planet moves through a displacement $\Delta\mathbf{r}$ in a short time Δt, then its new position is $\mathbf{r}(t + \Delta t) = \mathbf{r} + \Delta\mathbf{r}$. As illustrated in Fig. 13.7, the three vectors $\mathbf{r}(t)$,

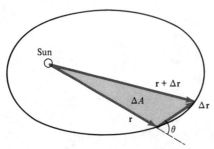

Figure 13.7 Position vector of a planet sweeping out an area ΔA in time Δt.

$\Delta\mathbf{r}$, and $\mathbf{r}(t + \Delta t)$ form a triangle. The area ΔA of this triangle is one-half the base r times the height, $\Delta r \sin \theta$; thus

$$\Delta A = \tfrac{1}{2} r (\Delta r \sin \theta)$$

where θ is the angle between $\mathbf{r}$ and $\Delta\mathbf{r}$.

Writing the area in this way suggests using the vector cross product to represent the area. Recall that in Chapter 5 we found that the magnitude of the cross product between two vectors is equal to the area of the parallelogram formed by the vectors, as Fig. 13.8 shows. And the area of the triangle is half the area of the parallelogram. In vector notation, we can write the area of the triangle as

$$\Delta A = \tfrac{1}{2} |\mathbf{r} \times \Delta\mathbf{r}|.$$

The direction of the cross product, given by the right-hand rule, is perpendicular to the plane of the vectors. So we can introduce the vector

$$\Delta\mathbf{A} = \tfrac{1}{2} \mathbf{r} \times \Delta\mathbf{r}$$

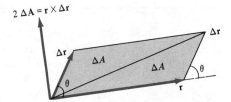

Figure 13.8 Area vector $\Delta\mathbf{A}$ as a vector cross product

and call it the area vector $\Delta\mathbf{A}$. Its length is the area of the triangle formed by $\mathbf{r}$ and $\Delta\mathbf{r}$.

The rate of change of this area vector, $d\mathbf{A}/dt$, is the limit of $\Delta\mathbf{A}/\Delta t$ as the time interval shrinks to zero:

$$\frac{d\mathbf{A}}{dt} = \lim_{\Delta t \to 0} \frac{\Delta\mathbf{A}}{\Delta t} = \lim_{\Delta t \to 0} \frac{\frac{1}{2}\mathbf{r} \times \Delta\mathbf{r}}{\Delta t} = \frac{1}{2}\mathbf{r} \times \lim_{\Delta t \to 0} \frac{\Delta\mathbf{r}}{\Delta t}.$$

But the last limit is simply the velocity $\mathbf{v}$ of the planet. Therefore we find that the rate of change of the area vector is

$$\frac{d\mathbf{A}}{dt} = \frac{1}{2}\mathbf{r} \times \mathbf{v}. \qquad (13.18)$$

Now the connection to Newton's dynamics becomes clear. Comparing Eqs. (13.18) and (13.3), we recognize that $d\mathbf{A}/dt$ is proportional to angular momentum:

$$\frac{d\mathbf{A}}{dt} = \frac{\mathbf{L}}{2m}. \qquad (13.19)$$

From Newton's second law we deduced that the angular momentum vector $\mathbf{L}$ is conserved in a central force field*; that is,

$\mathbf{L}$ *is constant in magnitude and direction.*

Thus, Kepler's law of equal areas follows from Newton's second law.

The result that $\mathbf{L}$ has constant *direction* also supplies part of Kepler's first law, which states that each orbit is an ellipse and hence lies in a plane. Since $\mathbf{L}/m$ is the cross product of $\mathbf{r}$ and $\mathbf{v}$, each of $\mathbf{r}$ and $\mathbf{v}$ must be perpendicular to $\mathbf{L}$. But $\mathbf{L}$ has constant direction, so both $\mathbf{r}$ and $\mathbf{v}$ must lie in a plane perpendicular to this direction. In other words, conservation of angular momentum implies a planar orbit.

*Strictly speaking, it is the center of mass of the sun-planet system, not the center of the sun, at which the origin of the central force field is located and about which $\mathbf{L}$ is conserved. The reason is that the center of the sun is always accelerated toward the planet, and the sun as well as the planet revolves about the center of mass. Although at first sight the center of the sun is also the origin of the central force, in fact the reference frame with origin fixed at the center of the sun is noninertial, entailing noninertial forces which prevent $\mathbf{L}$ from being conserved about this point. The center of mass, on the other hand, is unaccelerated and thus can be placed at the origin of an inertial frame. In practice, however, the sun is so much more massive than the planets that its center lies very near the center of mass, and one can ignore the difference between the two points for most purposes.

Example 3

At its closest approach (perihelion), the earth is 1.47×10^8 km from the sun and its speed is 30.2 km/s. What is the speed at its farthest (aphelion) point from the sun, a distance of 1.52×10^8 km?

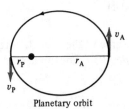

Planetary orbit

As illustrated with the ellipse above, when a planet is either at its closest or farthest point from the sun, its velocity is perpendicular to the radius vector. For these points Eq. (13.18) and the fact that dA/dt is constant imply that

$$\tfrac{1}{2} r_P v_P = \tfrac{1}{2} r_A v_A .$$

Solving for v_A, the speed of the planet at its farthest distance, we have

$$v_A = v_P(r_P/r_A) = (30.2 \text{ km/s})(1.47/1.52) = 29.2 \text{ km/s}.$$

13.6 VORTICES AND FIRESTORMS

We've just found that the law of conservation of angular momentum explains why planets orbiting the sun sweep out equal areas in equal times – an empirical law which Kepler discovered by examining voluminous data. The conservation of angular momentum applies not only to planets moving silently around the sun, but to any moving body for which $\mathbf{r} \times m\mathbf{v}$ is constant. One interesting application of this law is to vortices.

Suppose you have a huge bowl (a bathtub will suffice), which you fill with water through a hose, as shown in Fig. 13.9. As water pours into the bowl, it moves in some type of circulatory motion. If you turn off the water, that rotary motion will die out very slowly; it may not completely disappear for hours or even days. What happens if you pull the plug on the bottom of the bowl while the water is still rotating? Of course, the water will run out.

When you first pull the plug, the water flows straight out through the hole, but after a while, the water goes into a new kind of flow state. Instead of flowing straight out, it forms a whirlpool. Why? Aside from a small amount of viscosity from the walls of the bowl, there are no torques acting on the water. Consequently, the water conserves angular momentum as it runs out.

The angular momentum of a small portion of water at any instant as it spirals toward the center of the bowl is $\mathbf{L} = \mathbf{r} \times m\mathbf{v}$ about an axis through the center of the bowl.

Figure 13.9 Vortex formation in a bowl of water as a result of angular momentum conservation.

Because **r** and **v** are perpendicular, the magnitude of the angular momentum of this chunk is

$$L = mvr = \text{const.}$$

The precise value of the constant depends on the initial conditions of rotation.

As each small bit of water moves down toward the hole, it conserves angular momentum. So as the distance from the central axis becomes smaller, the velocity becomes increasingly larger, according to

$$v = \text{const}/mr.$$

When the water is moving in very small circles, it is moving very rapidly. But an inward force must keep this bit of water moving in a circle. That force which provides the centripetal acceleration is the tensile strength of the water – the ability of the water to hold itself together.

As the distance r becomes very small, the velocity correspondingly becomes extremely large. A large force is required to hold the water together. But when this force exceeds the tensile strength of the water, the water can no longer keep itself moving in a circle, so the surface ruptures and forms a hole. The hole in the center of the whirlpool is called a vortex.

The same dynamics occurs in other instances on a much larger scale. Whenever there is a large fire, such as a forest fire, great destruction can occur. This happened in the Chicago fire of 1871, and in the bombing of Hamburg, Germany, during the Second World War. The cause of the destruction was not a simple fire, but something more devastating, a firestorm. Firestorms are vortices similar to those in a bathtub, and are a result of conservation of angular momentum.

The heat of the fire causes the air to swell upward (just as we had water flowing downward). The resulting low-pressure region at the bottom draws in oxygen from the sides, so the fire burns faster. If there is any circulatory motion in the air, the air moves faster and faster near the center of the fire when the upwelling occurs. Since the air cannot provide the centripetal force to keep itself moving so rapidly, a vortex forms. Thus the same phenomenon occurs as in a bowl of water, except upside down. This vortex is a long-lived and violent state – the firestorm.

Meteorologists investigating historical records for the Great Fire of London in 1666 discovered that the air in the vicinity at the time had no distinct circulatory motion. Consequently, they believe that a firestorm never developed. Their conjecture is borne out by a comparison in the tolls: in the London fire, only four people died directly as a result of the fire and only 436 acres were burnt in 87 hours; on the other hand, in the Chicago fire, 2124 acres were burnt and 250 people killed in two days.

Another example of the same phenomenon occurring without a fire is upwellings of air in the atmosphere. If there is an upwelling of air, such as that above warm water, and if there is any circulatory motion, a vortex can form. In this case it is called a hurricane. Hurricanes can be long-lived (several days), persistent, and very destructive. The longest-lived hurricane we know of is on the planet Jupiter; it is the Great Red Spot of Jupiter, which has persisted for at least 300 years (we've only been able to observe it since the invention of the telescope in 1609). Figure 13.10 is a photograph of the Red Spot taken by *Voyager 1* on its flight past the planet. Years before the *Voyager* fly-by, it was suggested that the Red Spot was a huge hurricane, three times the size of the earth. Data from *Voyager* support this idea.

Figure 13.10 Photograph from *Voyager 1* of the Great Red Spot on Jupiter. (Courtesy JPL/NASA.)

13.7 CONSERVATION OF ANGULAR MOMENTUM AND ENERGY

When a particle moves in a gravitational or Coulomb force field, its direction changes continuously so momentum is not conserved. However, its angular momentum about the force center is conserved, as we have seen, and its total mechanical energy is also conserved.

Because of angular momentum conservation, the motion is confined to a plane perpendicular to the angular momentum vector. The situation is most readily analyzed by resolving the velocity vector into two perpendicular components, a radial component v_r and a transverse component v_θ, as indicated in Fig. 13.11.

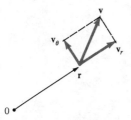

Figure 13.11 Components of velocity in perpendicular directions.

Since $v^2 = v_r^2 + v_\theta^2$ by the theorem of Pythagoras, the conserved energy can be written

$$E = U(r) + K = U(r) + \tfrac{1}{2}mv_\theta^2 + \tfrac{1}{2}mv_r^2 . \qquad (13.20)$$

For the gravitational or electrical applications we have in mind

$$U(r) = -\frac{D}{r}, \qquad (13.21)$$

where D is GMm or $-K_e q_1 q_2$, at large r, but $U(r)$ may have a less singular r dependence at small r if the source is spread out rather than being pointlike.

The angular momentum of the particle about the force center has magnitude

$$L = rmv_\theta ,$$

so we can rewrite the nonradial part of the kinetic energy as

$$\frac{mv_\theta^2}{2} = \frac{L^2}{2mr^2} . \qquad (13.22)$$

This is interpreted as the *rotational kinetic energy* of the particle. The total energy now has the form

$$E = U(r) + \frac{L^2}{2mr^2} + \frac{mv_r^2}{2} . \qquad (13.23)$$

If we call the first two terms on the right-hand side

$$U_{\text{eff}}(r) = U(r) + \frac{L^2}{2mr^2} \qquad (13.24)$$

and plot them as in Fig. 13.12, we find that $U_{\text{eff}}(r)$ rises rapidly at small r and exceeds the total energy E at some radius r_{min}. Because the radial kinetic energy $\tfrac{1}{2}mv_r^2$ must be positive, r_{min} is the closest the particle can approach the origin.

Another way to see that the particle cannot reach arbitrarily small r is by considering forces. Although the attractive gravitational or electrical force urges the particle to come closer, angular momentum conservation requires the particle to keep circulating as well.

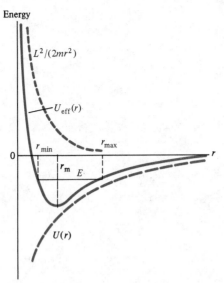

Figure 13.12 Potential, rotational kinetic, and total energy of a particle moving under a gravitational or electrical force.

The centripetal force

$$F_c = \frac{mv_\theta^2}{r} = \frac{(mv_\theta r)^2}{mr^3} = \frac{L^2}{mr^3} \tag{13.25}$$

required to keep the particle circulating grows faster at small r than the attractive force, which grows as $|F| = D/r^2$ for a point gravitational or electrical source or more slowly for a spread out source. The available attractive force is insufficient to supply the required centripetal force when

$$|\mathbf{F}| < \frac{L^2}{mr^3}, \tag{13.26}$$

as happens for example at

$$r < r_0 = \frac{L^2}{mD} \tag{13.27}$$

if $|\mathbf{F}| = D/r^2$. Beyond this point an inward-moving particle is slowed down, and eventually turns around.

If $E < 0$, that is, if a particle's initial radial velocity is less than its escape velocity, there is also a maximum distance r_{max} as pictured in Fig. 13.12. So a particle with $E < 0$ is constrained to shuttle back and forth between r_{min} and r_{max}, with the radial kinetic energy $\frac{1}{2}mv_r^2$ making up the difference between $U_{eff}(r)$ and the constant E at each r.

Although $U_{eff}(r)$ is a hybrid of $U(r)$ and rotational kinetic energy $L^2/2mr^2$, it depends only on r for constant L, and is often called the *effective potential* because of the potential-like role it plays in Fig. 13.12.

If there is other matter present to supply resistive or frictional forces, the particle will tend to lose radial kinetic energy and settle down into a circular orbit at the minimum or equilibrium point of $U_{eff}(r)$, located at r_m in Fig. 13.12. In determining r_m it is necessary to be careful about one point, though: the interaction with other matter will also tend to change the angular momentum of the particle we are considering. Nevertheless, its L is not likely to fall to zero, because the overall angular momentum of the system is conserved and must be shared among the particles. Thus the shape of $U_{eff}(r)$ for a typical particle will still resemble Fig. 13.12, although the particle's angular momentum will no longer have its initial value and the particle will still settle down into a circular orbit.

The process we have just gone through – setting up an idealized problem in which the energy and angular momentum of a particle in a spherically symmetric system are conserved, and then admitting that in reality dissipative forces change the energy and angular momentum of the particle (and even the spherical symmetry of the system), is typical of physics. The idealizations allow us to express the essence of the problem in terms of simple mathematics. Once this is done, the outlines of how realistic complications will change the answer can often be seen without having to calculate their effects in complete detail.

The conservation of overall angular momentum has interesting implications for a system that changes its size or shape. An especially fascinating application is something which not even the imaginative Kepler would have envisioned – the contraction of colossal clouds of gas and dust to form galaxies and solar systems.

Suppose initially the cloud is spherical, but has a rotation about a single axis. If a particle is initially on the "equator" near the outside of this mass distribution and sharing the general circulation, it will experience a $U_{eff}(r)$ of the qualitative nature we have discussed in Fig. 13.12, although $U(r)$ will have a somewhat different r dependence because the attracting matter is not all concentrated at the center.

In a condensing gas cloud, a particle spirals in, losing energy by collisions with other particles until it eventually reaches the lowest energy possible for its angular momentum. The stable orbit reached at this energy is a circle, as described earlier. As Fig. 13.13 indicates, particles with different values of angular momentum will have different minimum values of energy corresponding to circular orbits of different radii.

For a given $U(r)$ and L we can mathematically determine the radius of the circular orbit that corresponds to minimum energy; it is the value of r for which the slope of the curve is zero. In other words, the minimum energy occurs where $dU_{eff}(r)/dr = 0$. Taking the derivative of $U_{eff}(r) = U(r) + L^2/2mr^2$ in the approximation where the potential has the form $U = -GMm/r$, we obtain

$$\frac{dU_{eff}(r)}{dr} = -\frac{L^2}{mr^3} + G\frac{Mm}{r^2} .$$

Setting this equal to zero, and solving for r_m, the corresponding distance, we find

$$r_m = \frac{L^2}{GMm^2} . \tag{13.28}$$

Looking back at Eq. (13.27) we recognize r_m as r_0, the distance where the gravitational attraction supplies exactly enough centripetal force to maintain circular motion. The radius of the circular orbit can be found either by minimizing U_{eff} or from this force requirement.

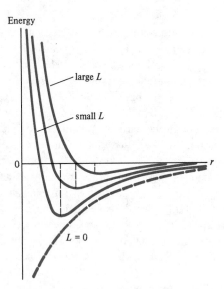

Figure 13.13 Stable orbit of a particle subject to dissipative collisions depends on the angular momentum.

We now have a picture of the evolution of a rotating gas cloud. As Fig. 13.14 depicts, the cloud begins to contract toward a point under the mutual attraction of gravity. As it contracts in the equatorial direction, the cloud can't move all the way in because of overall conservation of angular momentum. Particles will settle into circular orbits whose radii are prescribed by Eq. (13.28). However, in the vertical direction, contraction can take place without any increase in rotational kinetic energy and therefore continues after the limit to contraction in the equatorial direction has been reached. Consequently, the cloud tends to flatten out, forming galaxies like those in Fig. 13.15.

In addition to the pancakelike shape, some galaxies exhibit spiral arms. The simple explanation we found for the flattening is not sufficient to explain the structure of the spiral arms. Different phenomena are occurring. Likewise, when gas clouds condense to form solar systems, planets are formed, rather than a uniformly flattened structure as we have in galaxies. The same thing happens in planetary systems; moons are formed out of the gas and dust condensing into planets. Sometimes rings are formed. Since at least three planets in our solar system have rings, we know that ring formation is not unusual.

In all these phenomena, there presumably is no underlying physics involved other than Newton's laws. If we could solve the differential equations for all sorts of conditions, then we should be able to predict all the spectacular phenomena, such as Saturn's braided rings, which were unexpected before the *Voyager* exploration. There are solutions to Newton's simple equations which nobody ever guessed existed. Commenting on this, Nobel laureate Richard P. Feynman wrote in 1962:

> There are those who are going to be disappointed when no life is found on other planets. Not I. I want to be reminded and delighted and surprised once again through interplanetary exploration of the infinite variety and novelty of phenomena that can be generated from such simple principles. The test of science is its ability to predict. Had you never visited earth,

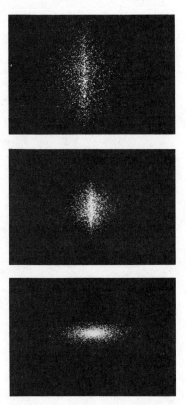

Figure 13.14 Successive stages in the contraction of a rotating sphere of gas.

could you predict the thunder storms, the volcanoes, the ocean waves, the auroras and the colorful sunsets? A salutory lesson it will be when we learn of all that goes on, on each of those dead planets – those eight or ten balls – each agglomerated from the same dust cloud and each obeying exactly the same laws of physics.

13.8   A FINAL WORD

We now have derived the third of the three great conservation laws in physics: conservation of energy, conservation of linear momentum, and conservation of angular momentum. All three laws came from Newton's law, $\mathbf{F} = m\mathbf{a}$. On the other hand, the conserved quantities involved are simple things which must be true *all* the time; they are useful bookkeeping devices for applying Newton's laws in complicated situations. For example, to have written Newton's laws for the water in the vortex bowl as it spirals down and out through the hole would have been exceedingly difficult. But the principle of why a hole develops in the middle is not difficult to understand, once we understand that angular momentum is conserved. This suggests that the conservation laws may be more than simply special ways of applying Newton's laws.

In the twentieth century the world witnessed a scientific revolution which is comparable in magnitude to the original scientific revolution of the period from Copernicus to Newton. We no longer believe that Newton's laws are a sufficient, adequate description

Figure 13.15 Pictures of galaxies exhibiting pancakelike structure.
(Palomar Observatory photograph.)

of the way the universe actually works. His laws are a special case, an approximation to deeper and more exact laws, known as the laws of quantum mechanics. Surprisingly, though, the three conservation laws of energy, linear momentum, and angular momentum exist in quantum mechanics also. Remember that they were derived from Newton's laws, which we no longer believe are universally correct. But in the most profound and correct theory that we have, the conservation laws themselves persist.

We've revealed some of the most profound aspects of nature – the three conservation laws, laws you'll never have to abandon. Hold on to them; they'll serve you well.

Problems

Torque and Angular Momentum

1. A worker finds it difficult to twist a stubborn bolt with a wrench, so he attaches a rope to the wrench as shown in (b), and pulls just as hard on the rope. How does the torque applied with the rope compare to that without the rope? How does the applied torque change if the worker runs the rope over a fixed pulley as shown in (c)?

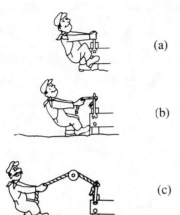

(a)

(b)

(c)

2. A 1.5-kg particle has position and velocity as shown with $r = 0.4$ m and $v = 3.0$ m/s. The magnitude of the force acting on it is 4.0 N. Relative to the origin find

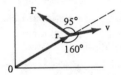

(a) the angular momentum of the particle,
(b) the torque acting on it.

3. A particle at position $\mathbf{r} = (3\hat{\mathbf{i}} + 6\hat{\mathbf{j}} - 2\hat{\mathbf{k}})$ m is acted upon by a force $\mathbf{F}$ equal to $(-4\hat{\mathbf{i}} - 3\hat{\mathbf{j}} + 5\hat{\mathbf{k}})$ N. Find the torque about the origin acting on the particle.

4. In a certain coordinate system, a 0.5-kg particle is at a position given by the vector $\mathbf{r} = (3\hat{\mathbf{i}} - 2\hat{\mathbf{j}} + \hat{\mathbf{k}})$ m moving with a velocity $\mathbf{v} = (-5\hat{\mathbf{i}} + 3\hat{\mathbf{j}} - 2\hat{\mathbf{k}})$ m/s.

(a) Find the angular momentum of this particle with respect to the origin of the coordinate system.
(b) Calculate the angular momentum relative to the point (1,1,1).

5. A particle of mass m is at the end of a rod of negligible mass and length R.

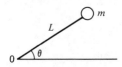

(a) If at any instant the rod makes an angle θ with the horizontal as it falls, calculate the torque about 0 acting on the mass as it falls.

(b) If the mass is started from the vertical postion ($\theta = \pi/2$) with initial speed v, find its angular momentum about 0 as a function of θ.

(c) Use your results to find the angular speed $d\theta/dt$ of the mass.

6. Two identical masses, each of mass m, move along straight lines with velocities $\mathbf{v}$ and $-\mathbf{v}$, respectively, as shown in the diagram.

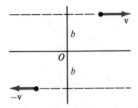

(a) Calculate the angular momentum of the system with respect to the origin 0.

(b) In this particular problem, $\mathbf{L}$ is the same with respect to any reference point, as you can readily check by considering some simple cases such as $\mathbf{L}$ about a point on the upper dotted line. Construct a proof that the angular momentum of two mass points is independent of the reference point whenever (as in the present problem) the total *linear* momentum $\Sigma\, m_i v_i$ of the system vanishes.

7. A body of mass m falls from rest in the earth's gravitational field according to Galileo's law, $z = z_0 - \frac{1}{2}\, gt^2$. Its horizontal coordinates are $x = x_0$, $y = 0$.

(a) Determine the position vector $\mathbf{r}$ and velocity vector $\mathbf{v}$ for this case.

(b) Find the angular momentum vector $\mathbf{L}$ about the origin as a function of time.

(c) Find the torque vector $\boldsymbol{\tau}$ about the origin.

(d) Show explicitly that $\boldsymbol{\tau} = d\mathbf{L}/dt$ and that the angular impulse equals the change in angular momentum.

8. A particle of charge e moving with velocity $\mathbf{v}$ in a magnetic field $\mathbf{B}$ feels a force $\mathbf{F} = e\mathbf{v} \times \mathbf{B}$. The field is uniform along the z axis, $\mathbf{B} = B_0\hat{\mathbf{k}}$.

(a) For what direction of $\mathbf{v}$ is momentum $\mathbf{p}$ conserved?

(b) If $\mathbf{v} = 3\hat{\mathbf{j}} + 5\hat{\mathbf{k}}$ and the particle is located at $\mathbf{r} = 2\hat{\mathbf{i}} + 2\hat{\mathbf{j}}$, what is the torque on the particle around the origin?

(c) At an instant when the particle is at $\mathbf{r} = 5\hat{\mathbf{i}}$, for what directions of $\mathbf{v}$ is $d\mathbf{L}/dt = 0$?

(d) If the particle has initial position $\mathbf{r} = 3\hat{\mathbf{j}}$ and initial velocity $\mathbf{v} = 5\hat{\mathbf{i}}$, in what plane does it move? Show that the motion is a circle of radius $mv/(eB_0)$ in that plane.

Angular Momentum Conservation

9. Suppose an object were acted upon by the force $\mathbf{F} = -k\mathbf{r}$, where $\mathbf{r}$ is the radius vector from some origin and k is constant. Would the angular momentum about the origin of the particle be conserved? Explain why.

10. A mass m is twirled in a circle of radius r_1 with a constant speed v_1. If the string is pulled so that the mass moves in a circle of radius r_2 as shown, find

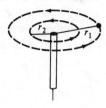

   (a) the speed of the particle,
   (b) the angular speed of the particle.

11. Two objects of mass 2.0 kg and 3.0 kg are connected by a light rod and move in a horizontal circle as shown. The speed of each is 1.5 m/s.

   (a) What is the total angular momentum of the objects about the center?
   (b) If the rod contracts uniformly to one-third of its original length, will the speed of the objects change? If so, by how much?

12. A mass m revolves at a distance of r_1 about an axis with an angular velocity ω_1. Suddenly the distance from the axis is decreased to r_2 by application of a radially inward force.

   (a) What is the new angular velocity?
   (b) By how much is the kinetic energy increased?
   (c) Show that your answer to part (b) is precisely the amount of work done by the centripetal force during the contraction from r_1 to r_2.

13. A toy train slides with coefficient of friction μ on a track fastened to a horizontal bicycle wheel. The wheel turns about its vertical axis without friction. Initially the train and wheel are at rest.

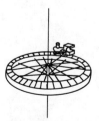

(a) The train is struck horizontally, setting it into clockwise motion. Describe the subsequent motion of the train and the wheel.

(b) If, after losing half its initial speed, the train is again struck horizontally, bringing it suddenly to rest while the wheel still rotates, describe the ensuing motion. Give the relation between the final wheel speed in this case and the final wheel speed achieved in part (a) when the wheel was not struck a second time.

Force and Torque

14. (a) Find the resultant for the unbalanced pair of forces $-F\hat{\mathbf{j}}$ applied at the origin and $(F + \Delta F)\hat{\mathbf{j}}$ applied at $x\hat{\mathbf{i}}$.

(b) Show that the point of application of this resultant moves to larger x as ΔF is decreased, and in fact moves to $x = \infty$ as ΔF approaches zero.

(c) In the limit as $\Delta F \rightarrow 0$ we are left with a couple $-\mathbf{F}$ and $\mathbf{F}$. Show that $-\mathbf{F}$ and $\mathbf{F}$ could point in any direction in the xy plane and still provide the same torque.

Kepler's Law of Equal Areas

15. Referring to the elliptical orbit shown in Fig. 13.6, where is the velocity of the planet a maximum? Where is the velocity a minimum? What does Kepler's second law imply about the speed of a planet moving in a circular orbit?

16. If the speed of the earth is 29.8 km/s when it is 1.49×10^8 km from the sun, what angle does the velocity vector make with its position vector? (Use data in Example 3 on p. 348.)

Vortices

17. In what way would you expect the radius of a hole in the center of a whirlpool of given angular momentum to depend on the tensile strength of the liquid?

18. In an attempt to provide a physical mechanism that causes the planets to orbit the sun, Descartes (working before Newton) formulated a vortex theory. In this model, the universe is filled with matter. The sun is located at the center of a whirlpool of air. The earth and other planets are embedded in the matter and hence are swept around the sun by the vortex motion of the surrounding medium. Based upon your understanding of vortices, can you find fault with this model?

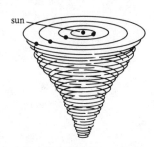

Conservation of Angular Momentum and Energy

19. The orbits of all the planets lie roughly in the same plane. If the planets were formed from a condensing gas cloud, why would you expect this feature?

20. Suppose two gas clouds initially have the same angular momentum but one is much more massive than the other. If these clouds eventually form galaxies, which will form a disk with larger radius?

21. In what type of orbit, elliptical or circular of the same angular momentum, does a planet have greater total energy?

22. A rocket fired off the earth with speed $v_e > 2GM_e/R_e$ will escape the earth no matter which direction it is fired in, as pointed out in Section 10.4. However, the maximum height attained by a rocket fired with speed $v < v_e$ is greater the more vertical the initial direction. Prove this by drawing the curve

$$U_{eff} = \frac{L^2}{2mr^2} + U(r)$$

for a vertical launch at energy

$$E = \frac{1}{2} mv_r^2 + U_{eff} < 0,$$

drawing U_{eff} for a nonvertical launch at the same E, and comparing the maximum distance r_{max} for the two cases. On the same plot, show that $E > 0$ leads to escape for any initial direction of firing.

CHAPTER 14

ROTATIONAL DYNAMICS FOR RIGID BODIES

If each element of a body be multiplied into the square of its distance from the axis 0A and all these products be collected into one sum and if this is put = *Mkk*, which I call the moment of inertia of the body with respect to the axis 0A, then the moment of force required to produce acceleration α will be *Mkk* · α.

• Leonhard Euler, in *Theoria Motus Corporum Solidorum seu Rigidorum* (1765)

14.1 ROTATION OF A RIGID BODY ABOUT A FIXED AXIS

The motion of bodies is determined by Newton's laws. The detailed application of these laws, however, to actual problems such as the motion of the earth under all forces exerted on it by the sun, the moon, and the other planets is a complex matter. One of the main difficulties is that the earth does not spin about a fixed axis in space. Another complication, which manifests itself in tides, is that the earth is not perfectly rigid.

We will tackle some of these complications in due course, but for the present we confine our attention to rigid bodies rotating about fixed axes. The rotation of a flywheel, for example, occurs about a fixed axis. The rolling of a wheel along a straight path involves rotation about an axis which, while moving in space, does not change its direction. These problems are relatively simple. Moreover, we have already noted that torque bears the same relation to rotation as force bears to translation, and in the case of rigid-body rotation we shall find a number of further analogies between rotational and translational quantities.

14.2 CENTER OF MASS OF A CONTINUOUS MASS DISTRIBUTION

Before entering into the details of rigid-body rotation, we need to become more familiar with the techniques for locating the center of mass of an extended body. We recall from Chapter 11 that the center of mass of any extended body, rigid or not, plays a basic role in mechanics as the point $\bar{\mathbf{r}}$ at which Newton's second law acts in the simple form

$$m\frac{d^2\bar{\mathbf{r}}}{dt^2} = \mathbf{F}_{\text{ext}}.$$

Later in this chapter we shall find that complicated motion involving rotation is best analyzed if we divide the problem mentally into a motion of the center of mass and a rotation about the center of mass. We shall also find that in rotational problems a new quantity called *moment of inertia* enters quite naturally and plays the role of inertia. Calculation of moment of inertia is closely related to that for determining center of mass. Therefore we turn now to the center of mass of a continuous mass distribution, generalizing the discussion initiated in Chapter 11.

In Eq. (11.8) we defined the center of mass of a system of n positive masses m_1, m_2, . . ., m_n located at discrete points $\mathbf{r}_1$, $\mathbf{r}_2$, . . ., $\mathbf{r}_n$ by the weighted sum

$$\bar{\mathbf{r}} = \frac{1}{M}\sum_{i=1}^{n} m_i\mathbf{r}_i , \qquad\qquad (11.8)$$

where $M = \Sigma_{i=1}^{n}\, m_i$ is the total mass of the system.

When we deal with a system whose total mass is distributed along all the points of an interval or throughout some region in the plane rather than at a finite number of discrete points, the concepts of mass and center of mass are defined by integrals rather than sums.

For example, consider a rod of length L made of material of varying density. Place the rod along the positive x axis with one end at the origin, and let $m(x)$ denote the mass of the portion of the rod from 0 to x. If there is a continuous function λ such that

$$m(x) = \int_0^x \lambda(x')\, dx',$$

then

$$\frac{dm}{dx} = \lambda(x)$$

and the function λ is called the *mass density* of the rod. The number $\lambda(x)$ is the *mass per unit length* at the point x.

The integral

$$\int_0^L x\lambda(x)\ dx,$$

also written as $\int_0^L x\ dm$, is called the *first moment* of the rod about 0. The *center of mass* is the point whose x coordinate is

$$\bar{x} = \frac{1}{m(L)} \int_0^L x\lambda(x)\ dx. \tag{14.1}$$

This formula is analogous to that in (11.8) except that the sums are now replaced by integrals.

Example 1

A rod of length L has mass per unit length $\lambda(x)$. Locate the center of mass if
  (a) $\lambda(x) = \lambda = $ const. (Such a rod is called *uniform*.)
  (b) $\lambda(x) = cx^2$, where c is constant.

The mass of the rod is $m(L) = \int_0^L \lambda(x)\ dx$ and the center of mass is located at

$$\bar{x} = \frac{1}{m(L)} \int_0^L x\lambda(x)\ dx.$$

  (a)   If $\lambda(x) = \lambda$ (a constant) then the mass is $m(L) = \lambda L$ and

$$\bar{x} = \frac{1}{\lambda L} \int_0^L \lambda x\ dx = \frac{1}{\lambda L} \frac{\lambda L^2}{2} = \frac{L}{2}.$$

The center of mass lies at the center of the rod, as expected.

  (b)   If $\lambda(x) = cx^2$ the mass is $m(L) = \int_0^L cx^2\ dx = \frac{1}{3}cL^3$ and

$$\bar{x} = \frac{3}{cL^3} \int_0^L cx^3\ dx = \frac{3}{cL^3} \frac{cL^4}{4} = \frac{3L}{4}.$$

The center of mass is three-quarters of the way toward the heavy end.

If mass is distributed throughout a plane region rather than on a line, the concepts of mass and center of mass are defined by double integrals.

For example, consider a thin plate having the shape of a region S. Assume that matter is distributed over this plate with a known mass density (mass per unit area). This means there is a nonnegative function ρ defined on S such that $\rho(x, y)$ is the mass per unit area at the point (x, y). The total mass $m(S)$ of the plate is defined by the double integral

$$m(S) = \iint_S \rho(x, y)\ dx\ dy.$$

By analogy with (14.1) the *center of mass* is the point $(\bar{x}, \bar{y})$ whose coordinates $\bar{x}$ and $\bar{y}$ are determined by the equations

$$\bar{x} = \frac{1}{m(S)} \iint_S x\rho(x, y)\, dx\, dy \tag{14.2}$$

and

$$\bar{y} = \frac{1}{m(S)} \iint_S y\rho(x, y)\, dx\, dy. \tag{14.3}$$

In order to understand these definitions it is necessary to explain the concept of the double integral. For the applications we have in mind, double integrals can be expressed in terms of the ordinary one-dimensional integrals that were introduced in Chapter 3. Specifically, suppose the region S consists of all points (x, y) lying between two curves $y = g_1(x)$ and $y = g_2(x)$, with $a \leq x \leq b$, as shown in Fig. 14.1. Then an integral of the form

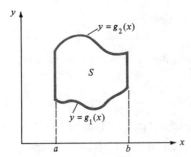

Figure 14.1 Region of integration S.

$$\iint_S F(x, y)\, dx\, dy$$

is defined as follows:

$$\iint_S F(x, y)\, dx\, dy = \int_a^b \left(\int_{g_1(x)}^{g_2(x)} F(x, y)\, dy \right) dx. \tag{14.4}$$

The right-hand side of Eq. (14.4) is to be interpreted as follows. First calculate the one-dimensional integral

$$\int_{g_1(x)}^{g_2(x)} F(x, y)\, dy$$

in which x is kept fixed and $F(x, y)$ is thought of as a function of y, integrated from $y = g_1(x)$ to $y = g_2(x)$. The result of this first integration is a number depending on x; hence it is a function of x which we then integrate from $x = a$ to $x = b$ to obtain the double integral. This definition is illustrated in the next example.

Example 2

An isosceles triangular plate has a constant mass density ρ per unit area and altitude h, as shown in the figure. Find its center of mass.

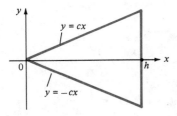

We place the triangle with its axis of symmetry along the x axis and one vertex at the origin. The two equal sides of the triangle have slopes c and $-c$, and the triangular region S consists of all points between the two curves,

$$y = -cx \quad \text{and} \quad y = cx \quad (0 \leqslant x \leqslant h).$$

Hence the mass of the plate is

$$m(S) = \iint_S \rho \, dx \, dy = \int_0^h \left(\int_{-cx}^{cx} \rho \, dy \right) dx = \int_0^h 2\rho cx \, dx = \rho ch^2.$$

The x coordinate of the center of mass is

$$\bar{x} = \frac{1}{m(S)} \iint_S \rho x \, dx \, dy = \frac{1}{\rho ch^2} \int_0^h \rho x \left(\int_{-cx}^{cx} dy \right) dx$$

$$= \frac{1}{\rho ch^2} \int_0^h 2\rho cx^2 \, dx \ = \ \frac{1}{\rho ch^2} \frac{2\rho ch^3}{3} = \frac{2h}{3}.$$

The same calculation for $\bar{y}$ gives $\bar{y} = 0$ because $\int_{-cx}^{cx} y \, dy = 0$. This is to be expected because the center of mass should lie on the axis of symmetry.

Example 3

Find the center of mass of a semicircular disk D of radius R and constant mass density ρ.

When D is placed as shown in the figure it lies between the two curves

$$y = 0 \quad\text{and}\quad y = \sqrt{R^2 - x^2} \quad (-R \leqslant x \leqslant R).$$

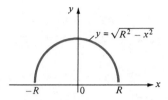

Its mass is

$$m(\text{D}) = \iint_{\text{D}} \rho \, dx \, dy = \int_{-R}^{R} \left(\int_{0}^{\sqrt{R^2 - x^2}} \rho \, dy \right) dx$$

$$= \int_{-R}^{R} \rho \sqrt{R^2 - x^2} \, dx = \rho(\text{area of D}) = \frac{\rho \pi R^2}{2}.$$

For the center of mass $(\bar{x}, \bar{y})$ we have $\bar{x} = 0$ by symmetry, and

$$\bar{y} = \frac{1}{m(\text{D})} \iint_{\text{D}} \rho y \, dx \, dy$$

$$= \frac{1}{m(\text{D})} \int_{-R}^{R} \left(\int_{0}^{\sqrt{R^2 - x^2}} \rho y \, dy \right) dx$$

$$= \frac{2}{\rho \pi R^2} \int_{-R}^{R} \frac{\rho}{2} (R^2 - x^2) \, dx$$

$$= \frac{2}{\rho \pi R^2} \frac{2 \rho R^3}{3} = \frac{4R}{3\pi}.$$

A number of properties of center of mass were discovered by Pappus of Alexandria, who lived around 300 A.D. and was one of the last geometers of the Alexandrian school of Greek mathematics. We mention two of his theorems, which often enable us to determine the center of mass without the use of integration.

One theorem states that the center of mass of a uniform region (constant mass density) lies on any line which is an axis of symmetry of the region.

Another theorem of Pappus states that a region C that is the union of two non-overlapping regions A and B has its center of mass on a line joining the center of mass of A and the center of mass of B.

To illustrate the use of these theorems we consider two examples shown in Figs. 14.2 and 14.3. In Fig. 14.2 a uniform rod C has been bent at its center to form an angle. The vertical dotted line is an axis of symmetry so by Pappus' first theorem the center of mass of C lies on this line. The left-hand portion A has its center of mass at its geometric

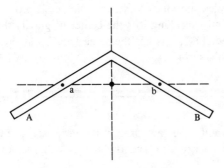

Figure 14.2 Determining the center of mass of a bent rod by using the two theorems of Pappus.

center a, and the right-hand portion B has its center of mass at its geometric center b. Therefore by Pappus' second theorem the center of mass of the entire rod C lies on the horizontal dotted line joining a and b. Hence the center of mass of C lies on the intersection of the two dotted lines as shown.

In Fig. 14.3 we use Pappus' theorems again to determine the center of mass of an L-shaped region C of constant mass density which does not have an axis of symmetry. We divide this region into the union of two rectangles in two different ways, as indicated. By Pappus' first theorem, the center of mass of each rectangle is at its geometric center. Draw a line through the centers a and b of A and B, and another through the centers a' and b' of A' and B', as indicated in the figure. Their intersection c is the center of mass of C.

The foregoing examples illustrate that the center of mass of a body might be outside the body. A surprising example of this phenomenon and of using the location of the

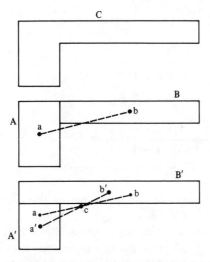

Figure 14.3 Determining the center of mass of an L-shaped region by using the two theorems of Pappus.

center of mass to analyze complex motions of extended bodies is provided by the pole vault in the next example. (Strictly speaking it is the center of gravity that is dealt with in this case, but as noted in Sec. 11.3, this has the same location as the center of mass for ordinary bodies on the earth's surface.)

Example 4

How high can an athlete pole vault if he can sprint at the very fast speed of 10 m/s and is able to convert all his kinetic energy into potential energy with the aid of a fiberglass pole?

The initial kinetic energy $\frac{1}{2}mv^2$ is converted at the peak of the vault into an increase $\Delta U = mgh$ in gravitational potential energy:

$$\tfrac{1}{2}mv^2 = mgh.$$

Solving for h we find

$$h = v^2/2g = (10 \text{ m/s})^2/(2 \times 9.8 \text{ m/s}^2) = 5.1 \text{ m} = 16.8 \text{ ft},$$

well short of the 1985 world record vault of 19 ft 8 in.

But note that the h we have calculated is not the height of the cross bar above the ground, but the height the vaulter's center of gravity has been raised. For it is his center of gravity **r** that the force of gravity acts upon in Eq. (10.24): $\Delta U = -\int \mathbf{F} \cdot d\mathbf{r}$. Taking into account that the runner's center of gravity is at least 3 ft above the ground in his initial vertical position, it can certainly be lifted close to 20 ft above the ground in our ideal vault.

Moreover, as shown in the illustration, much of the jumper's body dangles below the cross bar at the peak of his jump. As in Fig. 14.2, his center of gravity can lie outside of his body in this bent-over position, and it is even possible that his center of gravity passes slightly *under* the bar as he passes over! Thus the theoretical maximum for a pole vault by a 10 m/s sprinter is over 20 ft.

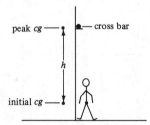

The pole vault is a highly complex event, and you may be able to think of additional ways a skillful performer adds height to his vault.

Passing one's center of gravity under the cross bar is also a useful strategy in the high jump. The modern "Fosbury flop" technique of jumping over backwards, with arms and legs arched downward, accomplishes this better than the older techniques such as the "straddle style," in which the jumper passes over parallel to the bar.

Fosbury flop   straddle

14.3 MOMENT OF INERTIA

We now return to the central concern of this chapter, rigid-body rotation. Consider a rigid-body rotating about a fixed axis, which we take to be the z axis, with angular speed ω. A particle of mass m_i, situated at radius r_i from the z axis and circulating with speed v_i about it, has angular momentum

$$L_i = r_i(m_i v_i) \tag{14.5}$$

about a point on the axis (Fig. 14.4). Since $v_i = \omega r_i$ for rotation about the axis, we can write

$$L_i = m_i r_i^2 \omega. \tag{14.6}$$

Figure 14.4 Circular motion of a particle about an axis normal to the page and passing through point P.

Summing over all particles in the body, we obtain the total angular momentum of the body about the axis of rotation:

$$L = \sum m_i r_i^2 \, \omega, \tag{14.7}$$

since in a rigid body all particles have the same angular speed ω.

The coefficient of ω in Eq. (14.7) is denoted by the symbol I:

$$\boxed{I = \sum m_i r_i^2.} \tag{14.8}$$

There is an exact analogy between the relation

$$\boxed{L = I\omega} \tag{14.9}$$

and $p = mv$: angular momentum L is the rotational analog of momentum p, angular speed ω is the analog of speed v, and I plays a role analogous to mass m.

Applying the law $\boldsymbol{\tau} = d\mathbf{L}/dt$ to rotation about the z axis, we obtain

$$\tau = \frac{d}{dt}(I\omega) = I\alpha \qquad (14.10)$$

where

$$\alpha = \frac{d\omega}{dt} \qquad (14.11)$$

is the *angular acceleration* about the z axis. Equation (14.10) is the fundamental dynamical law for rigid-body rotation about a fixed axis. Once again there is an exact analogy to the dynamical law $F = ma$ for linear motion, with torque τ, angular acceleration α, and I being the rotational analogs of force F, acceleration a, and m.

The quantity $I = \Sigma\, m_i r_i^2$ is called the *moment of inertia*. As usual, *inertia* denotes resistance to change. We see from Eq. (14.10) that I resists angular acceleration due to torque just as, in $F = ma$, m resists acceleration due to force. Thus both I and m are inertial quantities. The term *moment* arises from the fact that in the definition of I, the mass of each particle is multiplied or "weighted" by the square of its distance from the axis of rotation. Just as the quantity $\Sigma\, m_i x_i$ occurring in the center of mass is called the *first moment*, the quantity $\Sigma\, m_i r_i^2$ is called the *second moment*. The weighting makes rotational inertia depend on both the mass *and* its distribution.

Example 5

A flywheel (rotating disk) is attached to a horizontal axle of radius $r = 2$ cm, as shown in the figure. A cord is wrapped tightly about the axle. A mass $m = 2$ kg is attached to the cord and allowed to descend.

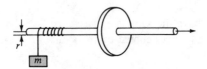

(a)   Find an expression for the angular acceleration α of the flywheel.

(b)   If the flywheel completes its first revolution in one second, find the total moment of inertia I of the flywheel and axle about their axis of rotation.

(a)   The linear acceleration of the mass is related to the tension T in the cord by

$$ma = mg - T.$$

The tension T exerts a torque Tr on the axle-flywheel combination, whose angular acceleration is controlled by the dynamical equation

$$Tr = I\alpha.$$

Since $a = \alpha r$, we have two equations relating α to T. Eliminating T, we find

$$m\alpha r = mg - \frac{I\alpha}{r},$$

so the angular acceleration is

$$\alpha = \frac{mgr}{mr^2 + I}.$$

In other words, the angular acceleration is in the form of torque divided by moment of inertia, and both the suspended mass and the axle-flywheel combination contribute inertia.

(b)   The angular acceleration is constant. Integrating, we find in complete analogy with constant acceleration that the angular speed ω and angle θ are

$$\omega = \alpha t + \omega_0$$

and

$$\theta = \tfrac{1}{2}\alpha t^2 + \omega_0 t + \theta_0.$$

The first revolution is completed when $\theta = 2\pi$. If we take $\theta_0 = \omega_0 = 0$, the time it takes for this is

$$t_1 = \sqrt{\frac{2\theta}{\alpha}} = \sqrt{\frac{4\pi}{\alpha}},$$

which implies

$$I = -mr^2 + \frac{mgr}{\alpha} = -mr^2 + \frac{mgrt_1^2}{4\pi}.$$

When $t_1 = 1$ s we have $\alpha = 4\pi$ and hence $I = 0.003$ kg m^2.

It is important to understand clearly the precise meaning of I. The quantity r_i appearing in its definition is the distance of the ith particle from the axis of rotation,

$$r_i = (x_i^2 + y_i^2)^{1/2}$$

(illustrated in Fig. 14.5), not its distance from the origin, which would be

$$(x_i^2 + y_i^2 + z_i^2)^{1/2}.$$

Because r_i is the distance from the axis of rotation, the numerical value of I depends on the axis chosen. For example, consider a cylindrical disk lying in the horizontal plane. Its I with respect to the vertical axis through its center differs from its I with respect to

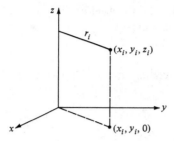

Figure 14.5 Distance r_i from the axis of rotation (taken to be the z axis in this example).

a vertical axis placed off-center. It has yet another value with respect to a *horizontal* axis passing through, say, its center.

The quantity

$$\overline{r^2} = \frac{\sum m_i r_i^2}{\sum m_i} \tag{14.12}$$

is a particular kind of average or mean value of the square of all particle distances from the axis of rotation. We see that the moment of inertia is equal to this mean value multiplied by the total mass M of the body. The quantity

$$k = \sqrt{\overline{r^2}} = \sqrt{\frac{I}{M}}, \tag{14.13}$$

which has the physical dimension of length, is called the *radius of gyration*. It represents the distance from the axis of rotation at which a point mass M would produce the same moment of inertia as the body.

14.4 CALCULATION OF MOMENTS OF INERTIA

The moment of inertia is a number associated with a body and a given axis about which the body could be rotated. For the simplest case of a single particle of mass m at distance r from the axis the moment of inertia is, by definition,

$$I = mr^2,$$

the mass times the square of the distance from the axis.

The moment of inertia (about a given axis) of a system of n particles is the sum of the moments of inertia of the individual particles (Fig. 14.6):

$$I = \sum_{i=1}^{n} m_i r_i^2. \tag{14.8}$$

If all the particles are the same distance $r_i = r$ from the axis then

$$I = r^2 \sum_{i=1}^{n} m_i = Mr^2$$

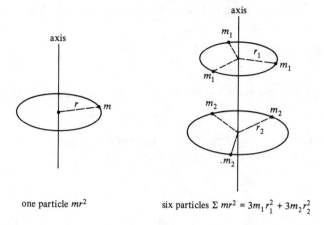

one particle mr^2 six particles $\Sigma \, mr^2 = 3m_1 r_1^2 + 3m_2 r_2^2$

Figure 14.6 Moment of inertia about an axis.

where M is the total mass of the system. In this special case the radius of gyration k equals r, but in general k lies between the smallest and largest r_i.

For a system whose total mass is distributed along a line or throughout some plane region or some solid body in space rather than at a finite number of discrete points, the moment of inertia is defined by an integral rather than a sum. The type of integral depends on the type of body but is always arrived at by the same process. The body is divided into a large number n of small pieces with masses $\Delta m_1, \, . \, . \, ., \Delta m_n$. Each piece of mass Δm_i is treated as though it were a particle with its center of mass at distance r_i from the given axis. According to (14.8) the moment of inertia of this system is the sum

$$\sum_{i=1}^{n} r_i^2 \, \Delta m_i.$$

As the number of pieces n increases without bound and the mass Δm_i of each piece approaches zero, the sum approaches a limiting value which we express symbolically in integral notation as follows:

$$I = \int_{\text{body}} r^2 \, dm. \tag{14.14}$$

To illustrate how such an integral should be interpreted and calculated we consider some simple examples.

Example 6

(a)   Express as an integral the moment of inertia of a rod of given mass density and length L about an arbitrary axis perpendicular to the rod. (b)   Calculate the integral when the rod is uniform (with constant density λ) and the axis passes through the center of the rod.

(a)   Place the rod along the positive x axis with one end at $x = 0$ and the other end at $x = L$, and assume the axis of rotation passes through the point $x = a$. If $m(x)$ denotes the mass of the portion of the rod from 0 to x then the mass density $\lambda(x)$ at x is the derivative

$$\lambda(x) = \frac{dm}{dx}.$$

Divide the rod into n pieces of lengths $\Delta x_1, \ldots, \Delta x_n$. The mass Δm_i of the ith piece is

$$\Delta m_i = \lambda_i \, \Delta x_i$$

where λ_i is the ratio $\Delta m_i / \Delta x_i$. If the center of mass of the ith piece is at x_i then the moment of inertia of this system of n pieces is, according to (14.8),

$$\sum_{i=1}^{n} \Delta m_i (x_i - a)^2 = \sum_{i=1}^{n} (x_i - a)^2 \, \lambda_i \, \Delta x_i.$$

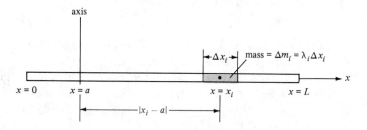

As we let the pieces shrink to zero and let n increase without bound the sum is replaced by the integral

$$I = \int_0^L (x - a)^2 \, \lambda(x) \, dx.$$

(b)   When $a = \frac{1}{2}L$ and the density is constant, $\lambda(x) = \lambda$, the integral becomes

$$I = \int_0^L \left(x - \frac{L}{2} \right)^2 \lambda \, dx = \lambda \int_{-L/2}^{L/2} x^2 \, dx = \lambda \frac{L^3}{12}$$

$$= \lambda L \frac{L^2}{12} = \frac{ML^2}{12},$$

where $M = \lambda L$ is the mass of the rod.

Example 7

Calculate the moment of inertia of a rotating flywheel of radius R and constant mass density ρ about its axis of rotation.

Let D denote the thickness of the flywheel. This time we divide the region into small pieces arranged in thin concentric rings about the axis. We find the moment of inertia of the pieces in each ring and then sum over all the rings to get the total moment of inertia.

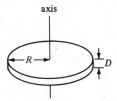

Each thin ring can be thought of as a system of n particles, all at the same distance r from the axis. A typical piece on the ring at distance r has volume $\Delta r\,\Delta L_i\,D$ as shown in the figure, so its mass is $\rho D\,\Delta r\,\Delta L_i$.

The total mass $M(r)$ of the thin ring is

$$M(r) = \rho D\,\Delta r \sum_{i=1}^{n} \Delta L_i = \rho D\,\Delta r\,2\pi r,$$

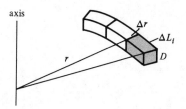

so the moment of inertia of the system in each ring is

$$M(r)r^2 = 2\pi\rho D r^3\,\Delta r.$$

Summing over all the rings and letting $\Delta r \to 0$ we are led to the integral

$$I = \int_0^R 2\pi\rho D r^3\,dr = \tfrac{1}{2}\,\pi\rho D R^4.$$

Since $\rho D\pi R^2 = M$, the mass of the flywheel, we find

$$I = \tfrac{1}{2}\,MR^2.$$

The positive number k such that $I = Mk^2$ is the radius of gyration. In this case we find $k = \sqrt{I/M} = R/\sqrt{2}$.

The *perpendicular-axis theorem* provides a shortcut that is sometimes useful in finding the moment of inertia of a *lamina*, i.e., a thin flat object of uniform thickness and density, such as is obtained by cutting a figure out of a sheet of cardboard. We place our lamina in the xy plane. If the lamina is made up of a finite system of particles we have

$$I_x = \sum m_i y_i^2 \quad \text{(moment of inertia about the } x \text{ axis)}$$

and

$$I_y = \sum m_i x_i^2 \text{ (moment of inertia about the } y \text{ axis).}$$

[Normally, $I_x = \sum m_i(y_i^2 + z_i^2)$, but for a sufficiently thin lamina in the xy plane, z_i^2 is so small it can be ignored. Similarly it can be ignored in I_y.] Adding the two equations for I_x and I_y we get

$$I_x + I_y = \sum m_i(x_i^2 + y_i^2) = \sum m_i r_i^2$$

where $r_i^2 = x_i^2 + y_i^2$ as shown in Fig. 14.7. But the last sum is the moment of inertia of the system about the z axis (perpendicular to the plane of the lamina). Thus we obtain the *perpendicular-axis theorem*, which states that for a lamina of arbitrary shape we have

$$I_x + I_y = I_z, \tag{14.15}$$

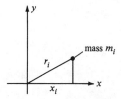

Figure 14.7 Distance of mass m_i from the z axis.

if the x and y axes are in the plane of the lamina. It is important to note that Eq. (14.15) does not apply to objects that are thick or nonplanar.

Example 8
Calculate the moment of inertia of a thin uniform circular disk about a transverse axis through its center (i.e., an axis in the plane of the disk).

   In Example 7 we learned that the moment of inertia of the disk about the z axis is $I_z = \frac{1}{2}MR^2$. Because of the circular symmetry of the disk, the moment of inertia is the same about any axis lying in the plane of the disk and passing through the center. In particular, $I_x = I_y$ so the perpendicular-axis theorem (14.15) implies that

$$I_x = \tfrac{1}{2}I_z = \tfrac{1}{4}MR^2.$$

Table 14.1 lists the moments of inertia for various uniform bodies.

14.5   THE PARALLEL-AXIS THEOREM

Often we know the moment of inertia of a body about an axis through the center of mass, as in Table 14.1, and wish to find it about some other axis parallel to the first one. At

Table 14.1 Moments of Inertia for Uniform Bodies

Body	Axis	I
Rod (length L)	Perpendicular axis through center	$\frac{1}{12}ML^2$
Thin ring (radius R)	Perpendicular axis through center	MR^2
Circular cylinder	Axis of cylinder	$\frac{1}{2}MR^2$
Thin disk	Transverse axis through center	$\frac{1}{4}MR^2$
Solid sphere	Any axis through center	$\frac{2}{5}MR^2$
Thin spherical shell	Any axis through center	$\frac{2}{3}MR^2$
Rectangular plate (length a, height b)	Axis through center perpendicular to the plate	$\frac{1}{12}M(a^2 + b^2)$

first sight the calculation may look complicated, but in fact the answer can be written down immediately with the aid of the parallel-axis theorem, which we'll now prove.

Figure 14.8 represents a section through the rigid body whose moment of inertia we wish to calculate. The axis of rotation, which is perpendicular to the plane of the figure, pierces this plane at point P, and C is the point where an axis passing through the center of mass and parallel to the axis of rotation pierces this plane. The distance between P and C is r; for convenience we choose the x axis so that it passes through both P and C. The moment of inertia of the body about the axis through P is

$$I_\text{P} = \sum m_i r_i^2.$$

From Fig. 14.8, the law of cosines gives

$$r_i^2 = r_i'^2 + r^2 + 2rr_i' \cos \theta_i, \tag{14.16}$$

and we have $r_i' \cos \theta_i = x_i'$, the x coordinate of m_i relative to the center of mass. Hence

$$I_\text{P} = \sum m_i r_i'^2 + \sum m_i r^2 + 2r \sum m_i x_i'. \tag{14.17}$$

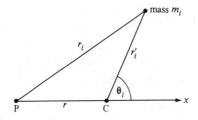

Figure 14.8 The parallel-axis theorem.

The first sum on the right is I_C, the moment of inertia about the center of mass; the second is M, the mass of the body, times r^2, the square of the distance between the two parallel axes. The last sum vanishes because $\Sigma\, m_i x_i'/M$ represents the distance of the x component of the center of mass from C, which is zero. Thus we obtain the *parallel-axis theorem*:

$$I_P = I_C + Mr^2. \tag{14.18}$$

The parallel-axis theorem is very useful in determining moments of inertia about axes that lie off-center in a body, or even outside the body. Note that the moment of inertia is always smaller about an axis through the center of mass than about any other parallel axis.

We can rediscover the parallel-axis theorem for a rod by using the result of Example 6, where we showed that the moment of inertia I_a of a rod about an axis through any point a of the rod is the integral

$$I_a = \int_0^L (x - a)^2\, \lambda(x)\, dx$$

where $\lambda(x)$ is the mass density. Since $(x - a)^2 = x^2 - 2ax + a^2$ we have

$$I_a = \int_0^L x^2 \lambda(x)\, dx - 2a \int_0^L x\lambda(x)\, dx + a^2 \int_0^L \lambda(x)\, dx.$$

The first integral on the right is I_0, the moment of inertia about the left end (where $a = 0$). The second integral is $\bar{x}M$, and the third is M, where $\bar{x}$ is the center of mass and M is the total mass of the rod. Therefore

$$I_a = I_0 - 2a\bar{x}M + a^2 M.$$

Similarly, for a parallel axis through any other point c we have

$$I_c = I_0 - 2c\bar{x}M + c^2 M.$$

Subtracting this from I_a we get

$$I_a - I_c = (a^2 - c^2)M - 2\bar{x}M(a - c).$$

If c is placed at the center of mass so that $\bar{x} = c$, the last equation becomes $I_a - I_c = (a^2 - 2ac + c^2)M$, or

$$I_a = I_c + M(a - c)^2,$$

which is the parallel-axis theorem once again. For a uniform rod (constant density) we can find I_0 by taking $a = 0$ in the above integral, or we can use the parallel-axis theorem with $c = \frac{1}{2}L$, $I_c = \frac{1}{12}ML^2$, and get

$$I_0 = \tfrac{1}{12}\, ML^2 + M\left(\tfrac{1}{2}L\right)^2 = \tfrac{1}{3}ML^2.$$

Example 9

Find the moment of inertia of a disk of radius R suspended by a string of length $L - R$, about an axis which passes through the point of suspension P and is perpendicular to the disk.

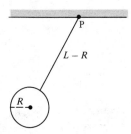

As we found in Example 7, the moment of inertia of the disk about a transverse axis through its center is $I_C = \frac{1}{2}MR^2$. By the parallel axis theorem, the moment of inertia about the axis through P is

$$I_P = \frac{1}{2} MR^2 + ML^2.$$

14.6 ENERGY AND WORK IN RIGID-BODY ROTATION

The kinetic energy of any moving body is the sum of the kinetic energy of its particles;

$$K = \frac{1}{2} \sum m_i v_i^2. \tag{14.19}$$

For a rigid body rotating with angular speed ω about an axis, we have $v_i = \omega r_i$, where r_i is the distance of the ith particle from that axis, so

$$K = \frac{1}{2} \sum m_i r_i^2 \omega^2.$$

But $\sum m_i r_i^2$ is just the moment of inertia about the axis of rotation, so

$$\boxed{K = \frac{1}{2} I \omega^2.} \tag{14.20}$$

Note the analogy to $K = \frac{1}{2}mv^2$ with I once again taking the place of mass and ω replacing v.

The work W done by a constant force F moving a body a distance s parallel to the force is

$$W = Fs. \tag{14.21}$$

Now consider the case in which the force provides a torque about an axis of a rigid body. For example, the suspended weight in Fig. 14.9 supplies a constant torque FR about the

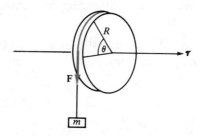

Figure 14.9 Work done by a constant torque.

central axis of the disk. As the disk turns through an angle θ, the weight descends through a distance $s = R\theta$. The work can be expressed in terms of the torque and the angle by

$$W = Fs = FR\theta = \tau\theta. \tag{14.22}$$

In the more general case in which the force is variable in magnitude, but still parallel to the displacement, Eq. (14.22) must be replaced by an integral

$$W = \int F \, ds = \int FR \, d\theta = \int \tau \, d\theta. \tag{14.23}$$

Rewriting Eq. (14.23) as

$$W = \int \tau \frac{d\theta}{dt} \, dt = \int \tau\omega \, dt \tag{14.24}$$

and applying the first fundamental theorem of calculus we find $dW/dt = \tau\omega$. But we recall that the general definition of power is $P = dW/dt$, so we obtain the formula

$$P = \frac{dW}{dt} = \tau\omega \tag{14.25}$$

for power spent in rotational motion. This should be compared with the formula Fv for power due to linear motion.

The dynamical relation (14.10), $\tau = d(I\omega)/dt$, can be inserted into Eq. (14.24) to obtain yet another formula for work,

$$W = \int \frac{d(I\omega)}{dt} \omega \, dt = \int \omega \, d(I\omega).$$

Since I is constant for a rigid body, the work done in passing from angular configuration 1 to 2 is

$$W_{12} = I \int_{\omega_1}^{\omega_2} \omega \, d\omega = \tfrac{1}{2} I\omega_2^2 - \tfrac{1}{2} I\omega_1^2. \tag{14.26}$$

The right-hand side of Eq. (14.26) is recognized as just the change in kinetic energy, so the work done in the rotational displacement from configuration 1 to 2 equals the change in kinetic energy, just as for linear motion of a particle.

14.7   ANALOGIES BETWEEN ROTATIONAL AND TRANSLATIONAL MOTION

We have repeatedly emphasized the general parallelism between the laws of motion as they apply to translation (nonrotating motion) and to rotation. The similarities involve both the *quantities* used to describe the motions and the *general laws* that govern them. We list the similarities systematically in Table 14.2. Keep in mind that the dynamical laws of rotation are not independent of Newton's laws; they are derived from the equation $\mathbf{F} = d\mathbf{p}/dt$.

Table 14.2 Analogies between Translation and Rotation

Corresponding Quantities		
Quantity	Translation	Rotation
Displacement	x	θ
Speed	$v = \dfrac{dx}{dt}$	$\omega = \dfrac{d\theta}{dt}$
Acceleration	$a = \dfrac{dv}{dt}$	$\alpha = \dfrac{d\omega}{dt}$
Inertia	m	$I = \sum mr^2$
Force	$\mathbf{F}$	$\boldsymbol{\tau} = \mathbf{r} \times \mathbf{F}$
Momentum	$\mathbf{p} = m\mathbf{v}$	$\mathbf{L} = \mathbf{r} \times \mathbf{p}$
Impulse	$\displaystyle\int \mathbf{F}\, dt$	$\displaystyle\int \boldsymbol{\tau}\, dt$
General Laws		
Law	Translation	Rotation
Newton's second law	$\mathbf{F} = \dfrac{d\mathbf{p}}{dt}$	$\boldsymbol{\tau} = \dfrac{d\mathbf{L}}{dt}$
Work	$W = \displaystyle\int F\, dx$	$W = \displaystyle\int \tau\, d\theta$
Power	$P = \dfrac{dW}{dt} = Fv$	$P = \dfrac{dW}{dt} = \tau\omega$
Impulse	$\displaystyle\int \mathbf{F}\, dt = \Delta\mathbf{p}$	$\displaystyle\int \boldsymbol{\tau}\, dt = \Delta\mathbf{L}$

There are further similarities for the *special laws* for various detailed kinds of motion. We list some of these in Table 14.3.

14.8 THE PHYSICAL PENDULUM

We have already studied the *simple pendulum*, i.e., a small bob suspended by an inextensible weightless cord or rod, from the point of view of forces. Let us now use torque to analyze the more general case of the *physical pendulum*, defined as an arbitrarily shaped rigid body, suspended as in Fig. 14.10 and swinging about a fixed horizontal support axis through O under the action of gravity. The concept of moment of inertia first arose in connection with this problem, which was solved by Christian Huygens, along with many other dynamical problems related to clock design, in his classic book *Horologium Oscillatorium* published in 1673.

In Fig. 14.10 gravity acts through the center of mass C, exerting a torque about O when C is displaced from its equilibrium position by an angle θ. If D is the distance between O and C, the torque about O is $mgD \sin \theta$. Applying the dynamical law of rigid-body rotation, $\tau = I\, d^2\theta/dt^2$ [Eq. (14.10)], we obtain

$$I \frac{d^2\theta}{dt^2} = -mgD \sin \theta \tag{14.27}$$

Table 14.3 Analogies between Translation and Rotation for Special Motions

Law	Translation	Rotation
	Special Laws	
Uniform motion	$a = 0$	$\alpha = 0$
	$v = $ const	$\omega = $ constant
	$x = x_0 + vt$	$\theta = \theta_0 + \omega t$
Uniformly accelerated motion	$F = $ const	$\tau = $ constant
	$a = $ const	$\alpha = $ constant
	$v = v_0 + at$	$\omega = \omega_0 + \alpha t$
	$x = x_0 + v_0 t + \frac{1}{2}at^2$	$\theta = \theta_0 + \omega_0 t + \frac{1}{2}\alpha t^2$
	$v^2 - v_0^2 = 2as$	$\omega^2 - \omega_0^2 = 2\alpha\theta$
Rigid-body motion	*of center of mass:*	*about fixed axis:**
	$p = mv$	$L = I\omega$
	$K = \frac{1}{2}mv^2 = \frac{1}{2}p^2/m$	$K = \frac{1}{2}I\omega^2 = \frac{1}{2}L^2/I$
	$F = ma$	$\tau = I\alpha$

*These laws also apply about an axis passing through the center of mass, even if the center of mass moves, provided the *direction* of the axis remains fixed.

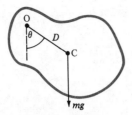

Figure 14.10 Physical pendulum.

where I is the moment of inertia of the physical pendulum about O. As in the case of the simple pendulum, this is not the equation for simple harmonic motion, but it becomes so in the small-angle approximation $\sin \theta \approx \theta$, which gives us

$$\frac{d^2\theta}{dt^2} = -\frac{mgD\theta}{I} . \tag{14.28}$$

This represents simple harmonic motion with angular frequency

$$\omega_0 = \sqrt{\frac{mgD}{I}} \tag{14.29}$$

and period

$$T = 2\pi \sqrt{\frac{I}{mgD}} . \tag{14.30}$$

According to the parallel-axis theorem [Eq. (14.18)], $I = I_C + mD^2$, where I_C is the moment of inertia of the rigid body about its center of mass. Therefore the period can be written

$$T = 2\pi \sqrt{\frac{I_C}{mgD} + \frac{D}{g}} . \tag{14.31}$$

In the special case of the simple pendulum all the mass is concentrated at the center of mass ($I_C = 0$), so the moment of inertia about the point of suspension is simply mD^2 and the period is $T = 2\pi\sqrt{D/g}$ as we found in Chapter 12. Any distribution of mass away from the center increases rotational inertia and makes the period longer.

Example 10
The pendulum in a grandfather clock consists of a bronze disk of mass m and radius R which is attached to the point of suspension O by a wooden rod of length $D = 10R$ whose weight we shall neglect. Find D so that the half-period $\frac{1}{2}T$ (the time between a "tick" and the next "tock") is exactly 1 s.

For a simple pendulum, $T = 2\pi\sqrt{D/g}$, and a 1-s half-period is obtained by choosing $D = g(T/2\pi)^2 = 0.994$ m.

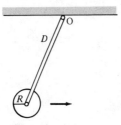

To find the correction due to the spatial extent of the disk, we insert the entry of Table 14.1 for the disk, $I_C = \frac{1}{2}mR^2$, into Eq. (14.31) and obtain

$$T = 2\pi\sqrt{\frac{\frac{1}{2}mR^2}{mgD} + \frac{D}{g}} = 2\pi\sqrt{\left(\frac{R^2}{2D^2} + 1\right)\frac{D}{g}} .$$

Solving for D we obtain, for a 1-s half-period,

$$D = g\left(\frac{T}{2\pi}\right)^2 \Big/ \left(1 + \frac{R^2}{2D^2}\right) = g\left(\frac{T}{2\pi}\right)^2 \Big/ \left(1 + \frac{1}{200}\right) = 0.988 \text{ m}.$$

The correction factor $1 + R^2/2D^2$ differs from 1 by $\frac{1}{2}\%$ in our case – a small amount, but significant for good timekeeping.

Example 11

How far from its center of mass should a rigid body be suspended to minimize the period of oscillation?

In Eq. (14.31), I_C is independent of the distance D between C and O. The minimum of T^2, found by taking the derivative of T^2 with respect to D,

$$\frac{d(T^2)}{dD} = (2\pi)^2\left(\frac{-I_C}{mgD^2} + \frac{1}{g}\right),$$

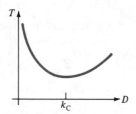

occurs when $d(T^2)/dD = 0$, which gives

$$D^2 = \frac{I_C}{m}.$$

Recalling the definition of the radius of gyration in Eq. (14.13), $k_C = \sqrt{I_C/m}$, we find that the minimum is located at

$$D = k_C$$

corresponding to the period

$$T_{min} = 2\pi\sqrt{\frac{2k_C}{g}}.$$

At larger D, T grows because the rotational inertia becomes large, and at smaller D, T grows because the restoring torque becomes small.

14.9 THE TORSION PENDULUM

When the bottom of a wire is twisted about its central axis through an angle θ relative to the top of the wire, the wire gives rise to a restoring torque $-c\theta$. The minus sign indicates that the torque is in a direction opposite to θ, and the linear dependence on θ is a manifestation of Hooke's law, accurate up to near the elastic limit of the wire. The constant c is an empirical parameter, analogous to the spring constant k, and is called the *torsion constant* of the wire.

The accurate linearity of the twisting of a wire subjected to moderate torsion makes it useful in various measuring devices. We have already remarked on the use of a torsion fiber in the Cavendish balance for determining the gravitational constant G. A similar application of torsion is made in ammeters and voltmeters.

In the *torsion pendulum*, a solid disk is hung from a wire as shown in Fig. 14.11. When the disk is rotated through an angle θ and then released, the ensuing motion is described by $\tau = I\,d^2\theta/dt^2$ [Eq. (14.10)] with the torque given by $-c\theta$:

$$I\frac{d^2\theta}{dt^2} = -c\theta. \tag{14.32}$$

Here I is the moment of inertia of the disk about its central axis, $I = \frac{1}{2}mR^2$. The motion

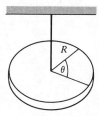

Figure 14.11 Torsion pendulum.

is simple harmonic, in this case not only for small angles but for the entire range of angles over which Hooke's law $\tau = -c\theta$ is satisfied. The angular frequency is $\omega = \sqrt{c/I}$ and the period of oscillation is

$$T = 2\pi \sqrt{\frac{I}{c}}. \tag{14.33}$$

This formula provides a convenient method for measuring the torsion constant of a wire.

14.10 COMBINED TRANSLATIONS AND ROTATIONS

The action of external forces on any body causes its center of mass to move according to the law

$$\mathbf{F} = m \frac{d^2\bar{\mathbf{r}}}{dt^2}. \tag{11.14}$$

The torque applied by external forces on any body causes it to rotate about an arbitrary point according to the law

$$\tau = \frac{d\mathbf{L}}{dt}. \tag{13.4}$$

Both Eqs. (11.14) and (13.4) are true for general motion.

In problems involving rigid-body rotation about a fixed axis, we have found it convenient to focus on Eq. (13.4), or more precisely on its fixed-axis version $\tau = I\,d^2\theta/dt^2$ [Eq. (14.10)]. Equation (11.14) is also valid, but the center-of-mass motion is already known from the overall rotation of the body about the fixed axis. And to evaluate Eq. (11.14) would require consideration of the reaction force applied to the rotating body by the support at the fixed axis. In a fixed-axis problem we would normally make explicit use of Eq. (11.14) only if we wished to find this reaction force.

Nevertheless it is often of interest to separate motion into rotation about the center of mass and translation of the center of mass. We have already remarked, for example, on the parallel-axis theorem which expresses I in terms of moment of inertia about the center of mass plus a contribution representing motion of the center of mass about another axis. The parallel-axis theorem has an immediate consequence for the kinetic energy of a body that rotates about an axis through an off-center point P with angular speed ω. The kinetic energy is

$$K = \tfrac{1}{2}I_\text{P}\omega^2 \tag{14.20}$$

and the parallel-axis theorem tells us that

$$I_\text{P} = I_\text{C} + mr^2 \tag{14.18}$$

where r is the distance from P to C. Therefore

$$K = \tfrac{1}{2}I_\text{C}\omega^2 + \tfrac{1}{2}mr^2\omega^2.$$

But $r\omega$ is the speed $\bar{v}$ with which the center of mass C moves about P. Thus

$$K = \tfrac{1}{2}I_C\omega^2 + \tfrac{1}{2}m\bar{v}^2; \qquad\qquad (14.34)$$

the kinetic energy is the sum of the *rotational* energy $\tfrac{1}{2}I_C\omega^2$ about the center of mass, and the kinetic energy $\tfrac{1}{2}m\bar{v}^2$ of *linear* motion of the center of mass with the entire mass considered to be at the center of the body.

In the same vein, we may consider the angular momentum of a system of particles about an arbitrary origin O:

$$\mathbf{L} = \sum \mathbf{r}_i \times m_i\mathbf{v}_i. \qquad\qquad (14.35)$$

Expressing $\mathbf{r}_i$ in terms of center-of-mass coordinates $\mathbf{r}'_i$ and the position $\bar{\mathbf{r}}$ of the center of mass, we have $\mathbf{r}_i = \mathbf{r}'_i + \bar{\mathbf{r}}$ and $\mathbf{v}_i = \mathbf{v}'_i + \bar{\mathbf{v}}$. Substituting these relations into Eq. (14.35), we find

$$\mathbf{L} = \sum (\mathbf{r}'_i + \bar{\mathbf{r}}) \times m_i(\mathbf{v}'_i + \bar{\mathbf{v}})$$

$$= \sum \mathbf{r}'_i \times m_i\mathbf{v}'_i + \left(\sum \mathbf{r}'_i m_i\right) \times \bar{\mathbf{v}}$$

$$+ \bar{\mathbf{r}} \times \sum m_i\mathbf{v}'_i + \bar{\mathbf{r}} \times \sum m_i\bar{\mathbf{v}}.$$

Fortunately the second term is zero because $\sum m_i\mathbf{r}'_i$ is proportional to the position of the center of mass, which vanishes in center-of-mass coordinates. Similarly, the third term is zero because

$$\sum m_i\mathbf{v}'_i = \sum m_i\frac{d\mathbf{r}'_i}{dt} = \frac{d}{dt}\left(\sum m_i\mathbf{r}'_i\right)$$

again depends on the vanishing quantity $\sum m_i\mathbf{r}'_i$. The first term is the angular momentum with respect to the center of mass, $\mathbf{L}_C$; the last term is $\bar{\mathbf{r}} \times M\bar{\mathbf{v}}$, the angular momentum of the center of mass (with all masses considered concentrated there) about O. So we are left with

$$\mathbf{L} = \mathbf{L}_C + \bar{\mathbf{r}} \times M\bar{\mathbf{v}}. \qquad\qquad (14.36)$$

For example, it is convenient to split the angular momentum of the earth into two parts: the "spin" associated with the earth's daily rotation about its own center of mass, and the "orbital" contribution associated with the earth's yearly passage about the sun. A similar splitup of angular momentum into spin and orbital parts is extremely useful in describing the motion of electrons around the nucleus in atomic physics.

Example 12
A yo-yo of radius R and mass m is lowered by its string. At a given instant it rotates about its center of mass with angular speed ω_C and its center of mass falls with speed

v_C, where $v_C = R\omega_C$. If we treat the yo-yo as a uniform solid disk, what is its kinetic energy, and what is its angular momentum L about an axis normal to the disk and passing through the point O where the yo-yo is held?

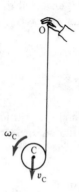

The disk has moment of inertia $I = \frac{1}{2}mR^2$. Using Eq. (14.34) and $v_C = R\omega_C$ we find that the kinetic energy is

$$K = \frac{1}{2}\left(\frac{1}{2}mR^2\right)\omega_C^2 + \frac{1}{2}mv_C^2 = \frac{3}{4}mv_C^2.$$

From Eq. (14.36) and $v_C = R\omega_C$ the angular momentum about O is

$$L = I\omega_C + Rmv_C = \frac{1}{2}mR^2\omega_C + Rmv_C = \frac{3}{2}Rmv_C.$$

(Note that in the orbital term, $|\bar{\mathbf{r}} \times m\mathbf{v}_C| = Rmv_C$ independent of the height of O above C.) In this example the "spin" of the yo-yo adds 50% to the translational kinetic energy and angular momentum about O.

When a rigid body rotates about a fixed axis, the alternative viewpoint of separating the motion into rotation about the center of mass and translation of the center of mass represents something of a complication. But when a rigid body has no fixed axis and tumbles freely through space, then its center of mass is the point it rotates about, so separating the motion into rotation about the center of mass and translation of the center of mass is the natural and simple way to analyze the motion. In this case both Eqs. (11.14) and (13.4) must be used. We shall give some illustrations in Examples 13 and 14.

Example 13

An irregularly shaped lamina has been cut out of a Styrofoam sheet and its center of mass marked with an orange spot for a lecture demonstration. Describe the motion of the lamina when it is tossed spinning through the air with the flat side vertical and the spin axis perpendicular to the flat side.

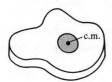

Under the action of gravity the center of mass moves in a Galilean parabola. The orange spot makes the parabola readily visible. Gravity exerts no torque about the center of mass, so the angular momentum about the center of mass is conserved and the lamina spins with constant angular speed.

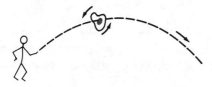

There are unique advantages to analyzing the motion with reference to the center of mass in this problem. About any other point, gravity exerts a torque and $\mathbf{L}$ is not conserved. Moreover, a spot placed on any other point would follow a less regular trajectory.

Example 14

A uniform thin bar of mass M and length $2R$ lies at rest on a frictionless table. A putty ball of mass m with horizontal velocity v normal to the bar hits the bar at one end, sticking to it. Describe the subsequent motion.

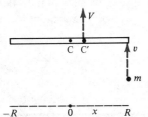

After the impact, the center of mass C' of the combined system travels with uniform velocity V in a straight line, since no further horizontal forces act. Moreover, the system rotates with uniform angular speed ω about its center of mass, because no further torques act. Note that C, the geometrical center of the bar, moves along a more complicated path. (What is this path?)

The collision is most simply treated by using momentum and angular momentum conservation. Momentum conservation,

$$mv = (m + M)V, \tag{1}$$

determines the final velocity $V = mv/(m + M)$. Angular momentum conservation about the center of mass C' relates the initial angular momentum $(R - x_{C'})mv$ (see figure) to the final angular momentum $I_{C'}\omega$ as follows:

$$(R - x_{C'})mv = I_{C'}\omega. \tag{2}$$

To evaluate ω we need to know the position $x_{C'}$ of C'. As measured from the geometrical center of the bar, it is

$$x_{C'} = \frac{mR}{M + m}.$$

The moment of inertia about this point is

$$I_{C'} = \tfrac{1}{12}M(2R)^2 + Mx_{C'}^2 + m(R - x_{C'})^2$$

where the first two terms give the contribution for the bar alone, and the last term is the contribution from the putty. (Note that the parallel-axis theorem has been used in the first two terms to find the moment of inertia of the bar around the off-center point C'.) Using the value for $x_{C'}$ we get

$$I_{C'} = \frac{MR^2}{3} + M\left(\frac{mR}{M + m}\right)^2 + mR^2\left(1 - \frac{m}{M + m}\right)^2,$$

which reduces after a bit of algebraic manipulation to

$$I_{C'} = \frac{MR^2}{3} + \frac{MmR^2}{M + m}.$$

Finally, substituting $x_{C'}$ and $I_{C'}$ into the angular momentum conservation relation (2) above we find, after more algebra,

$$\omega = \frac{R - x_{C'}}{I_{C'}} mv = \frac{3mv}{R(M + 4m)}.$$

14.11 KINEMATICS OF THE ROLLING WHEEL

An important special case of combined translation and rotation is the motion of a rolling wheel. There are two ways, both equally correct, in which this can be described. (i) One can describe the wheel as rotating with angular speed ω_C about a *moving* axis through the center of mass C, while C progresses with a linear speed v_C (Fig. 14.12). (ii) Alternatively, one can say that the wheel rotates with angular speed ω_P about an axis through the point of contact P, an axis which is *instantaneously at rest*! This is physically plausible from the fact that by "rolling" we mean a motion without slipping, and if the part of the wheel at the point of contact is not slipping, it cannot be moving relative to the surface below.

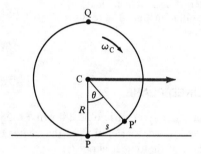

Figure 14.12 The rolling wheel.

With regard to the first point of view, as the point of contact moves from P to P′, the wheel rotates through an angle θ and progresses a distance s. From Fig. 14.12, $s = R\theta$. Differentiating, we find for the linear speed of C

$$v_C = \frac{ds}{dt} = R\frac{d\theta}{dt} = R\omega_C \tag{14.37}$$

and for the acceleration of C

$$a_C = \frac{dv_C}{dt} = R\frac{d\omega_C}{dt} = R\alpha. \tag{14.38}$$

We can immediately verify that point P has instantaneous speed zero as claimed. We find v_P (see Fig. 14.13) by adding to v_C the velocity of rotation of P about C, namely $-R\omega_C$, and this gives zero by Eq. (14.37).

What is the relation between ω_P and ω_C? Surprisingly, they are equal. We have just shown that $v_C = R\omega_C$, whereas from the second point of view C has speed $v_C = R\omega_P$ as a consequence of its rotation about P. Consistency of the two results requires

$$\omega_C = \omega_P. \tag{14.39}$$

As a check, let us calculate the speed of the point Q at the top of the wheel by both methods. In the first method (Fig. 14.13) we find v_Q by adding to v_C the speed of rotation of Q about C, namely $R\omega_C$, and with the help of Eq. (14.37) this gives $2v_C$. In the second method $v_Q = (2R)\omega_P$, and this also equals $2v_C$.

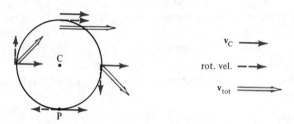

Figure 14.13 Vector addition of center-of-mass velocity and rotational velocity about the center in rolling motion. The result is equivalent to instantaneous rotation about P.

14.12 ROLLING DOWN AN INCLINED PLANE

The rolling of a cylinder, sphere, or other symmetric object down a rough inclined plane (Fig. 14.14) is conveniently analyzed as an acceleration of the center of mass C along the plane and a simultaneous rotation of the object about C.

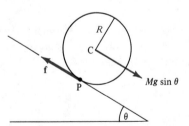

Figure 14.14 Rolling down an inclined plane.

We have previously shown in Eq. (11.14) that C moves as if all the external force were applied at C. Normal to the plane the component $Mg \cos \theta$ of gravity is evidently canceled by the normal reaction force exerted by the plane, for there is no acceleration in that direction. Along the plane the component $Mg \sin \theta$ of gravity is opposed by the frictional force **f** produced by the rough plane. Although we have not yet determined the magnitude of **f**, the linear acceleration a_C of C for an object of mass M on an incline of angle θ is related to f by Newton's second law,

$$Ma_C = Mg \sin \theta - f. \tag{14.40}$$

We also recall that the object will rotate about C according to the law $\tau = I\alpha$ [Eq. (14.10)]. Since gravity acts at C and the normal reaction force points directly toward the center of the object, neither exerts any torque about C. The torque due to f is fR, exerted clockwise. Therefore

$$fR = I\alpha. \tag{14.41}$$

If the object rolls without slipping, $a_C = R\alpha$ [Eq. (14.38)], and Eq. (14.41) gives

$$f = \frac{Ia_C}{R^2}. \tag{14.42}$$

When this is introduced into Eq. (14.40), we find

$$a_C = \frac{Mg \sin \theta}{M + I/R^2}. \tag{14.43}$$

The physical interpretation of this result becomes especially transparent if we write $I = bMR^2$, where $0 \leq b \leq 1$. In terms of b, Eq. (14.43) reads simply

$$a_C = \frac{g \sin \theta}{1 + b}. \tag{14.44}$$

This tells us that the acceleration of the rolling object does not depend on M or R; only the *distribution* of the mass (as represented by b) affects the result. We note that $b = k^2/R^2$, where k is the radius of gyration.

For the particular example of a homogeneous cylinder, $b = \frac{1}{2}$ and we find

$$a_C = \tfrac{2}{3} g \sin \theta. \tag{14.45}$$

The acceleration is only two-thirds of what it would be in frictionless sliding. The frictional force f has the magnitude $\frac{1}{3} Mg \sin \theta$. If the coefficient of friction is μ, the maximum value of f is given by μ times the normal reaction force, i.e., $\frac{1}{3}(Mg \sin \theta) \leq \mu Mg \cos \theta$. Because the frictional force cannot exceed this value, rolling can only occur when $\mu > \frac{1}{3} \tan \theta$. When this inequality is not satisfied, the cylinder will slip.

Rolling down an inclined plane can also be analyzed by means of energy considerations. When the object rolls a distance s, its potential energy decreases by

$$Mgh = Mgs \sin \theta.$$

If it started from rest, Eq. (14.34) tells us that it will have gained kinetic energy of amount

$$\tfrac{1}{2} I \omega_C^2 + \tfrac{1}{2} M v_C^2 = \tfrac{1}{2} b M R^2 \omega_C^2 + \tfrac{1}{2} M v_C^2.$$

Since $R\omega_C = v_C$, the kinetic energy of the rolling object can be expressed succinctly as $\frac{1}{2}(1 + b)Mv_C^2$. Conservation of mechanical energy then requires that

$$Mgs \sin \theta = \tfrac{1}{2}(1 + b)Mv_C^2,$$

or $v_C^2 = [2/(1 + b)]gs \sin \theta$. Comparing this with the general relation $v^2 = 2as$ for uniformly accelerated motion we find

$$a_C = \frac{g \sin \theta}{1 + b},$$

in agreement with the result (14.44) obtained by analyzing the forces.

At first sight it may appear inconsistent to apply mechanical energy conservation to a problem involving frictional forces. However, we recall from the kinetics of the rolling wheel that the point of contact P, where the frictional force is applied, is momentarily at rest. This provides justification for assuming that friction does no work in our idealized treatment of rolling motion.

14.13   CORIOLIS FORCES

As we have seen in Chapter 9, the extension of Newton's second law to accelerated frames requires that extra "inertial forces" be introduced in addition to any applied (i.e., real) forces that may be acting. In a frame that accelerates at a rate $\mathbf{a}_0$ relative to an inertial frame, the inertial force is $\mathbf{f} = -m\mathbf{a}_0$.

The rotation of a rigid body offers a natural instance of an accelerated frame of reference. We live on a slowly rotating rigid body, the earth, and occasionally experience rapidly rotating ones, such as merry-go-rounds. We have already discussed the fact that a point located at distance r from the axis of a frame rotating with uniform angular speed ω is accelerated at a rate $\omega^2 r$ toward the axis, and therefore a particle at that point feels a centrifugal force $\mathbf{f} = m\omega^2\mathbf{r}$ away from the axis in that frame. Recognizing the effects of centrifugal force on the latitude variation of the gravitational acceleration g, we were able to demonstrate the earth's rotation in a way that was not known to Galileo.

When a particle *moves* with respect to a rotating frame, the centrifugal force is supplemented by another inertial force called the *Coriolis force*. The effect is relatively easy to visualize at the axis of rotation, where the centrifugal force is negligible, so let us begin with that case. Consider a puck crossing the axis of a rotating horizontal disk. If friction can be ignored, the puck is free of horizontal forces and therefore moves in a straight line (the solid line of Fig. 14.15a) with constant velocity $\mathbf{v}$ relative to the laboratory (which will serve sufficiently well as an inertial frame for our present purposes). As seen from the laboratory, the rotating disk turns, say, counterclockwise with angular speed ω. But as seen from a frame fixed in the disk, it is the laboratory that rotates, and it rotates with the same angular speed in the opposite sense, clockwise. In this frame the puck's trajectory turns clockwise, following the curved path indicated by the *dashed line* in Fig. 14.15b. Thus there must be an inertial force in the rotating frame to provide the curvature that was not present in the inertial frame, and we call that force the Coriolis force.

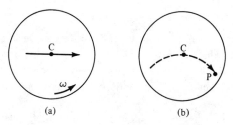

Figure 14.15 Motion of a frictionless puck passing over the rotation axis C, as seen from above in (a) an inertial frame (solid line) and (b) the rotating frame (dashed line).

To find the direction and magnitude of the Coriolis force we argue as follows. In a travel time Δt from the axis C to point P a small distance $r = v\,\Delta t$ away, the puck is deflected through an angle $\Delta\theta = \omega\,\Delta t$ in the rotating frame (Fig. 14.16). The distance Δs it is deflected from a straight line out to radius r is therefore $\Delta s = r\,\Delta\theta = r\omega\,\Delta t$. And because $r = v\,\Delta t$, we have

$$\Delta s = \omega v (\Delta t)^2. \tag{14.46}$$

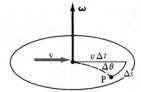

Figure 14.16 Kinematics of puck motion through the axis in the rotating frame.

Comparing this expression with the constant-acceleration law $\Delta s = \frac{1}{2}a(\Delta t)^2$, we see that the magnitude of the acceleration is $a = 2\omega v$ and that the associated Coriolis force has magnitude

$$f_C = ma = 2m\omega v. \tag{14.47}$$

To determine its direction, we define an *angular velocity vector* $\boldsymbol{\omega}$ with magnitude ω directed along the axis of rotation (Fig. 14.16). We see from Fig. 14.16 that the deflection is in a direction perpendicular to both $\mathbf{v}$ and $\boldsymbol{\omega}$. Therefore we can express the Coriolis force as a vector cross product of $\mathbf{v}$ and $\boldsymbol{\omega}$, and using the right-hand rule to find the direction we obtain

$$\mathbf{f}_C = -2m\boldsymbol{\omega} \times \mathbf{v}.$$

Though we have only given an argument for the form of $\mathbf{f}_C$ in a special case, the above equation turns out to be true in general provided we replace $\mathbf{v}$ by $\mathbf{v}'$, the velocity in the *rotating* frame:

$$\mathbf{f}_C = -2m\boldsymbol{\omega} \times \mathbf{v}'. \tag{14.48}$$

In the above example, $\mathbf{v} = \mathbf{v}'$ at the moment when the puck moves radially through the axis, so we did not notice the distinction. Equation (14.48) can be derived by working out the acceleration relative to an inertial frame of a particle moving with velocity $\mathbf{v}'$ relative to a rotating frame. We shall not give the derivation here, but will provide further examples and physical discussion.

The direction of the Coriolis force can often be seen from angular momentum arguments. For example, the frictionless puck in Fig. 14.15a has no angular momentum with respect to the center at the instant when it passes through the center. To conserve angular momentum in the inertial frame as it subsequently moves outward, it must maintain zero angular speed in the inertial frame. But this requires that it curve to the right at angular speed ω in the rotating frame. In any example of this type, one can deduce either from the Coriolis force in the rotating frame or from angular momentum conservation in the inertial frame that the trajectory curves to the right in a frame rotating counterclockwise and to the left in one rotating clockwise.

The magnitude of the Coriolis force can be appreciable on a turntable or merry-go-round. For example, if ω is one radian per second and v' is five meters per second, the Coriolis acceleration $2\omega v'$ is 10 m/s^2, equal to the acceleration g of gravity.

Away from the axis of a turntable, Coriolis and centrifugal forces are both present in the rotating frame. For example, consider an experiment in which a frictionless puck is initially tethered by a string of length r to the axis of a rotating disk (Fig. 14.17a). From the point of view of an inertial frame, it circles with the disk at its angular speed ω under the centripetal force $m\omega^2 r$ supplied by the string tension. From the point of view of a frame fixed in the rotating disk, the puck is in equilibrium under the opposing pulls of the centrifugal force and the string tension. Then at time $t = 0$ the string is broken, releasing the puck. It now travels in a straight line in the inertial frame, being acted upon by no horizontal forces in that frame. Relative to the rotating frame, it is at rest until $t = 0$, then accelerates radially outward under the action of the centrifugal force. As it acquires velocity, it also feels the Coriolis force, which deflects it to the right (Fig. 14.17b).

An example in which we can readily check the Coriolis force law (14.48) quantitatively is provided by a particle which remains at rest in the *inertial* frame. In a rotating frame, the particle rotates backward with speed $v' = \omega r$ at fixed radius r. To keep it

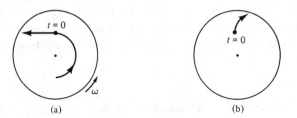

Figure 14.17 Motion of a frictionless puck (a) as seen in the laboratory, (b) relative to the turntable.

circling at fixed radius r in the rotating frame, the *net* force must be of magnitude $m\omega^2 r$ directed *inward*. Since the particle (being at rest in the inertial frame) is evidently subject to no net true forces, and the centrifugal force is known to be $m\omega^2 r$ directed *outward*, the Coriolis force must supply $2m\omega^2 r$ inward. But we see from Fig. 14.18 that this is indeed what we get from the vector

Figure 14.18 A particle at rest in the laboratory has this motion relative to a rotating disk.

$$\mathbf{f}_C = -2m\boldsymbol{\omega} \times \mathbf{v}'$$

where $\boldsymbol{\omega}$ is the angular velocity of the rotating frame relative to the inertial frame, $\mathbf{v}'$ is the velocity of the particle relative to the rotating frame, and the magnitude of $\mathbf{v}'$ is ωr.

The Coriolis force associated with the *earth's* rotation is much weaker than the effects considered above because the earth rotates only once per day, corresponding to an angular speed $\omega \approx 2\pi \times 10^{-5}$ rad/s. Even at projectile velocities of meters per second, the Coriolis acceleration $2\omega v'$ is only on the order of 10^{-2} m/s^2, far less than g. That is why the Coriolis force is not intuitively familiar.

When the Coriolis force associated with the earth's rotation acts over a sufficient period of time, however, it can have striking effects. Trajectories curve slowly but surely to the right in the Northern Hemisphere since, as seen from above the North Pole, the earth rotates counterclockwise like the turntables in the preceding figures. (In the Southern Hemisphere these directions are reversed.) A classic example is the *Foucault pendulum*. If a pendulum were placed at the North Pole, and suspended in such a way that no torque was exerted about the vertical axis, the plane of oscillation would remain fixed with respect to an inertial frame such as that provided by the fixed stars. As the earth turned under it, this plane would precess with respect to the earth with a period of one day. The precession would be clockwise when viewed from above (Fig. 14.19). The sideways acceleration $2\omega v'$ to the right is tiny, but acting over the 10^5 seconds in a day it turns the pendulum around completely. At latitudes other than the North Pole the precession of a Foucault pendulum is somewhat harder to visualize, but it is always described by the acceleration $-2\boldsymbol{\omega} \times \mathbf{v}'$ and thus provides evidence of the earth's rotation.

A number of important large-scale phenomena on the earth are driven by Coriolis forces acting over a span of hours or days. For example, the air in a high-pressure zone tends to flow outward in all directions (dashed lines in Fig. 14.20a). If we look at a high-pressure zone in the Northern Hemisphere from a vantage point above, the Coriolis force tends to deflect the air currents to their right (solid lines in Fig. 14.20a). The result is that the high-pressure zones known as *anticyclones* which periodically move across the temperate zone, bringing generally clear weather, normally rotate in the clockwise di-

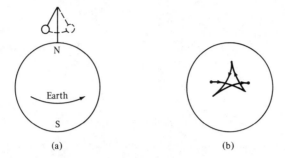

(a)   (b)

Figure 14.19 Path of Foucault pendulum bob at the North Pole, as seen from (a) side in inertial frame, (b) above in frame corotating with earth. (The precession is greatly exaggerated.)

rection in the Northern Hemisphere. As they generally move from west to east, we see from Fig. 14.20a that their approach is announced by west or northwest winds.

Conversely, air flows toward a low-pressure zone (dashed line in Fig. 14.20b). The Coriolis force deflects these winds to their right in the Northern Hemisphere (solid lines in Fig. 14.20b). Therefore the low-pressure weather zones known as *cyclones* which periodically traverse the temperate zone, bringing stormy weather as disparate air masses converge, generally rotate counterclockwise in the Northern Hemisphere. Since lows, like highs, generally move from west to east in the temperate zone, we see from Fig. 14.20b that the approach of the stormy weather they bring is announced by east or southeast winds. The somewhat more localized but more violent subtropical storms known as *hurricanes* or *typhoons* also rotate counterclockwise in the Northern Hemisphere. In the Southern Hemisphere, all these rotations are reversed.

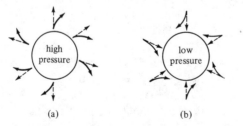

(a)   (b)

Figure 14.20 Wind circulation in the Northern Hemisphere about (a) high-pressure area, (b) low-pressure area.

In Chapter 13 we mentioned the formation of a vortex when water goes down a bathtub drain, a phenomenon in some ways like an upside-down version of a hurricane. However, the direction in which the bathtub vortex starts to rotate is usually determined by whatever circulation pattern, even a small one, it has to begin with, plus the way the plug is pulled, etc. The Coriolis effect is normally negligible on this scale, and the water is as likely to rotate one way as the other. Only under specially controlled conditions where outside disturbances are eliminated, and the water is allowed to settle down over a considerable period of time, can the Coriolis effect be demonstrated in a tub.

14.14   A FINAL WORD

From a certain point of view, rigid-body rotation is merely an application of Newton's laws – no new basic principles are used. So why spend time on this complicated subject?

One answer is that the applications are very important. Understanding rigid-body rotation, scientists of the seventeenth and eighteenth centuries could begin to deal quantitatively with wheels, pulleys, and clocks in the real world, the very objects man's growing use of natural forces was relying upon. And one could learn subtle details of forces on the rotating earth itself, such as the Coriolis force which helps shape the pattern of winds and ocean currents.

Moreover, the study of rigid-body rotation, though complicated, has its own organizing principles and quantities. An example is the use of moment of inertia, a central quantity in the equations and a rotational analog of mass. Another is center of mass, already encountered in Chapter 11, but growing in importance with continuing use. It has been successively revealed as the point in a body that moves under the action of purely external forces (all other parts of a body can be jerked about in a more complicated way by internal forces), and as the point about which the rotation of a body in free flight is most conveniently analyzed. By no means restricted to rigid bodies, it was also discussed in Chapter 11 as a point relative to which the motion of two colliding or otherwise interacting bodies can conveniently be analyzed.

The center of mass remains an important tool today. In treating the collision of an electron and a proton, it is still useful and natural to analyze the motion and discuss the angular momentum in terms of the center of mass. Like energy conservation and angular momentum conservation, center of mass turns out to be a concept which, though first encountered in the context of Newton's laws of classical mechanics, remains valid and important in the quantum mechanics of the twentieth century.

Problems

Center of Mass

1. Why does a hurdler bend his upper torso down as he passes over the hurdle?

2. The density λ of a thin rod OA of length L varies linearly with the distance x from the end O according to $\lambda = \lambda_0 x/L$. Find the distance from O to the center of mass of the rod.

3. A fork of uniform density ρ has dimensions as shown on p. 402:

   (a) How far from the left end is the center of mass?
   (b) How far does the center of mass shift if the handle [area $(2h)(2L)$] is replaced by wood of density $\rho/2$?

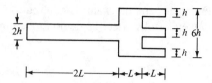

4. A thin uniform bar of mass M is bent into a right angle of edges L and $2L$. It is at rest while propped against a smooth wall as shown, making an angle θ with the floor, which has coefficient of friction $\mu \neq 0$.

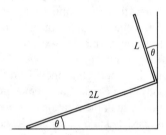

(a) Draw a free body diagram for the bar.
(b) Determine the horizontal and vertical forces of the floor on the bar, and the force of the wall on the bar.
(c) Determine the interval of values of θ for which the bar will not slide or fall away from the wall.

Calculation of Moments of Inertia

5. Derive the formula for the moment of inertia of a thin rectangular plate about an axis through its center perpendicular to the plate, as given in Table 14.1, by using the perpendicular-axis theorem together with the result given in the same table for a rod.

6. Show that if the fork of Problem 3a has its three prongs removed, the radius of gyration k of the resulting figure about a perpendicular axis passing through the left end of the handle is given by

$$k^2 = \frac{29}{15}h^2 + \frac{13}{3}L^2.$$

7. The density λ of a thin rod OA of length L varies linearly with the distance x from the end at O according to $\lambda = \lambda_0 x/L$. Find the moment of inertia of the rod about each of the following axes:

(a) An axis through O perpendicular to the rod.

(b) An axis through A perpendicular to the rod.

8. An object is made by welding together a solid disk of mass M and radius R, a hoop of mass M and radius $2R$, and a bar of length $4R$ and mass $2M$, as shown in the figure. Assume each object has negligible thickness and lies in the same plane.

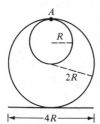

(a) Locate the center of mass of the object relative to point A.

(b) Calculate the moment of inertia of the object about an axis that passes through A and is perpendicular to the plane of the object.

Rigid Body Rotation about a Fixed Axis

9. A disk undergoes a constant angular acceleration that speeds it up from rest to 1800 rpm in 5 s. How long does it take the disk to rotate from rest through the first 90°?

10. A turntable of radius 1 m rotates freely at 5 rad/s about a fixed vertical axis. Its moment of inertia is 4 kg m². A 1-kg mass falls vertically onto it halfway out to the rim and is observed to slide on the surface until it slides off the edge with a speed of 1 m/s directed tangent to the rim and in the direction of rotation. Find the angular velocity of the turntable just after the mass leaves it.

11. A thin wire of mass M is bent to form the sides of a square of edgelength L. This rigid square is mounted on a vertical axis along one edge. A constant horizontal force $\mathbf{F}$, making a constant angle θ with the plane of the square, is continually applied at the center of the edge opposite the vertical axis. Neglect all friction.

(a) Find the object's moment of inertia about the axis.

(b) Calculate the magnitude of the torque about the axis due to $\mathbf{F}$.

(c) Determine how long it will take to make the square rotate through five revolutions, starting from rest.

12. Two pulleys with radii R_1 and R_2, are connected by a thin belt of negligible mass. They rotate about parallel axes with moments of inertia I_1 and I_2, respectively. A force of constant magnitude F is applied to pulley 1 at a distance r from its axis, causing angular acceleration. The force is maintained continually at an angle β

from the radial direction. Neglect friction at the axles and assume the belt does not slip.

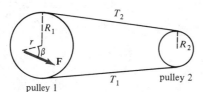

pulley 1
pulley 2

(a) If pulley 1 rotates through an angle θ_1, find the angle through which pulley 2 rotates.

(b) Find a relation between the angular accelerations of the two pulleys.

(c) Find the tangential acceleration of a point on the belt.

(d) The figure shows tensions T_1 and T_2 in those parts of the belt between the pulleys. Determine whether T_1 is greater than, equal to, or less than T_2.

(e) By applying the rotational analog of Newton's second law to each pulley, determine the angular acceleration of pulley 1 in terms of the given quantities.

(f) Calculate the difference of the tensions, $T_2 - T_1$, in terms of the given quantities.

13. A uniform board of length L and weight mg rests horizontally on supports at its endpoints A and B. Suddenly the support at the right endpoint B is removed. At that instant, calculate each of the following:

(a) The torque acting on the board about endpoint A.

(b) The magnitude of the angular acceleration about A.

(c) The magnitude of the linear acceleration of the center of mass.

(d) The force exerted by the support at A.

Energy and Work in Rigid Body Rotation

14. A flywheel with a horizontal axis has moment of inertia 120 kg m^2 about its axis. It starts from rest at time $t = 0$, and a constant torque of 2400 N m is continually applied to it for 10 s. Friction is negligible.

(a) Find the magnitude of its angular acceleration.

(b) How many revolutions does it make between $t = 5$ s and $t = 10$ s?

(c) Find its angular momentum at $t = 10$ s.

(d) How much work is done on the flywheel during those 10 s?

15. A wheel is spinning at 10 rad/s on a horizontal axle that has negligible friction at its supports. The wheel has an outer radius of 0.6 m and a moment of inertia about the axle of 15 kg m^2.

(a) Determine the magnitude of the constant force F, applied tangent to the rim and in the plane of the wheel, required to produce an opposite spin of 6 rad/s in 10 s.

(b) How many revolutions will the wheel make in those 10 s?

(c) Calculate the work done on the wheel in those 10 s.

16. Two cylindrical pucks, each with radius R, mass M, and uniform density, are sliding on a frictionless air table. They approach one another from opposite directions, each with speed v, and with impact parameter $2R$ between their centers of mass (i.e., a grazing collision so that they barely touch). If they stick together, determine

(a) the angular speed ω with which each puck rotates about the common center of mass,

(b) the fraction of initial energy lost to heat in the collision.

17. Two weights, one of mass M and the other of mass $3M$ are connected by a weightless string that passes over a wheel mounted on a fixed horizontal axle of negligible friction. The wheel has radius R and moment of inertia I about its axis. Initially the smaller mass rests on the ground and the larger mass is at a distance h above the ground, as shown in the figure. Then the masses are released from rest.

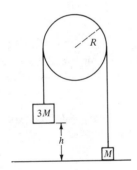

(a) Find the speed the larger mass will have just before it strikes the ground.

(b) Calculate the maximum height the smaller mass will rise above the ground.

18. A uniform circular disk of radius R and mass M is initially rotating counterclockwise with angular speed ω_0 about a horizontal frictionless axle through its center. A point of mass M falls vertically along a line tangent to the disk and becomes attached to the rim of the disk, striking it in a direction opposite to the rotation.

(a) If the speed of the falling point mass is $2R\omega_0$ just before the collision and the time of the collision is very short, find the angular speed of the composite disk-point mass system just after the collision takes place.

(b) What amount of heat energy is produced during the collision?

(c) The disk-point mass system continues to rotate after the collision. Use work and energy considerations to find the angular speed of rotation when the point mass is at its lowest point.

Physical Pendulum and Torsion Pendulum

19. A thin, uniform flat metal sheet, cut in the shape of a rectangle of diagonal length L, is pivoted about a horizontal, frictionless axis perpendicular to the sheet at one corner. Determine the period for small oscillations.

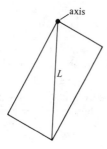

20. A thin uniform hoop of diameter d hangs on a nail. It is displaced from equilibrium through a small angle in its own plane and then released. Show that if the hoop does not slip on the nail, its period of oscillation is the same as that of a simple pendulum of length d.

21. A thin disk of radius $R = 10$ cm is suspended by a wire of length $L = 20$ cm. Determine the angular frequency of oscillation when

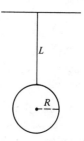

(a) the disk moves in the plane of the figure.

(b) the disk moves perpendicular to the plane of the figure.

(c) the center of mass doesn't move, but the disk twists about it.

22. A hollow rod of mass m and length L is hung from a pivot at one end. The rod is filled with sand having mass $2m$ (so the total mass is $3m$).

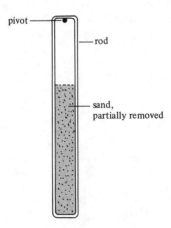

pivot

rod

sand, partially removed

(a) What is the period of the pendulum thus formed?

(b) What is the period of the pendulum if the upper half of the sand is removed?

(c) Determine qualitatively how the period of the pendulum changes as various amounts of sand are removed. Make a rough sketch of the period as a function of the amount of sand removed. (Do not attempt to find exact formulas. Just find interesting features by considering the system.)

Combined Translation and Rotation

23. A uniform bar of mass M and length L, initially at rest on a smooth floor, is struck at one end with impulse $\mathbf{P}$ perpendicular to the bar and tangent to the floor. Determine how far the center of mass will slide before it completes one revolution.

24. A uniform rod of mass M and length L is hung from a pivot at one end. An impulse P is applied perpendicular to the side of the rod while the rod is at rest in its equilibrium position. At what distance along the rod (measured from the pivot point) must the impulse be applied so that the pivot does not feel the impact? (This point is called the *center of percussion*.)

25. A wooden hoop of mass 1 kg and radius 0.5 m, lying flat on a frictionless table, is struck tangentially by a 1-kg pellet, moving at a speed of 4 m/s, that becomes imbedded in the wood.

(a) Find the center of mass of the final system relative to the center of the hoop.

(b) Calculate the velocity of the center of mass of the final system, the moment of inertia of the final system, its energy, and the energy lost in the collision.

26. Two equal, uniform, thin rigid rods of length L and mass M are free to move on a horizontal, frictionless table top. Initially, one rod is at rest and the other is moving toward it, in pure translation, at a velocity **v** perpendicular to both rods. The rods collide and stick together, moving thenceforth as a composite rod of total length $3L/2$. (One half of each rod coincides with one half of the other.) Determine:

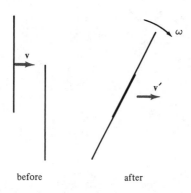

before   after

(a) The velocity of the composite rod after the collision.
(b) The moment of inertia of the composite rod about a perpendicular axis through its center of mass.
(c) The angular velocity of the composite rod about its center of mass after the collision.
(d) The amount of energy dissipated in the collision.

Rolling Motion

27. A sphere of mass M and radius R starts from rest at the top of an inclined plane tilted at an angle θ. The sphere is allowed to roll a distance L along the plane without twisting, curving, or slipping. Find each of the following:

(a) Its linear speed after it has rolled a distance L.
(b) Its angular momentum at the same time as in (a).
(c) The angular acceleration of the sphere about its center of mass.

28. A cart released from rest rolls without slipping down a plane inclined at an angle θ. The cart has four identical wheels, each of mass M, radius R, and moment of inertia I about its axis. The rest of the cart also has mass M. Find the speed of the center of mass of the cart after it has rolled a distance s.

29. A solid homogeneous disk of radius R starts from rest and rolls without slipping down the roof of a house along a plane inclined at angle θ with the horizontal. Its center falls a vertical distance $h \gg R$, at which point the disk rolls off the roof, and its center falls an additional distance $H > h$ before the disk strikes the ground.

(a) Find the translational speed of the disk (the magnitude of its velocity) just before it leaves the roof.

(b) Find its translational speed just before its strikes the ground.

(c) Calculate the amount of time it spent rolling on the roof.

30. The cushions around the edges of a billiard table come into contact with a ball somewhat above its center in order to reverse the spin as well as the normal velocity of a rolling ball as it rebounds. What is the ideal height H above the surface of the table that will accomplish this? (Express this height in terms of the radius R of the ball. Assume that the force of the cushion on the ball acts horizontally.)

31. Four bodies: a solid cylinder, a hoop, a solid sphere, and a hollow sphere, race down a smooth inclined plane.

(a) What are the ratios of the linear speeds they acquire in rolling the same distance down the incline without slipping? Compare these results to the linear speeds acquired by sliding the same distance down the same incline.

(b) If the plane makes an angle θ with the horizontal, find, for each body, the minimum coefficient of friction required to make it roll without slipping.

32. A yo-yo of mass 200 g and outer radius 3 cm is pulled horizontally on a surface without slipping by a string wrapped around its inner cylinder of radius 1 cm. The moment of inertia of the yo-yo about its rotation axis is 750 g cm^2 and its linear acceleration is 10 cm/s^2.

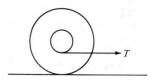

(a) Calculate the tension in the string.

(b) Find the minimum coefficient of friction required for the yo-yo not to slip.

33. A billiard ball of radius R is struck at its center, causing it to skid with no angular speed initially.

(a) Make a sketch indicating the manner in which the ball starts to spin as friction acts on it.

(b) Discuss whether friction increases or decreases the kinetic energy of the ball while it is "spinning up" (gaining angular speed).

(c) Now suppose the ball is struck below its center, giving it an initial underspin (a tendency to rotate backward). Describe qualitatively how friction changes the angular velocity and translational velocity of the ball, up to the point where the motion changes from skidding to rolling.

(d) Finally, suppose the ball is struck so high above its center that it receives an initial overspin ($\omega R > v_C$). While friction is "spinning down" the ball, does it increase or decrease the translational velocity of the ball?

34. A bowling ball is thrown down an alley with an initial speed v_0. Initially it slides without rolling. Show that it begins rolling without sliding when friction has slowed it down to a speed $v_1 = \frac{5}{7} v_0$.

35. A solid sphere of uniform density and radius R is initially at rest on the smooth icy surface of a frozen lake. Someone gives it a sudden horizontal kick at height $H < R$ above the surface and the center of the sphere moves to the right with speed v_0 while spinning about a horizontal axis through the center. The coefficient of friction between the sphere and smooth ice is zero.

   (a) Determine, in terms of R, v_0 and H, the angular speed of the sphere about its center of mass just after the kick. Indicate the direction of rotation.

   After traveling some distance, the sphere comes to rough ice where the coefficient of friction is $\mu > 0$. The sphere slips for a while on the rough ice before it starts rolling.

   (b) Determine the speed of the center of the sphere at time t after reaching the rough ice, while it is still slipping.
   (c) Find the angular speed of the sphere at this time t.
   (d) Find the time t_1 at which the sphere achieves a state of pure rolling motion on rough ice.
   (e) Calculate the ratio of the kinetic energy the sphere has when it achieves pure rolling motion to that it had just after the kick.

Coriolis Force

36. Use the Coriolis force to explain why the circulation pattern of surface ocean currents moves clockwise in both the North Atlantic (moving up the American coast and across to England) and the North Pacific (moving past Japan, the Aleutians, and the American Northwest). In which direction do you think surface currents circulate in the South Atlantic?

37. A space station spins as indicated in the figure to provide its inhabitants with a sense of artificial "gravity" when afloat in space. An astronaut in the space station stands firmly on the floor and drops a ball from chest height. *Relative to the astronaut's body*, does the ball fall straight down (toward B), backward (toward A), or forward (toward C)?

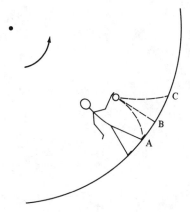

38. An artillery shell is fired due north from the South Pole, inclined at an angle of 45°. It follows a Galilean trajectory and lands 20 km away. In which direction does the Coriolis force deflect its path from due north, and by how many meters?

39. A stone is dropped from a tower of height y located on the Equator. While the stone is falling the main component of its velocity vector is vertically downward, but the Coriolis force also acts, causing a small movement x eastward. Find x in terms of y, g, and the earth's angular speed ω_E. Calculate the numerical value of x corresponding to $y = 100$ m.

CHAPTER 15

GYROSCOPES

To those who study the progress of exact science, the common spinning top is a symbol of the labours and the perplexities of men who had successfully threaded the mazes of planetary motions. The mathematicians of the last age, searching through nature for problems worthy of their analysis, found in this toy of their youth, ample occupation for their highest mathematical powers.

No illustration of astronomical precession can be devised more perfect than that presented by a properly balanced top, but yet the motion of rotation has intricacies far exceeding those of the theory of precession.

James Clerk Maxwell, "On a Dynamical Top" (1857)

15.1 AN ANCIENT QUESTION

In ancient times, people much more familiar with the night sky than we are helped themselves memorize its configurations by seeing heroes and creatures in clusters of stars. These constellations were patterns formed by stars fastened to a great sphere which surrounded the earth and formed the boundary of the universe. This celestial globe rotated on an axis through the earth, causing the stars to move along circular paths across the sky.

Figure 15.1 Time exposure of the night sky. (Lick Observatory photograph.)

Likewise the life-giving sun was fixed to its own sphere whose rotation made the sun seem to travel across the sky each day, rising in the east and setting in the west. But unlike the stars, the sun gradually changed its path each day, rising and setting more northerly in the summer and more southerly in the winter. Ancient astronomers accounted for the yearly motion of the sun by an extra annual rotation of the sun's sphere about an axis tilted by $23\frac{1}{2}°$. They named the plane of the sun's annual orbit the ecliptic, and the zodiac was the circular zoo of constellations through which the sun moved on its yearly journey around the stationary earth. Figure 15.2 illustrates the ecliptic and zodiac.

Figure 15.2 Celestial sphere with ecliptic and zodiac indicated.

Twice a year, once in the spring and once in the autumn, the orbit of the sun passes through the equatorial plane of the earth. On these dates, called the *vernal* and *autumnal equinoxes*, the sun rises due east and sets due west, and the lengths of day and night are equal. The points were marked by the position of the sun in the zodiac.

Through amazingly careful observations in the second century B.C., the Greek astronomer Hipparchus discovered that the position of the equinoxes in the zodiac slowly drifts westward. For this phenomenon, called the *precession of the equinoxes*, he reported a value of 36 seconds of arc per year. But to him the precession of the equinoxes merely represented an empirical fact, like so many astronomical facts, which had to be considered in the compilation of calendars for planting and harvesting.

Not until 1543 was an explanation proposed for the precession of the equinoxes. In his book *De Revolutionibus*, Copernicus conjectured that the earth orbits the sun and that the earth spins on an axis tilted by $23\frac{1}{2}°$ to the ecliptic. In his model of the universe, the precession of the equinoxes was due to the fact that the axis of the earth slowly traces out a circle, as shown in Fig. 15.3. As the axis shifts, the star it seems to point at, the pole star, also seems to drift until it's replaced by another. Copernicus concluded that the precessional rate is about 52 seconds of arc per year, which corresponds to a precessional period of 26,000 years.

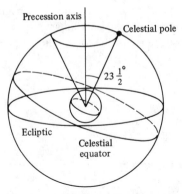

Figure 15.3 Precession of the equinoxes is caused by the rotation of the earth's axis.

By the time of Newton, this descriptive account of the precession of the equinoxes was well documented. However, the physical cause remained a mystery until Isaac Newton answered this great astronomical problem, the key to which lies in the gyroscope.

15.2 THE GYROSCOPE

Before we can explain the precession of the equinoxes, we first need to understand the underlying physics of a type of motion called *gyroscopic precession*. The simplest gyroscope is a wheel which is free to spin on an axle, which we'll call the *spin axis*. A spinning top is another example of a gyroscope.

Figure 15.4a shows a nonspinning bicycle wheel of mass M at the center of a horizontal axle 0P of negligible mass along the x axis. Initially the axle is supported at both end points 0 and P. If the support at P is removed, the torque due to the weight Mg

causes the wheel to fall and, in so doing, to rotate counterclockwise about the y axis as shown in Fig. 15.4b, where the wheel is viewed from along the positive y axis.

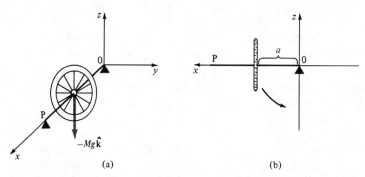

(a)   (b)

Figure 15.4 (a) Axle supported at both ends. (b) Support at P removed.

If the same wheel is set spinning rapidly on its axle, as indicated in Fig. 15.5, an extraordinary phenomenon occurs when the support is removed at P. The wheel does not fall as before but instead the axle remains almost horizontal and begins to revolve, or *precess*, about the z axis, as shown in Fig. 15.5. This apparently paradoxical motion, called *gyroscopic precession*, is a consequence of the torque due to gravity acting on the angular momentum of the spinning wheel.

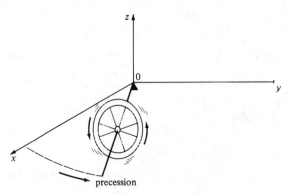

precession

Figure 15.5 A rapidly spinning wheel does not fall but exhibits precession.

As the nonspinning wheel in Fig. 15.4 begins to topple we can calculate its torque about 0 as follows. Let $a\hat{\mathbf{i}}$ denote the vector from 0 to the center of the mass of the wheel. When the support at P is removed the torque has initial value

$$\boldsymbol{\tau} = (a\hat{\mathbf{i}}) \times (-Mg\hat{\mathbf{k}}) = aMg(\hat{\mathbf{k}} \times \hat{\mathbf{i}}) = aMg\hat{\mathbf{j}}.$$

When the axle has dropped through an angle θ from the x axis, where $0 \le \theta \le \pi/2$, the corresponding torque about 0 is

$$\boldsymbol{\tau} = [(a \cos \theta)\hat{\mathbf{i}}] \times (-Mg\hat{\mathbf{k}}) = (aMg \cos \theta)\hat{\mathbf{j}}.$$

This torque is the rate of change of some angular momentum vector $\mathbf{L}_0$,

$$\boldsymbol{\tau} = \frac{d\mathbf{L}_0}{dt},$$

where $\mathbf{L}_0$ has the same direction as $\boldsymbol{\tau}$ (parallel to the plane of the wheel) since $\mathbf{L}_0 = \mathbf{0}$ when $t = 0$.

Now suppose you start the wheel spinning with a constant angular speed in the counterclockwise direction, as shown in Fig. 15.5, before the support at P is removed. This gives the wheel angular momentum about its center which we denote by $\mathbf{L}$, a vector of constant length along the axle. We call this the *spin momentum* of the wheel. The faster the wheel spins, the longer the vector $\mathbf{L}$. Let's analyze what happens now if you remove the support at P.

The torque $\boldsymbol{\tau}$ exerted by gravity about 0 is in the horizontal plane, since $\mathbf{r} \times \mathbf{F}$ must be perpendicular to $\mathbf{F}$. If the spinning wheel begins to fall under this torque, its angular momentum will be the vector sum $\mathbf{L} + \mathbf{L}_0$. Suppose for the moment that $\mathbf{L}$ is big enough (because the wheel is spinning very rapidly) that we can neglect $\mathbf{L}_0$ compared to $\mathbf{L}$. Then the torque is related to $\mathbf{L}$ by the equation

$$\boldsymbol{\tau} = \frac{d\mathbf{L}}{dt}. \tag{13.4}$$

So $\mathbf{L}$, a vector of constant length along the axle, must change in the direction of $\boldsymbol{\tau}$, which is horizontal. Thus the axle precesses horizontally rather than falling: $\mathbf{L}$ moves in a circle. As the axle moves about this circle in response to the torque, the direction of $\mathbf{L}$ changes and the direction of $\boldsymbol{\tau}$, being perpendicular to $\mathbf{L}$, changes as well. This is described by saying that $\mathbf{L}$ tries to follow $\boldsymbol{\tau}$, which leads it around in a circle, as suggested in Fig. 15.6. The resulting motion is uniform gyroscopic precession.

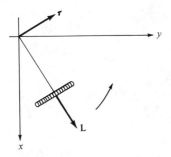

Figure 15.6 The angular momentum vector $\mathbf{L}$ tries to follow the torque vector $\boldsymbol{\tau}$.

The relationship between the vectors $\boldsymbol{\tau}$ and $\mathbf{L}$ in this motion is strikingly similar to the relationship between the vectors $\mathbf{F}$ and $\mathbf{p}$ in uniform circular motion, as illustrated in Fig. 15.7.

Uniform Circular Motion   Uniform Gyroscopic Precession

$$F = dp/dt$$ $$\tau = d\mathbf{L}/dt$$

p and F are constant L and τ are constant

$F \perp \mathbf{p}$, rotates $\mathbf{p}$ $\tau \perp \mathbf{L}$, rotates $\mathbf{L}$

Figure 15.7 Analogy between uniform circular motion and uniform gyroscopic precession.

Example 1

A gyroscope is spinning in space with angular momentum **L**, where it is acted on by two equal but opposite forces **F**, shown in the diagram, equidistant from the center of mass at 0.

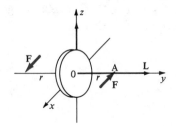

(a)   What is the torque acting on the gyroscope about point 0?
(b)   For the instant shown in the diagram, in what direction is end point A moving as a result of precession?

(a)   From $\boldsymbol{\tau} = \mathbf{r} \times \mathbf{F}$, we can find the direction of the torque created by each force using the right-hand rule. Gravity creates no torque about 0. Each force **F** creates a torque $rF\hat{\mathbf{k}}$ in the z direction. Therefore the total torque is

$$\boldsymbol{\tau} = 2rF\hat{\mathbf{k}}.$$

(b)   As the angular momentum vector **L** tries to follow the torque, end point A will move along with **L**. The precessional motion is such that **L** describes a circle in the yz

plane; so A moves counterclockwise when viewed from the positive x axis. In other words, the horizontal forces make the gyroscope axis pop up into the vertical direction!

Note that although the net force acting on the gyroscope is zero, the net torque is nonzero. We recall that a pair of forces equal in magnitude and oppositely directed is known as a couple; it produces rotation without acceleration of the center of mass.

Now that we have some idea of the cause of gyroscopic motion, let's consider the angular momentum once more as well as the energy associated with this motion. These considerations will provide an alternative viewpoint that yields a more general behavior than the uniform precession obtained above, including further subtle details of the motion.

In the foregoing description of gyroscopic precession we neglected L_0 and assumed that all the angular momentum of the gyroscope is purely spin momentum lying in the horizontal xy plane. But when you remove the support at P of a real gyroscope you observe that the axle does fall a bit, tilting down so that L now has a vertical component $-L_z$ as well as horizontal components. Since there is no torque to produce angular momentum in the z direction, any angular momentum in that direction must be conserved. How does that take place?

To conserve angular momentum in the vertical direction, the gyroscope rotates to create angular momentum that balances the downward vertical component $-L_z$. The resulting rotation of the axle about the z axis is just the gyroscopic precession obtained previously by a different argument. In accordance with our previous result (Fig. 15.5), it is directed so that the angular momentum associated with it is in the positive z direction. (This direction is obtained if L points outward along the axle as in Fig. 15.5; if the wheel is spun in the reverse direction the precession is reversed.)

The energy of the gyroscope must also be conserved. The precession of the gyroscope has a kinetic energy associated with it. That kinetic energy must come from somewhere, and the source is a change in gravitational potential energy of the center of mass. When the center of mass drops slightly as the gyroscope initially falls, the gravitational potential energy of the center of mass decreases. That decrease in potential energy appears as kinetic energy of the precession of the gyroscope.

The simple gyroscope displays even more intricacies in its motion. Consider what happens to the center of mass. Originally, the center of mass lies in the xy plane of Fig. 15.5. Once the gyroscope is released, the center of mass falls a bit as precession begins. The new (lower) height of the center of mass is one of stable equilibrium. But the gyroscope didn't start off in a stable equilibrium position. We recall from Chapter 12 that a system slightly disturbed from its stable equilibrium position oscillates. That's what happens to the center of mass of the gyroscope: the center of mass oscillates about the stable equilibrium position. This oscillatory motion which takes place in addition to the precession is called *nutation* (after the Latin word for *nodding*). Figure 15.8 illustrates the nutation we've just described, where the curved, cusplike path represents the motion of the center of mass. Eventually these oscillations are damped out by friction at the pivot point and air resistance, and the gyroscope's motion turns into uniform precession.

Let's briefly recapitulate the motion of a gyroscope. There are three qualitatively different kinds of motion occurring simultaneously. First, there is rotation about the gyro's spin axis; this *spin* is at some angular velocity which we'll call ω. Second, there

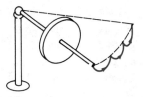

Figure 15.8 In nutation, the center of mass of a gyroscope traces a
path like the one shown.

is precession about the point of suspension. The *precession* has a different angular velocity
which we'll call Ω (capital omega), the angular velocity of precession. The third kind
of motion is the oscillation out of the plane of precession, and that's called *nutation*.

Example 2

Suppose that instead of releasing one end of a spinning horizontally oriented gyroscope,
you impart a slight horizontal velocity to the end as you release it. Describe the nutation
you would expect.

By imparting a slight velocity to the gyroscope when releasing it, you are adding
kinetic energy to it and giving the gyroscope a net vertical component of angular momentum L_z. At first, you might think that this added energy and angular momentum
would enable precession to occur without tilting down or nutation. However, the motion
of the released gyroscope must still conserve energy and L_z, and if the initial horizontal
velocity is very small that velocity cannot by itself supply enough energy or L_z for
precession. Therefore the gyroscope must still drop down a little and will nutate for the
reasons cited before. If the initial velocity is in the same direction as the precession, then
the nutation will consist of oscillations about the stable equilibrium position which are
"stretched out" more along the path of the center of mass, as illustrated below:

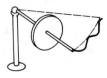

If the initial horizontal velocity is increased somewhat, the gyroscope does not need
to tilt down as far and the oscillations have less amplitude. With further increase, a
specific horizontal velocity is reached which gives the gyroscope exactly the kinetic
energy and L_z needed to precess without any nutation.

Example 3

This example describes another apparent paradox related to gyroscopic motion. Take the bicycle wheel in Fig. 15.4a, mount the supports at 0 and P on a turntable as shown below, then set the bicycle wheel spinning at a constant angular speed. The wheel acquires spin momentum **L** along the axle.

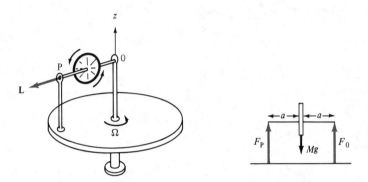

If the table does not rotate, this vector **L** remains constant, and the net torque $d\mathbf{L}/dt$ is **0**. This is to be expected, because if the axle has length $2a$, the contact forces F_0 and F_P are each equal to $\frac{1}{2}Mg$ and the net torque about the center of the axle is

$$F_0 a - F_P a = 0.$$

Now suppose we rotate the table with a constant angular speed Ω about the z axis, as suggested by the figure. A new paradox arises. The contact forces F_0 and F_P are no longer equal! The reason is that the wheel's angular momentum changes because of the rotation of the turntable, and to maintain this change in angular momentum a torque is required, which means $F_0 a - F_P a$ is no longer 0, so $F_0 \neq F_P$. We will show that as Ω increases, F_0 increases while F_P decreases and, in fact,

$$F_0 = \tfrac{1}{2}Mg + \tfrac{1}{2}\frac{L\Omega}{a} \qquad \text{and} \qquad F_P = \tfrac{1}{2}Mg - \tfrac{1}{2}\frac{L\Omega}{a} . \qquad (15.1)$$

As the turntable rotates, the angular momentum vector **L** of the spinning wheel rotates with the same angular speed but maintains its constant magnitude L. In a small time interval Δt the table turns through an angle $\Delta\phi$ and the vector **L** changes to **L** + Δ**L**, where ΔL, the length of Δ**L**, is approximately $L\,\Delta\phi$, as suggested by the accompanying figure.

Therefore

$$\frac{\Delta L}{\Delta t} = L\,\frac{\Delta\phi}{\Delta t} ,$$

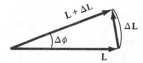

and as $\Delta t \to 0$ we find

$$\tau = \frac{dL}{dt} = L\,\frac{d\phi}{dt} = L\Omega.$$

Since $\mathbf{L}$ has constant length, $d\mathbf{L}/dt$ is perpendicular to $\mathbf{L}$ so the torque $\boldsymbol{\tau}$ is perpendicular to the axle, with magnitude $L\Omega$. On the other hand, this torque has magnitude equal to $(F_0 - F_P)a$, and hence $(F_0 - F_P)a = L\Omega$, so

$$F_0 - F_P = \frac{L\Omega}{a} \qquad \text{and} \qquad F_0 + F_P = Mg.$$

Adding and subtracting these equations we obtain (15.1). This shows that when the table is turning there are unequal contact forces at 0 and at P.

Now suppose we choose Ω so that $L\Omega/a = Mg$. Then (15.1) shows that the support at P has zero contact force and can therefore be removed. We then have a situation similar to gyroscopic precession as described earlier. If Ω is increased so that $L\Omega/a > Mg$, the contact force at P becomes negative while the support at 0 presses down on the turntable with a force greater than Mg.

The foregoing example shows that if an angular momentum vector of constant length L changes its direction at a constant angular speed Ω, a torque $L\Omega$ is required to maintain this motion. It does not matter where the torque comes from. It could be produced from the contact forces as above or from some other external source, but there must be a torque somewhere.

15.3   ANGULAR VELOCITY OF PRECESSION

Let's now find out what determines the angular velocity of precession of a gyroscope. The argument connects the rate of precession to the torque, and is related to Example 3.

Earlier we described precession as the rotation of the angular momentum vector $\mathbf{L}$, as shown in Fig. 15.9. Due to a constant torque from gravity, a gyroscope precesses with a certain precessional angular velocity $\boldsymbol{\Omega}$. Consider the angular momentum vector at times t and $t + \Delta t$, as shown in Fig. 15.9. The magnitude of the small change in the angular momentum vector $|\Delta\mathbf{L}|$ is given very nearly by

$$|\Delta\mathbf{L}| = (L \sin \theta)\,\Delta\phi$$

where $\Delta\theta$ is the angle the tip of **L** moves through in time Δt, θ is the angle between **L** and the axis of precession, and $L \sin \theta$ is the radius of the circle traced out by the tip of **L**. Therefore we have

$$\left| \frac{\Delta \mathbf{L}}{\Delta t} \right| = (L \sin \theta) \frac{\Delta \phi}{\Delta t}.$$

Taking the limit $\Delta t \rightarrow 0$, we find that the rate of change of the magnitude of the angular momentum vector is

$$\left| \frac{d\mathbf{L}}{dt} \right| = \lim_{\Delta t \rightarrow 0} \left| \frac{\Delta \mathbf{L}}{\Delta t} \right| = \lim_{\Delta t \rightarrow 0} L \sin \theta \frac{\Delta \phi}{\Delta t} = L \sin \theta \frac{d\phi}{dt}. \tag{15.2}$$

The direction of this vector is perpendicular to **L** and at any instant is tangent to the circle traced out by the tip of **L**.

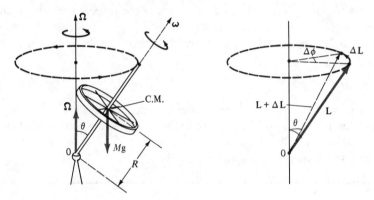

Figure 15.9 Precession of a gyroscope.

The quantity $d\phi/dt$ is the rate at which the angular momentum vector precesses and is the magnitude of what we have called the angular velocity of precession $\boldsymbol{\Omega}$. The direction of this vector is taken to be along the axis of rotation in such a way that if you curl the fingers of your right hand around in the direction the angular momentum vector is precessing, your thumb points in the direction of $\boldsymbol{\Omega}$.

With the direction of $\boldsymbol{\Omega}$ as specified above, we see that the angle between $\boldsymbol{\Omega}$ and **L** is θ. Thus $(L \sin \theta) \, d\phi/dt = |\mathbf{L}| \, |\boldsymbol{\Omega}| \sin \theta = |\boldsymbol{\Omega} \times \mathbf{L}|$ and Eq. (15.2) shows that $|d\mathbf{L}/dt| = |\boldsymbol{\Omega} \times \mathbf{L}|$. This suggests that we can write $d\mathbf{L}/dt$ as a vector cross product:

$$\frac{d\mathbf{L}}{dt} = \boldsymbol{\Omega} \times \mathbf{L}. \tag{15.3}$$

This is easily verified because $d\mathbf{L}/dt$ is perpendicular to both $\boldsymbol{\Omega}$ and **L** and is at any instant tangent to the circle traced out by the tip of **L**. Therefore the vectors $d\mathbf{L}/dt$ and $\boldsymbol{\Omega} \times \mathbf{L}$ have the same length and the same direction.

Example 4

By considering a point on a sphere of radius r which is rotating with an angular velocity $\boldsymbol{\omega}$, show that

$$\frac{d\mathbf{r}}{dt} = \boldsymbol{\omega} \times \mathbf{r},$$

where $\mathbf{r}$ is the position vector of the particle.

Comparing with Fig. 15.9, we see that our present problem of uniform rotation is completely analogous kinematically to uniform gyroscopic precession, with $\mathbf{r}$ the analog of $\mathbf{L}$ and $\boldsymbol{\omega}$ the analog of $\boldsymbol{\Omega}$. The result is therefore the analog of (15.3):

$$\frac{d\mathbf{r}}{dt} = \boldsymbol{\omega} \times \mathbf{r}.$$

For example, if $\mathbf{r}$ lies on the "equator," the magnitude of $d\mathbf{r}/dt$ is $v = \omega r$, a result familiar from uniform circular motion, and its direction is perpendicular to both $\boldsymbol{\omega}$ and $\mathbf{r}$.

More generally, the relation $d\mathbf{A}/dt = \boldsymbol{\omega} \times \mathbf{A}$ holds for any vector $\mathbf{A}$ that undergoes uniform rotation with angular velocity $\boldsymbol{\omega}$.

Torque is equal to the rate of change of angular momentum, so the equation which describes the precession of a gyroscope is

$$\boldsymbol{\tau} = \frac{d\mathbf{L}}{dt} = \boldsymbol{\Omega} \times \mathbf{L}. \tag{15.4}$$

From Fig. 15.9 we see that the torque due to gravity about the pivot is

$$\boldsymbol{\tau} = \mathbf{R} \times M\mathbf{g},$$

where $\mathbf{R}$ is the vector from the pivot to the center of mass and M is the mass of the wheel. The magnitude of this torque is

$$\tau = MgR \sin \theta.$$

As we saw earlier, the magnitude of $d\mathbf{L}/dt$ is

$$\left| \frac{d\mathbf{L}}{dt} \right| = \Omega L \sin \theta.$$

Because these two expressions are equal, we can solve for the magnitude of the precessional angular velocity:

$$\Omega = \frac{MgR}{L}. \tag{15.5}$$

Note that this result was already obtained for a special case in Example 3.

The denominator L in (15.5) can be expressed in terms of its angular speed ω and the moment of inertia I of the gyroscope about its spin axis,

$$L = I\omega, \tag{14.9}$$

so (15.5) becomes

$$\Omega = \frac{MgR}{I\omega}. \tag{15.6}$$

For example, the moment of inertia of a bicycle wheel of radius r about its spin axis is approximately

$$I = Mr^2,$$

because nearly all the mass is in the rim and tire. Thus, for an idealized bicycle wheel gyroscope, we can express the rate of precession as

$$\Omega = \frac{gR}{\omega r^2}.$$

In many practical applications a gyroscope is suspended about its center of mass so that the gravitational torque won't make it precess. When torques are eliminated, the angular momentum **L** of the balanced gyroscope holds its direction relative to an inertial frame of reference. This is one of the principles used in the *gyrocompass* and in *inertial guidance systems*.

In practice, perfect balance cannot be achieved, and the gyroscope will drift. To minimize Ω, besides the obvious strategy of making the distance R from the pivot to the center of mass as small as possible, we can make ω as large as possible, that is, have the gyroscope spin extremely fast. Furthermore, we can make I/M large by concentrating the mass as far as possible from the axis of rotation.

How good are the best gyroscopes? The exact answer is probably a military secret, but it is estimated that the best inertial guidance gyroscopes precess at about 100 seconds of arc (about 0.03°) per day.

15.4   THE EARTH AS A GYROSCOPE

Among the numerous and perplexing mysteries Isaac Newton removed from the pages of history was the precession of the equinoxes. In the *Principia*, Newton gives an explanation of the observed precession based on his dynamics. His penetrating insight was to realize that the earth itself acts like a gyroscope.

We usually think of the earth as being approximately spherical. If it were spherical, the sun (or moon) could not exert a torque on it around its center because the force of the sun (or moon) acts on the center of mass at the center of the earth. To this approximation, the angular momentum of the spinning earth is always in the same direction.

However, if we analyze the earth more closely, we find it is not a perfect sphere on account of its rotation, which causes a bulge at the equator, as exaggerated in Fig. 15.10. The width of each bulge at the equator is about 1/300 the polar radius of the earth; this corresponds to a deviation of 13 mi (21 km) at the equator from the radius of a perfect sphere.

Figure 15.10 Forces on equatorial bulges in (a) the summer and (b) the winter. The torque produced is in the same direction in both cases.

In addition, the earth's spin axis makes an angle of $23\frac{1}{2}°$ with respect to the plane of its orbit. Consequently, the bulge is oriented asymmetrically as shown in Fig. 15.10. Although the sun creates no torque on the spherical distribution of mass, it does exert a torque about the center of the bulges. The bulge nearer to the sun is more strongly attracted than the bulge which is farther away. This results in a torque τ pointed up from the plane of Fig. 15.10a. Because the precession occurs very slowly – taking 26,000 years to complete one circle – by the time the sun gets to the other side of the earth, the bulges have hardly moved at all. So when the sun is on the opposite side of the earth six months later as shown in Fig. 15.10b, the force on the bulge nearer the sun is greater and creates a torque in the *same* direction. (You can easily verify this with the right-hand rule.) The moon exerts a similar torque on the earth (in nearly the same direction, because the moon moves nearly in the plane of the ecliptic).

Since the earth rotates from west to east, its spin angular momentum is directed toward the North Pole. The torque acting on the spinning earth causes a precession in which the tip of the angular momentum vector traces out a circular path from east to west, as shown in Fig. 15.11, observed as the precession of the equinoxes. In the *Principia*, Newton, in addition to explaining the direction of the precession, estimated its rate to be 51 seconds of arc per year, a result within 10% of the observed value.

15.5 A FINAL WORD

The gyroscope is a strange, amusing, and important device. What is perhaps even stranger and more beguiling is the fact that Isaac Newton himself understood how it worked and managed to explain the precession of the equinoxes. For most of us, that little detail is lost in the magnitude of Newton's grander accomplishments. But if some other person who did nothing else in his lifetime had managed to do that, we would remember him today as an important scientist.

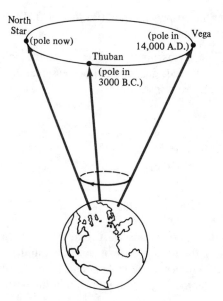

Figure 15.11 Precession of the equinoxes.

One day in 1684, Newton casually told a young friend named Edmund Halley that he had discovered that a $1/r^2$ force law leads to orbits in the form of conic sections. It's difficult to imagine what Halley felt at that moment. Having Newton as a friend cannot have been very relaxing at best, but Newton had just told him that he had discovered the key to the universe.

Halley prevailed upon his friend to publish the result, which Newton ultimately did. And indeed it *was* the key to the universe, although the known universe was smaller then than it is today. In the next two chapters, our job is to turn that key.

Problems

Gyroscopic Precession

1. **(a)** Suppose you had two gyroscopes of identical dimensions with one twice as massive as the other. Would you expect the rate of precession of the more massive gyroscope to be greater than, less than, or equal to that of the lighter one? Why?

   (b) Consider the two gyroscopes shown in the diagram. They have equal masses, and spin about their axes at the same rate, but the center of mass of gyroscope

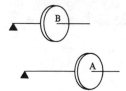

A is twice as far from the suspension point as that of B. How do their angular velocities of precession compare?

2. The angular momentum of the gyroscope shown below points along the negative y axis. If the support at B is removed, describe the precessional motion. What happens if the support at A is removed instead?

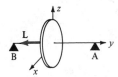

3. Imagine a gyroscope spinning in outer space. At one instant, its intrinsic angular momentum is directed along the negative x axis as shown in the diagram. Two equal but oppositely directed forces act at ends A and B, respectively. From the resulting precession, end A is moving in the xy plane in the positive y direction.

    (a) In what direction is the torque about 0 acting on the gyroscope?
    (b) In what directions are the forces at A and B acting?

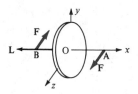

4. Shown below is a photograph of the motion of the end of a nutating gyroscope taken with a stroboscope. As the end moves around the loops, it doubles back as indicated by the arrows. This nutation was achieved by imparting a small velocity to the end of the gyroscope as it was released. Relative to the way the end is precessing, in what direction was the initial velocity? Explain your reasoning.

[From Kleppner, D., and Kolenkow, R. J. *An Introduction to Mechanics*, McGraw-Hill Book Co., New York (1973). By permission of the publisher.]

5. Everyone knows that to turn a bicycle to the right, you lean in that direction. In terms of torque and precession of the bicycle wheel explain how leaning makes the bicycle turn.

Using Spin Stability

6. Why is a spinning Frisbee more stable in flight than one that is not spinning? (For the same reason, bullets are given a spin as they leave the gun barrel.)

7. Suppose a playful physicist loads her suitcase with a heavy flywheel that is rotating. If a bellhop were to take such a suitcase as shown, with its spin angular momentum pointing along the negative x direction, and suddenly make a turn to the left (in the negative x direction), what would happen to the suitcase?

8. According to Galileo and the principle of inertia, uniform motion in a constant direction cannot be detected. This means that if you were in a sealed room with no windows you could not tell by any experiment done totally within that room whether you were at rest or moving with a constant velocity. But if you had a gyroscope in the room, you would be able to detect the rotation of the earth. How?

Angular Velocity of Precession

9. A toy gyroscope precesses faster as friction acting on the axle causes its spinning to slow down. Explain why.

10. A simple gyroscope consisting of a 3.0-kg wheel having a 0.2-m radius spins at 100 rad/s. The distance from the point of suspension to the center of mass is 0.6 m. Find the magnitude of the angular velocity of precession if

   (a) all the mass of the wheel is concentrated at its rim,
   (b) the wheel is a solid disk of uniform mass density.

11. Refer to Example 3 of Section 15.2. What changes must be made in the analysis if the support at 0 is not at the center of the table? What if the two supports lie symmetrically about the center?

The Earth as a Gyroscope

12. If the angular momentum of the earth about its axis is 7.1×10^{33} kg m^2/s, and the observed rate of precession of the equinoxes is 46.74 seconds of arc per year, what torque is exerted on the earth?

Let us make an oversimplified model in which this torque is entirely due to the action of the sun on two patches attached to the equator of an otherwise spherical earth. The patches are identical, one on the side nearest the sun, and the other on the far side. If the patches are separated by 13,000 km (the diameter of the earth), estimate the mass each patch must have to give the torque you found above. If each patch is a circular disk of thickness 20 km and density 5 g/cm^3, find its radius.

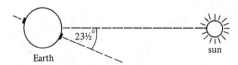

CHAPTER 16

KEPLER'S LAWS AND THE CONIC SECTIONS

I was almost driven to madness in considering and calculating the matter. I could not find out why the planet [Mars] would rather go on an elliptical orbit. Oh ridiculous me! As if the libration on the diameter could not also be the way to the ellipse. So this notion brought me up short, that the ellipse exists because of the libration. With reasoning derived from physical principles agreeing with experience, there is no figure left for the orbit of the planet except for a perfect ellipse. . . .

Why should I mince my words? The truth of Nature, which I had rejected and chased away, returned by stealth through the back door, disguising itself to be accepted. That is to say, I laid [the original equation] aside, and fell back on ellipses, believing that this was a quite different hypothesis, whereas the two, as I shall prove in the next chapter, are one and the same. . . . I thought and searched, until I went nearly mad, for a reason why the planet preferred an elliptical orbit. . . . Ah, what a foolish bird I have been!

Johannes Kepler, *Astronomia Nova* (1609)

16.1 THE QUEST FOR PRECISION

Not long after Copernicus published his revolutionary book, Tycho Brahe (1546–1601) provided a multitude of new observations that, despite his own intentions, provided crucial support for the Copernican hypothesis. At that time the furious debate between the Copernican and Ptolemaic systems was no longer waged in words alone; observations and exact measurements carried a new significance. Tycho realized the need for more precise astronomical observations and the instruments to make them. He built giant measuring devices on the island of Hveen, Denmark. Before Tycho's time, the locations of the heavenly bodies were known with a precision of about ten minutes (10') of arc. Tycho's painstakingly careful measurements reduced the uncertainty to about 2' of arc. His contribution as an astronomer was based on a method of observation that would become obsolete just ten years after his death, when the telescope was invented, transforming astronomy forever.

Tycho was a foul-tempered Danish lord who tongue-lashed kings, tormented peasants, sported a silver nose (his own having been lost in a youthful duel over mathematics), and kept a clairvoyant dwarf as his court jester and a tame elk that got drunk one night, fell down stairs, broke a leg, and died. Yet Tycho was a measuring maniac, a fussily precise man who opened a new age of observation in science.

The princes of Europe vied to acquire the services of Tycho as court astronomer and astrologer. King Frederick II of Denmark provided Tycho with the island of Hveen off the coast of his kingdom. There Tycho built Uraniborg, his hilltop observatory, which housed a windmill, a paper mill, fishponds, and a prison for unruly peasants, as well as workshops for the artisans who built his magnificent devices. His instruments were constructed on a huge scale. One of his quadrants, for example, was 38 ft in diameter and contained a life-size portrait of Tycho seated within its arc.

At Uraniborg, Tycho held court like a lord rather than a scholar. He ate and drank excessively, and at odd moments would pop off his gleaming metal nose to rub ointment on what lay beneath. When his feudal tenants complained of mistreatment, he tossed them into jail. And when the young King Christian IV reduced his benefits, Tycho, indignant, left the country. But at Uraniborg, he had so precisely determined the position of 777 stars and refined the measurements of Mars that today's measurements have only fractionally refined them.

Figure 16.1 Tycho seated at his giant quadrant. (Courtesy of the Archives, California Institute of Technology.)

In Prague in 1597, Rudolph II, King of Bohemia and Emperor of Germany, welcomed Tycho to his court as Imperial Mathematician with the fattest salary in the realm. On January 1, 1600, the impoverished Protestant Kepler left the Roman Catholic town of Graz for Prague to join the great Tycho. Tycho and Kepler knew that they needed each other. Kepler, in order to perfect his theoretical cosmology, the work that was his mission in life, needed Tycho's superb astronomical data. He was insolent and resentful when Tycho would only provide him with faint hints of key observations. And Tycho, in order to get his data organized into a useful form, needed Kepler's mathematical genius; he sensed that his hope for lasting fame lay in Kepler's penetrating intelligence. This tenuous relationship lasted for 18 months, but Tycho kept his secrets to the end.

At a banquet, Tycho overdrank and then held back his water beyond the demands of courtesy. An acute urinary infection set in, and within 11 days Tycho died. His last words to Kepler were, "Let me not seem to have lived in vain."

16.2 KEPLER'S LAWS

During the titanic period when the Aristotelian world was being replaced by the Copernican universe, Tycho believed in neither. He had his own model of the universe – the Tychonic theory, illustrated in Fig. 16.2. Tycho's theory was that the earth is stationary at the center of the universe and that the sun orbits the earth, with all of the planets orbiting the sun. Tycho fervently believed in his model and hoped that Kepler would build the universe on it. Kepler was to do just the opposite, to use Tycho's data to establish the Copernican universe.

Figure 16.2 The Tychonic model of the universe. (Courtesy of the Archives, California Institute of Technology.)

Immediately after Tycho's death, Kepler stealthily took Tycho's prized data for fear that they would be lost in the settlement of his estate. Rudolph appointed Kepler to the vacant post of Imperial Mathematician, and Kepler at last could settle down and work with the data he so much needed. Kepler eventually published in 1627 a full set of tables generated from Tycho's data, the *Rudolphine Tables*.

Kepler devoted the six years following Tycho's death to a battle with the planet Mars. Because the observed irregularities of its motion were greater than those of any other planet, the motion of Mars could not easily be described in terms of the Platonic ideal of uniform circular motion. Kepler, fully accepting a heliocentric universe, sought the smooth, continuous curve in which the planets orbited the sun. The problem he faced was that he had to find this curve on the basis of observations from a moving platform, the earth, which itself orbits the sun in some nonuniform way.

In his book, *Astronomia Nova* (*New Astronomy*), Kepler discusses the ingenious method by which he determined first the path of the earth itself. Kepler knew that the length of the Martian year (the time needed to complete one orbit) is 687 days. He used this information to identify the dates on which Mars would return to a given point in its orbit. The particular point chosen was point M in Fig. 16.3, when the earth at point E_0 was on a straight line between the sun and Mars. In the 687 days it takes Mars to return to point M, the earth moves $687/365 = 1.88$ revolutions, or through an angle of 677°. In other words, the earth moves 43° less than two complete revolutions. Thus, when Mars is again at point M, the earth is at point E_1. One Martian year later, Mars will again be at point M, but the earth will be at point E_2, which is 43° from E_1. By locating the earth's position on successive Martian years, Kepler succeeded in constructing a plot of the earth's orbit. He found this plot to be indistinguishable from a circle, except that the sun was slightly displaced from the center.

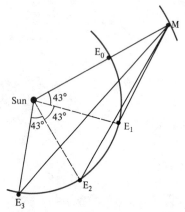

Figure 16.3 Kepler's determination of the earth's orbit from knowledge of the position of Mars on successive Martian years.

Kepler's plot of the earth's orbit also revealed that the earth moves fastest when it is nearest the sun. From an analysis of the speed of the earth at various points of its orbit, Kepler formulated what became known as his *second law*:

> A line directed from the sun to a planet
> sweeps out equal areas in equal times.

As we've seen in Chapter 13, this law can be understood as a consequence of the conservation of angular momentum.

Knowing the orbit and timetable of the earth, Kepler then reversed his analysis to find the shape of Mars' orbit as seen from the sun. Again he used observations of the position of Mars separated by one Martian year. By using the position of the earth at the same stage of successive Martian years, Kepler could triangulate to find a point on the orbit of Mars, as depicted in Fig. 16.3 where point M was fixed by sighting along the lines E_0M and E_1M. Then he could choose a second point, for example M′, the position of Mars the next time the earth – now at E_0' – was on a straight line between the sun and Mars (Fig. 16.4). When Mars returned to M′ after one further Martian year, the earth would be at E_1' as shown in Fig. 16.4, allowing a second triangulation to fix M′. Tycho's data, amassed over a quarter of a century, allowed Kepler to fix 12 points on the orbit of Mars in this manner. He could not force the points to fit a circular orbit, but rather he found the orbit of Mars to be an oval, as shown in Fig. 16.5. The disagreement between the data and the best circular path was about 8′ of arc. Here the improvement from the 10′ uncertainty in the data available to Copernicus to the 2′ uncertainty in Tycho's data proved crucial. Had Copernicus not tried to use epicycles but attempted instead to fit the orbit of Mars to a circle in this way, he would have succeeded in matching the best observations available to him. But Kepler, working with Tycho's improved data, could not. Faced with the choice of giving up the Platonic ideal of circular motion or violating Tycho's magnificent observations, he chose to believe the observations. After months of agonizing calculations, Kepler realized that the orbit of Mars could be fitted by an elegant curve whose special properties had been known for hundreds of years. The orbit of Mars, he realized, is an ellipse, as is, indeed, that of every planet. Thus he formulated what is now called his *first law*:

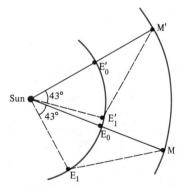

Figure 16.4 Kepler's triangulation to determine the orbit of Mars.

> Each planet orbits the sun along an elliptical path with the sun at one focus.

What made the orbit of Mars resist description as a circle is its large *eccentricity*. The eccentricity is a measure of how distant the sun is from the center of the ellipse. As

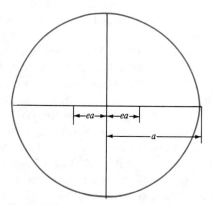

Figure 16.5 The elliptical orbit of Mars.

depicted in Fig. 16.5, if the length of the semimajor axis (half the longer dimension of the ellipse) is a, then the sun is located at a point (called a focus) at a distance ea from the center, where e is the eccentricity. If $e \neq 0$, there are two foci equidistant from the center. If $e = 0$, the foci coincide with the center of the curve, which then becomes a circle. As indicated in Table 16.1, Mars has the greatest eccentricity of those planets which Kepler could observe.

Table 16.1 Eccentricities of Planetary Orbits Known Today and Comments on Observational Possibilities in Kepler's Time

Planet	Eccentricity	Comment
Mercury	0.206	Too near the sun for observations
Venus	0.007	Very nearly circular
Earth	0.017	Small eccentricity
Mars	0.093	Largest eccentricity of planets known to Kepler
Jupiter	0.048	Slowly moving (12-yr period)
Saturn	0.056	Slowly moving (30-yr period)
Uranus	0.047	Not discovered until 1781
Neptune	0.009	Not discovered until 1846
Pluto	0.249	Not discovered until 1930

Kepler discovered many laws; three of them turn out to be correct. Buried within the pages of his book *Harmony of the World*, amid a long list of speculations, is what we call his *third law*:

> The square of the period of a planet is proportional to the cube of its semimajor axis.

In other words, $T^2 = ka^3$, where T is the period (the time to go once around the ellipse) and k, a constant of proportionality, is the same for all planets. Our task is to show why Kepler's laws are true in the heavens. In Chapter 17 we will deduce them from Newton's laws of motion and gravity.

16.3 CONIC SECTIONS

Before we can show that Kepler's laws follow from Newton's laws, we need a mathematical description of the ellipse. The ellipse is one of a family of curves that can be formed by the intersection of a cone with a plane. The curves obtained by slicing a cone with a plane not passing through the vertex are called *conic sections*, or simply *conics*. If the cutting plane is parallel to the side of the cone, as in Fig. 16.6a, the conic is called a *parabola*. Otherwise the intersection is called an *ellipse* or a *hyperbola*, according as the plane cuts just one or both nappes (portions of the cone), as depicted in Figs. 16.6b and 16.6c.

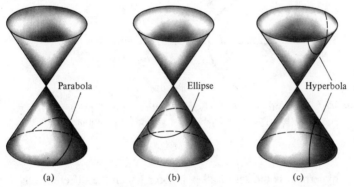

Figure 16.6 Conic sections formed by the intersection of a cone and plane: (a) parabola, (b) ellipse, (c) hyperbola.

The study of conic sections is believed to have originated in ancient Greece in an attempt to solve a puzzle posed by the oracle of Delos. When suffering citizens appealed to the oracle to halt a plague that ravaged Athens, they were instructed to double the size of Apollo's cubical altar. Their attempts to do so using a straightedge and compass were doomed to failure because, as we know now, the cube cannot be doubled in this way. The pestilence worsened. About 340 B.C. Menaechmus found two solutions using conic sections, one by intersecting a parabola and hyperbola, the other by intersecting two parabolas. It is also possible that the early Greeks considered conic sections in connection with sundials.

Apollonius of Perga (262–200 B.C.) wrote the first comprehensive treatise on conics. He used a double cone to generate the hyperbola and varied the tilt of the intersecting plane to generate the conics. He also coined the names ellipse, parabola, and hyperbola. For this work, he earned the title of "the great geometer."

There are other ways to introduce conic sections without referring to sections of a cone. One method refers instead to special points known as *foci*. The singular is *focus*, a word adopted by Kepler from Latin (meaning *fireplace*) for the place of the sun in the elliptical orbits.

An ellipse can be defined as the set of all points in a plane the sum of whose distances r and r' from two fixed points F and F' (the foci) is a constant. Figure 16.7a illustrates this definition of the ellipse. If the foci coincide, the ellipse reduces to a circle.

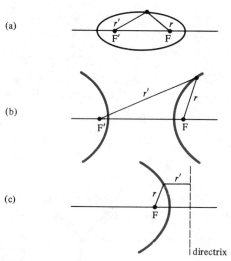

Figure 16.7 Focal properties of the conic sections: (a) ellipse: $r + r' = \text{const}$; (b) hyperbola: $|r - r'| = \text{const}$; (c) parabola: $r = r'$.

A hyperbola is the set of all points for which the difference $|r - r'|$ is constant, as shown in Fig. 16.7b. A parabola is the set of all points in a plane for which the distance from a fixed point F (the focus) is equal to the perpendicular distance from a given line (called the *directrix*).

Each of the conics has special properties which make it technologically important and interesting. For example, automobile headlights have parabolic reflectors because all light emitted from the focus will be reflected in the same direction, forming a beam. And projectiles, we know, follow parabolic paths. Elliptical domes allow sound generated from one focus to be entirely reflected to the other focus, forming a ''whispering gallery,'' convenient for listening in on conversations. And Kepler taught us that the orbit of a planet is an ellipse. The hyperbola is the basis for several navigational systems, one of which is called LORAN (LOng RAnge Navigation). A LORAN receiver on a ship records the time differences between arrival of signals sent simultaneously from several radio broadcasting stations located at known positions. Multiplication of the time difference $t_1 - t_2$ by the speed of light gives the difference $|r_1 - r_2|$ between the distances to stations 1 and 2. The ship lies somewhere on the hyperbola with that $|r_1 - r_2|$, and with foci at stations 1 and 2. Use of the distance differences $|r_1 - r_3|$ and $|r_2 - r_3|$ for the pairs of stations 1, 3 and 2, 3 determines further hyperbolas, and the intersection of the hyperbolas gives the location of the ship.

16.4 THE ELLIPSE

From the definition of an ellipse as a set of points for which the sum of distances from the foci $r + r'$ is constant, we have a handy method for drawing an ellipse. Take a piece of string of length $2a$ with its ends attached to two pins at the foci. By moving a pencil so as to keep the string taut you can draw an ellipse like the one in Fig. 16.8. The equation

$$r + r' = 2a \tag{16.1}$$

is satisfied for each point on the ellipse.

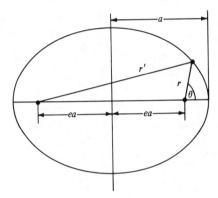

Figure 16.8 Quantities that specify an ellipse.

Presently we will describe another equation relating one of the focal distances r to the angle θ shown in Fig. 16.8, but first we give some definitions.

An ellipse has a center of symmetry midway between its foci. A line through the foci intersects the ellipse at two points called *vertices* of the ellipse, each at distance a from the center. The segment joining the vertices is called the *major axis* and has length $2a$. And, as already mentioned in connection with the orbit of Mars, the distance from the center to either focus is some fraction of a, usually denoted by ea, where e is a number between 0 and 1, $0 < e < 1$, called the *eccentricity* of the ellipse.

16.5 THE CONICS AND ECCENTRICITY

The concept of eccentricity can be used to give a unified treatment of all the conics. A conic section can be defined as a curve traced out by a point moving in a plane in such a way that the ratio of its distance from a fixed point (a focus) and a fixed line (a directrix) is constant. This constant ratio is called the eccentricity and is denoted by e. In Example 3, Section 16.6, we'll show that for an ellipse this definition of e implies the one given earlier in Section 16.4.

If $0 < e < 1$, the conic is an *ellipse*; if $e = 1$, it is a *parabola*; and if $e > 1$, it is a *hyperbola*.

In Fig. 16.9, F denotes the focus, the vertical line is the directrix, P is any point on the conic, Q is the nearest point to P on the directrix, and the eccentricity is the ratio e = FP/QP.

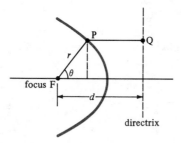

Figure 16.9 Geometry for obtaining a polar equation for conics.

From this definition we can easily find a polar equation for any conic with a vertical directrix. Let d be the distance from the focus to the directrix, and introduce r and θ as shown in Fig. 16.9. If both P and F are to the left of the directrix, as shown, we have

$$FP = r$$

and

$$QP = d - r \cos \theta,$$

so the relation FP = eQP becomes

$$r = e(d - r \cos \theta).$$

Solving for r we obtain

$$r = \frac{ed}{1 + e \cos \theta}. \tag{16.2}$$

The three types of conics are shown in Fig. 16.10. Because $\cos(-\theta) = \cos \theta$, all three conics are symmetric about the horizontal axis. On an ellipse, because $0 < e < 1$, the distance from the focus to a point on the curve is always less than the distance from the directrix to that point. In this case the curve crosses the horizontal axis at two points, when $\theta = 0$ and when $\theta = \pi$. Because of the symmetry, the curve is closed as shown in Fig. 16.10a.

What happens if $e = 1$? The curve is a parabola, and to visualize it we examine its polar equation, which now becomes

$$r = \frac{d}{1 + \cos \theta}.$$

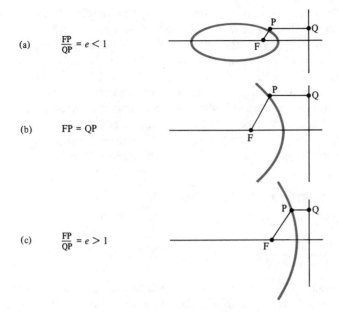

(a) $\dfrac{FP}{QP} = e < 1$

(b) $FP = QP$

(c) $\dfrac{FP}{QP} = e > 1$

Figure 16.10 The conic sections: (a) the ellipse has $0 < e < 1$, (b) the parabola has $e = 1$, (c) the hyperbola has $e > 1$.

When $\theta = 0$ the parabola cuts the axis at $r = \tfrac{1}{2}d$, the point equidistant from the focus and directrix. But it never again crosses the axis. In fact, as θ increases toward π, the denominator $1 + \cos\theta$ approaches 0 and the distance r becomes infinite. In other words, the curve spreads out to arbitrarily large distances from the axis as θ increases toward π, as shown in Fig. 16.10b. The symmetric lower half corresponds to values of θ in the interval $-\pi < \theta < 0$.

When $e > 1$ the polar equation (16.2) shows that when $\theta = 0$ the hyperbola crosses the axis at $r = ed/(1 + e)$. It, too, never crosses the axis again, as we can see by writing the polar equation in the form

$$r = \frac{d}{1/e + \cos\theta} \tag{16.3}$$

where we have divided top and bottom of the right-hand side of (16.2) by e. The term $1/e$ in the denominator is now less than 1, so there is a specific angle α between $\pi/2$ and π whose cosine is $-1/e$. Hence Eq. (16.3) can be written as

$$r = \frac{d}{\cos\theta - \cos\alpha}$$

where $\cos\alpha = -1/e$. As θ approaches α, the denominator approaches 0 and again r becomes infinite. The hyperbola is also open ended, but it spreads out in a different manner from the parabola. As r becomes larger, the angle θ increases but never reaches the value α.

The circle, $e = 0$, is a limiting case of the ellipse,* and the parabola, $e = 1$, can be viewed as a limiting case of the hyperbola. But the transition from ellipse to parabola at $e = 1$ is a surprising phenomenon in which variation of a continuous parameter e causes an abrupt jump from a closed to an open curve. This is not an isolated case; for example, the orbit of a mass moving in the potential $U = -GMm/r$ changes character abruptly as the total enegy $E = U + \frac{1}{2}mv^2$ is varied from negative values, for which planetary orbits confined to limited r occur, to positive values, for which the orbits are unbounded and take objects out to arbitrarily large values of r. As we shall see in the next chapter, the jumps in behavior as eccentricity and total energy increase through the values 1 and 0, respectively, are linked in a remarkable way.

16.6 PROPERTIES OF THE ELLIPSE

If $0 < e < 1$ the polar equation

$$r = \frac{ed}{1 + e \cos \theta} \tag{16.2}$$

represents an ellipse with eccentricity e, where d is the distance from the focus F to the directrix. The following examples show how easy it is to extract information about the ellipse from this equation.

Example 1
If we imagine a planet orbiting a fixed sun (located at F) the point on the ellipse nearest the sun is called the *perihelion* and the point farthest from the sun is called the *aphelion*. (See Fig. 16.11.) Show that the perihelion distance r_p and the aphelion distance r_a are given by

$$r_p = \frac{ed}{1 + e} \tag{16.4}$$

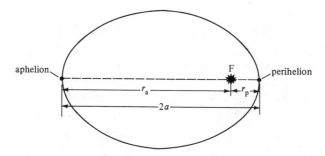

Figure 16.11 Perihelion and aphelion on an elliptical orbit about the sun.

*A circle of radius a is the limiting case of an ellipse in which $e \rightarrow 0$ and $d \rightarrow \infty$ in such a way that $ed \rightarrow a$. The limiting form of Eq. (16.2) is then simply $r = a$.

and

$$r_a = \frac{ed}{1 - e} . \tag{16.5}$$

Because r is smallest when $\cos \theta$ is largest ($\theta = 0$), Eq. (16.2) gives us (16.4) when $\theta = 0$; and because r is largest when $\cos \theta$ is smallest ($\theta = \pi$) we obtain (16.5) when $\theta = \pi$.

Example 2

Show that the length a of the semimajor axis is related to d and e as follows:

$$a = \frac{ed}{1 - e^2} \tag{16.6}$$

or

$$ed = a(1 - e^2). \tag{16.7}$$

The sum $r_p + r_a$ is $2a$, the length of the major axis; hence from Example 1 we find

$$2a = r_p + r_a = \frac{ed}{1 + e} + \frac{ed}{1 - e} = \frac{2ed}{1 - e^2} ,$$

from which we obtain (16.6) and (16.7).

When (16.7) is used in (16.4) and (16.5) we obtain alternative formulas for the perihelion and aphelion distances:

$$r_p = \frac{a(1 - e^2)}{1 + e} = a(1 - e) = a - ae \tag{16.4a}$$

and

$$r_a = \frac{a(1 - e^2)}{1 - e} = a(1 + e) = a + ae. \tag{16.5a}$$

Example 3

Let C denote the center of the major axis. The chord through C perpendicular to the major axis is called the *minor axis*. Let B denote the point of the minor axis directly above C and let $b = BC$. Show that

$$CF = ae, \tag{16.8}$$

$$BF = a, \tag{16.9}$$

and

$$b = a\sqrt{1 - e^2} . \tag{16.10}$$

From Fig. 16.12 we see that $CF = a - r_p$ and from (16.4a) we find $a - r_p = ae$, which proves (16.8). This result shows that the definition of eccentricity given in Sec. 16.4, $CF = ae$, follows from the definition in Sec. 16.5, where e is the ratio of the distances from a focus and from the directrix.

The point B is at a distance $ae + d$ from the directrix, so

$$BF = e(ae + d) = ae^2 + ed = ae^2 + a(1 - e^2) = a,$$

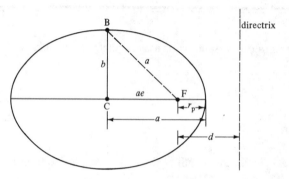

Figure 16.12 Calculations of distances CF, BF, and BC.

which gives (16.9). To prove (16.10) we refer to the right triangle FCB in Fig. 16.12. Its hypotenuse BF has length a, and one side CF has length ae, so by the theorem of Pythagoras we have

$$b^2 = a^2 - (ae)^2,$$

which implies (16.10).

Example 4

Let F' be the second focus, obtained by reflecting F through the center C, and let r' be the distance from F' to any point P on the ellipse, as shown in Fig. 16.13. Show that the definition of an ellipse given in Sec. 16.4,

$$r + r' = 2a, \tag{16.1}$$

(i.e., the string construction) follows from the definition in terms of a directrix and eccentricity.

We shall derive (16.1) from (16.2), the polar equation for an ellipse with a vertical directrix. First, applying the theorem of Pythagoras to the large right triangle in Fig. 16.13, we find

$$r'^2 = (r\cos\theta + 2ae)^2 + (r\sin\theta)^2 = r^2 + 4ear\cos\theta + 4a^2e^2 \tag{16.11}$$

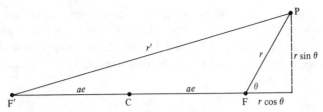

Figure 16.13 Proof that $r + r' = 2a$ on an ellipse.

because $\sin^2 \theta + \cos^2 \theta = 1$. The basic equation (16.2) together with (16.7) gives us

$$r(1 + e \cos \theta) = ed = a(1 - e^2),$$

which, when multiplied by $4a$ and rearranged, becomes

$$4ear \cos \theta + 4a^2 e^2 = 4a^2 - 4ar.$$

Using this in (16.11) we find

$$r'^2 = r^2 - 4ar + 4a^2 = (2a - r)^2.$$

Taking the positive square root of each member we obtain $r' = 2a - r$, which proves (16.1).

The basic polar equation for an ellipse can now be written, using (16.7), as

$$r = \frac{a(1 - e^2)}{1 + e \cos \theta}. \tag{16.12}$$

It tells us that the shape of the ellipse is governed by its eccentricity e. Figure 16.14 shows several ellipses with the same rightmost focus and the same value of a, but with various eccentricities. From Eq. (16.12) we see that $r \rightarrow a$ as $e \rightarrow 0$, which tells us that the smaller the eccentricity the rounder the ellipse. This is illustrated in Fig. 16.14.

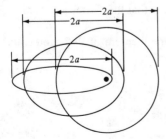

Figure 16.14 For fixed a, the shape of an ellipse is determined by its eccentricity.

As the eccentricity approaches zero, the distance between foci approaches zero and the ellipse approaches a circle of radius a. On the other hand, the ellipse becomes flatter as the eccentricity approaches 1 because $b = a \sqrt{1 - e^2}$ approaches 0 as $e \to 1$.

One other property, which we state without proof, is that the area A of the region enclosed by an ellipse is

$$A = \pi ab. \tag{16.13}$$

When $a = b$ this becomes the familiar formula for the area of a circle.

16.7 CARTESIAN EQUATIONS FOR CONIC SECTIONS

In our treatment of Kepler's laws it is convenient to use polar equations for analyzing the conic sections. But in discussing trajectories of projectiles it is more natural to use rectangular coordinates to discuss the Galilean parabola. This section describes briefly how all conics can be described in rectangular coordinates. The material in this section is not needed for the later chapters and can be considered as optional.

After the advent of analytic geometry in the seventeenth century, the conic sections were studied by algebraic methods. It was shown that no matter how the coordinate axes are chosen, the rectangular coordinates (x, y) of every point on a conic section satisfy a quadratic equation of the form

$$Ax^2 + Bxy + Cy^2 + Dx + Ey + G = 0$$

where A, B, C, D, E, and G are constants. We will illustrate this in Example 5 for the ellipse and in Example 6 for all three types of conics through the origin. If the conic has eccentricity e it can be shown that the quantity $4AC - B^2$ has the same algebraic sign as $1 - e$, so the type of conic can be recognized from the equation. It is an ellipse, parabola, or hyperbola, depending on whether $4AC - B^2$ is positive, zero, or negative.

Example 5

The equation of an ellipse is often seen in the form

$$\frac{x^2}{a^2} + \frac{y^2}{b^2} = 1 \tag{16.14}$$

where a and b are the lengths of the semiaxes. Take the x axis through the foci of an ellipse and the origin at its center and show that the Cartesian coordinates (x, y) of each point on the ellipse satisfy Eq. (16.14).

Referring to Fig. 16.13 we see, by the theorem of Pythagoras, that

$$r'^2 = (x + ae)^2 + y^2$$

and

$$r^2 = (x - ae)^2 + y^2.$$

Adding and subtracting these two equations we find

$$r^2 + r'^2 = 2(x^2 + y^2 + a^2e^2) \tag{16.15}$$

and

$$r'^2 - r^2 = (x + ae)^2 - (x - ae)^2 = 4axe. \tag{16.16}$$

But $r'^2 - r^2 = (r' + r)(r' - r) = 2a(r' - r)$ because

$$r + r' = 2a; \tag{16.1}$$

hence (16.16) implies $2a(r' - r) = 4axe$, or

$$r' - r = 2xe. \tag{16.17}$$

Adding and subtracting (16.1) and (16.17) we find

$$r' = a + xe \qquad \text{and} \qquad r = a - xe.$$

Putting these values in (16.15) and rearranging terms we obtain

$$x^2(1 - e^2) + y^2 = a^2(1 - e^2),$$

which gives (16.14) after we divide by $b^2 = a^2(1 - e^2)$.

Example 6

A conic of eccentricity e with its focus on the x axis has a vertical directrix and passes through the origin. Show that the rectangular coordinates (x, y) of each point on the conic satisfy an equation of the form

$$y^2 = (e^2 - 1)x^2 - cx \tag{16.18}$$

where c is a constant depending on the conic.

   If we let the line $x = p$ be the directrix, then the focus will be at $(-ep, 0)$ because the origin lies on the curve. If the directrix is to the right of the origin, as shown here, p is positive. Squaring the fundamental defining relation $PF = ePQ$, we have

$$(PF)^2 = e^2(PQ)^2.$$

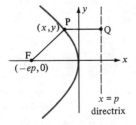

From the figure, we see that

$$(PF)^2 = (x + ep)^2 + y^2 \qquad \text{and} \qquad (PQ)^2 = (x - p)^2,$$

so the defining relation gives us

$$(x + ep)^2 + y^2 = e^2(x - p)^2.$$

Solving for y^2 and simplifying we find

$$y^2 = (e^2 - 1)x^2 - 2ep(1 + e)x.$$

This is Eq. (16.18) with $c = 2ep(1 + e)$.

16.8 A FINAL WORD

The English language contains words and constructions similar in form and definition to the conic sections. The parabola corresponds to the word *parable*, the ellipse corresponds to the *ellipsis*, and the hyperbola corresponds to the word *hyperbole*. This is no accident, because as any dictionary will reveal, these words are actually derived from the same words as the respective conic sections.

An *ellipsis* is a construction in which a grammatically necessary element in a sentence is omitted because it can be understood. For example, *Physics is more fun than chemistry*. This is an ellipsis, because a correct statement would be *Physics is more fun than chemistry is*. By removing part of a sentence, the result is less than the whole. Mathematically this corresponds to $e < 1$.

A *parable* is a short story that has some hidden meaning, such as a moral lesson. An example would be the legend of Newton and the apple, or Adam and Eve and the other apple. Mathematically, parabolas correspond to $e = 1$ and are formed by the intersection of a cone with a plane parallel to the generator of the cone. The parable represents a parallel to life.

The word *hyperbole* means an extravagant exaggeration or overstatement. The hyperbola corresponds to $e > 1$. An example of hyperbole is the following statement: *Physics is the greatest course in the world*.

In any case, it is a breathtaking fact that planets flying around in space follow particular mathematical curves with special properties. This brings us face to face with the great mystery that has awed everyone from Galileo down to Albert Einstein: mathematical relationships describe the laws of nature.

Problems

Elliptical Orbits

1. Using the data in Table 16.1, find the ratio of the perihelion distance to aphelion distance for (a) the earth and (b) Mars.

2. On a certain ellipse it is 5.0 cm from one focus to the farthest point and 2.0 cm from the same focus to the nearest point. Find (a) the eccentricity, (b) the length of the semimajor axis, (c) the length of the semiminor axis, and (d) the area of the ellipse.

3. Knowing that the perihelion distance of Mercury is 45.8×10^6 km and its eccentricity is 0.206, calculate the aphelion distance.

4. The elliptical orbit of a satellite about the earth is described by the polar equation

$$r = \frac{(8000 \text{ km})}{1 + 0.4 \cos \theta}.$$

Find (a) the eccentricity, (b) the length of the semimajor axis, and (c) the length of the semiminor axis.

5. Using Eq. (16.12) and conservation of angular momentum, derive an expression for the ratio of the speed of a planet at aphelion to that at perihelion. For what eccentricity is this ratio smallest?

6. A satellite in elliptical orbit around the earth is 7500 km from the center of the earth at perihelion and has a speed of 8000 m/s there. Its aphelion distance is 12000 km. Find (a) the eccentricity of the orbit, (b) the length of the semimajor axis, and (c) the speed at aphelion.

Polar Equations for Conics

Each of Problems 7–11 gives a polar equation for a conic section with focus F at the origin and a vertical directrix lying to the right of F. In each case, determine the eccentricity e, the type of conic, and the distance d of F from the directrix.

7. $r = \dfrac{2}{1 + \cos \theta}.$

8. $r = \dfrac{6}{3 + \frac{1}{2} \cos \theta}.$

9. $r = \dfrac{4}{\frac{1}{2} + \cos \theta}.$

10. $r = \dfrac{1}{1 + 3 \cos \theta}.$

11. $r = \dfrac{4}{6 + 3 \cos \theta}.$

12. The orbit of Halley's comet can be approximated by a parabolic orbit having a closest approach to the sun of 0.4 AU. Assuming that the earth's orbit is a circle of radius 1 AU, find the angle θ shown in the figure.

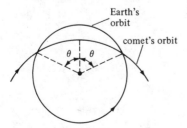

Cartesian Equation for Conics

13. Use the method of Example 6 to show that a parabola has a Cartesian equation of the form

$$y = ax^2$$

if the focus is on the y axis, the directrix is parallel to the x axis, and the curve passes through the origin. Make a sketch of the curve and indicate the geometric meaning of a.

14. Using Eq. (16.18) for an ellipse through the origin, show that the curve also crosses the x axis when $x = 2ep/(e-1)$.

15. Modify the argument in Example 5 as needed to show that each point (x,y) on a hyperbola of eccentricity e satisfies the equation

$$x^2(1 - e^2) + y^2 = a^2(1 - e^2)$$

where ae is the distance from the origin to the focus. This can also be written as

$$\frac{x^2}{a^2} - \frac{y^2}{b^2} = 1$$

where $b^2 = (e^2 - 1)a^2$. Since the equation remains unchanged when (x, y) is replaced by $(-x, -y)$, the hyperbola is symmetric about the origin. Make a sketch showing the symmetry and indicate the geometric meanings of a and b on your sketch.

16. The point $(2,2)$ is a focus and the line $x + y = 2$ is a directrix of a hyperbola with eccentricity $e = \sqrt{2}$. Proceed directly from the definition in terms of eccentricity to show that each point (x,y) on the hyperbola satisfies the equation

$$xy = 2.$$

CHAPTER 17

SOLVING THE KEPLER PROBLEM

Therefore, during the whole time of their appearance, comets fall within the sphere of activity of the circumsolar force, and hence are acted upon by its impulse and therefore (by Corollary 1, Proposition XII) describe conic sections that have their foci in the center of the sun, and by radii drawn from the sun describe areas proportional to the times. For that force propagated to an immense distance, will govern the motions of bodies far beyond the orbit of Saturn.

Isaac Newton, *Principia* (1687)

17.1 SETTING THE STAGE

In 1543 Corpernicus published his famous book; a generation later Kepler formulated his three laws; and then 150 years after Copernicus's book, Isaac Newton took Kepler's third law and used it to deduce the law of universal gravitation. From the law of gravitation and his dynamics, Newton was able to deduce Kepler's other two laws.

The task of deducing all three of Kepler's laws from Newton's laws is called the *Kepler problem*. Its solution is one of the crowning achievements of Western thought.

It is part of our cultural heritage just as Beethoven's symphonies or Shakespeare's plays or the ceiling of the Sistine Chapel are part of our cultural heritage. But it differs from a symphony or a play or a painting in an important way. It is a living idea. It is not something to be executed or performed by others and merely admired by us. We can absorb it, penetrate it, master it, and it becomes our very own, to take with us forever. For the same reason it is not necessary to try to do it in the same way Newton did it for himself, and we shall not. To make the task easier, we'll use ideas and techniques Newton didn't have: energy and vectors. We have been carefully preparing ourselves for this, and now we are on the threshold of the great discovery.

What we will do is prove that the differential equation we get from Newton's second law and the law of gravitation is satisfied only by the conic sections – the ellipse, parabola, or hyperbola. The orbits of the planets turn out to be ellipses, but other heavenly bodies such as meteors or comets can travel along ellipses, hyperbolas, or even parabolas. The solution of the differential equation does not, by itself, reveal which type of conic the orbit will be. Energy considerations will help us discover which orbits are ellipses and which are hyperbolas or parabolas.

Our task is not as formidable as it may seem because we have already solved part of the problem when we derived the law of equal areas in Chapter 13. Recall that the rate of change of the area vector of any object orbiting the sun by the action of a central force is

$$\frac{d\mathbf{A}}{dt} = \frac{1}{2}\mathbf{r} \times \mathbf{v} = \frac{\mathbf{L}}{2M} = \text{const} \qquad\qquad (13.18)$$

where $\mathbf{A}$ is the area vector, $\mathbf{v}$ the velocity, M the mass of the moving object, and $\mathbf{L}$ its angular momentum. The vector $\mathbf{L}/2M$ is always constant, because the sun acting on the object by a central force can apply no torque to it, and therefore the angular momentum of the object (planet, meteor, or comet) is conserved, as is its mass.

Because $d\mathbf{A}/dt$ is a constant vector it has constant magnitude and constant direction. Constant magnitude implies Kepler's second law, and constant direction tells us that the orbit lies in a plane perpendicular to that direction.

17.2 POLAR COORDINATES AND THE UNIT VECTORS r̂ AND θ̂

We have already used polar coordinates to study circular motion and to describe the conic sections. Since a heavenly body orbiting the sun must lie in a plane, polar coordinates can be used to describe its orbit.

We recall that the polar coordinates r and θ of a point in a plane are related to the rectangular coordinates (x, y) by the equations

$$x = r \cos \theta, \qquad y = r \sin \theta.$$

For any plane curve, the position vector $\mathbf{r} = x\,\hat{\mathbf{i}} + y\,\hat{\mathbf{j}}$, shown in Fig. 17.1, is given by

$$\mathbf{r} = r(\cos \theta)\hat{\mathbf{i}} + r(\sin \theta)\hat{\mathbf{j}} = r[(\cos \theta)\hat{\mathbf{i}} + (\sin \theta)\hat{\mathbf{j}}]$$

where $r = |\mathbf{r}|$. We will consider both r and θ as functions of time.

The vector $(\cos \theta)\hat{\mathbf{i}} + (\sin \theta)\hat{\mathbf{j}}$ is a vector of unit length having the same direction as $\mathbf{r}$. We denote this unit vector by our customary notation $\hat{\mathbf{r}}$, so we have

$$\mathbf{r} = r\hat{\mathbf{r}}$$

where

$$\hat{\mathbf{r}} = (\cos \theta)\hat{\mathbf{i}} + (\sin \theta)\hat{\mathbf{j}}. \tag{17.1}$$

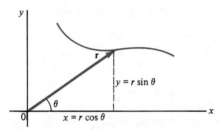

Figure 17.1 Position vector expressed in terms of polar coordinates r and θ.

We also introduce a unit vector $\hat{\boldsymbol{\theta}}$ which is perpendicular to $\hat{\mathbf{r}}$. Because $\hat{\mathbf{r}}$ has constant length, its derivative is perpendicular to it, and in fact we define $\hat{\boldsymbol{\theta}}$ to be the derivative of $\hat{\mathbf{r}}$ with respect to θ,

$$\hat{\boldsymbol{\theta}} = \frac{d\hat{\mathbf{r}}}{d\theta} = -(\sin \theta)\hat{\mathbf{i}} + (\cos \theta)\hat{\mathbf{j}}. \tag{17.2}$$

Like the unit vectors $\hat{\mathbf{i}}$ and $\hat{\mathbf{j}}$, both $\hat{\mathbf{r}}$ and $\hat{\boldsymbol{\theta}}$ are dimensionless. You can easily verify that

$$\hat{\mathbf{r}} \cdot \hat{\boldsymbol{\theta}} = 0,$$

which implies that $\hat{\mathbf{r}}$ and $\hat{\boldsymbol{\theta}}$ are indeed perpendicular. Another property is that

$$\frac{d\hat{\boldsymbol{\theta}}}{d\theta} = -(\cos \theta)\hat{\mathbf{i}} - (\sin \theta)\hat{\mathbf{j}} = -\hat{\mathbf{r}}.$$

The unit vectors $\hat{\mathbf{r}}$ and $\hat{\boldsymbol{\theta}}$ play the same role in polar coordinates as $\hat{\mathbf{i}}$ and $\hat{\mathbf{j}}$ play in Cartesian coordinates. It must be kept in mind, however, that unlike $\hat{\mathbf{i}}$ and $\hat{\mathbf{j}}$, which are fixed in space, the directions of $\hat{\mathbf{r}}$ and $\hat{\boldsymbol{\theta}}$ change as we move along the curve, as illustrated in Fig. 17.2.

Observe that, by the chain rule, we have

$$\frac{d\hat{\boldsymbol{\theta}}}{dt} = \frac{d\hat{\boldsymbol{\theta}}}{d\theta} \frac{d\theta}{dt} = -\hat{\mathbf{r}} \frac{d\theta}{dt}, \tag{17.3}$$

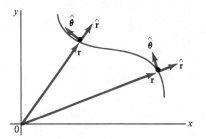

Figure 17.2 Unit vectors $\hat{\mathbf{r}}$ and $\hat{\boldsymbol{\theta}}$ attached to a curve at two points.

so we get the formula

$$-\hat{\mathbf{r}} = \frac{1}{d\theta/dt}\frac{d\hat{\boldsymbol{\theta}}}{dt}, \tag{17.4}$$

which we will use in solving the Kepler problem.

The vectors $\hat{\mathbf{r}}$ and $\hat{\boldsymbol{\theta}}$ share another property with $\hat{\mathbf{i}}$ and $\hat{\mathbf{j}}$. We recall that $\hat{\mathbf{i}} \times \hat{\mathbf{j}} = \hat{\mathbf{k}}$. Similarly, we have

$$\hat{\mathbf{r}} \times \hat{\boldsymbol{\theta}} = \hat{\mathbf{k}}, \tag{17.5}$$

which is illustrated geometrically in Fig. 17.3 and can be proved algebraically from the formula

$$\hat{\mathbf{r}} \times \hat{\boldsymbol{\theta}} = \begin{vmatrix} \hat{\mathbf{i}} & \hat{\mathbf{j}} & \hat{\mathbf{k}} \\ \cos\theta & \sin\theta & 0 \\ -\sin\theta & \cos\theta & 0 \end{vmatrix} = (\cos^2\theta + \sin^2\theta)\hat{\mathbf{k}}.$$

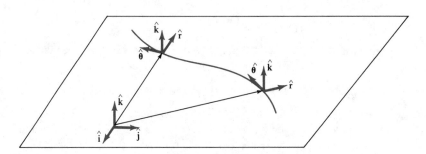

Figure 17.3 Unit vectors $\hat{\mathbf{r}}$, $\hat{\boldsymbol{\theta}}$, and $\hat{\mathbf{k}}$ at two points.

We will now express the velocity vector of any plane motion as a combination of the unit vectors $\hat{\mathbf{r}}$ and $\hat{\boldsymbol{\theta}}$. The velocity is given by

$$\mathbf{v} = \frac{d\mathbf{r}}{dt} = \frac{d}{dt}(r\hat{\mathbf{r}}).$$

Because both r and $\hat{\mathbf{r}}$ may change with time we apply the product rule to get

$$\mathbf{v} = \frac{dr}{dt}\hat{\mathbf{r}} + r\frac{d\hat{\mathbf{r}}}{dt} .$$

We can write the derivative $d\hat{\mathbf{r}}/dt$ in terms of $\hat{\boldsymbol{\theta}}$ by applying the chain rule,

$$\frac{d\hat{\mathbf{r}}}{dt} = \frac{d\hat{\mathbf{r}}}{d\theta}\frac{d\theta}{dt} = \frac{d\theta}{dt}\hat{\boldsymbol{\theta}},$$

where we used Eq. (17.2) for $d\hat{\mathbf{r}}/d\theta$. Substituting this into our expression for the velocity, we obtain the formula we were seeking:

$$\mathbf{v} = \frac{dr}{dt}\hat{\mathbf{r}} + r\frac{d\theta}{dt}\hat{\boldsymbol{\theta}}. \tag{17.6}$$

The scalar factors dr/dt and $r\, d\theta/dt$ multiplying $\hat{\mathbf{r}}$ and $\hat{\boldsymbol{\theta}}$ are called, respectively, the *radial* and *transverse components* of velocity.

Since $\hat{\mathbf{r}}$ and $\hat{\boldsymbol{\theta}}$ are perpendicular unit vectors, we can easily determine the *speed*:

$$v = \sqrt{\mathbf{v}\cdot\mathbf{v}} = \sqrt{\left(\frac{dr}{dt}\right)^2 + \left(r\frac{d\theta}{dt}\right)^2} . \tag{17.7}$$

Notice that the terms under the square root are not second derivatives but the squares of first derivatives.

Example 1

Express the acceleration for arbitrary circular motion ($r = $ const, $\omega = |d\theta/dt|$ allowed to vary with time) in terms of $\hat{\mathbf{r}}$ and $\hat{\boldsymbol{\theta}}$.

For circular motion, Eq. (17.6) for the velocity simplifies to

$$\mathbf{v} = r\frac{d\theta}{dt}\hat{\boldsymbol{\theta}}$$

with r constant. Differentiating this by the product rule and using Eq. (17.3) we find

$$\mathbf{a} = \frac{d\mathbf{v}}{dt} = r\frac{d}{dt}\left(\frac{d\theta}{dt}\right)\hat{\boldsymbol{\theta}} + r\frac{d\theta}{dt}\frac{d\hat{\boldsymbol{\theta}}}{dt}$$

$$= r\frac{d}{dt}\left(\frac{d\theta}{dt}\right)\hat{\boldsymbol{\theta}} - r\left(\frac{d\theta}{dt}\right)^2\hat{\mathbf{r}}.$$

If θ increases with t then $d\theta/dt$ is positive and the relations $\omega = d\theta/dt$ and $v = r\omega$ allow us to express the acceleration in the alternative forms

$$\mathbf{a} = r\frac{d\omega}{dt}\hat{\boldsymbol{\theta}} - r\omega^2\hat{\mathbf{r}}$$

$$= r\frac{d\omega}{dt}\hat{\boldsymbol{\theta}} - \frac{v^2}{r}\hat{\mathbf{r}}.$$

(17.8)

This establishes the result that we applied to the loop-the-loop ride in Sec. 7.6: The centripetal component of acceleration in circular motion equals v^2/r even when the motion is nonuniform. The new features in nonuniform circular motion are that the magnitude $v = r\omega$ varies around the circle, and that a tangential component of acceleration occurs, $r\, d\omega/dt$.

Example 2
Show that for any plane motion we have

$$\mathbf{r} \times \mathbf{v} = r^2\frac{d\theta}{dt}\hat{\mathbf{k}}.$$

(17.9)

Taking the cross product of $\mathbf{r} = r\hat{\mathbf{r}}$ with $\mathbf{v}$ as given by (17.6) we find

$$\mathbf{r} \times \mathbf{v} = r\frac{dr}{dt}\hat{\mathbf{r}} \times \hat{\mathbf{r}} + r^2\frac{d\theta}{dt}\hat{\mathbf{r}} \times \hat{\boldsymbol{\theta}}.$$

But, $\hat{\mathbf{r}} \times \hat{\mathbf{r}} = \mathbf{0}$ and in (17.5) we showed that $\hat{\mathbf{r}} \times \hat{\boldsymbol{\theta}} = \hat{\mathbf{k}}$ so this proves (17.9).

The accompanying figure shows why this result could have been anticipated. Only the component of $\mathbf{v}$ normal to $\mathbf{r}$, with magnitude $r|d\theta/dt|$, contributes to $\mathbf{r} \times \mathbf{v}$, so $\mathbf{r} \times \mathbf{v}$ should have magnitude $r(r|d\theta/dt|) = r^2|d\theta/dt|$ and should be perpendicular to the plane of $\mathbf{r}$ and $\mathbf{v}$.

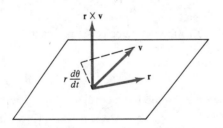

In Chapter 13 we introduced the angular momentum vector

$$\mathbf{L} = M\mathbf{r} \times \mathbf{v}$$

(13.3)

for any moving object of mass M. We can now obtain a simple formula for $\mathbf{L}$ in polar coordinates when the motion is in a plane. Using Eq. (17.9) we have

$$\mathbf{L} = Mr^2 \frac{d\theta}{dt} \hat{\mathbf{k}}. \tag{17.10}$$

This tells us that for any plane motion the angular momentum vector has a fixed direction perpendicular to the plane of the motion and that its length is equal to $Mr^2\omega$, where $\omega = |d\theta/dt|$ is the angular speed.

In Chapter 13 we also showed that for any plane motion the vector $\frac{1}{2}\mathbf{r} \times \mathbf{v}$ is the rate of change of the area vector,

$$\frac{1}{2}\mathbf{r} \times \mathbf{v} = \frac{d\mathbf{A}}{dt}. \tag{13.18}$$

Equation (17.9) now tells us that

$$\frac{d\mathbf{A}}{dt} = \frac{1}{2} r^2 \frac{d\theta}{dt} \hat{\mathbf{k}}, \tag{17.11}$$

so Kepler's law of equal areas is equivalent to the statement that

$$r^2 \frac{d\theta}{dt} = \text{const.}$$

From (17.10) we deduce that the absolute value of this constant is L/M. If we measure θ so that it increases with t, then

$$r^2 \frac{d\theta}{dt} = \frac{L}{M}. \tag{17.12}$$

This property will be used to help us solve the Kepler problem.

17.3 SOLUTION OF THE KEPLER PROBLEM

We now have all the machinery in place to solve the Kepler problem. Assume we have a fixed sun of mass M_0 and a moving body of mass M attracted to the sun by a gravitational force $\mathbf{F}$. Newton's law of gravity states that

$$\mathbf{F} = -G\frac{MM_0}{r^2}\hat{\mathbf{r}}$$

where $\mathbf{r} = r\hat{\mathbf{r}}$ is the radius vector from the sun to the body. Newton's second law of motion describes the acceleration due to that force,

$$\mathbf{F} = M\frac{d\mathbf{v}}{dt}.$$

Equating the two expressions for $\mathbf{F}$ and canceling M, we find

$$\frac{d\mathbf{v}}{dt} = -G\frac{M_0}{r^2}\hat{\mathbf{r}}.$$

Using (17.4) and then Eq. (17.12), assuming that θ is measured so that it increases with t, we can write this as

$$\frac{d\mathbf{v}}{dt} = \frac{GM_0}{r^2(d\theta/dt)}\frac{d\hat{\boldsymbol{\theta}}}{dt} = \frac{GMM_0}{L}\frac{d\hat{\boldsymbol{\theta}}}{dt} = \frac{D}{L}\frac{d\hat{\boldsymbol{\theta}}}{dt}$$

or as

$$\frac{L}{D}\frac{d\mathbf{v}}{dt} = \frac{d\hat{\boldsymbol{\theta}}}{dt},$$

where L/D is a constant and $D = GMM_0$. This last differential equation can be integrated at once to give

$$\frac{L}{D}\mathbf{v} = \hat{\boldsymbol{\theta}} + \mathbf{C}$$

where $\mathbf{C}$ is a constant vector which depends on the initial conditions. Let's measure t so that at time $t = 0$ the body is closest to the sun. Then $dr/dt = 0$ at $t = 0$ because r has a minimum there. Hence $\mathbf{v}(0)$ has the same direction as $\hat{\boldsymbol{\theta}}(0)$ since the radial component of $\mathbf{v}(0)$ is zero. Because we are measuring θ so that it increases with t, we have $\hat{\boldsymbol{\theta}}(0) = \hat{\mathbf{j}}$, as indicated in the diagram.

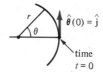

Hence $\mathbf{C}$ is a scalar times $\hat{\mathbf{j}}$. We call this scalar e and write $\mathbf{C} = e\hat{\mathbf{j}}$. (We shall see below that e is the eccentricity.) The basic equation for $\mathbf{v}$ becomes

$$\frac{L}{D}\mathbf{v} = \hat{\boldsymbol{\theta}} + e\hat{\mathbf{j}}.$$

Taking the dot product with $\hat{\boldsymbol{\theta}}$ we have

$$\frac{L}{D}\mathbf{v}\cdot\hat{\boldsymbol{\theta}} = \hat{\boldsymbol{\theta}}\cdot\hat{\boldsymbol{\theta}} + e\,\hat{\mathbf{j}}\cdot\hat{\boldsymbol{\theta}} = 1 + e\cos\theta. \qquad (17.13)$$

But from (17.6) and (17.12) we have

$$\mathbf{v}\cdot\hat{\boldsymbol{\theta}} = r\frac{d\theta}{dt} = \frac{1}{r}\left(r^2\frac{d\theta}{dt}\right) = \frac{1}{r}\frac{L}{M}$$

hence (17.13) becomes

$$\frac{L^2}{DM}\frac{1}{r} = 1 + e\cos\theta. \qquad (17.14)$$

This equation implies that e is positive. To see this, let θ take the values 0 and π. Then L^2/DMr takes the values $1 + e$ and $1 - e$, respectively. Because r has a minimum when $\theta = 0$, this implies $1 + e > 1 - e$, so $e > 0$ as asserted. When we solve (17.14) for r we get

$$r = \frac{L^2}{DM\,(1\ +\ e\cos\theta)}\,, \tag{17.15}$$

the polar equation for a conic with eccentricity e.

For an ellipse the polar equation was shown in Chapter 16 to be

$$r = \frac{a(1\ -\ e^2)}{1\ +\ e\cos\theta} \tag{16.12}$$

where a is the length of the semimajor axis. Comparing this with (17.15) we see that for an elliptical orbit we have

$$\frac{L^2}{DM} = a(1\ -\ e^2). \tag{17.16}$$

It should be realized that the foregoing solution of the Kepler Problem is based on simplifying assumptions which are not exactly true in the real solar system. We have assumed that the sun is fixed (which it is not) and that the only force acting on the body is the gravitational attraction of the sun. In reality, all planets and other objects in the solar system also exert gravitational forces on the body, but these are negligible compared to the massive attraction of the sun. In a solar system such as ours with one huge sun and a small number of little planets (called a "Keplerian system") these simplifications seem reasonable because the predicted orbits agree with the actual observed orbits to a remarkable degree of accuracy.

17.4 CELESTIAL OMENS: COMETS

Ancient astronomers earned their keep and kept their heads by making accurate predictions of the arrival of the seasons and of such troubling celestial events as solar and lunar eclipses. As their technical expertise improved, astronomers learned to predict even the wandering motion of the planets. Yet at times, interlopers, such as meteors, appeared in the night sky. Although meteors seemed as unpredictable as the weather (and were thought to be related to it, which is why their name shares the same Greek root with the science of weather – meteorology), even meteor showers were observed to occur regularly. For example, the most spectacular meteor showers occur every year in mid-August.

Nevertheless, objects that are not planets or meteors occasionally and mysteriously appear in the heavens. Trailing plumes of cold fire, these objects were named *comets*, from the Greek word meaning *thing with hair*. Because their appearance was unpredictable, comets were interpreted as omens of impending disaster. The Bayeaux Tapestry, which tells the story of the Norman conquest of England, shows a comet that appeared in 1066. Later in that year Harold, the Anglo-Saxon king, was defeated by William the Conqueror at the Battle of Hastings. Shakespeare places a comet in the sky the night before the murder of Julius Caesar. And according to legend, Montezuma, the Aztec leader, fell into such a depression at the foreboding appearance of a comet that he could not lead his people against the invading conquistadors.

For ages comets remained a perplexing anomaly of the heavens. Neither the system of Ptolemy nor the heliocentric theory of Copernicus nor even the ellipses of Kepler made the appearance of comets understandable and predictable. In 1682 a spectacular comet

Figure 17.4 Part of the Bayeux Tapestry depicting the comet of 1066, known today as Halley's comet. (Courtesy Science Graphics.)

blazed across the sky, and among the astronomers who charted its position was Isaac Newton. He saw in the orbit of the comet the same force and dynamics at work that govern the motion of the planets. Some comets, Newton realized, could swing past the sun in open curves – parabolas and hyperbolas – and so would never return. But other comets should move along elliptical paths like the planets' except on much longer leashes. Newton's penetrating insight revealed that comets are members of the solar system, and thus he cast off the superstition that enshrouded them.

17.5 ENERGY AND ECCENTRICITY

We have shown that a planet, comet, meteor, or any heavenly body that orbits the sun must move along a conic section with polar equation

$$r = \frac{L^2}{DM(1 + e \cos \theta)} \tag{17.15}$$

where L is the angular momentum of the body, M is its mass, $D = GMM_0$, and M_0 is the mass of the sun. To determine whether the orbit is an ellipse, parabola, or hyperbola we relate the eccentricity e to the total energy E of the moving body.

The energy E consists of two parts,

$$E = K + U,$$

where $K = \frac{1}{2}Mv^2$ is the kinetic energy and $U = -D/r$ is the potential energy associated with the gravitational force. To calculate K in polar coordinates we use the equation preceding Eq. (17.13) and write the velocity in the form

$$\mathbf{v} = \frac{D}{L} (\hat{\boldsymbol{\theta}} + e\hat{\mathbf{j}}).$$

Taking the square of the length of each member we get

$$v^2 = \frac{D^2}{L^2} (\hat{\boldsymbol{\theta}} + e\hat{\mathbf{j}}) \cdot (\hat{\boldsymbol{\theta}} + e\hat{\mathbf{j}})$$

$$= \frac{D^2}{L^2} (1 + 2e \cos \theta + e^2)$$

because $\hat{\boldsymbol{\theta}} \cdot \hat{\mathbf{j}} = \cos \theta$, so the kinetic energy is

$$K = \tfrac{1}{2} Mv^2 = \frac{D^2 M}{2L^2} (1 + 2e \cos \theta + e^2). \qquad (17.17)$$

To calculate the potential energy $U = -D/r$ we use Eq. (17.14), which gives us

$$\frac{1}{r} = \frac{DM}{L^2} (1 + e \cos \theta);$$

hence

$$U = -\frac{D}{r} = -\frac{D^2 M}{2L^2} (2 + 2e \cos \theta). \qquad (17.18)$$

The total energy $E = K + U$, obtained by adding (17.17) and (17.18), is given by

$$E = \frac{D^2 M}{2L^2} (e^2 - 1). \qquad (17.19)$$

The formula for the total energy is suddenly much simpler! Both r and θ have been eliminated and the total energy is constant, as it should be. It is expressed in terms of the masses M and M_0 (because $D = GMM_0$), the angular momentum L, and the eccentricity e of the orbit. This formula for the energy will give us important insight into the shapes of the orbits that we couldn't get from the equation of the conics alone.

For an elliptical orbit Eq. (17.16) gives us

$$\frac{L^2}{DM} = a(1 - e^2), \qquad (17.16)$$

so the energy in this case is simply

$$E = -\frac{D}{2a} = -\frac{GMM_0}{2a}. \qquad (17.20)$$

This formula is useful because it expresses the total energy of an elliptical orbit entirely in terms of the mass of the planet, the mass of the sun, and the length of the major axis of the ellipse.

17.6 ORBITS AND ECCENTRICITY

A planet, comet, or satellite orbiting the sun has an initial energy E, an initial angular momentum L, and a definite value of $D = GMM_0$. These numbers should completely determine the type of orbit, and indeed they do, for if we solve Eq. (17.19) for the eccentricity we find

$$e = \sqrt{1 + \frac{2L^2 E}{D^2 M}} \, . \tag{17.21}$$

Let's explore the connection between energy and eccentricity.

Case I: E > 0.

For positive total energy, the object always has more kinetic energy than potential energy. In this case $e > 1$ so the orbit is a hyperbola. This can be interpreted to mean that the object starts its motion at an infinite distance from the sun and slowly falls toward it, with its change in potential energy appearing as kinetic energy. It makes one pass by the sun at some minimum separation distance and whips away along the hyperbola, never to return. Some comets have been seen with hyperbolic orbits. (This requires careful measurement. The orbits of comets can only be observed when they are near the sun and, as Fig. 17.5 shows, hyperbolic, parabolic, and highly eccentric elliptic orbits all look qualitatively similar there.)

Case II: E = 0.

For zero energy, the potential energy is exactly equal to the kinetic energy and the eccentricity is 1 so the object moves along a parabola. It, too, makes one pass by the sun and moves away, never to return. Parabolic orbits are theoretically possible but highly unlikely because they would require a perfect balance between the negative potential energy and the positive kinetic energy.

Case III: E < 0.

If E is negative the quantity under the radical sign in (17.21) is less than 1, hence $e < 1$, and the orbit is an ellipse. In this case the gravitational potential energy is always greater in magnitude than the positive kinetic energy. The object never gains enough kinetic energy to escape so it remains trapped forever in a closed elliptical orbit about the sun as shown in Fig. 17.5.

If $E = -D^2 M/2L^2$ the eccentricity is zero and the orbit is a circle, a Platonic orbit. The orbits of most planets have a small eccentricity and are very nearly circular. Presumably interactions with nearby debris orbiting the sun in the early solar system reduced

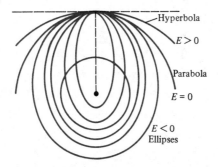

Figure 17.5 Possible orbits about the sun, with common point of nearest approach.

the eccentricities of the planetary orbits to their present small values (we have not taken account of this effect here but will consider it briefly in Sec. 17.8). However, an exactly circular orbit would be a surprising coincidence.

The structure of the solar system, the motion of planets, comets, and satellites, has finally been revealed. We've seen that Newton's dynamics lead to Kepler's first two laws. Next we consider the third law.

17.7 KEPLER'S THIRD LAW

In 1618 Kepler published the *Harmony of the Worlds*, the culmination of his lifelong obsession. In this book he attempted to reveal the ultimate secret of the universe in a synthesis of geometry, music, astronomy, and astrology – an ambitious undertaking which had not been attempted since Plato. The harmonies Kepler refers to are certain geometric proportions which he finds everywhere reflecting the universal order from which the planetary laws, the harmonies of music, the variations of the weather, and the fortunes of man are derived.

Hidden among the luxuriant fantasies of the *Harmony of the Worlds* is Kepler's third law of planetary motion, which states that the square of the period of revolution of a planet is proportional to the cube of the length of its semimajor axis. Mathematically we can write this law as

$$T^2 = ka^3$$

where k is a constant, the same for all planets. Kepler searched for such a law because he thought the universe would be hopelessly disharmonious without such a correlation. If the sun had the power to govern a planet's motion, then that motion must somehow depend on its distance from the sun. But how?

Veiled in the third law is the clue that led Newton to the universal law of gravity and the edifice of the solar system. No small achievement of Newton's was that he spotted the three laws in Kepler's writings and plucked them from the numerous other laws. Kepler never realized their real importance.

For the special case of uniform circular motion, the derivation of Kepler's third law from Newton's law of gravity is relatively simple, and has already been presented in Section 7.5. Let's now treat the general case of elliptic orbits by a different method, deriving Kepler's third law from his first and second laws. The second law states that the radius vector from the sun to a planet sweeps out area at a constant rate. From (17.11) and (17.12) we can express this in the form

$$\frac{dA}{dt} = \frac{L}{2M} . \tag{17.22}$$

If T denotes the time to complete one elliptical orbit, then integrating (17.22) from $t = 0$ to $t = T$ we find

$$A = \frac{L}{2M} T$$

where A is the area of the region enclosed by the ellipse. Solving for T and squaring both sides we obtain

$$T^2 = \left(\frac{2M}{L}\right)^2 A^2.$$

Using the formula $A = \pi ab$ for the area of an ellipse we get

$$T^2 = \left(\frac{2M}{L}\right)^2 \pi^2 a^2 b^2 = \left(\frac{2M}{L}\right)^2 \pi^2 a^4 (1 - e^2)$$

because $b^2 = a^2(1 - e^2)$. But from (17.16) we have

$$\frac{L^2}{DM} = a(1 - e^2)$$

so we obtain

$$T^2 = \pi^2 a^3 \frac{4M^2}{L^2} \frac{L^2}{DM} = 4\pi^2 \frac{M}{D} a^3.$$

Because $D = GMM_0$, we have Kepler's third law:

$$T^2 = \frac{4\pi^2 a^3}{GM_0}. \tag{17.23}$$

From our derivation we see that the constant of proportionality $k = 4\pi^2/GM_0$ depends only on the mass of the sun, and therefore is the same for all the planets. Since Kepler's third law is a consequence of the universal law of gravity and Newton's laws, it holds not only for planetary orbits but also for the elliptical orbits of moons or satellites about planets. For the motion of satellites, the mass of the attracting planet replaces M_0 in Eq. (17.23).

17.8 PLANETARY MOTION AND EFFECTIVE POTENTIAL

We can gain additional insight by analyzing the motion of heavenly bodies on the basis of conservation of energy and momentum. We use the fact that the total energy

$$E = K + U = \tfrac{1}{2} M v^2 + U(r)$$

is constant, and that the angular momentum is constant. We use Eq. (17.7) for v, and then Eq. (17.12) to replace $r^2 \, d\theta/dt$ by L/M, and finally $U(r) = -D/r$ to obtain

$$E = \frac{1}{2} M \left(\frac{dr}{dt}\right)^2 + \frac{1}{2} \frac{L^2}{Mr^2} - \frac{D}{r}. \tag{17.24}$$

This expresses the total energy in polar coordinates and also incorporates the angular momentum as part of the kinetic energy term.

In Chapter 13 we discussed the effective potential of a particle. We now apply this idea specifically to the Kepler problem. We define the *effective potential* function as

$$U_{\text{eff}}(r) = \frac{L^2}{2Mr^2} - \frac{D}{r}, \tag{17.25}$$

and denote the radial component of velocity dr/dt by v_r. Then the total energy in (17.24) is given by

$$E = \tfrac{1}{2} M v_r^2 + U_{\text{eff}}(r),$$

which remains constant as r and v_r change.

Figure 17.6 illustrates the effective potential and helps us further understand the motion which we described analytically earlier in this chapter. The precise shape of the curve depends on the angular momentum, as we saw in Chapter 13. Suppose a particle has some energy E which is negative, implying that the gravitational potential energy is greater than the kinetic energy. Because E is constant, the amount of energy between the horizontal line of constant height E and the effective potential curve is the kinetic energy $\tfrac{1}{2} M v_r^2$, as indicated on Fig. 17.6. Now what happens at the points where $E = U_{\text{eff}}(r)$? Whenever the total energy is entirely potential energy, that is, at the points where $E = U_{\text{eff}}(r)$, the kinetic energy $\tfrac{1}{2} M v_r^2$ is zero. That's precisely what the graph indicates. These points, labeled as $r_{\min}$ and $r_{\max}$, are the *turning points* at which v_r changes sign, indicating that the particle changes its direction relative to the center of its motion located at the origin.

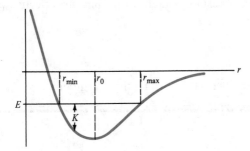

Figure 17.6 Effective potential function for planetary motion.

What does this imply about the motion of planets? A negative total energy tells us that the motion of the planet is bounded by a minimum and maximum distance from the sun. And for a $1/r$ gravitational potential, this corresponds to an elliptical orbit. The points $r_{\min}$ and $r_{\max}$ correspond to the perihelion and aphelion, respectively.

Figure 17.6 also indicates that there is a minimum in the effective potential. In Chapter 13 we found that this minimum separation distance corresponded to circular motion, and so it does here as well. The graph tells us that if a particle has an energy $E_{\min}$ equal to the minimum of the effective potential, then the kinetic energy term $\tfrac{1}{2} M v_r^2$ is always zero. This means that the particle has zero radial velocity (but it still has a transverse component of velocity) and moves at a constant distance from the center of motion – in other words, it moves in a circle. As discussed in Chapter 13, when a particle in an elliptical orbit suffers dissipative collisions it loses energy and tends to settle down into less eccentric, nearly circular orbits near the minimum of U_{eff}.

Example 3

What is the energy of a particle in a circular orbit in terms of its angular momentum?

We can use Eq. (17.24) to find the energy at any r, but first we need to know what the radius of the circular orbit is. Because a circular orbit occurs for E_{min} equal to the minimum of the effective potential, we can find the corresponding value of r by realizing that

$$\frac{d}{dr} U_{eff} = 0$$

at the minimum. Using our expression for $U_{eff}(r)$ in (17.25), we find

$$\frac{d}{dr} U_{eff} = -\frac{L^2}{Mr^3} + \frac{D}{r^2}.$$

Setting this equal to zero and solving for r_0, the radius of a circular orbit, we have

$$r_0 = \frac{L^2}{MD}.$$

Because $E_{min} = U_{eff}(r_0)$, we can substitute for r_0 into the effective potential and get (after some algebra)

$$E_{min} = -\frac{MD^2}{2L^2}.$$

This agrees with the result obtained by setting the eccentricity $e = 0$ in (17.21).

From Fig. 17.6 we also see that if the energy is zero or positive the motion is no longer bounded. In these cases there is a minimum value of r, which occurs where $E = U_{eff}(r)$, but there is no maximum value. Thus if a body moves toward the center of force, the separation will decrease until some minimum value, the turning point, where v_r changes sign. Then the body continues to move away from the center and never returns. As we've already learned, zero and positive values of the total energy correspond, respectively, to orbits which are parabolas and hyperbolas. In these cases the total energy is the limit the kinetic energy approaches at very large distances. By contrast, when the total energy is negative its magnitude $|E|$ is a *binding energy*, the amount of *extra* kinetic energy the body would have to be given to escape the gravitational potential to arbitrarily large r.

17.9 APPLICATIONS OF ORBITAL DYNAMICS

We illustrate applications of the results of this chapter through a number of examples.

Example 4

A satellite is fired from the surface of a spherical, nonrotating planet of mass M_0 and radius R, which has no atmosphere, with a speed v_0 at an angle of 30° from the radial direction. In its subsequent orbit, the satellite reaches a maximum distance of $\frac{5}{2}R$ from the center of the planet. Using conservation of energy and angular momentum, find v_0 in terms of M_0, R, and G.

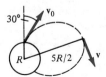

Take the mass of the satellite to be m, and calculate its energy at point A, where it is fired off, and at point B in its orbit.

$$E_A = \frac{1}{2} mv_0^2 - G\frac{mM_0}{R}, \qquad E_B = \frac{1}{2} mv^2 - G\frac{mM_0}{\frac{5}{2}R}.$$

We know that $E_A = E_B$, but we have two unknowns, namely, v_0 and v. The additional information needed to solve the problem is provided by the equation stating conservation of angular momentum. Recalling that $L = mvr$, where v is the component of $\mathbf{v}$ perpendicular to $\mathbf{r}$, we have

$$L_A = mRv_0 \sin 30°, \qquad L_B = mv\left(\tfrac{5}{2}R\right).$$

Because $L_A = L_B$, we can solve for v in terms of v_0:

$$v = \tfrac{1}{5}v_0.$$

Substituting this value into E_B and setting E_A equal to E_B, we get

$$\frac{1}{2}mv_0^2 - G\frac{mM_0}{R} = \frac{1}{2}m\left(\frac{v_0}{5}\right)^2 - G\frac{2mM_0}{5R}.$$

After some algebra, we find that

$$v = \sqrt{\frac{5GM_0}{4R}}.$$

Example 5

Show that Galileo's parabolic trajectory is approximately a small segment of an ellipse with the earth's center at the more distant focus.

Because the energy and angular momentum determine the type of orbit, we begin by evaluating the energy for a typical object projected on the surface of the earth. If v_0 is the speed of the object of mass m, the total energy is

$$E = \frac{1}{2} mv_0^2 - G\frac{Mm}{R}$$

where M is the mass of the earth (6×10^{24} kg) and R is its radius (6×10^6 m). The angular momentum of the object about the center of the earth is

$$L = mv_0 R \sin \theta$$

where, as shown in the sketch, θ is the angle from the vertical.

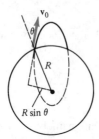

According to Eq. (17.21), the eccentricity of the orbit is given by

$$e = \sqrt{1 + \frac{2L^2 E}{D^2 m}} \qquad\qquad (17.21)$$

with $D = GMm$. Substituting for E and L and factoring terms we can write the eccentricity of a projectile as

$$e = \sqrt{1 + \frac{2v_0^2 \sin^2 \theta \left(\frac{1}{2}v_0^2 - GM/R\right)}{(GM/R)^2}}.$$

Now typical initial speeds of most projectiles are in the range of hundreds of meters per second at the most, so the term $\frac{1}{2}v_0^2$ is on the order of 10^4 m^2/s^2. On the other hand, the term $GM/R = 7 \times 10^7$ m^2/s^2. This means that we can ignore the $\frac{1}{2}v_0^2$ term compared to the term GM/R in the numerator of the expression for the eccentricity, so

$$e \approx \sqrt{1 - \frac{2v_0^2 \sin^2 \theta}{GM/R}}.$$

This last equation indicates that the eccentricity is less than 1, which means that the "orbit" is part of an ellipse. The center of the earth is at the more distant focus of the ellipse, as shown above in the sketch, because the distance from the center is maximum when the highest point of the orbit is reached, and the highest point lies only slightly above the earth's surface for a typical projectile of low kinetic energy. In addition, since the speed is comparatively small, the eccentricity is very nearly 1; that is, the elliptical orbit is extremely elongated. Taking $v_0 = 100$ m/s and $\theta = 45°$, we find that

$$1 - e \approx 7 \times 10^{-5},$$

so the eccentricity of the orbit differs from that of a parabola by only seven parts in 100,000.

Example 6

In Chapter 10 we found that the escape velocity from a planet of mass M and radius R is given by $v_0 = (2GM/R)^{1/2}$. What type of orbit will an object follow if projected tangentially with speed v_0 from the surface of a planet without an atmosphere?

To find the type of orbit, let's calculate the eccentricity, which is given by Eq. (17.21),

$$e = \sqrt{1 + \frac{2L^2E}{D^2m}} \, . \tag{17.21}$$

If m is the mass of the object, then its energy is

$$E = \frac{1}{2}mv_0^2 - G\frac{Mm}{R} = \frac{1}{2}m\frac{2GM}{R} - G\frac{Mm}{R} = 0.$$

(This is the meaning of the term *escape velocity*: a velocity just sufficient to get the object out to an infinite distance with no energy "left over.") Since zero energy corresponds to $e = 1$, we see that the orbit will be a parabola. In the special case of tangential launch, the object starts at the vertex of the parabola, the point nearest the focus.

17.10 CALCULATING THE ORBIT FROM INITIAL CONDITIONS

Now that we understand the factors that determine the orbit of a planet, a comet, or a satellite, let's indicate just how the orbit can be found if some initial conditions are known. Suppose a satellite of mass m is launched, for example, from a space shuttle a distance r_0 from the center of the earth (of mass M) as indicated in Fig. 17.7. As part of the initial conditions the initial speed v_0 relative to the earth, as well as the angle of launch ϕ are also known. How do we find the size, shape, and orientation of the orbit?

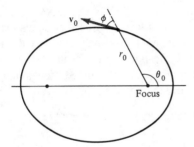

Figure 17.7 Initial conditions which determine the size, shape, and orientation of an orbit.

First, we know that the orbit will be closed only if the energy is negative. The energy, which remains constant, is specified by

$$E = \frac{1}{2} m v_0^2 - G \frac{Mm}{r_0}.$$

The first step is to verify that the energy is negative, which means the initial conditions must satisfy the relation $v_0^2 < 2GM/r_0$.

According to Eq. (17.20), this energy is also given by

$$E = -G \frac{Mm}{2a} \tag{17.20}$$

where $2a$ is the length of the major axis if the orbit is closed. Because E is known from the initial speed and distance, Eq. (17.20) determines the length of the major axis. That tells us the size of the orbit.

The angular momentum about the center of the earth likewise remains constant and equal to its initial value, which is given by

$$L = m v_0 r_0 \sin \phi.$$

Because the energy and angular momentum are known, the eccentricity of the orbit can be determined from Eq. (17.21):

$$e = \sqrt{1 + \frac{2L^2 E}{D^2 m}} \ . \tag{17.21}$$

Once the eccentricity is known, the perigee (nearest point to the earth) and apogee (farthest point from the earth) can be found. Their distances from the focus are, respectively,

$$r_\mathrm{p} = a(1 - e) \quad \text{and} \quad r_\mathrm{a} = a(1 + e).$$

So far, we've determined the size (given by a) and shape (given by e) of the orbit. The orientation of the orbit in space is specified by the line from the focus to the perigee. To find this line it suffices to find the angle θ_0 between this line and the known vector $\mathbf{r}_0$, as indicated in Fig. 17.7. The angle θ_0 can be determined from the polar equation of an ellipse.

$$r = \frac{L^2}{Dm(1 + e \cos \theta)} \tag{17.15}$$

when the values of r_0, e, m, and L are inserted. Note that θ_0 also specifies the initial position of the satellite along the orbit. (Actually, we have only determined $\cos \theta_0$, so there remains a sign ambiguity in θ_0 corresponding to whether we are departing from the perigee or approaching it. The ambiguity can be resolved by examining Fig. 17.7. We find that $\phi < 90°$ when departing from the perigee, and $\phi > 90°$ when approaching it.)

Example 7

As 5×10^3 kg satellite is launched in space with an initial speed $v = 4000$ m/s at distance $r_0 = 6R = 3.6 \times 10^7$ m from the center of the earth, at an angle of 30° from the radial direction. Calculate (a) the length of the semimajor axis, (b) the angular momentum, (c) the eccentricity, (d) the orientation, and (e) the perigee and apogee distances of the orbit.

Using Eq. (17.20),

$$E = -G \frac{Mm}{2a} . \tag{17.20}$$

we can determine the length of the semimajor axis. We know that the initial energy is

$$E = \frac{1}{2} mv_0^2 - G \frac{Mm}{r_0} = -1.6 \times 10^{10} \text{ J}$$

when we insert the values of v_0, r_0, m, M, and G. Equating this to Eq. (17.20) and solving for a, we find that $a = 6.4 \times 10^7$ m, which is about 10.4 earth radii.

The angular momentum is simply $L = mv_0r_0 \sin \phi$, which turns out to be equal to 3.6×10^{14} kg m^2/s. Once we know both E and L, we can use Eq. (17.21),

$$e = \sqrt{1 + \frac{2L^2E}{D^2m}} , \tag{17.21}$$

to calculate the eccentricity. Substituting values, we find that $e = 0.89$.

The orientation of the orbit comes from taking the equation of an ellipse, Eq. (17.15),

$$r = \frac{L^2}{Dm(1 + e \cos \theta)} , \tag{17.15}$$

and solving for $\cos \theta_0$ when $r = r_0$:

$$\cos \theta_0 = \frac{1}{e} \left(\frac{L^2}{Dmr_0} - 1 \right).$$

Here we find that $\theta_0 = 136°$. As stated before, this result gives both the orientation of the orbit in space and the position of the satellite along the orbit.

Finally, the perigee and apogee distances are given by

$$r_p = a(1 - e) = 7.0 \times 10^6 \text{ m} = 1.2R,$$

$$r_a = a(1 + e) = 1.2 \times 10^8 \text{ m} = 20R.$$

The orbit is illustrated below:

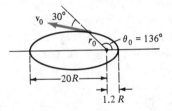

17.11 A FINAL WORD

Aside from Newton himself, the British astronomer Edmund Halley was the first person entrusted with knowledge of the structure of what we now call the solar system. Halley realized that although the planets had nearly circular orbits, Newton's results implied that all bodies in space will have elliptical orbits if they are captured by the sun at all. Some of them, for example comets that pass by infrequently, might have highly eccentric orbits.

In 1682 an awesome comet dominated the sky for months, and Halley made repeated measurements of its path. Later he applied Newton's method to find its orbit. Although his observations covered only a small segment of the complete orbit, Halley calculated that the comet had a highly elliptical orbit with a semimajor axis 20 times the earth's and a period of 76 years. Looking back into astronomical records, he found observations of comets in 1607 and 1531 so nearly identical that he concluded that they had to be the same object. The comet of the Bayeux Tapestry, it turns out, was also Halley's comet. Then Halley made the crucial test, which, of course, was to predict the comet's next return. He said that it would be seen at the end of 1758, a prediction that neither he nor Newton lived to see fulfilled.

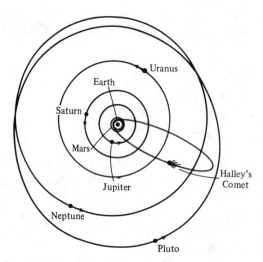

Figure 17.8 Orbit of Halley's comet.

Halley's comet was next seen at Christmas 1758. It has not stopped making its rounds. At the comet's nearest approach to the sun, it travels inside the orbit of Venus and shines brightly. At its most distant point, it goes beyond the orbit of Neptune, as shown in Fig. 17.8. The most recent round trip brings it to this part of the solar system in 1910, and again, just as Halley predicted, in 1986.*

---

*Note that Fig. 17.8 indicates that the motion of the comet around its orbit is retrograde, that is, in the direction opposite to that of the planets. This feature increases the difficulty and expense of observing the comet at close range when it is near the sun. Unless special measures are taken, a spacecraft tends to approach the comet head-on at high relative velocity and has only a single brief pass in which to collect data on it.

> Of all the comets in the sky,
> There's none like comet Halley.
> We see it with the naked eye,
> And periodically.

> – Anon

Problems

Plane Motion in Polar Coordinates

1. A particle moving in a plane traces out a spiral given by

    $$r(t) = r_0 e^t, \qquad \theta(t) = \omega_0 t$$

    where r_0 and ω_0 are constants. Determine the velocity and speed of the particle.

2. **(a)** How does the angular speed of a planet vary with its distance from the sun?

    (b) If an object has constant mass and angular momentum in a circular orbit, what must be true for $d\theta/dt$?

3. By differentiating the formula for velocity in polar coordinates show that the acceleration vector $\mathbf{a} = d\mathbf{v}/dt$ can be expressed as

    $$\mathbf{a} = \left[\frac{d^2 r}{dt^2} - r \left(\frac{d\theta}{dt} \right)^2 \right] \hat{\mathbf{r}} + \left(r \frac{d^2\theta}{dt^2} + 2 \frac{dr}{dt} \frac{d\theta}{dt} \right) \hat{\boldsymbol{\theta}}.$$

    The factors multiplying $\hat{\mathbf{r}}$ and $\hat{\boldsymbol{\theta}}$ are called, respectively, the *radial* and *transverse components* of acceleration.

4. Refer to the formula for acceleration in Problem 3. Show that the transverse component is equal to

    $$\frac{1}{r} \frac{d}{dt} \left(r^2 \frac{d\theta}{dt} \right).$$

    Deduce that the acceleration is radial (parallel to $\mathbf{r}$) if and only if the position vector $\mathbf{r}$ sweeps out area at a constant rate.

5. For circular motion in polar coordinates the vector

    $$\boldsymbol{\omega} = \frac{1}{r^2} \mathbf{r} \times \mathbf{v}$$

    is called the angular velocity vector. Show that

    (a) $\boldsymbol{\omega} = \dfrac{d\theta}{dt} \hat{\mathbf{k}}$,

    (b) $\mathbf{v} = \boldsymbol{\omega} \times \mathbf{r}$,

    (c) $\mathbf{v} \times \boldsymbol{\omega} = \dfrac{v^2}{r^2} \mathbf{r}$,

    (d) $\boldsymbol{\omega} \times (\boldsymbol{\omega} \times \mathbf{r}) = -\omega^2 \mathbf{r}$, where $\omega = \left| \dfrac{d\theta}{dt} \right|$.

Energy, Angular Momentum, and Orbits

6. Knowing that the aphelion and perihelion distances for the earth are given by 1.47 $\times 10^{11}$ m and 1.53×10^{11} m, respectively, calculate the energy associated with the earth's orbit. (Use data in Appendix D.)

7. Show that the ratio of the kinetic energy at aphelion to that at perihelion is given by

$$\frac{K_a}{K_p} = \frac{r_p^2}{r_a^2},$$

where r_p and r_a are the perihelion and aphelion distances.

8. The elliptical orbit of a 2500-kg satellite about the earth is described by the polar equation

$$r = \frac{8600 \text{ km}}{1 + 0.4 \cos \theta}.$$

(a) What is the eccentricity of the orbit?
(b) What is the total energy of the orbit?
(c) What is the angular momentum of the orbit?

(Use data in Appendix D for the mass of the earth.)

9. Can the total energy of a planet be negative? If so, what does this mean? Can kinetic energy ever be negative?

10. A satellite of mass m is traveling at a speed of v_0 in a circular orbit of radius r_0 about a planet. Show that the total energy of the satellite is $-\frac{1}{2}mv_0^2$.

11. A satellite of mass m has a circular orbit around a planet of mass M_0. The angular momentum of the satellite is L. Find the total energy of the satellite in terms of m, M_0, and L.

12. What is the eccentricity of an orbit with zero angular momentum? Is the motion in this case really an "orbit"?

13. Using data in Appendix D for values of eccentricity and semimajor axis, calculate the energy and angular momentum of the earth in its orbit.

14. The absolute magnitude of the energy of a meteor approaching the earth is given by

$$|E| = \frac{G^2 M^2 m^3}{L^2},$$

where M is the mass of the earth, m the mass of the meteor, and L its angular momentum. What are the types and eccentricities of possible orbits?

15. A satellite of mass m orbits the earth (of mass M and radius R). The perigee distance (distance of closest approach) measured from the center of the earth is $1.5R$, and the apogee distance (farthest distance) is $2.5R$.

(a) What is the eccentricity of the orbit?

(b) What is the energy of the satellite?

(c) What is its angular momentum in terms of G, M, m, and R?

16. A spherical, nonrotating planet with no atmosphere has mass M_0 and radius R. A spacecraft is fired from the surface with a speed

$$v_0 = \tfrac{3}{4}\sqrt{2GM_0/R} \ .$$

By considering conservation of energy and angular momentum, calculate the farthest distance it reaches from the center of the planet if it is fired off (a) radially and (b) tangentially.

17. Repeat Problem 16 for the case of $v_0 = \sqrt{2GM_0/R}$.

18. A rocket is fired from Cape Canaveral with an initial speed v_0 at an angle θ from the horizon as shown. Neglecting air resistance and the earth's rotation, calculate the maximum distance from the center of the earth that the rocket reaches in terms of the mass and radius of the earth, M and R, v_0, θ, and G.

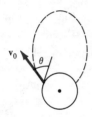

Effective Potential

19. Draw on your understanding of harmonic motion and effective potentials to explain qualitatively the motion of a particle which is slightly disturbed while it is in a circular orbit.

20. Qualitatively explain how the shape of the effective potential changes if the angular momentum is increased while all other factors remain the same. How does this change alter the turning points r_{min} and r_{max}?

21. Verify your answer for Problem 16 qualitatively. That is, show by sketching and using the curve for $U_{eff}(r)$ whether it is radial or tangential takeoff that produces the larger r_{max}.

Applications of Orbital Dynamics

Use Appendix D for data on masses and distances.

22. Determine the mass of the sun from the period and semimajor axis of the earth's orbit.

23. The period of the moon's orbit about the earth is 27.3 days and the semimajor axis has a length of 3.85×10^8 m. From this information calculate the mass of the earth.

24. What is the shortest possible period for an earth satellite?

25. A communications satellite is placed in a circular "synchronous orbit" about the equator where it remains overhead, having a period of one day. Determine the radius of this orbit.

26. The moon Ganymede of Jupiter orbits the planet with a period of 7.16 days and a semimajor axis of length 1.07×10^9 m. From this information calculate the mass of Jupiter.

27. Knowing that the eccentricity of the orbit of Halley's comet is 0.967 and the length of its semimajor axis is 20 times that of the earth's orbit, use additional data in Appendix D as needed to calculate the following:

    (a) the perihelion and aphelion distances for the comet,
    (b) its period.

28. A satellite of mass m is launched with a speed $v_0 = (1.5GM/R)^{1/2}$ from a distance of $4R$ from a planet for mass M and radius R as shown in the sketch. In terms of G, m, M, and R find the following for the orbit: (a) the energy, (b) the type, and (c) the eccentricity.

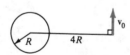

29. From a distance of $5R$ from the center of a planet of mass M and radius R, a satellite of mass m is launched with a speed $v_0 = (0.2GM/R)^{1/2}$ in the direction shown in the sketch. In terms of G, m, M, and R, determine the following quantities for the orbit: (a) the energy, (b) the angular momentum, and (c) the eccentricity.

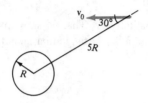

30. Using conservation of energy and angular momentum, calculate the speed of the satellite in Problem 29 at both its perigee and apogee in terms of G, M, and R.

CHAPTER 18

NAVIGATING IN SPACE

The initial shock [of acceleration] is the worst part of it, for he is thrown upward as if by an explosion of gun powder. . . . Therefore he must be dazed by opiates beforehand; his limbs must be carefully protected so that they are not torn from him and the recoil is spread over all parts of his body. Then he will meet new difficulties: immense cold and inhibited respiration. . . . When the first part of the journey is completed, it becomes easier because on such a long journey the body no doubt escapes the magnetic force of the earth and enters that of the moon, so that the latter gets the upper hand. At this point we set the travellers free and leave them to their own devices: like spiders they will stretch out and contract, and propel themselves forward by their own force – for, as the magnetic forces of the earth and moon both attract the body and hold it suspended, the effect is as if neither of them were attracting it – so that in the end its mass will by itself turn toward the moon.

Johannes Kepler, *Somnium*, published posthumously in 1634

18.1 FREEWAYS IN THE SKY

Not many years ago, the only conceivable use of the beautiful celestial mechanics developed over hundreds of years was to compute the positions of bodies in the heavens. Today that situation has changed radically. Many of the objects floating around in space are not natural at all; they were launched either by us or by our friends on the other side of this tiny planet we inhabit. And if we somehow avoid blowing each other up, there will be many more objects leaving this planet in the future.

In 1973 *Mariner 10* traveled to cloud-cloaked Venus and to Mercury. In 1975, the *Viking* landers arrived on the red planet Mars. And in 1977 *Voyager 2* was launched for a rendezvous to uncover the secrets of the giant planets – Jupiter, Saturn, Uranus, and Neptune. These exquisite machines are our scientific eyes and ears – the explorers we send across the new ocean of space. Sometimes they shatter old legends and illusions, but just as often they create new ones. Even though they are logical extensions of our five senses, these machines remain the metal slaves of our human minds and imaginations.

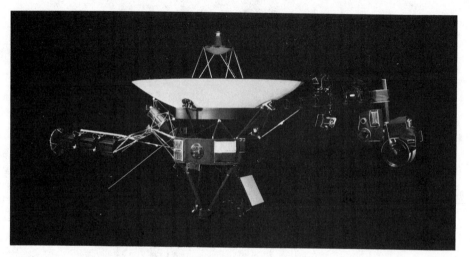

Figure 18.1 Spacecraft *Voyager 1*. (Courtesy JPL/NASA.)

How do interplanetary probes reach those strange and distant worlds? How do they navigate across the vast ocean of the solar system without the aid of so much as a lighthouse? As we are about to learn, there are no free rides in space. Every trip requires some expenditure of energy. But there are ways to go that minimize the cost.

18.2 NAVIGATING IN SPACE

How do we navigate to other planets? There are a number of possible solutions. The first method that comes to mind is to use brute force. We could build giant rockets to hurl a spacecraft directly toward a planet, then use a blast of the rockets to slow the craft down when it arrives at its destination. However, there is a much more elegant and practical method: to make use of the sun's gravitational field and Kepler's ellipses.

The first step is to launch the spacecraft into a temporary orbit about the earth. From there we place the spacecraft into a Hohman *transfer orbit*, as shown in Fig. 18.2. This is a semielliptical orbit about the sun which is tangent to both the earth's orbit and the orbit of the target planet. Once the spacecraft is in the transfer orbit, its rockets need not be fired again until it reaches the orbit of the target planet. On the way, its energy and angular momentum remain constant. The gravitational force of the sun does all the work and this part of the ride is free. When it reaches the orbit of the target planet, the rockets must be fired to remove it from the transfer orbit and place it in the orbit of the target. This mode of travel requires the least amount of fuel, and that's much more important when you're traveling from Earth to Saturn than when you're driving across town.

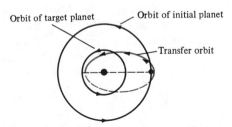

Figure 18.2 A transfer orbit is an elliptical orbit, with the sun at one focus, which is tangent to the orbits of both the initial and target planets.

Transfer orbits put constraints on space travel. We can't send a probe to any planet at any time by this method. Instead the launch must take place when the earth and the target are in the correct relative positions, a happening known as a *launch opportunity*. The earth must be at one end of the major axis of the transfer orbit at launch, and the target planet must arrive at the other end simultaneously with the spacecraft. By this method, we can send spacecraft to Venus every 19 months, to Mars every 780 days, and to Jupiter every 13 months. The opportunities are a consequence of the different orbital periods of the planets. According to Kepler's third law, each planet has an orbital period which is related to the length of its semimajor axis by

$$T^2 = ka^3,$$
(17.23)

where $k = 4\pi^2/GM_0$ is a constant of proportionality which depends on the mass M_0 of the sun, but is the same for all planets.

When a launch opportunity occurs, a spacecraft is initially placed in a temporary orbit about the earth known as a *parking orbit*. To enter a transfer orbit a spacecraft must somehow leave its parking orbit and escape the earth's gravity. Rocket thrusts supply a spacecraft with energy to escape, but when and how the thrusts are made depends on the destination.

Even during an opportunity the craft must be launched from the right place in its parking orbit. That place is called a *launch window*. If the spacecraft is headed for one of the inner planets (Mercury or Venus), the launch window occurs as the craft emerges onto the sunlit side of the earth, as shown in Fig. 18.3a. To launch toward the outer planets, a craft must leave its parking orbit as it approaches the dark side of the earth, as shown in Fig. 18.3b. Here's why: The initial velocity of the craft includes a contribution due to the earth's orbital velocity about the sun in addition to the craft's orbital velocity about the earth. (The parking orbit around the earth is always in the same sense as the earth's rotation. Why?) When the craft is on the sunlit side of the earth, these contributions are in opposite directions (see Fig. 18.3). An additional rocket thrust in the direction of the craft's orbital motion about the earth allows the craft to escape from its earth orbit. But since the velocity of the craft relative to the sun now is smaller than the earth's, the craft falls into an orbit closer to the sun – a transfer orbit. Although the craft takes off in a hyperbolic (escape) orbit relative to the earth, its orbit relative to the sun (once it becomes essentially free of the earth's influence) is elliptical.

To enter a transfer orbit to an outer planet, a spacecraft is launched from its parking orbit while approaching the dark side of the earth. When on the dark side the craft's orbital velocity about the earth is in the same direction as the earth's about the sun. Thus the two contributions will add, and a rocket blast causes the craft to escape from its Earth orbit and move into a larger orbit about the sun. (See Fig. 18.3b.)

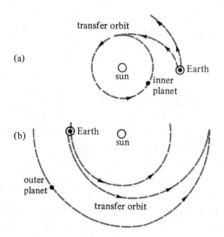

Figure 18.3 Launch window for a journey to (a) an inner planet, (b) an outer planet.

18.3 TRANSFER ORBITS

Let's now mathematically describe the steps involved in calculating a transfer orbit from, say, the earth to another planet. To simplify the calculations somewhat and emphasize the physical ideas, we'll assume the orbits of the planets are circles. We'll also ignore the need to escape from the earth's gravity and the speed of the spacecraft in its parking orbit around the earth. As discussed earlier, a transfer orbit is semielliptical with the two planets at perihelion and aphelion. In plotting a course for such an orbit, the first step is to find the length $2a$ of the major axis of the ellipse. As shown in Fig. 18.4, if r_1 and r_2 are the radii of the planetary orbits, then the length of the major axis of the transfer orbit is

$$2a = r_1 + r_2.$$

Next we need to know the speed which the satellite must have to move into the transfer orbit. Recalling that the energy of any elliptical orbit about the sun is

$$E = -G\,\frac{mM_0}{2a}\,,\tag{17.20}$$

where M_0 is the mass of the sun and m is the mass of the spacecraft, we see that once $2a$ is known, the energy of the orbit is determined. The energy of the spacecraft is also given by

$$E = \tfrac{1}{2}mv_1^2 - G\,\frac{mM_0}{r_1}\,,$$

so we can solve for v_1, the speed needed to begin the transfer orbit. We ignore the gravitational potential energy from the earth because it is much smaller than that of the sun. From the orbital speed of the earth, we can then calculate precisely the increase or decrease in speed necessary to send the spacecraft into the transfer orbit.

If the probe travels to an inner planet, it gains speed as it falls toward the sun. On the other hand, if it is traveling to an outer planet, it loses speed. But in either case, the speed of the spacecraft will have to be changed at the end of its trajectory to match that of the planet in its orbit about the sun. To make this change when the spacecraft reaches the orbit of the target planet, we need to know its arrival speed v_2. Since angular momentum is conserved along the transfer orbit, we can easily calculate v_2. Conservation of angular momentum, applied to the two points of the transfer orbit indicated in Fig. 18.4, tells us that

$$mv_1r_1 = mv_2r_2$$

because the velocity is perpendicular to the radius vector from the sun at both points. For this equation we can solve for v_2.

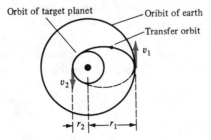

Figure 18.4 Quantities determining a transfer orbit.

From Kepler's third law, we can find the time for the spacecraft to travel along the transfer orbit; it's one-half of the period T. By using Kepler's third law $T^2 = ka^3$ twice, the constant k can be eliminated. Thus we have

$$\frac{T^2}{T_E^2} = \frac{a^3}{a_E^3}$$

where T_E and a_E are the period (one year) and semimajor axis of the earth's orbit, while T and a are those of the transfer orbit. The semimajor axes are measured in astronomical units, 1 AU being 1.496×10^{11} m, the mean sun-to-earth distance.

Finally, the launch opportunity can be found by deciding where the target planet should be relative to the earth at launch, so that the spacecraft will intercept it upon arrival at the orbit of the target planet. From the time of travel t and the orbital period of the planet T_p, the fraction $(360°)t/T_p$ tells us the angle in degrees the planet will move through while the spacecraft is on its way. From this angle we can determine the opportunity for launch. The following examples illustrate these calculations.

Example 1

For a transfer orbit between Earth and Venus, assume that both orbits are circular. The radius of Venus' orbit is 0.72 AU and its orbital period and speed are 225 days and 35.0 km/s, respectively.

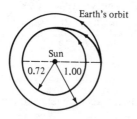

Determine the following: (a) the length of the major axis, (b) the speed needed to propel a spacecraft into the transfer orbit, (c) the change in speed required when the spacecraft reaches Venus, (d) the time of the trip, and (e) the relative positions of Venus and Earth at launch.

Making the approximations described in the text, we assume that initially the spacecraft has the same speed as the earth in its orbit around the sun. The energy of a spacecraft moving in that orbit is

$$E_0 = -G \frac{mM_0}{2r_E}$$

where M_0 is the mass of the sun, m the mass of the spacecraft, and r_E the radius of the earth's orbit. Equating this energy to

$$E = \tfrac{1}{2}mv_0^2 - G \frac{mM_0}{r_E},$$

we find that the speed of the satellite as it moves with the earth is

$$v_0 = (GM_0/r_E)^{1/2} = 29.8 \text{ km/s}.$$

We'll express the other speeds in terms of this speed.

As indicated in the above diagram, the length of the major axis of the orbit is $2a = 0.72$ AU $+ 1.00$ AU $= 1.72$ AU $= 1.72 r_E$. The energy of the transfer orbit is determined by $2a$, so we have

$$E = -G \frac{mM_0}{1.72 r_E} = \tfrac{1}{2} m v_1^2 - G \frac{mM_0}{r_E}$$

where v_1 is the speed needed for the spacecraft to enter the transfer orbit beginning at the earth. Solving for v_1, we find

$$v_1 = (0.84 GM_0/r_E)^{1/2} = 0.91 v_0 = 27.2 \text{ km/s}.$$

Therefore a rocket blast is needed to slow down the craft by 2.6 km/s, the difference between 29.8 and 27.2.

By Kepler's third law, the period of the transfer orbit can be found from the ratio

$$\frac{T^2}{T_E^2} = \frac{a^3}{a_E^3}.$$

This yields a travel time (one-half the period) $t = 0.4$ yr $= 146$ days.

Using conservation of angular momentum, we can calculate v_2, the speed of the spacecraft when it reaches the orbit of Venus. Because

$$m v_1 r_E = m v_2 r_V$$

where r_V is the radius of the orbit of Venus, we find $v_2 = 37.8$ km/s. The speed of Venus is 35.0 km/s, so the spacecraft needs to be slowed down by 2.8 km/s.

While the spacecraft spends 146 days traveling to Venus, the planet moves through $(146/225)360° = 234°$. Therefore, as indicated in the diagram below, Venus should be at point V_{launch}, which is 234° from the arrival point V_{arr}. During the trip, the earth moves through 144° and, as you can easily verify, is at the point E_{arr} on the diagram.

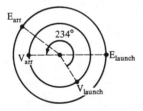

Example 2
Plan a transfer orbit to Mars, knowing that the radius of the planet's orbit (assumed to be circular) is $1.52 r_E$ and that the orbital speed of Mars is 24.1 km/s.

The calculations for this voyage are identical to those in Example 1, with the exception that the spacecraft travels to a higher orbit. This means that at launch and arrival it will need to speed up.

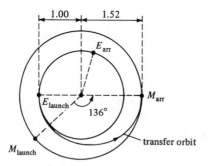

As indicated in the diagram above, the length of the major axis is 2.52 AU $= 2.52r_E$. Using this to determine the energy of the transfer orbit and setting that energy equal to

$$E = \tfrac{1}{2}mv_1^2 - G\frac{M_0 m}{r_E},$$

we find that the necessary speed for the transfer orbit is

$$v_1 = 1.10v_0 = 32.8 \text{ km/s}.$$

The orbital speed of the earth is 29.8 km/s, so the spacecraft needs to have its speed boosted by 3.0 km/s.

Conservation of angular momentum allows us to calculate the speed v_2 when it arrives at the orbit of Mars:

$$mv_1 r_E = mv_2 r_M,$$

which tells us that $v_2 = 0.72v_0 = 21.6$ km/s. Therefore, to attain the speed of Mars, the craft needs to be boosted by 2.5 km/s when it reaches this point.

The travel time is one-half of the period of the transfer orbit, and is calculated from Kepler's third law:

$$t = \tfrac{1}{2}(1.26)^{3/2} \text{ yr} = 0.71 \text{ yr} = 259 \text{ days}.$$

While the spacecraft is voyaging to Mars, that planet moves through $(259/687)360° = 136°$ along its orbit. Therefore the launch opportunity occurs when Mars is at the point M_{launch} in its orbit, as indicated in the above diagram. You can show that when the spacecraft arrives, the earth is at the point E_{arr}, at an angle of 255° from its position at launch.

18.4   GRAVITY ASSIST

In 1977 *Voyager 2* was launched for a once in a lifetime chance to tour the four outer gaseous giant planets – Jupiter, Saturn, Uranus, and Neptune, as illustrated in Fig. 18.5. Once every 175 years these planets line up so that one spacecraft can visit all of them. Mission navigators used the gravitational fields of these planets themselves to provide extra boosts to *Voyager 2* in a technique known as *gravity assist*. Through a gravity assist from Jupiter, *Voyager 2* would visit Saturn, Uranus, and Neptune within 12 years (Fig. 18.5). If the spacecraft had been launched directly to Saturn, that part of the voyage alone would have taken more than 6 years, instead of 4 years with gravity assist. A trip to Uranus without gravity assist requires 16 years. And *Voyager 2* could never have reached Neptune at all.

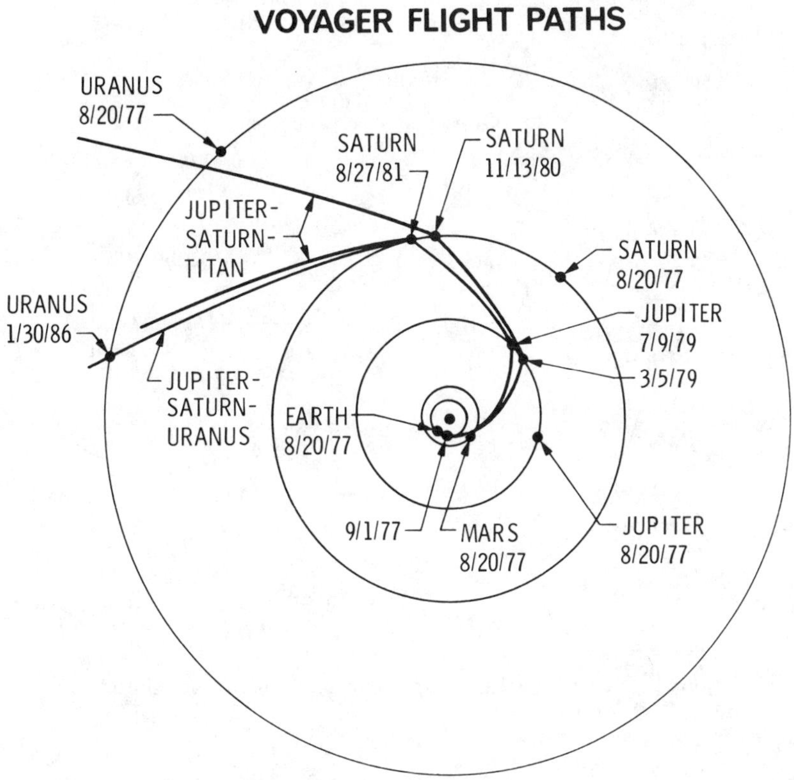

VOYAGER FLIGHT PATHS

Figure 18.5 Path of *Voyager 2* on its grand tour of the outer giant planets. (Courtesy JPL/NASA.)

Imagine a spacecraft traveling close to a large planet, for example, Jupiter. In this case, we actually need to take into account the gravitational force of both the sun and Jupiter. But if the satellite passes close enough to Jupiter, Jupiter's attraction will temporarily become much stronger than that of the sun; therefore the force from the sun can be temporarily ignored. (This simplifies the problem from one involving three bodies to a solvable problem of two bodies – the spacecraft and Jupiter.) Let's see what happens from the perspective of Jupiter.

The spacecraft approaches Jupiter from very far away with some small velocity. According to the expression for its total energy,

$$ E = \tfrac{1}{2}mv^2 - G\,\frac{Mm}{r}\,, $$

that energy is either zero or slightly positive, because the kinetic energy is larger than the gravitational potential energy. Thus the orbit around Jupiter will be open – either a parabola or hyperbola – as indicated in Fig. 18.6.

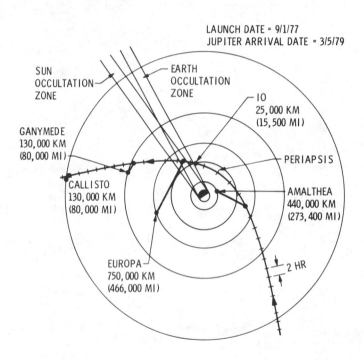

Figure 18.6 Gravity assist to a spacecraft from Jupiter. (Courtesy JPL/NASA.)

Consequently, the spacecraft has a change in velocity; the velocity at point 2 in Fig. 18.6 is in a different direction from the velocity at point 1. Although the velocity changes, the speed at points 1 and 2 in Fig. 18.6 is the same. This is what happens as viewed from Jupiter.

Looking at the interaction from the viewpoint of the sun, which is what must be done to determine the trajectory of a spacecraft navigating the solar system, we realize that something else has happened because Jupiter is moving. In order for the spacecraft to describe a smooth parabola or hyperbola around Jupiter as seen from Jupiter, it had to be dragged along with Jupiter's motion around the sun. This means that with respect to the sun, the spacecraft has acquired an extra component of velocity, namely Jupiter's velocity, which changes its kinetic energy. Although its potential energy hasn't changed because the spacecraft's distance from the sun has hardly changed, its kinetic energy is increased by the gravity assist of Jupiter. Therefore gravity assist makes use of Jupiter's motion to increase the total energy of a spacecraft.

Gravity assist is analogous to hitting a baseball with a bat. The bat not only reverses the direction of the velocity of the ball, but also increases the ball's speed; the stronger the swing, the better the chance for a home run. Similarly, planets with large kinetic energy can give large gravity assists to spacecraft.

Since the total energy of the solar system must be conserved, the energy the spacecraft gains from Jupiter comes from the potential energy of the planet as it orbits the sun. So as a result of this generous gravity assist, Jupiter orbits the sun a hair closer than before.

There are two ways to make a gravity assist. One way is the method we've just described where all motion takes place in the plane of the ecliptic, which you might recall is the plane of the orbits of most of the planets. This method of increasing the energy of an orbit in the plane of the ecliptic is called *pumping*.

Another possible form of gravity assist is to send a spacecraft in a different direction with respect to Jupiter, in such a way that the gravitational force of Jupiter applies a torque on the spacecraft about the sun. This torque can tilt the angular momentum vector of the spacecraft. Since the angular momentum vector is perpendicular to the plane of the orbit of the spacecraft, tilting the plane of the orbit means that the orbit is tilted out of the plane of the ecliptic. This type of gravity assist is called *cranking*. Both methods are standard tricks of the navigational trade.

18.5   A FINAL WORD

One of the spectacular revelations of the *Voyager 2* mission was the intricacies of the rings of Saturn, shown in the photograph of Fig. 18.7. There is an outer ring, called the

Figure 18.7   *Voyager 2* photograph of the rings of Saturn. (Courtesy JPL/NASA.)

F ring, which is a very thin, well-organized ring separate from the other rings. Even before *Voyager 2* flew past Saturn, calculations and theories described how such a well-organized, thin ring might have come about.

In order for the F ring to have become so narrow there must be two moons orbiting Saturn, one just inside the ring and one just outside it. How do the two moons shape the F ring? Suppose we examine the interaction between the F ring and one moon. If two bodies pass alongside each other in space, they interact gravitationally, and if one body is moving faster, it loses kinetic energy, whereas the slower one gains kinetic energy. We've just seen this in gravity assists. It is also true that the closer a moon is to a planet, the faster it moves. Therefore the inside moon travels faster than the outer moon. The ring material moves at an intermediate speed.

The material in the ring interacts with the fast inner satellite and gains energy. The increase in energy pumps the material into a higher orbit. So through this interaction with the inner moon, material on the inner edge of the ring is pushed outward, away from Saturn, toward the center of the ring. On the other hand, material on the outer edge of the ring interacts with the outer moon, which is moving slower than it. Consequently, the particles lose energy to the slower satellite and move down to a lower orbit, toward the center of the ring.

The presence of the two moons on the outside and inside of the ring tends to compress the ring together into a narrower ring. The moons act like shepherds keeping their flock together; they keep the particles in a narrow ring gravitationally.

This theory was suggested by Dr. Peter Goldreich of Caltech before the *Voyager 2* fly-by of Saturn. The theory predicted that there would be two moons, which hadn't been observed yet, shepherding the F ring. And when *Voyager 2* flew past the rings, the two moons were found (Fig. 18.8), just as predicted – a natural example of gravity assist.

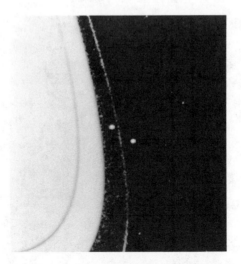

Figure 18.8 *Voyager 2* photograph of the F ring of Saturn and two shepherding moons. (Courtesy JPL/NASA.)

Problems

Navigating in Space

1. A satellite is placed into orbit around the earth before voyaging to Jupiter. Explain how by firing its rockets both the energy and angular momentum of the satellite can be changed.

2. One method proposed for space travel is to use ion propulsion. By constantly burning fuel and ejecting ions, a spacecraft could gain speed smoothly rather than in spurts from the firing of a rocket. How does this method of propulsion, in comparison to rocket blasts, complicate calculating transfer orbits?

3. Show that the number of days T_L between launch opportunities for any planet is given by

$$ T_L = \left| \frac{T_{earth} T_{planet}}{T_{earth} - T_{planet}} \right| $$

where T_{earth} and T_{planet} are the orbital periods about the sun. Find the launch opportunity period for Saturn and the earth.

4. Launch opportunity is associated with a physical alignment of two planets and the sun. Imagine a planet on each of the two hands of a clock. If the correct alignment happened to be that the two planets should be directly opposite each other, how many minutes would pass between successive favorable positions?

Transfer Orbits

5. For the transfer orbit of Example 1 calculate the energy that must be supplied to a 3000-kg spacecraft (a) to enter the transfer orbit and (b) to slow down to match the speed of Venus.

6. Some science fiction stories refer to a mysterious sister planet of the earth, which shares the same orbit as the earth but is always opposite the sun and hence remains unobserved. Approximating the earth's orbit as a circle, qualitatively explain how a spacecraft could be sent from the earth to the sister planet.

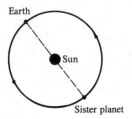

7. A satellite is in a circular orbit of radius $2R$, where R is the radius of the earth. The spacecraft is to be boosted to a higher orbit having a radius $4R$. Using data in

Appendix D for the mass and radius of the earth, and taking the mass of the satellite to be 2500 kg, find the following for the transfer orbit:

(a) length of the major axis,
(b) change in speed at the beginning and end of the orbit.

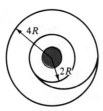

8. A satellite is in a circular orbit of radius r_0 about the earth. The velocity of the satellite is changed by firing its rockets tangent to its orbit in such a way that the speed is increased to 1.1 times its original speed. Find

(a) the eccentricity of the new orbit,
(b) the apogee distance in terms of r_0.

9. Using necessary data in Appendix D determine the following for a transfer orbit between the earth and Mercury:

(a) length of the major axis,
(b) change in speed to enter orbit,
(c) speed of spacecraft when it arrives at planet,
(d) travel time, and
(e) position of Mercury for launch opportunities.

10. Repeat Problem 9 for a voyage from the earth to Jupiter.

11. Dr. Lee DuBridge, president of Caltech, posed the following question in an after-dinner speech to the American Physical Society on April 27, 1960: Suppose that two spacecraft are in the same circular orbit around the earth, but one is a few hundred yards behind the other. An astronaut in the rear wants to throw a ham sandwich to his partner in the other craft. How can he do it? Qualitatively describe the possible paths of transfer.

CHAPTER 19

TEMPERATURE AND THE GAS LAWS

There are however innumerable other local motions which on account of the minuteness of the moving particles cannot be detected, such as the motions of the particles in hot bodies, in fermenting bodies, in putrescent bodies, in growing bodies, in the organs of sensation and so forth. If any one shall have the good fortune to discover all these, I might almost say that he will have laid bare the whole nature of bodies so far as the mechanical causes of things are concerned.

Isaac Newton, in *Unpublished Papers of Isaac Newton*

19.1 TEMPERATURE AND PRESSURE

Everybody talks about the weather, and that usually means the temperature, an inescapable part of our environment. Yet Newton's laws of mechanics tell us nothing about temperature. Is there any connection between mechanics and temperature?

In Chapter 10 we saw a connection. If you drop a block from above a table, its potential energy first turns into kinetic energy, and then is transformed into thermal energy when the block hits the table. After a while the only evidence that those events occurred is a slight warming of the surroundings, that is, a small increase in temperature.

What really happens is that the kinetic energy of the falling block is turned into the energy of motion of atoms and molecules. The energy is still there, but the motions are in random directions, not the organized motion of a whole block of matter. The energy of those random motions is internal energy, and the evidence of its existence is a change in temperature.

Before we go any further, let's make a short digression on temperature scales. Most of the world uses the Celsius scale, but in the United States the Fahrenheit scale is predominant. By *temperature* we mean a number assigned on a definite scale so that we can tell whether it was hotter today in Nairobi or in Calgary. Temperature scales offer a means of comparing the temperature of an object to standards set by fixed calibration points.

On the Fahrenheit scale, water freezes at 32° and boils at 212°. Temperatures on this scale are followed by units of degrees Fahrenheit, abbreviated °F. On the Celsius scale the freezing and boiling points of water are given by 0° and 100°, respectively. The abbreviation °C is used for this scale. In Section 19.5 we'll discuss how these common temperature scales came into being.

Example 1

If the temperature at a seaside resort is 86°F, what is the corresponding temperature in °C?

To find the conversion between Fahrenheit and Celsius temperature scales, let T_C denote the temperature in degrees Celsius, and T_F in degrees Fahrenheit. The relation between T_F and T_C is linear, as illustrated in the figure,

$$T_F = AT_C + B$$

where A and B are constants.

To find these constants we use the fact that water freezes at $T_F = 32$ and $T_C = 0$. This immediately gives $B = 32$, so

$$A = \frac{T_F - 32}{T_C}.$$

But water boils when $T_F = 212$ and $T_C = 100$, so

$$A = \frac{212 - 32}{100} = \frac{180}{100} = \frac{9}{5}.$$

Hence

$$T_F = \tfrac{9}{5} T_C + 32.$$

We can also solve this last equation for T_C in terms of T_F to find

$$T_C = \tfrac{5}{9}(T_F - 32).$$

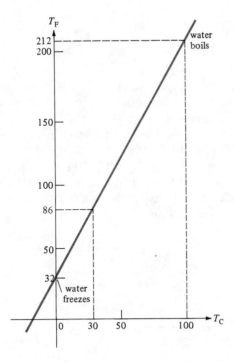

In particular, when $T_F = 86°F$ we get $T_C = 30°C$.

For most purposes, the best way to make this conversion is to remember a few fixed values, and interpolate between them. Unless great precision is needed, this can be done quickly without writing anything down. Here are some handy reference points:

°C	°F
0	32
10	50
20	68
30	86
40	104

Now to continue our search for a connection between mechanics and temperature. If we want to learn something about the temperature of the air, we should analyze the random motions of the molecules it contains.

The random motion of gas molecules is related not only to temperature, but also to another mechanical property, pressure. Pressure is defined as force per unit area. If an object of area A has a force F applied over that area, the corresponding pressure is

$$P = \frac{F}{A}.$$ (19.1)

In SI units, a pressure of one newton per square meter (1 N/m^2) is one pascal (1 Pa). A commonly used unit of pressure is the atmosphere, 1 atm = 1.01×10^5 Pa. Other units and conversions are listed in Appendix B.

Some of the greatest minds of the nineteenth century investigated the relationship between temperature and pressure. One fruitful idea is to imagine the motion of molecules in a closed container like that shown in Fig. 19.1. The constant, rapid drumbeat of molecules bouncing off the walls of the container exerts a steady average force on the walls. Let's calculate how that force is related to quantities which specify the motion of the molecules, namely, mass and velocity.

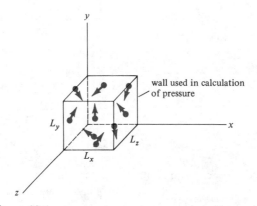

Figure 19.1   As a result of collisions, the molecules of a gas exert pressure on the container walls.

Consider first a single collision. A molecule of mass m approaches the wall with velocity **v**, and bounces off like a rubber ball in an elastic collision. In other words, the molecule bounces off with the same speed it had before the collision. As illustrated in Fig. 19.2, the angle of incidence is equal to the angle of reflection.

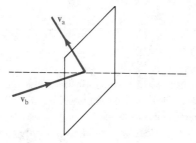

Figure 19.2   A gas molecule undergoes a change in momentum in a collision with a wall of a container.

Although the speed of the rebounding molecule is the same, its velocity is changed. As a result of the collision the molecule undergoes a change in its momentum $m\mathbf{v}$. The velocity $\mathbf{v}$ has components v_x, v_y, v_z, and if the direction perpendicular to the wall is the x direction, as shown in Fig. 19.2, the molecule has a positive x component of velocity before it hits the wall and a negative x component on rebound. In other words, the x component is exactly reversed as a result of the collision; the other components remain unchanged. Therefore the momentum $\mathbf{p}_b$, of the molecule before collision is given by

$$\mathbf{p}_b = m(v_x\hat{\mathbf{i}} + v_y\hat{\mathbf{j}} + v_z\hat{\mathbf{k}})$$

whereas the momentum $\mathbf{p}_a$ after collision is

$$\mathbf{p}_a = m(-v_x\hat{\mathbf{i}} + v_y\hat{\mathbf{j}} + v_z\hat{\mathbf{k}}).$$

According to Newton's second law the force on the molecule is equal to the rate of change of momentum, $\mathbf{F} = d\mathbf{p}/dt$, which we can approximate as

$$\mathbf{F} = \frac{\Delta\mathbf{p}}{\Delta t},$$

where $\Delta\mathbf{p}$ is the change in momentum and Δt is the change in time. By Newton's third law, this force is equal and opposite to the force the molecule exerts on the wall. Our formulas for $\mathbf{p}_a$ and $\mathbf{p}_b$ show that the change in momentum of the molecule is given by

$$\Delta\mathbf{p} = \mathbf{p}_a - \mathbf{p}_b = -2mv_x\hat{\mathbf{i}}.$$

Now suppose for a moment that the gas is extremely rarefied, so a typical molecule travels back and forth across the box many times before colliding with other molecules. Let's say a particular molecule returns and hits the wall once every t_c seconds. Then it transfers momentum to the wall at the rate $-\Delta p_x/t_c$. To calculate t_c, note that between one collision with the wall and the next, the molecule has to cross the box, hit the wall, and come back, traveling a distance $2L_x$ in the x direction (twice the length of the box in the x direction) at speed v_x. So the time between collisions is given by

$$t_c = \frac{2L_x}{v_x}.$$

All in all, then, the rate of momentum transfer from the molecule to the wall is

$$\frac{-\Delta p_x}{t_c} = 2mv_x\frac{v_x}{2L_x} = \frac{mv_x^2}{L_x}.$$

Each of the N molecules in the box has its own velocity, and therefore its own travel time and momentum transfer. Let v_{xi} denote the x component of the velocity of the ith molecule, where $i = 1, 2, \ldots, N$. Summing over all the molecules in the box, we obtain the total rate of momentum transfer to the wall, and thus the force on the wall:

$$F = \sum_{i=1}^{N} \frac{mv_{xi}^2}{L_x} = \frac{1}{L_x}\sum_{i=1}^{N} mv_{xi}^2. \tag{19.2}$$

Although the momentum transfers from a single molecule are intermittent, the total momentum transfer is a drumbeat of impulses that provides a steady push on the wall.

If the gas is not extremely rarefied, it's not quite so simple – the molecules may collide with one another before completing the round trip. But these collisions turn out to conserve kinetic energy and momentum, so the net impact on the wall is the same as if the particles had not collided with one another.

In Eq. (19.2) we recognize the term mv_{xi}^2 as twice the contribution to the kinetic energy from the x component of velocity. Since the motion of the molecules is random, the contributions from the y and z components have the same value, so we can write

$$\sum_{i=1}^{N} mv_{xi}^2 = \frac{1}{3} \sum_{i=1}^{N} (mv_{xi}^2 + mv_{yi}^2 + mv_{zi}^2) = \frac{1}{3} \sum_{i=1}^{N} mv_i^2$$

where

$$v_i^2 = v_{xi}^2 + v_{yi}^2 + v_{zi}^2$$

is the square of the speed of the ith molecule. But

$$\frac{1}{3} \sum_{i=1}^{N} mv_i^2 = \frac{2}{3} \sum_{i=1}^{N} \frac{1}{2} mv_i^2 = \frac{2}{3} K,$$

where K is the total kinetic energy of all the molecules. Using this in (19.2) we obtain

$$F = \frac{2}{3} \frac{K}{L_x}$$

for the total force on the wall from all the molecules in the box.

Now the pressure on the wall is the average force divided by the area. We just found the force, and we know that the area of the wall on which we are calculating the pressure is $L_y L_z$. Therefore we find the pressure to be

$$P = \frac{F}{A} = \frac{2}{3} \frac{K}{L_x L_y L_z}.$$

The volume V of the box is $L_x L_y L_z$, so we can cast the result into the simpler form

$$P = \frac{2}{3} \frac{K}{V}. \tag{19.3}$$

Finally, since the total kinetic energy K equals the average kinetic energy $\overline{K}$ of a single molecule times the number N of molecules, we can write Eq. (19.3) as

$$P = \frac{2}{3} \frac{N\overline{K}}{V}. \tag{19.4}$$

This argument was first presented in a simplified version by James Prescott Joule. From Eq. (19.4) we see that as the random motion of the gas molecules becomes more vigorous, the average kinetic energy increases, resulting in correspondingly greater pressure on the containing walls. Furthermore, we've found a relationship between the macroscopic pressure and the microscopic average kinetic energy of the individual molecules of the gas.

19.2 THE GAS LAWS OF BOYLE, CHARLES, AND GAY-LUSSAC

By connecting two mechanical quantities, pressure and kinetic energy, we found a relationship between heat and pressure. But how does this connection aid us in finding a relationship between heat and temperature? What we need is an understanding of the behavior of gases.

A major advance in understanding gases was provided in the seventeenth century by Robert Boyle. A staunch advocate of careful, thorough experimentation, Boyle discovered experimentally a relation between the volume of a gas and its pressure: for a given sample of gas, as long as the temperature and mass remained unchanged, the pressure was inversely proportional to the volume. In other words, if you squeeze a gas, its pressure rises proportionately; gases act like springs.

Boyle's experiments suggested that for a given sample of gas at a fixed temperature, the product of the pressure P and the volume V is constant. This empirical relation,

$$PV = \text{const} \qquad \text{(at fixed temperature)}, \tag{19.5}$$

is called *Boyle's law*, and it can be compared with formula (19.4) derived in the last section. If we rewrite (19.4) as

$$PV = \tfrac{2}{3} N\overline{K} \tag{19.6}$$

it has a striking resemblance to Boyle's law. In fact, if the average kinetic energy of a gas doesn't change at a fixed temperature, then for a given sample of gas both N and $\overline{K}$ are constant, so PV is constant.

According to Boyle's law, twice as much gas at a given pressure will occupy twice the volume. Similarly, for half as much gas, the volume is also half. Equation (19.6) shows that this is reasonable, since the quantity PV is proportional to the number of gas molecules N and otherwise depends only on the average kinetic energy $\overline{K}$, which presumably depends only on temperature. This suggests that we rewrite (19.6) as follows:

$$\frac{PV}{N} = \text{a function of temperature only.} \tag{19.7}$$

It is not hard to guess something about this function. Because gases expand when heated, as temperature increases so does the quantity PV, so we anticipate that this function should increase with temperature.

The exact dependence on temperature was revealed by further experiments into the nature of gases carried out by Jacques Alexandre César Charles, an eighteenth-century French scientist and hot-air balloon enthusiast. Charles's curiosity about the behavior of gases led him to the important discovery that *all* gases expand by the same amount with a given rise in temperature. For each 1°C rise in temperature, the volume of any gas expands by 1/273 of the volume it would have at 0°C. Similarly, for each 1°C decrease in temperature, any gas contracts by 1/273 of its volume.

By extrapolating the behavior of a gas as the temperature is progressively decreased, we conclude that at −273°C, the volume would reach zero, as illustrated in Fig. 19.3. Actually, gases liquefy or solidify before they reach −273°C, so the volume never quite reaches zero. But if the law did hold, there could be no lower temperature than −273°C. This temperature, seemingly the lowest attainable, is designated as *absolute zero* on

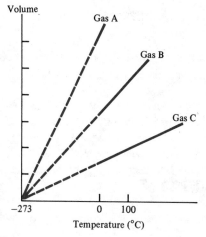

Figure 19.3 A volume versus temperature graph of the behavior of different gases when the temperature is lowered.

another temperature scale called the *absolute system*. The units in this system are called kelvins (K) in honor of William Thomson, Lord Kelvin:

$$T = T_C + 273.$$

Thus Charles proposed that the volume V is proportional to T, where T is the absolute temperature in kelvins. Although Charles's investigations were never published, studies made 15 years after his work by Joseph Louis Gay-Lussac, another French scientist and balloon enthusiast, confirmed that to a good approximation the volume of a given amount of any gas is proportional to the absolute temperature when the pressure is held constant. We can write this as an equation,

$$\frac{V}{T} = \text{const} \qquad \text{(for constant pressure)} \tag{19.8}$$

where T is the absolute temperature. This is called the gas law of Charles and Gay-Lussac.

The findings of Boyle, Charles, and Gay-Lussac amount to saying that gas behaves in much the same way as does mercury or alcohol in a thermometer. When heated, the mercury in a thermometer expands and rises up the tube. A gas also expands when heated, but with an important difference: all gases expand by approximately the same amount, 1/273 of their volume at 0°C for each 1°C of rise in temperature. Often scientists conveniently speak of an *ideal* gas whose behavior is described exactly by Boyle's law and the law of Charles and Gay-Lussac. Liquids and solids on the other hand, expand by amounts varying more noticeably with the type of substance when heated. For example, when a jar lid is hard to open, you place it in hot water. The metal lid expands more than the glass, thereby making the lid looser fitting and easier to open.

A significant conclusion drawn from Boyle's experiments was that gases must be composed of discrete particles separated by void since the gas is compressible. This interest in the behavior of gases marked a revival of the ancient Greek conjecture that matter is composed of incessantly moving particles called atoms. The original idea of atoms is credited to the fifth-century B.C. philosopher Leucippus and his student Democritus. Although many qualitative ideas of atomic theory were developed by the ancient Greeks, controlled, quantitative investigations were not carried out until the seventeenth and eighteenth centuries.

19.3 THE IDEAL-GAS LAW

Because the quantity pressure times volume divided by the number of molecules, PV/N, increases with temperature, the easiest way to define temperature would be to set that quantity equal to the temperature. The only trouble with this scheme is that the temperature would depend on the units we choose to measure pressure and volume. To compensate for this we insert a constant which may be different for each set of units of pressure and volume, leaving temperature always numerically the same. Then we can write down an equation which describes the state of a gas,

$$PV = NkT,$$ (19.9)

where k is the constant. The constant universally adopted (called Boltzmann's constant) has the value

$$k = 1.38 \times 10^{-23} \quad \text{J/K}.$$

As we have seen, the resulting unit of temperature, called the kelvin, is the same size as 1°C. Figure 19.4 compares some temperatures on the three different scales.

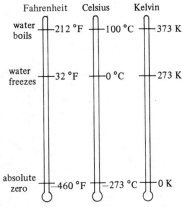

Figure 19.4 Temperature on the Fahrenheit, Celsius, and Kelvin temperature scales.

Equation (19.9), known as the *ideal-gas law*, is extremely useful in physics and chemistry. It is, of course, much more than a mere definition of the Kelvin temperature scale. Also known as an equation of state, this equation combines both Boyle's and Charles's laws, and the value of the constant k also tells us the number of molecules of a gas in a given volume at known temperature and pressure.

Boyle himself had noticed small departures from the constancy of the product of pressure and volume. But, because of its simplicity and general usefulness, he ignored deviations from his law. Why don't gases follow the ideal-gas law for all ranges of temperature and pressure? One reason is that gas molecules are not point particles but have nonzero volumes. Consequently, the amount of space available for each molecule is not the entire volume of the container. Instead each molecule has a volume less than V available to move in. In addition, the forces that the molecules exert on one another are not entirely negligible. When the density of the gas is great enough for the molecules to be relatively close together, the pressure is reduced as a result of the attraction of molecules to one another.

Numerous attempts have been made to deduce a general equation holding for real gases. The most celebrated is that developed by Johannes van der Waals in 1873. Van der Waals assumed that the effect of the finite size of molecules could be taken into account by using $V - Nb$ for the volume, where b is experimentally determined for each gas. Furthermore, the attractive forces between the molecules were represented by replacing P by the quantity $P + aN^2/V^2$, where the coefficient a is also experimentally determined for each gas. Thus we have the Van der Waals equation

$$(P + aN^2/V^2)(V - Nb) = NkT. \tag{19.10}$$

19.4 TEMPERATURE AND ENERGY

Let's summarize what we've learned so far. Heat, the random kinetic energy of molecules, is related to the pressure of a gas through

$$PV = \tfrac{2}{3} N\overline{K}, \tag{19.6}$$

which gives

$$\overline{K} = \frac{3}{2}\frac{PV}{N}. \tag{19.11}$$

On the other hand, studies of the dependence of volume and pressure of an ideal gas on temperature suggest the ideal-gas law

$$PV = NkT. \tag{19.9}$$

Solving Eq. (19.9) for PV/N and inserting that into Eq. (19.11), we are led to another result:

$$\boxed{\overline{K} = \tfrac{3}{2} kT.} \tag{19.12}$$

We've discovered that the average kinetic energy of an atom or molecule in a gas is directly proportional to the absolute temperature. This result agrees pleasantly with the starting point of this chapter: warmth is evidence of the disorganized motions of atoms and molecules. We have succeeded in finding the connection between heat, temperature, and energy.

One final question. Is the average kinetic energy of a gas, $\frac{3}{2} kT$, all of its energy, or can it have other forms of energy? One way to investigate this question is through an experiment. Let's theorize that the energy of a gas is $\frac{3}{2} NkT$. To test this theory, take an isolated container of gas of fixed volume and add it to a known amount Q (say, one joule) of heat energy. Because the volume is fixed, nothing moves, so no work flows in or out, and because the system is isolated, no heat, other than what we've added, can flow in or out. Since Q is the change in energy, our theory states that the rate of change in energy with respect to temperature, dQ/dT, must be

$$\frac{dQ}{dT} = \frac{3}{2} Nk.$$

In other words, the temperature is predicted to rise by exactly $2Q/(3Nk)$.

Experimentally, this prediction holds for certain gases, for example, the noble gases helium and argon. But for air, we would find $dQ/dT = \frac{5}{2} Nk$ approximately. What does this imply? Other gases have somewhere other than translational kinetic energy to store energy. In fact, all gases which are composed of simple atoms are consistent with our theory, while those which are made of more complicated molecules are not; air, for example, is composed mainly of N_2 and O_2 molecules.

Where is the extra energy stored? What happens is that molecules both vibrate and tumble, and do more of both as they heat up. And that is where the energy goes. Technically we say that the molecules are *internally excited*. Atoms are simpler and tougher structures and do not tend to get excited as easily as the more complicated molecules.

At room temperature and above, the internal excitation energy of gas molecules, like kinetic energy, is proportional to the temperature. This allows us to write a simple equation for the total energy U of a gas:

$$\boxed{U = qNkT.} \tag{19.13}$$

For simple atomic gases like helium, q is found to be equal to $\frac{3}{2}$. For molecular gases it is larger.

19.5   A FINAL WORD

A morning shower is steamy hot, a refreshing glass of lemonade is icy cold. But how hot is *hot*, or cold *cold*? What exactly do these two terms mean? How do we measure the degree of warmth, or lack of it, in a body? In other words, how can we specify the temperature of something? In order to discuss temperature, a scale on which to compare various objects is needed.

An intuitive feeling for temperature comes from the sense of touch. But whereas the length and mass of an object can be measured in terms of another object (and that measurement will be the same for every judge), the sense of touch does not provide an objective means for measuring temperature. What one might call "a comfortable shower," another might call too hot or too cold. Like force, temperature can only be measured by its effects – and a dependable, reproducible scale is needed on which changes can be measured.

One of the earliest thermometers was invented in 1602 by our old friend Galileo Galilei. Called a thermoscope, Galileo's thermometer merely consisted of a glass bulb containing air and having a long, open-ended stem which was placed in water. When the temperature changed, the air inside the bulb expanded or contracted and, correspondingly, the water in the stem fell or rose. But Galileo's apparatus lacked one vital ingredient needed to understand the changes in temperature: a visible, fixed scale.

One of the first temperature scales was established by the Danish astronomer Olaus Roemer, who lived in the eighteenth century. Roemer chose two convenient, useful, and reliable fixed points: the melting point of snow and the boiling point of water. He designated the melting point of snow as $7\frac{1}{2}°$, while the boiling point of water he set at 60°. The former, seemingly arbitrary point was chosen by Roemer for meteorological temperature so as to position one-eighth of his entire scale below freezing. As a result, zero degrees on Roemer's scale approximated the temperature of an ice and salt mixture, which was widely believed to be at the lowest possible attainable temperature, and so all readings on his thermometer were assumed to be positive.

Roemer did not publish anything about his thermometer, and its existence went largely unnoticed save by a very few of his contemporaries, including a young Polish scientific instrument maker by the name of Daniel Fahrenheit, who sensed a future in calibrated thermometers. In 1709, Fahrenheit watched Roemer graduate several thermometers. Later he described what he saw in a letter to Hermann Boerhaave, a colleague.

> I found that he had stood several thermometers in water and ice, and later he dipped these in warm water, which was at blood-heat and after he had marked these two limits on all the thermometers, half the distance between them was added below the point in the vessel with ice, and the whole distance divided into $22\frac{1}{2}$ parts, beginning with 0 at the bottom, then $7\frac{1}{2}$ at the point in the vessel with ice and $22\frac{1}{2}$ degrees for that at blood-heat.

For ease of construction Fahrenheit multiplied each degree by four, so that the upper point became 90° and the lower one 30°. Later he changed the upper point to 96° and the lower to 32° for a very good instrument maker's reason: the 64° interval between them could then be engraved into single degrees by successively dividing it in half six times. With this temperature scale, Fahrenheit (somewhat inaccurately) found the temperature of boiling water to be 212°. With 212° fixed as the boiling point of water, this scale is the one popular in the United States. As a result of the adjustment in the scale, normal body temperature became 98.6° instead of Fahrenheit's 96°. The range 0 to 100°F is the range of common weather temperatures.

In 1742 Anders Celsius, a Swedish astronomer, proposed a centigrade system in which the temperature scale is made up of 100 equal degrees spanning the liquid state of water: Celsius called the freezing point of water 100° and the boiling point of water 0°. Shortly after his death, Celsius's colleagues at the Uppsala Observatory inverted the scale to make the freezing point 0° and the boiling point 100°, giving us the scale generally used today; it is really Celsius's scale inverted, so we should call a temperature on this scale ℧° rather than °C as an abbreviation.

Problems

Temperature and Pressure

1. What temperature is the same on both the Fahrenheit and Celsius temperature scales?

2. Cite a couple of everyday examples that illustrate that heating a gas increases its pressure.

3. A spherical balloon of radius 10.0 cm contains helium gas at a pressure of 1.50×10^5 Pa. How many helium atoms are contained in the balloon if each atom has a kinetic energy of 4.2×10^{-22} J?

4. Compute the average kinetic energy per atom of three moles of argon gas in a cylindrical container of radius 5.0 cm and height 20.0 cm under a pressure of 3 atm. (One mole is 6.02×10^{23} molecules, or in the case of a monatomic gas like argon, 6.02×10^{23} atoms.)

5. A machine gun fires two rubber bullets per second at a speed of 5.0 m/s directly at a square plate with sides 0.10 m long. The bullets, each of mass 15.0 g, bounce back elastically along their initial direction. What is the pressure exerted on the plate by the stream of bullets?

6. What would be the relation between the pressure and average kinetic energy per molecule if the box in Figure 19.1 contained a mixture of gases?

7. A mass of 5.0 kg of a gas is in a container whose volume is 0.30 m³ at a pressure of 7.0×10^5 Pa. What is the square root of the average velocity squared of a molecule? This quantity is often referred to as the root mean square of the velocity – the rms value.

Boyle's Law and Charles's Law

8. If the volume occupied by a gas is decreasing, can you conclude that the temperature must be decreasing? Explain your reasoning.

9. If a helium-filled balloon is placed in a freezer, must its volume decrease? Explain.

10. The pressure in a bicycle tire is increased from 30 lb/in.² to 60 lb/in.², yet the volume doesn't decrease by half. Why?

11. A gas is placed in a vessel which maintains a constant temperature at a pressure of 2.0 atm and a volume of 3.0 L. Overnight a leak develops and the pressure is found to be 1.75 atm and the volume 2.4 L. What fraction of gas has escaped?

The Ideal-Gas Law

12. On a cool day when the temperature is 16°C the pressure in a car tire is 49 lb/in.^{2} (3.3 × 10^5 Pa). What is the pressure on a hot day when the temperature is 36°C?

13. Find the pressure of 0.70 g of argon in a volume of 5.0 L at a temperature of 25°C. (Argon has atomic weight 40. The mass of a single hydrogen atom, of atomic weight 1, is 1.67 × 10^{-24} g.)

14. The state of a gas is initially specified by T = 350 K, P = 6.0 atm, and V = 8.5 L. If the pressure is lowered to 4.0 atm and the volume increased to 10.0 L, what is the new temperature of the gas?

15. A sample of 100 g of helium gas is at a temperature of 0°C and a pressure of 500 atm. Calculate the volume of the gas according to (a) the ideal-gas law, and (b) the van der Waals equation with a = 1.4 L^2 atm/mol^2 and b = 0.03 L/mol. (The mass of one helium atom is 6.64 × 10^{-24} g.)

Temperature and Energy

16. What is the average kinetic energy per atom of helium gas at a temperature of 23°C?

17. What is the root-mean-square speed of a helium atom (mass 6.64 × 10^{-24} g) in the preceding question?

18. One mole (N = 6.02 × 10^{23}) of hydrogen gas at 80°C has a total energy of 7300 J. What is the value of q?

19. If 4.0 J of energy is added to a one mole (N = 6.02 × 10^{23}) of argon gas, then how much will the temperature of the gas increase?

20. You wake up one cold morning, turn up your thermostat, and heat the air in your house from 285 to 293 K. What is the change in the total internal energy of the air in your house? (*Note:* The answer is zero – why?)

CHAPTER

20

THE ENGINE OF NATURE

Everybody knows that heat can cause movement, that it possesses great motive power; steam engines so common today are a vivid and familiar proof of it. . . . The study of these engines is of the greatest interest, their importance is enormous, and their use increases every day. They seem destined to produce a great revolution in the civilized world. . . .

Despite studies of all kinds devoted to steam engines, and in spite of the satisfactory state they have reached today, the theory of them has advanced very little and the attempts to improve them are still directed almost by chance.

Sadi Carnot, "The Motive Power of Heat" (1824)

20.1 THE AGE OF STEAM

The age of steam is past. The steam engine is a curiosity, an object of nostalgia that has been replaced by diesel engines, electric motors, turbine engines, and gasoline engines to drive the wheels of civilization. Nonetheless, steam did have its day. The steam engine not only caused the Industrial Revolution, which changed our lives; it also led to discoveries in physics so profound that they changed the way we think. How did investigations into the nature of steam engines lead to a deeper understanding of the universe?

First, we need to understand how a steam engine operates. In essence, a steam engine is a device which heats water in a closed container, a boiler, thereby converting it to steam. As the steam and water mixture becomes hotter the pressure rises; the steam engine controls and takes advantage of the force applied by that pressure. When the high-pressure steam is released through a valve to a cylinder where it can push a movable piston, it produces work. So a steam engine starts with heat and produces work.

More than one stroke of an engine is needed to drive a civilization. After the first stroke of the piston the low-pressure steam that remains can be expelled from the cylinder through another valve, and the process can be repeated over and over again.

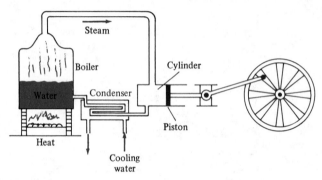

Figure 20.1 Schematic representation of a steam engine.

In the 1700s engineers were building steam engines that worked by suction rather than positive pressure, but these early devices left much to be desired. The age of steam had not yet arrived because crucial ideas were missing. The discovery that made the steam engine into a practical device was patented in 1769 by a Scottish instrument maker named James Watt. When asked to repair a model of a steam engine that was used as a lecture demonstration, Watt conducted a series of experiments that led him to realize that the engine wasted most of its heat in warming up the walls of the cylinder. Watt's idea was to condense the spent steam by cooling it outside the cylinder, so that the cylinder walls could always stay hot. Later engineers worked up the courage to use the potentially explosive power of positive pressure. These were among a series of developments and improvements that made the steam engine a practical source of power. The age of steam had arrived.

For the next half century, engineers devised ways to make steam engines more efficient. Their goal was to get more work out of each ton of coal used to heat the boiler. Like James Watt, these engineers were concerned with practical results, not theory. And by the 1820s steam engines worked well: there were steamships plodding across the seas and trains chugging across continents.

In 1824 a young French military engineer, Nicholas Léonard Sadi Carnot (1796–1832), published a remarkable essay on steam engines. After reviewing the industrial, political, and economic importance of the steam engine, Carnot raised the question: Is there an assignable limit to the efficiency of such engines? Although he never succeeded in making steam engines work better, his ideas had a profound influence in another direction.

James Watt, whose purpose had been to make money, invented a more efficient steam engine, and thereby revolutionized society. Sadi Carnot, whose purpose was to make steam engines more efficient, gave birth to thermodynamics, and thereby revolutionized physics. Thermodynamics is the science that deals with phenomena involving heat.

Figure 20.2 Nicolas Léonard Sadi Carnot. (Courtesy of the Archives, California Institute of Technology.)

Carnot's approach was radically different from that of steam engineers before him. Instead of tinkering with knobs, valves, and piston strokes of engines, Carnot developed an abstract theory of how engines work. He wanted to formulate the underlying principles governing the ideal engine. His question was not how to extract a little more work from a ton of coal, but what is the maximum amount of work that can be extracted from a ton of coal.

20.2 WORK AND THE PRESSURE–VOLUME DIAGRAM

Carnot undertook his quest unaware of the law of conversation of energy, which had not yet been formulated. For him, heat was not a form of energy that could be transformed into other forms, but rather a colorless, weightless fluid, called caloric, which was conserved and not consumed by the engine. So Carnot never thought that the most efficient engine would be one that turned all the available heat energy into work. That is just as well, because it is the wrong answer.

As discussed in Section 10.9, Carnot visualized caloric running a steam engine in much the same way as water runs a waterwheel. As water falls from greater to lesser height, it drives the waterwheel, but is itself not consumed. The ''spent'' water must be carried away or the wheel will soon be flooded and stop functioning. By analogy, Carnot imagined that a steam engine runs by the flow of caloric from the higher temperature of the boiler to the lower temperature of the surroundings. In fact a steam engine would not function unless heat were extracted at the lower temperature; that was precisely the job of the condenser that James Watt invented.

The fact that heat must be extracted doesn't depend in any way on steam. Air can similarly be used to drive an air engine. For example, air can be heated in a large can, raising it to high pressure. Then a valve opens, allowing the high-pressure air to push on a piston which does work because a force is applied through a distance. Heating air pressurizes it because as it becomes hotter, if it can't expand, its pressure rises. The reverse is also true: cooling air lowers the pressure. Expanding the volume of a gas lowers the pressure and tends to cool it, whereas compressing tends to warm it. These are the essential points in Carnot's analysis of the efficiency of an engine.

Let's express this idea quantitatively by assuming that air obeys the ideal-gas law (which it very nearly does). Then, as shown in Chapter 19, we can describe the state of the gas by

$$PV = NkT, \qquad\qquad\qquad (19.9)$$

and we also can write the internal kinetic energy plus the energy stored in the molecules or atoms as

$$U = qNkT. \qquad\qquad\qquad (19.13)$$

Imagine a gas confined to a cylinder which has a movable piston at one end as shown in Fig. 20.3. If we compress the gas by pushing in the piston, the volume of the gas changes and either the pressure or temperature or both must change. As a result of this compression, the gas will not be in an equilibrium state; the pressure near the piston initially will be greater than that far away from the piston. Consequently we cannot define the state of the gas, that is, specify T, P, or U, until the gas settles down. However, if we push the piston slowly in small steps, waiting for equilibrium to be reestablished after each step, we can compress the gas so that it is never too far from an equilibrium state. This kind of process is known as a *quasistatic* process, and in practice it can be achieved fairly well.

With that aside, let's return to the quantitative analysis. Consider a piston in a circular cylinder, closed at the left end, as shown in Fig. 20.3. If a variable force F moves the piston from 0 to x, the work done by this force is expressed by the integral

$$W = \int_0^x F(x')\, dx'.$$

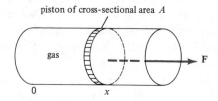

Figure 20.3 The work done by a gas depends on its pressure and volume.

Differentiating this equation we obtain, by the first fundamental theorem of calculus,

$$\frac{dW}{dx} = F(x).$$

If the force F is applied by a gas inside the cylinder under pressure P, then

$$F = PA,$$

where A is the cross-sectional area of the piston, so we get

$$\frac{dW}{dx} = PA.$$

Now the state of a gas is usually described by its pressure P, its volume V, and its temperature T. In this case the volume is $V = Ax$ so $dV/dx = A$ because A is constant, and the equation for dW/dx becomes

$$\frac{dW}{dx} = P \frac{dV}{dx}.$$

We eliminate x from this equation by using the chain rule,

$$\frac{dW}{dx} = \frac{dW}{dV} \frac{dV}{dx}$$

and obtain the fundamental relation

$$\frac{dW}{dV} = P, \tag{20.1}$$

which relates the work W to the pressure P and the volume V. To calculate the work W_{12} done *by the gas* in expanding the cylinder from initial volume V_1 to final volume V_2 we integrate this relation and obtain

$$W_{12} = \int_{V_1}^{V_2} P \, dV. \tag{20.2}$$

To carry out the integration we need to know how the pressure P varies as a function of the volume V.

Note that this is the work done by an *expanding* gas. If an equal but opposite force $-F$ *compresses* the gas, then the gas gains energy (or does negative work $-W_{12}$) as a result of the compression.

Recall that the pressure, volume, and temperature of an ideal gas completely specify the state of the gas. Since these three variables are related by the ideal-gas law, knowledge of two alone specifies the state. When work is performed by a gas, P and V are the natural variables to define the state, which can be represented by a point (V,P) on a pressure–volume diagram.

Figure 20.4 illustrates this idea. As the gas is expanded from a smaller volume V_1 to a larger volume V_2, it defines a set of points (V,P) which we call a *PV curve*. Since

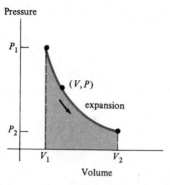

Figure 20.4 The state of a gas is represented by a point on a pressure–volume graph and the work done by the expanding gas is equal to the area of the region under the curve.

the work done by the gas is an integral, it is equal to the area of the region under the curve and over the interval $[V_1, V_2]$.

The actual shape of the PV curve will depend on how P and V are related to each other and to the temperature T. Different dependencies will produce different curves, or paths, that represent how the pressure and volume were related during the expansion or compression process. Three such paths are illustrated in Fig. 20.5. The work done by the expanding gas is different along each of these paths because the corresponding areas are different. So the work done in going from one state (V_1, P_1) to another (V_2, P_2) depends not only on the initial and final states, but also on the path taken between the states.

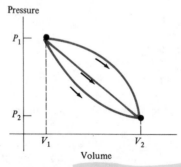

Figure 20.5 The work done by an expanding gas is equal to the area of the region under the curve in a PV diagram and depends on the path taken.

Example 1

A frictionless piston compresses a gas in a chamber that keeps the pressure constant, a process known as an isobaric compression. If (V_1, P_1) specifies the initial state of the gas and (V_2, P_1) the final state, find the work done by the gas in this compression if P_1 = 3 atm, V_1 = 5, and V_2 = 1 (in liters).

Pressure (atm)

expansion

3.0

1.0   5.0

Volume (L)

As the gas is compressed it gains energy and does a negative amount of work. So let's calculate instead the corresponding positive work W_{12} done by the gas if it *expands* from 1 L to 5 L at a constant pressure of 3 atm. According to Eq. (20.2) this work is

$$W_{12} = \int_1^5 P \, dV = 3 \int_1^5 dV = 3 \times 4 = 12 \text{ L atm}$$

$$= 12 \times 10^{-3} \text{ m}^3 \times 10^5 \text{ N/m}^2 = 1200 \text{ J}.$$

The work done by the gas in compression is -12 L atm. In this example the PV curve is a horizontal line segment and the work W_{12} is equal to the area of a rectangle.

Suppose a gas is initially compressed from a pressure P_1 and volume V_1 to another state specified by P_2 and V_2 along path A in Fig. 20.6. Next imagine that the gas is allowed to expand and follows path B back to the initial state, where B lies above path A. Has any net work been done? As indicated in Fig. 20.6 the net work done in the complete cycle is the difference between the area of the region under curve B and that under curve A. This difference is the area of the shaded region between the two paths. The work done by the gas is positive if the upper path is an expansion and the lower

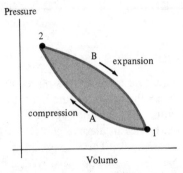

Pressure

2

B   expansion

compression   A

1

Volume

Figure 20.6 The net work done by a gas in a complete cycle is equal to the area of the region enclosed by the path.

path a compression. In other words, if the cycle is clockwise on the *PV* diagram, the gas does positive work.

Example 2

Suppose the gas in Example 1 is compressed from initial state (5,3) to (1,3) as before, and then is allowed to expand linearly to a third state (5,4) (volume 5 L, pressure 4 atm) before being cooled at constant volume to return to its initial state (5,3) as shown in the figure below. How much work is done by the gas in this cycle?

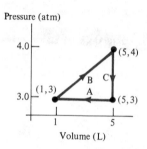

First, we recognize that the total work done is the sum of the work done along each of the three straight-line paths A, B, and C:

$$W_{tot} = W_A + W_B + W_C.$$

From Example 1, we already know that the work done along path A is $W_A = -12$ L atm. Because path B taken from the state (1,3) to the state (5,4) is a straight line on the *PV* diagram, the work done along B is the area of a trapezoid,

$$W_B = \text{base} \times \text{average height} = (5 - 1)\frac{4 + 3}{2} = 14 \text{ L atm.}$$

For path C the work done is zero because the volume does not change. Thus we find the total work done by the gas to be equal to 2 L atm. Of course, this is also equal to the area of the triangle enclosed by the complete path.

20.3   THE FIRST LAW OF THERMODYNAMICS

Everyday experiences show us that whenever a hot body is placed in contact with a cool body, heat flows from the hot body to the cool one until the two reach a common equilibrium temperature. Numerous ingenious experiments in the eighteenth and nineteenth centuries indicated that for two bodies in thermal contact, when heat leaves one body an equal amount enters the other body. These experiments led to the idea of conservation of caloric. Caloric was neither created nor destroyed, but merely transferred from one body to another.

The idea of conservation of caloric was eventually abandoned when experiments showed that it is not conserved. Late in the eighteenth century, Benjamin Thompson, also known as Count Rumford, studied the origin of heat in the production of a cannon. As director of the Bavarian arsenal, he supervised the boring of cannons for that kingdom. He observed that the cooling water had to be continually replaced during the boring process because the heat from the boring tool boiled it away. According to the caloric theory, the metal chips formed in the boring process released caloric to the water, causing it to boil. Thompson noted that even when chips were not produced but the boring tool was turning, heat was still produced. In other words, caloric could be created by friction and could be produced endlessly; evidently, caloric was not conserved. Thompson's experiments indicated a close connection between the work done by the boring tool and the heat produced.

In the 1840s a series of careful experiments performed by the British scientist James Prescott Joule once and for all destroyed the principle of conservation of caloric. As discussed in Section 10.6, Joule's experiments used a slowly descending weight that turned paddle wheels in a container of water. By precise measurements of the increase in temperature of the water and the work done by the falling weight, Joule uncovered the relationship between heat and work. In these experiments, the container of water was carefully insulated to prevent heat from entering or leaving the system. Such a system is said to be adiabatically shielded, and the transfer of work into internal energy, when no heat is allowed to enter or leave the system, is called an *adiabatic* process. In adiabatic processes, the path taken from one state to another is unique, unlike the more general processes discussed in the previous section, and so the work done depends only on the initial and final states of the system.

Work and heat are both forms of energy in the process of transfer. We define heat as that form of energy which is transferred from one body to another solely by virtue of differences in their temperature. Joule's experiments showed that neither heat nor mechanical energy is conserved independently. Rather, his experiments indicated that the mechanical energy lost from a system is converted into heat. Thus the *total* quantity of mechanical energy and heat is conserved.

Joule's contribution to thermodynamics was discovering the law of conservation of energy: the heat energy added to a system is equal to the sum of the work done by the system and the change in internal energy of the system. Known as the *first law of thermodynamics*, this result can be stated mathematically as

$$Q = W + \Delta U$$ (20.3)

where Q is the *heat added*, W is the work done *by* the system, and ΔU is the change in internal energy of the system.

To understand the distinction between internal energy and heat and work, consider the analogy of a bank account. You can add money to your account by depositing either cash or checks. Likewise, you can withdraw money from your account in the form of cash or checks. The change in the amount of money in the account over any period is equal to the algebraic sum of all the deposits and withdrawals. The total balance shown in your monthly statement does not depend on which transactions were in cash and which were by check; only the algebraic sum of the deposits and withdrawals is reflected in the balance. So it is with the internal energy function: although heat and work contribute separately, only the algebraic sum of the two matters.

Example 3

One hundred grams of a certain ideal diatomic gas contains 3.57 mol.

(a)   What is the internal energy of the gas at 35°C?

(b)   If the gas is heated from 35°C to a temperature T with the pressure and volume changing linearly as shown, find the change in internal energy for this process.

(c)   Find the heat required to produce the change in internal energy in (b).

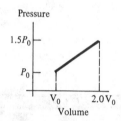

(a)   From Eq. (19.13) we know that the internal energy of a gas is given by

$$U = qNkT \qquad\qquad (19.13)$$

where $q = \frac{5}{2}$ for a diatomic gas and T is the absolute temperature. Using the fact that 6.02×10^{23} atoms are in one mole, we can substitute to find

$$U = \tfrac{5}{2}\,(3.57 \text{ mol})\,(6.02 \times 10^{23} \text{ atoms/mol}) \times (1.38 \times 10^{-23} \text{ J/K})\,(308 \text{ K}),$$

$$U = 2.28 \times 10^4 \text{ J}.$$

(b)   The change in internal energy is given by $\Delta U = qNk(T_f - T_i)$, so we need to find the final temperature of the gas. According to the ideal-gas law, $PV = NkT$, we can find the final temperature by forming a ratio:

$$\frac{P_i V_i}{P_f V_f} = \frac{T_i}{T_f}.$$

From the graph we know that $P_i = P_0$, $P_f = 1.5P_0$, $V_i = V_0$, $V_f = 2V_0$, and $T_i = 308$ K. Solving, we find $T_f = 924$ K. Therefore the change in internal energy is $\Delta U = 4.57 \times 10^4$ J.

(c)   According to the first law of thermodynamics, $Q = W + \Delta U$, where Q is the heat added and W is the work done by the gas. We can calculate the work done by the gas from the PV diagram because the work done is equal to the area of a trapezoid. Breaking the trapezoid into a triangle and a rectangle, we have

$$W = \tfrac{1}{2}\,V_0(0.5P_0) + P_0 V_0 = \tfrac{5}{4}\,P_0 V_0.$$

By the ideal-gas equation, $P_0 V_0 = NkT_i$, and the work done by the gas is $W = \frac{5}{4} NkT_i$ $= 1.14 \times 10^4$ J. Therefore the heat added is 5.71×10^4 J.

20.4   ADIABATIC AND ISOTHERMAL PROCESSES

All kinds of heat engines basically work because the force of a compressed gas can be used to make something move. When that's done, gas must be compressed again to start the next engine cycle. Thus the expansion and compression of gases are the crucial elements of a working engine. To analyze engines, we need to ask how the pressure of a gas behaves while its volume is changing.

The answer depends on how the change is accomplished. For example, it's possible to keep the gas at constant pressure and make it expand by heating it. On the other hand, if the temperature is carefully kept constant, then according to $PV = NkT$ the pressure will decrease as the gas expands. In other words, the relation between P and V depends on what T is doing.

For analyzing engines, two processes are particularly important. One is the case of constant temperature, called an *isothermal* process. In the other, the system is isolated so that no heat can flow in or out. Then the behavior of the temperature, and therefore the relation between P and V, can only be deduced indirectly by invoking the first law of thermodynamics (or in other words, the conservation of energy). This second kind of process is called *adiabatic*. We discuss the adiabatic case first.

In an adiabatic process, the gas is expanded or contracted in a container which is so well insulated from its surroundings that no heat flows in or out of the system. Mathematically, this means that $Q = 0$ in the first law of thermodynamics (20.3). Hence for an adiabatic process the first law becomes

$$W + \Delta U = 0 \quad \text{— ADIABATIC} \tag{20.4}$$

where W is the work done by the system, and ΔU is the change in internal energy. If W is positive, then ΔU is negative, so the work done by the gas equals the decrease in internal energy.

We now investigate the implications that Eq. (20.4) has for the relations between pressure, volume, and temperature of an ideal gas.

Let U_0 denote the initial value of the internal energy. Then the change in internal energy is $\Delta U = U - U_0$, and Eq. (20.4) gives us

$$W = -U + U_0.$$

Differentiating with respect to volume V we obtain

$$\frac{dW}{dV} = -\frac{dU}{dV}.$$

But by Eq.(20.2) we have $dW/dV = P$, so the previous equation becomes

$$\frac{dU}{dV} = -P. \quad \text{for ADIABATIC PROCESS} \tag{20.5}$$

We also know that the internal energy is given by

$$U = qNkT,$$ (19.13)

which can be combined with the ideal-gas law $NkT = PV$ to give us

$$U = qPV$$

where q is a positive constant. Differentiating this equation with respect to V, using the product rule, we get

$$\frac{dU}{dV} = qP + qV\frac{dP}{dV}.$$

Equating this to (20.5) and rearranging terms we obtain

$$\frac{1 + q}{q}P\frac{dV}{dP} + V = 0,$$

a differential equation relating P and V. The factor $(1 + q)/q$ is a constant greater than 1 which we denote by γ, and the differential equation becomes

$$\gamma P\frac{dV}{dP} + V = 0.$$

To solve this equation we first multiply each member by $V^{\gamma - 1}$, which gives us

$$P\gamma V^{\gamma - 1}\frac{dV}{dP} + V^{\gamma} = 0.$$

The reason for doing this is that now we recognize the left-hand side of the equation as simply the derivative of the product PV^{γ}. In other words, the differential equation now reads

$$\frac{d}{dP}(PV^{\gamma}) = 0,$$

which means that PV^{γ} must be a constant. Thus we have shown that *in an adiabatic process the pressure and volume of an ideal gas are related as follows:*

$$\boxed{PV^{\gamma} = C}$$ (20.6)

where C and γ are constants, with $\gamma > 1$. Therefore the pressure as a function of volume is given by

$$P = CV^{-\gamma},$$ (20.7)

whereas the volume as a function of pressure is

$$V = \left(\frac{C}{P}\right)^{1/\gamma} = \text{const} \times P^{-1/\gamma}.$$ (20.8)

The corresponding PV diagram is given in Fig. 20.7.

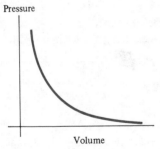

Figure 20.7 PV diagram for an adiabatic process: $PV^\gamma = C$.

We can also incorporate the ideal-gas law $PV = NkT$ to express the temperature directly in terms of P or of V. The results are

$$T = \frac{PV}{Nk} = \frac{C}{Nk} V^{1-\gamma} = \text{const} \times V^{1-\gamma} \qquad (20.9)$$

and

$$T = \frac{C}{Nk} \left(\frac{C}{P}\right)^{(1-\gamma)/\gamma} = \text{const} \times P^{(\gamma-1)/\gamma}. \qquad (20.10)$$

Because $\gamma > 1$, the power of V in (20.9) is negative, so T is a decreasing function of V. In other words, when a gas expands adiabatically, its temperature falls as the volume increases. In (20.10) the exponent of P is positive, so T is an increasing function of P. Thus, when a gas is compressed adiabatically the temperature rises as the pressure increases.

For a monatomic ideal gas we know from Chapter 19 that $U = \frac{3}{2} PV$ so $q = \frac{3}{2}$ and $\gamma = (1 + q)/q = \frac{5}{3}$.

Example 4

Suppose that 2 mol of helium gas are compressed along the path from A to B adiabatically as shown in the PV diagram. At point B the temperature is 194 K and the pressure is 3.2 atm.

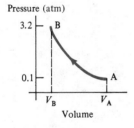

(a) Find the volume of the gas at point B.

(b) Assuming the pressure at A is $P_A = 0.1$ atm, find the volume of the gas at A.

(c) Find the change in internal energy of the gas.

(d) Find the amount of work done on the gas during compression.

(a) From the ideal-gas law $PV = NkT$ we find the volume at B to be $V_B = 0.01$ m^3.

(b) The compression is adiabatic, so we know that the quantity $PV^\gamma = $ const, where $\gamma = \frac{5}{3}$ for a monatomic gas. Therefore we have $P_A V_A^{5/3} = P_B V_B^{5/3}$, which tells us that $V_A = V_B (P_B/P_A)^{3/5}$, or $V_A = 0.08$ m^3.

(c) According to Eq. (19.13), $U = qNkT$ specifies the internal energy of a gas, where for a monatomic gas, $q = \frac{3}{2}$. The change in internal energy is given by $\Delta U = U_B - U_A = \frac{3}{2} Nk(T_B - T_A)$. We are given the temperature T_B and can use the ideal-gas law to find that the temperature at A is 49 K. Substituting back into our equation, we find that $\Delta U = 3.6 \times 10^3$ J.

(d) The compression is adiabatic, so no heat is transferred to the gas, that is, $Q = 0$. By the first law of thermodynamics, $W + \Delta U = 0$, where W is the work done by the gas. Therefore the work done *on the gas* is equal to $-W = \Delta U = 3.6 \times 10^3$ J; since the gas was compressed, the work done on it is positive.

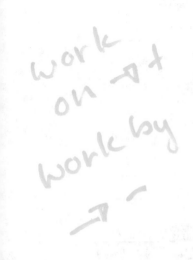

Next we consider an *isothermal* process. Here a gas is expanded or contracted in a container which is immersed in a heat reservoir, a large body whose temperature is constant. Because $U = qNkT$, if the temperature is held constant the internal energy U also remains constant, and the first law of thermodynamics states that

$$Q = W$$

because $\Delta U = 0$. The ideal-gas law $PV = NkT$ implies that in an isothermal process the product of pressure and volume is constant, so the PV curve (called an *isotherm*) is a hyperbola like the one shown in Fig. 20.8.

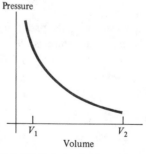

Figure 20.8 In an isothermal process the PV curve is a hyperbola, $PV = C$.

To calculate the work done by a gas in isothermal expansion we use the integral (20.2)

$$W_{12} = \int_{V_1}^{V_2} P \, dV \qquad\qquad (20.2)$$

where now $P = NkT/V$ with T constant. Therefore

$$W_{12} = NkT \int_{V_1}^{V_2} \frac{dV}{V} = NkT \ln\left(\frac{V_2}{V_1}\right). \qquad\qquad (20.11)$$

For an expanding gas, $V_2 > V_1$ and W_{12} is a positive quantity, so the gas does a positive amount of work which must equal the amount of heat Q added to the system because the internal energy remains constant. To keep the temperature of the gas constant, this added heat must come from the reservoir.

On the other hand, if the gas is isothermally compressed, then the work done by the gas is negative so Q is also negative, which means that heat must leave the system. In this case the heat is absorbed by the reservoir.

20.5 THE SECOND LAW OF THERMODYNAMICS

Now that we've explored the behavior of gases, let's return to the air engine. In the cycle of an engine, one stroke of the piston allows the gas to expand and cool. This is called the power stroke. To keep the engine going two things must be accomplished: the piston must return to where it started, and the same amount of air must replace the air in the boiler that was lost. One obvious way to achieve this would be to push the piston back in and let it recompress the air in the cylinder.

To push the piston back in would require exactly the same amount of work as the piston performed on the way out (ignoring friction), hence no net work would be achieved. In the real frictional world the situation is worse: more work would have to be added to the machine than was extracted, just to keep it going.

The situation can be remedied, however, by simply cooling the air after the outward piston stroke. Cooling the air in the cylinder causes it to contract and so the piston is pulled inward. This means that less force, and hence less work, is required to return the piston to its original position. One more step also needs to be included. The cool, dense air is heated back to boiler temperature and pressure, then injected back into the cylinder. At this point the engine is back to its initial condition and a net amount of work has been done by the machine.

The crucial step that made the engine run was to cool the air after the power stroke, in other words, to remove heat at lower temperature. This is the "runoff" from Carnot's waterwheel analogy. Watt's invention, the condenser, serves precisely this purpose. That's also why your car engine must have an efficient cooling system. Figure 20.9 illustrates a schematic representation of a heat engine.

Carnot's waterwheel analogy, as we mentioned, is not quite accurate. Heat is not a fluid but a form of energy. What an engine does is turn heat into work. Nevertheless, what we learn from Carnot remains true. The subtle and remarkably profound insight of Carnot's analogy is that not all of the heat, but only part of it, can be turned into work. The remaining heat must be discarded.

The initial and final states of the gas in an engine are the same, so the change in internal energy must be zero. Now if the heat absorbed into the boiler each cycle is Q_i, and the heat extracted per cycle is Q_o, then the net heat absorbed is

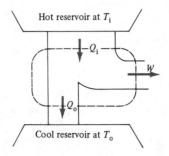

Figure 20.9 Schematic representation of a heat engine.

$$Q = Q_i - Q_o.$$

Then the first law of thermodynamics requires that

$$W = Q_i - Q_o \tag{20.12}$$

where W is the net work done by the engine.

The purpose of any engine is to do as much work as possible for a given expenditure of heat extracted from a source, such as burning coal or ignited gasoline. The efficiency e of a machine is defined as the ratio of the work done to the amount of heat put in:

$$e = \frac{W}{Q_i}. \tag{20.13}$$

Using Eq. (20.12), we can write the efficiency as

$$e = 1 - \frac{Q_o}{Q_i}. \tag{20.14}$$

This indicates that for optimum efficiency we want to discard as small an amount of heat as possible. For a perfectly efficient engine, e would equal one. In that case, all the heat absorbed from the higher temperature reservoir would be converted into work, and no heat would be discarded into a lower-temperature reservoir. But the efficiency of any engine, even an ideal frictionless engine, is always less than one. And a real engine, having friction, will perform less than the ideal amount of work, thereby making Q_o even larger, and the efficiency even less than ideal.

The experimental observation that it is impossible to make an engine that is one hundred percent efficient is known as the *second law of thermodynamics*:

A process whose *only* net result is to take heat from a reservoir and convert it completely to work is impossible.

We haven't yet answered the question Carnot set out to answer: Does an ideal engine exist? What is the highest efficiency a perfect, frictionless engine can possibly have? So far we've only argued that the efficiency must be less than one. But how much less?

Although the question seems to be a problem in engineering, the answer has implications far beyond the realm of engines. One of the implications concerns the properties of all matter. The other relates to the flow of time itself.

The implications for the properties of matter come from the fact that the result cannot depend on what the working fluid of an engine is. Otherwise it would be possible to design an engine that circumvents Carnot's arguments. We therefore arrive at general principles applicable to all matter.

The second point, having to do with the flow of time, stems from the fact that, once heat has flowed "downhill" from high temperature to low temperature, it is very difficult to run it "uphill" again. If some amount of heat Q_i flows into a perfect, frictionless engine, producing work W and depositing heat Q_o at lower temperature, the very best that can ever be done is to use exactly the same amount of work W to retrieve Q_o from the low temperature, depositing Q_o and W together as high-temperature heat Q_i. But that ideal, the net result of which is that nothing happens, is unattainable in the real world.

Once any real engine has used heat, the best any other real engine can do to reverse the process is return to high temperature a little less heat than was originally extracted, leaving a little more waste at low temperature; that's what the second law of thermodynamics says. Once this has happened the universe has been irreversibly changed forever.

20.6   THE CARNOT ENGINE

Although we have tasted the flavor of Carnot's reasoning, we have not yet followed enough of it to see what it is based on, nor to understand exactly where it leads. Let's remedy that situation by examining the logic of one of his central arguments concerning the efficiency of engines. Carnot invented an engine of his own, similar to the air engine we've discussed, and no more practical, but with a few further refinements.

The Carnot engine consists of a high-temperature source of heat at temperature T_i (i for input) and a lower temperature reservoir at temperature T_o (o for output), plus a sealed cylinder containing a gas and having no valves, but with a movable piston. In one cycle the engine uses the following processes: First the piston is as far in as it will go, so the gas is compressed to its smallest volume of the cycle. This corresponds to point 1 on the PV diagram in Fig. 20.10. At this point the temperature of the gas is T_i, the same as that of the high-temperature reservoir. Next the cylinder is placed in contact with the heat source – the boiler – and the gas expands, pushing the piston partially out. Although the gas would tend to cool in the expansion, its temperature is held constant since the cylinder is in contact with the heat reservoir. In other words, the gas moves along an isotherm from point 1 and 2 in Fig. 20.10. As we discussed in Sec. 20.4, an amount of heat is added to the gas during the isothermal expansion. So in the first step, some heat Q_i is extracted from the reservoir at T_i and some work is done by the engine.

In the second step of the Carnot cycle, the cylinder is removed from the heat source and the gas is allowed to expand further, doing more work. This time, however, the cylinder is not connected to any source of heat, so the gas cools as it adiabatically expands. In Fig. 20.10 this corresponds to the adiabat from point 2 to point 3. The gas expands until its temperature drops to T_o, the temperature of the cool reservoir.

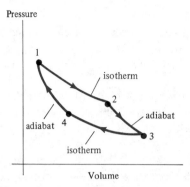

Figure 20.10 The Carnot cycle illustrated on a PV diagram.

For the next step in the cycle, the cylinder is placed in contact with the cool reservoir and the cylinder is pushed in partially. This compression would tend to heat the gas, but since the gas is in contact with the cool reservoir, its temperature remains constant at T_o, while heat flows out of the gas and into the reservoir. The path from point 3 to point 4 in Fig. 20.10 illustrates this isothermal compression. During this step some work is done on the gas, but since the temperature is held constant, the pressure remains lower than it could otherwise have been. This is accomplished by allowing some amount of heat Q_o to flow out into the cool bath.

During the final step, the cylinder is removed from contact with the cool bath, and work is done on it to push the piston further in. This pressurizes the gas, and because heat can't escape, the gas warms. At the end of this stroke, represented by the adiabat from point 4 to point 1 in Fig. 20.10, the compressed gas reaches the initial temperature T_i. And the cycle is ready to begin again. From the discussion of Sec. 20.2, the amount W of net work done is equal to the area of the region enclosed by the cycle.

The entire cycle can operate in reverse without violating any physical principles because each individual step has been designed to work equally well backward and forward. If we run the Carnot cycle forward, an amount of heat Q_i is extracted at high temperature T_i, an amount of net work W is done, and the remaining heat $Q_i - W = Q_o$ is rejected at the low temperature T_o. If we run the cycle backward, net work must be put into the machine. Heat is extracted from the low temperature, and energy equal to the heat in plus the work is discarded at high temperature. That's the kind of machine we use to cool our houses and food on a hot day. It is the principle of a refrigerator. Carnot prescribed a machine that can run equally well backward or forward. In other words, the Carnot engine is a reversible engine.

Carnot's result for an ideal engine states that no engine that works between two heat reservoirs can be more efficient than a reversible engine that works between the same reservoirs. We can prove this statement as follows: Suppose you have a Carnot engine that can run backward or forward with an efficiency

$$e = 1 - \frac{Q_o}{Q_i} = \frac{W}{Q_i}.$$

(20.14)

However, knowing that Carnot engines are not really practical, you obtain another device from a competing company. The manufacturer claims that this engine has an even better efficiency:

$$e^* = 1 - \frac{Q_o^*}{Q_i} = \frac{W^*}{Q_i}.$$

In other words, for a given amount of heat Q_i extracted from the boiler, this engine will do more work $W^* > W$ and dump less heat $Q_o^* < Q_o$ than the impractical Carnot device.

How can we test the manufacturer's claim? Suppose we use the practical engine to run the Carnot device backward as a refrigerator. The practical engine extracts heat Q_i, does work W^*, and deposits heat Q_o^*. The Carnot device uses up W, less than the available W^*, to extract heat Q_o, which is of greater magnitude than Q_o^*, from the low-temperature bath, and returns the same amount Q_i to the boiler. The net result is that the boiler doesn't need any fuel, since we return to it all the heat that we used. The combined machine still does useable work $W^* - W$ at the expense of the cool bath, which is losing heat $Q_o - Q_o^*$. If we install the engines in a ship, we could use them to drive the ship across the ocean by extracting heat from the ocean, which we use as the low-temperature reservoir. We've invented a device to make use of the limitless energy stored in the ocean.

That's impossible. The starting point of Carnot's reasoning is that work cannot be done by extracting heat over and over again from a single temperature source; that was in his waterwheel analogy. But that is just what we've done. Making heat run "uphill," with no other net effect, or running an engine from a single temperature source, are logically equivalent, and both equally impossible. That is the second law of thermodynamics, and it cannot be proved or deduced from other laws. Its logical consequences are Sadi Carnot's great gift to us. In the view of many physicists, the second law is perhaps the most profound of all the laws of physics.

An engine more efficient than the Carnot engine would violate the second law of thermodynamics. In other words, no engine can be more efficient than the Carnot engine. The crucial property of the Carnot engine is that it is reversible. In fact, any engine operating between temperatures T_i and T_o, that can run with equal efficiency backward and forward, must by the same arguments have the same efficiency as the Carnot engine.

What is the ideal value of the efficiency of a Carnot engine? The efficiency is the same for different kinds of machines with different working fluids, so it cannot depend on any of the details of the machine. The efficiency can only depend on the things that are the same for all, namely, the temperatures of the reservoirs T_i and T_o.

Let's calculate the efficiency of the Carnot cycle illustrated in Fig. 20.11. During the isothermal expansion from V_1 to V_2, the heat added is equal to the work done by the gas. By Eq. (20.11), we know that this amount of heat is given by

$$Q_i = W_{12} = NkT_i \ln(V_2/V_1). \tag{20.11}$$

The second stroke expands the cylinder further, from V_2 to V_3. Then the third stroke, this time with the gas in contact with the bath at T_o, squirts out heat by contracting from V_3 and V_4. Along the isotherm from point 3 to point 4, the heat extracted is given by

$$Q_o = NkT_o \ln(V_3/V_4).$$

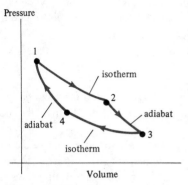

Figure 20.11 The Carnot cycle.

Now we could find the ratio Q_o/Q_i and therefore the efficiency if we knew how the volumes were related to each other.

Since strokes two and four are adiabats, we know that $PV^\gamma = \text{const}$, and because $P = NkT/V$, we have

$$NkTV^{\gamma-1} = \text{const}$$

for adiabats. Since Nk doesn't change, it follows that

$$T_i V_2^{\gamma-1} = T_o V_3^{\gamma-1} \quad \text{(from 2 to 3)},$$

$$T_o V_4^{\gamma-1} = T_i V_1^{\gamma-1} \quad \text{(from 4 to 1)}.$$

Taking a ratio of these equations to eliminate the temperature, we get

$$\frac{V_3}{V_4} = \frac{V_2}{V_1}.$$

Using this result for the ratio of the volumes and inserting it into the expressions for Q_i and Q_o, we find

$$\frac{Q_o}{Q_i} = \frac{T_o}{T_i}. \tag{20.15}$$

Therefore the efficiency of a Carnot engine is given by

$$\boxed{e = 1 - \frac{T_o}{T_i}.} \tag{20.16}$$

This astonishingly simple result is one of the most important in all of thermodynamics and raises momentous issues for both engineering and physics.

Example 5

A reversible engine operates with an efficiency of 40% and rejects 150 cal of heat to a reservoir at 300 K each cycle.

(a)   What is the temperature of the hot reservoir?
(b)   How much work is done each cycle?

(a)   For a reversible or Carnot engine the efficiency is given by Eq. (20.16), $e = 1 - (T_o/T_i)$. Solving this equation for T_i we find $T_i = 500$ K.

(b)   The efficiency is defined as $e = W/Q_i$, and $Q_o/Q_i = T_o/T_i$, so we have $W = eQ_oT_i/T_o$, and we find $W = 100$ cal.

20.7   A FINAL WORD

The efficiency of real engines is an important problem, intimately related to the so-called energy crisis, problems of the environment, and so on. Every analysis of those questions leads us back to Sadi Carnot. Nevertheless, we will be more interested in yet other consequences of his reasoning that have implications for the flow of time and the properties of matter. How did the focus of our discussion shift from grubby, clanking steam engines to the universe itself? Carnot's reasoning gently led us from arguments about engines to arguments about principles. His imaginary devices are no longer idealized steam engines, they are idealizations of the human will, the magical but physical law-abiding machines that allow us to find the outer limits of the extent to which we can ever aspire to alter the world, and by extension, the universe.

Within Carnot's own brief lifetime he was respected, but his work was largely ignored. One reason was that it was abstract and difficult to understand. Another was that he was an engineer and his ideas had no practical engineering significance at the time. His work, however, was rescued from obscurity by theoretical physicists who discovered it decades after it was done.

The first of these was a French engineer named Émile Clapeyron, who cleverly simplified Carnot's arguments by doing what Carnot did not: using the "law" of conservation of caloric. Because this idea was wrong, many believed that Carnot's deductions were based on faulty arguments. They were not; only Clapeyron's arguments were. The next physicist who realized that there was something entirely new at the heart of Carnot's arguments was Rudolph Clausius. The conservation of energy, although true and profoundly important, just wasn't enough to explain a world in which every machine would eventually grind to a halt if it weren't wound up or supplied with new fuel. Another idea was needed, something even more subtle and profound than the conservation of energy. Clapeyron and Clausius found that fruitful idea in Carnot's work, and Clausius called it *entropy*. That is what the next chapter of our story is about.

Problems

Work and the Pressure–Volume Diagram

1. When a cold can of soda pop is opened, a thin fog forms near the opening. Formulate a possible explanation.

2. Why do you think the valve on a bicycle pump gets hot when you pump up a tire? Once you've answered that, then explain why the valve on the compressed air at a gas station doesn't heat up.

3. Cite the underlying physics behind the formation of a mushroom cloud after a large bomb explodes.

4. At an altitude of 30,000 ft the air temperature is $-30°F$, yet passenger jets flying at this altitude use air conditioners to cool the air. Explain why.

5. For an isobaric compression, show that the work done by the gas is also equal to $Nk(T_2 - T_1)$, where T_2 and T_1 are the final and initial temperatures of the gas, respectively.

6. One mole ($N = 6.02 \times 10^{23}$) of an ideal gas is in the state given by point A in the PV diagram.

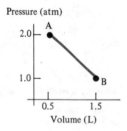

(a) What is the temperature of the gas at A?
(b) If the gas expands from A to B, how much work does it do?

7. Suppose N molecules of an ideal diatomic gas are compressed from a volume V_1 to a volume V_2 in such a way that the temperature varies according to $T = T_1(V_1/V)^{3/2}$. Calculate the work done by the gas in this compression in terms of N, k, T_1, V_1, and V_2.

The First Law of Thermodynamics

8. Can a gas absorb heat without any change in internal energy?

9. A monatomic ideal gas starts with pressure $P_1 = 1.0 \times 10^7$ Pa, volume $V_1 = 0.1$ m^3, and temperature $T_1 = 1500$ K. First its pressure is decreased to $P_2 = 3 \times 10^6$ Pa, with the volume kept constant. Then the gas expands to a volume $V_2 = 0.25$ m^3, with the pressure constant, as shown in the diagram.

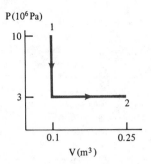

(a) Determine the change in internal energy of the gas in going from point 1 to point 2.

(b) Calculate the work done in the process.

(c) Find the heat added to the gas in the process.

10. For the process in Problem 7, determine the change in internal energy and the heat added.

11. An ideal monatomic gas initially at 300 K and occupying a volume of 2.0×10^{-4} m^3 is allowed to expand to a volume of 4.0×10^{-4} m^3 in such a way that the pressure (in Pa) is given by

$$P = 1.0 \times 10^5 - (1.0 \times 10^{-3})/V^2$$

where V is measured in m^3.

(a) Find the initial and final pressures of the gas.

(b) Determine the final temperature of the gas.

(c) Calculate the work done by the gas during the expansion.

(d) Find the amount of heat added to the gas during the expansion.

Adiabatic and Isothermal Processes

12. Gently blow across your knuckles with your mouth wide open. Compare that feeling to that from blowing across your knuckles with your lips close together. Why does the air feel cooler the second time?

13. One mole ($N = 6.02 \times 10^{23}$) of an ideal gas for which $\gamma = \frac{5}{3}$ is at 300 K and 1.0 atm. Find the initial and final energies of the gas and the work done by the gas when 800 J of heat energy is added at

(a) constant volume,

(b) constant pressure.

14. Suppose that 2.0 mol of helium gas (6.02×10^{23} atoms/mol) are compressed isothermally along the path AB as shown in the PV diagram on p. 528. At point A the volume is 6.0 L and the pressure is 2.5 atm.

(a) Find the temperature of the gas at point B.

(b) Assuming the volume at point B to be 3.0 L, find the pressure.

(c) Calculate the work done on the gas during the compression.

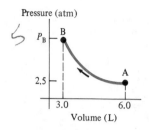

(d) How much heat was added to the gas during the compression?

15. An ideal gas at pressure P_1 and volume V_1 is compressed adiabatically to a volume V_2 and pressure P_2. Show that the work done by the gas is given by $W = (P_2V_2 - P_1V_1)/(\gamma - 1)$.

16. A monatomic gas initially at a pressure of 1.5 atm and a volume of 0.3 m³ is compressed adiabatically to a volume of 0.1 m³.

(a) Find the pressure of the gas in the final state.
(b) Determine the work done by the gas in compression.
(c) Calculate the final temperature of the gas. (Assume 1 mole of gas, 6.02×10^{23} atoms.)

The Second Law of Thermodynamics

17. Suppose that an inventor approached you with an idea to make an engine that would extract heat from the ocean and use it to power a ship without the need to expel heat at a lower temperature. Would you invest in such an idea?

18. An engine has an output of 800 J/cycle and an efficiency of 40%. How much heat is absorbed and how much is rejected in each cycle?

19. Logically, could a device satisfy the first law but violate the second law of thermodynamics? Devise an example to explain your answer.

The Efficiency of Carnot and Other Cycles

20. Why do designers of power plants try to increase the temperature of the steam fed into engines as much as possible?

21. An engine works between reservoirs at 450 and 300 K, extracting 100 J from the hot reservoir during each cycle.

(a) What is the maximum possible efficiency for this machine?
(b) What is the greatest amount of work it can perform during each cycle?

22. An ideal gas having $\gamma = \frac{5}{3}$ follows the cycle shown in the PV diagram at the top of p. 529; the temperature at point 1 is 300 K. Determine the efficiency of this cycle. (*Note:* The efficiency is defined as $e = W/Q_i$ irrespective of whether the heat input occurs at a single temperature or over a whole range of temperatures.)

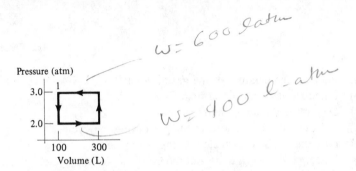

$W = 600 \ \ell\text{-atm}$

$W = 400 \ \ell\text{-atm}$

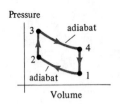

Pressure (atm)

3.0 — 1

2.0 —

100   300

Volume (L)

23. A diatomic gas is used in a heat engine. Starting at point 1 with $P_1 = 1.0 \times 10^5$ N/m², $V_1 = 1.0 \times 10^{-3}$ m³, and $T_1 = 400$ K, the cycle follows that for an idealized gasoline engine, as shown in the following diagram. For the compression 1 to 2, the volume decreases to 10^{-4} m³, and for the pressure increase in 2 to 3, $P_3 = 2P_2$.

Pressure

3 adiabat

4

2

adiabat   1

Volume

(a) Fill in the following table with the values of pressure and temperature for the various points indicated:

	1	2	3	4
P	1.0×10^5 N/m²			
V	1.0×10^{-3} m³	1.0×10^{-4} m³	1.0×10^{-4} m³	1.0×10^{-3} m³
T	400 K			

(b) Calculate the heat added and indicate in which part of the cycle this occurs.
(c) Determine the heat extracted and indicate in which part of the cycle this occurs.
(d) Find the work done in one cycle.
(e) Calculate the efficiency of the engine.

24. A certain monatomic gas is taken through the cycle shown on the PV diagram below. At point A the gas has pressure 4.0×10^5 Pa, temperature 1200 K, and volume 2.56×10^{-4} m³. It expands adiabatically to a volume 5.00×10^{-4} m³ at point B. From B to C the pressure is constant, and from C to A the volume is constant.

Pressure

A

adiabat

C   B

Volume

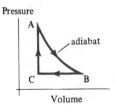

(a) Find the temperature of the gas at point B.

(b) Calculate the pressure at point B.

(c) Determine the temperature at point C.

(d) Calculate the heat added to the gas during the cycle.

(e) Find the net heat extracted during the cycle.

(f) Calculate the work done and the efficiency of the cycle.

CHAPTER 21

ENTROPY

For the present I will limit myself to quoting the following result: if we imagine the same quantity, which in the case of a single body I have called its entropy, formed in a consistent manner for the whole universe (taking into account all the conditions), and if at the same time we use the other notion, energy, with its simpler meaning, we can formulate the fundamental laws of the universe corresponding to the laws of the mechanical theory of heat in the following simple form:

1. The energy of the universe is constant.
2. The entropy of the universe tends to a maximum.

Rudolph Clausius in *Annalen der Physik*, **125** (1865)

21.1 TOWARD AN UNDERSTANDING OF ENTROPY

In this chapter we turn our attention to the *entropy principle*, a concept which, like Newton's second law, is an organizing principle for understanding the world. The principle is relatively simple to state, but understanding its meaning is more challenging.

Through theoretical studies of Carnot's work in 1865, the German physicist Rudolph Clausius introduced a new physical quantity closely linked to energy. He called it *entropy*, a word which sounds like "energy" and comes from the Greek word for "transformation." The use of entropy provides a way to analyze the behavior of energy in transformation.

To obtain an intuitive feeling for the concept of entropy, let's start with a familiar mechanical system we've used many times: Galileo's experiment with a ball rolling down and up two inclined planes. If friction is ignored, the ball rolls down one plane and back up the other, conserving mechanical energy. But if friction is *not* ignored, as time goes on the ball loses energy. This energy is transformed to heat which warms the ball and the material on the inclined planes. Without friction, all of the energy in the system can be accounted for by describing the motion of a single object, the ball itself. But with friction, the energy is shared among all the atoms and molecules that have been warmed by friction. As time goes on, even remote parts of the apparatus are warmed through conduction of heat. As a result, the number of objects sharing the energy continues to grow. When the ball finally comes to rest at the lowest point of its travels, a huge number of atoms have gained roughly the same fraction of the energy originally available. The number of objects sharing this energy has increased dramatically, from one object (the ball) to a number on the order of 10^{26} (the number of atoms and molecules in the ball, the material of the planes, their supports, and everything else in the system). Consequently, everything in the system is slightly warmer. Presently (in Example 2) we shall show that everything warms up by about 0.002°C, not enough for anyone to notice.

When all the energy resides in the ball, the ball can perform a positive amount of work. But at the end of the experiment, when the energy has been distributed among 10^{26} atoms and molecules in the form of heat, no useful work can be extracted. The entropy of the system is a measure of the amount of energy *unavailable* for work during the process. For this particular experiment the amount of energy transformed to heat keeps increasing as more and more subunits share the energy. Very crudely, the entropy tends to increase in proportion to the number of subunits sharing the energy.

The example of the rolling ball illustrates two important points. First, the transfer of energy in this process is *irreversible*. We cannot make the ball gain energy and roll back up the inclined planes by cooling the system. Second, the entropy S increases as more and more subunits come to share the available energy. But once all the atoms have roughly equal shares of the energy, the entropy can increase no further.

These two points are contained in the following statement of the entropy principle: *In an irreversible process, the total entropy of a system always increases until it reaches a maximum value.* After that, nothing else happens.

Example 1

If the experiment just described is carried out with a 0.50-kg ball which is initially lifted to a height of 1.0 m, estimate the final change in the average energy per atom in the system.

Initially the system has potential energy

$$E = mgh = (0.50 \text{ kg}) (9.8 \text{ m/s}^2) (1.0 \text{ m}) = 4.9 \text{ J}.$$

The change in average energy per atom is roughly $E/N \approx 4.9 \times 10^{-26}$ J.

Example 2

By the end of the foregoing experiment, everything in the system is slightly warmer. How much warmer?

For this estimation, we treat the system as though it were an ideal gas consisting of $N = 10^{26}$ atoms, and we assume that all the initial potential energy E (calculated in Example 1 to be 4.9 J) is equally distributed as heat among the N atoms. Then the increase in energy of each atom is E/N. Now in Chapter 19 we found that the average kinetic energy $\overline{K}$ of an atom in an ideal gas at absolute temperature T is given by

$$\overline{K} = \tfrac{3}{2} kT, \tag{19.12}$$

so the change in absolute temperature ΔT caused by a change $\Delta \overline{K}$ in the kinetic energy is

$$\Delta T = \frac{\Delta \overline{K}}{3k/2} .$$

In this example, $\Delta \overline{K} = E/N$ so

$$\Delta T = \frac{E/N}{3k/2} .$$

Substituting the values $E/N = 4.9 \times 10^{-26}$ J and $k = 1.38 \times 10^{-23}$ J/K we find the temperature of the system rises by a meager 2.4×10^{-3} K, or about 0.002°C. (Most solid materials have more energy per atom than an ideal gas at the same temperature, so the temperature change is actually smaller – perhaps half this size.)

Example 3

For the system in Example 1, what is the average speed of the center of mass of the ball at the end of the experiment?

The ball doesn't actually come to rest; it continues to jiggle around with the same fraction of the original energy of the system that each atom ends up with, namely E/N. The average speed $\overline{v}$ of any particle of mass M in the system at the end of the experiment satisfies

$$\tfrac{1}{2} M\overline{v}^2 = E/N.$$

Hence for the center of mass of the ball we have

$$\overline{v} = \sqrt{\frac{2E}{NM}} = \sqrt{\frac{2(4.9 \text{ J})}{(10^{26})\,(0.50 \text{ kg})}} = 4.4 \times 10^{-13} \frac{\text{m}}{\text{s}} .$$

Now that's splitting hairs!

21.2 ENGINES AND ENTROPY

There is a quantitative meaning for entropy that arises out of an analysis of reversible processes. In studying the efficiency of a Carnot engine we learned that if an ideal engine absorbs an amount of heat Q_i at a temperature T_i and then discards heat Q_o at temperature T_o, the quantities are related by Eq. (20.15), $Q_o/Q_i = T_o/T_i$, or

$$\frac{Q_o}{T_o} = \frac{Q_i}{T_i}. \tag{21.1}$$

In other words, in a reversible cycle, the quantity Q/T is conserved, and therefore has physical significance. For individual stages in a reversible process the ratio

$$\Delta S = \frac{Q}{T} \tag{21.2}$$

is also called *the change in entropy* and has units joules per kelvin (J/K).

In a reversible cycle Q/T stays constant, but in an irreversible process the ratio Q/T increases. Increasing Q/T has the same effect as friction increasing the entropy of a system like the rolling ball described earlier. To better understand the relation between this meaning of entropy and the qualitative description in Sec. 21.1 we compare an ideal Carnot engine with a nonideal engine.

At a high temperature T_i a Carnot engine has entropy change $\Delta S_i = Q_i/T_i$, and at a low temperature T_o its entropy change is $\Delta S_o = Q_o/T_o$. For the ideal Carnot engine, Eq. (21.1) states that

$$\Delta S_o = \Delta S_i \quad \text{(ideal)}.$$

But a less efficient (nonideal) engine starting with the same Q_i and T_i produces less work and therefore deposits more heat Q_o' at temperature T_o. Hence its entropy change at the low temperature T_o is

$$\Delta S_o' = \frac{Q_o'}{T_o}.$$

This change in entropy is greater in magnitude than that of ΔS_o, that is, $\Delta S_o' > \Delta S_o$, because more heat is extracted by a nonideal engine. Because $\Delta S_o = \Delta S_i$ for an ideal engine, the magnitudes of the changes in entropy obey

$$\Delta S_o' > \Delta S_i \quad \text{(nonideal)}.$$

So ideal engines keep the entropy constant, whereas nonideal engines increase it.

Now we can see the analogy to friction. The ideal Carnot engine doesn't create any entropy. That's why a Carnot engine connected to a Carnot refrigerator could operate forever, but do nothing besides run itself. It would be exactly analogous to a frictionless ball in Galileo's experiment that rolls up and down the inclined planes forever. In the real world there is always friction and rolling balls always come to rest. Likewise, in the real world there are no ideal Carnot engines and entropy is not conserved.

What property of the Carnot cycle makes the change in entropy zero? The key is found in the reversibility of the cycle. In order for a process to be reversible, friction must be eliminated and the process must be quasistatic. Most processes in nature, however, are irreversible. To prove that the entropy increases in irreversible processes, let's develop a theoretical method to handle such processes.

At this point it is convenient to introduce the convention that heat extracted is negative and heat input is positive. Thus $-Q_o = |Q_o|$ and, for a Carnot engine, Eq. (21.1) can be written

$$\frac{Q_i}{T_i} = \frac{-Q_o}{T_o},\qquad\qquad(21.1)$$

which becomes

$$\frac{Q_i}{T_i} + \frac{Q_o}{T_o} = 0.\qquad\qquad(21.3)$$

This was the starting point of Clausius's derivation of the entropy principle: *In a reversible Carnot cycle, the entropy of the system does not change.* What about *any* reversible engine?

Clausius realized that any reversible engine operating between two heat reservoirs can be approximated as accurately as desired by a series of alternating quasistatic adiabatic and isothermal paths. Here's how: first a number of adiabats are drawn, then adjacent adiabats are connected with two isotherms. The temperatures of the isotherms correspond to the temperatures at the top and bottom of the strip. This idea is illustrated in Fig. 21.1.

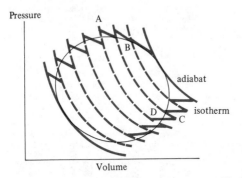

Figure 21.1 Any reversible cycle on a *PV* diagram can be approximated as well as desired by a series of alternating adiabats and isotherms.

The lines A, B, C, D, for example, constitute a Carnot cycle in Fig. 21.1. In other words, an arbitrary cycle can be approximated by a series of Carnot cycles. No heat is absorbed or rejected in the adiabatic parts of the cycle, so all the isothermal parts may be paired off into Carnot cycles. And we know how to analyze Carnot cycles. The approximation can be made as close to the actual cycle as we wish by making the mesh of adiabatic and isothermal lines still finer. For each of the small Carnot cycles, we know that

$$\frac{\Delta Q_i}{T_i} + \frac{\Delta Q_o}{T_o} = 0. \tag{21.3}$$

Summing over all the isothermal paths along which heat ΔQ_j is either absorbed or rejected at temperature T_j, we obtain

$$\sum_j \frac{\Delta Q_j}{T_j} = 0. \tag{21.4}$$

By letting the heat ΔQ_j become arbitrarily small and the number of such subcycles become arbitrarily large, the sum approaches an integral:

$$\oint \frac{dQ}{T} = 0, \tag{21.5}$$

where the symbol $\oint$ is that for a line integral taken around a complete cycle. In other words, the change in entropy for *any* reversible system is zero. This implies that for a reversible cycle, the entropy difference between two states depends only on those states and not on the particular path connecting them – a result Clausius first published in 1854.

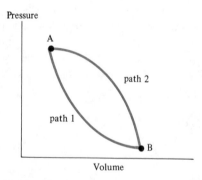

Figure 21.2 A reversible process from a state A to a state B made along two different paths in the *PV* diagram.

To understand why the entropy is independent of the path, consider the system moving reversibly along path 1 from A to B as shown in Fig. 21.2, then back to A along a different reversible path. Equation (21.5) implies that

$$\int_{A}^{B} \frac{dQ}{T} + \int_{B}^{A} \frac{dQ}{T} = 0.$$
$$\text{path 1} \qquad \text{path 2}$$

Because path 2 is reversible,

$$\int_{B}^{A} \frac{dQ}{T} = -\int_{A}^{B} \frac{dQ}{T},$$
$$\text{path 2} \qquad \text{path 2}$$

so

$$\int_{A}^{B} \frac{dQ}{T} = \int_{A}^{B} \frac{dQ}{T} .$$
$$\text{path 1} \qquad \text{path 2}$$

Because this is true of an arbitrary cycle through points A and B, the change in entropy is independent of the path taken, provided that path is reversible. This result is surprising because the heat entering or leaving the substance does depend on the path, yet the entropy change doesn't.

From this analysis it seems reasonable to define the change in entropy between any two states A and B connected by a reversible process as

$$\Delta S = S_{B} - S_{A} = \int_{A}^{B} \frac{dQ}{T} . \tag{21.6}$$

The change in entropy depends *only* on the initial and final equlibrium states of a system and not on the path joining them. The sign of the entropy change is determined by the heat flow direction (+ if added, − if extracted).

What is the change in entropy for an irreversible process? An irreversible process cannot be represented by a continuous path on a PV diagram because it doesn't move through a series of equilibrium states. However, we can join the ends of an irreversible process on a PV diagram by a reversible process, as shown in Fig. 21.3, forming a closed

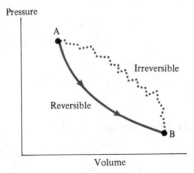

Figure 21.3 The initial and final equilibrium states of an irreversible process can be connected by a reversible process between the same points.

cycle. Then we can approximate the entire cycle by a sequence of small Carnot cycles. Consider a portion of the cycle operating between T_{ij} and T_{oj} in which heat ΔQ_{ij} is extracted from the hot reservoir and an amount ΔQ_{oj} is discarded at low temperature. The irreversible cycle is less efficient than a reversible cycle operating between the same two temperatures, so we have

$$1 - \frac{-\Delta Q_{oj}}{\Delta Q_{ij}} < 1 - \frac{T_{oj}}{T_{ij}} ,$$

or by rearranging terms,

$$\frac{\Delta Q_{ij}}{T_{ij}} + \frac{\Delta Q_{oj}}{T_{oj}} < 0.$$

Summing over all such cycles, and passing to the limit we obtain

$$\oint \frac{dQ}{T} < 0 \qquad\qquad (21.7)$$

for an irreversible cycle. What has happened is that the irreversible part of the cycle has generated more heat than it should have, and that extra heat has been extracted from the system to bring it back to its starting point. That's why the integral $\oint dQ/T$ is negative. The result in (21.7) is known as the *Clausius inequality* and it allows us to prove that in any irreversible process the entropy of an isolated system always increases.

Here's how: Imagine that the irreversible process shown in Fig. 21.3 proceeds from the equilibrium state A to B. For the complete cycle, first going from A to B in an irreversible process, then returning from B to A by a reversible path, Clausius's inequality implies

$$\underset{\substack{\text{irreversible}}}{\int_A^B \frac{dQ_I}{T}} + \underset{\substack{\text{reversible}}}{\int_B^A \frac{dQ_R}{T}} < 0,$$

or

$$\int_A^B \frac{dQ_I}{T} < -\int_B^A \frac{dQ_R}{T}$$

or

$$\int_A^B \frac{dQ_I}{T} < \int_A^B \frac{dQ_R}{T}.$$

From the definition of entropy for the reversible path, we know

$$\Delta S = S_B - S_A = \int_A^B \frac{dQ_R}{T}, \qquad\qquad (21.6)$$

so

$$\int_A^B \frac{dQ_I}{T} < S_B - S_A.$$

In particular this last inequality applies to an irreversible process taking the isolated system adiabatically from equilibrium state A to B. For an adiabatic process, no heat is exchanged and the left-hand side of the equation is zero, implying that $S_B > S_A$. In other words, for an isolated system, the entropy of the system increases after any irreversible process.

This result makes a profound statement about the behavior of the universe. Because the universe is an isolated system, the entropy of the universe must increase in time. As a consequence, the universal usefulness of energy decreases.

Example 4

An ideal gas is allowed to expand freely and adiabatically from a volume V_1 to a volume V_2. What is the change in entropy for this irreversible process? (A free expansion is one in which no work is done, as when a balloon bursts in a vacuum.)

Because no heat is exchanged with the surroundings, we might think that the change in entropy is zero. However, something irreversible does happen. We never observe the reverse process in which the air in a room freely contracts. On the other hand, because neither work nor heat is exchanged, the temperature of the gas does remain constant, because the temperature of an ideal gas depends only on its energy U. We compute the change in entropy by considering a *reversible* process connecting the two states: a quasi-static expansion from V_1 to V_2 at constant T. In this case, the change in entropy will be given by Q/T, where Q is the heat absorbed by the gas. Because the internal energy is constant in this process, the work done by the gas arises from heat absorbed by it. From Eq. (20.2) we calculate the work done by the gas as

$$Q = W = \int_{V_1}^{V_2} P \, dV = NkT \int_{V_1}^{V_2} \frac{dV}{V} = NkT \ln\left(\frac{V_2}{V_1}\right).$$

Therefore the change in entropy for the irreversible process is

$$\Delta S = Q/T = Nk \ln(V_2/V_1).$$

21.3 ENTROPY AND THE SECOND LAW OF THERMODYNAMICS

Clausius's indelible contribution to thermodynamics was his analysis of irreversible processes. He derived his inequality (21.7) from the observation that heat never flows by itself from low temperature to high temperature. For example, an ice cube placed in water never causes the water to boil. He noted that a negative value of the change in entropy for an isolated system would correspond to heat flowing from cold to hot. Consequently, the change in entropy for any cycle must be zero, if it is reversible, and positive, if it is irreversible. Later he capsulized his ideas in the lines quoted at the beginning of this chapter: "The energy of the universe is constant. The entropy of the universe tends to a maximum."

Let's look into another simple, but crucial, example of how the entropy principle works when there is no friction involved. Suppose a body consists of two parts that are not at the same temperature. One body is at temperature T_i and the other is at T_o. Now imagine momentarily connecting the two parts by a piece of copper wire, as in Fig. 21.4. Copper is a good conductor of heat, so heat flows from the hot piece at T_i to the cold piece at T_o. The entropy change of the hot piece is

$$\Delta S_i = \frac{-Q}{T_i}. \tag{21.2}$$

No work is done, and all the heat flows into the cool body at T_o. The entropy of the cooler body is increased by

$$\Delta S_o = \frac{Q}{T_o}.$$

The process has thereby increased the entropy of the two pieces combined by

$$\Delta S = \Delta S_o + \Delta S_i = \frac{Q}{T_o} - \frac{Q}{T_i},$$

which is positive because $T_o < T_i$.

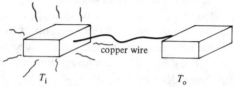

Figure 21.4 Heat conduction from a hot body to a cool body.

The flow of heat warmed the cooler piece and cooled the warmer piece. The two temperatures are now closer together, but so long as the temperatures are not equal we can reconnect the copper wire and repeat the process. Each time we do so, the entropy of the combined body will increase slightly, and the two temperatures will become closer to each other. We can repeatedly reconnect the two bodies until the two temperatures become equal. Then no heat will flow, and the entropy will no longer increase. At that point, without changing the mass, volume, or energy of the system, we have made the entropy as large as it can possibly be. And at that point the combined body has reached thermal equilibrium. This situation is closely analogous to the ball at the end of Galileo's experiment, which is in a state of mechanical as well as thermal equilibrium.

Example 5

A copper wire connecting two large pieces of metal conducts 40 J of heat from one piece at 400 K to the other at 350 K. What is the change in entropy for this process?

The hot piece of metal loses an amount of energy $Q = 40$ J. Because this occurs at a constant temperature $T_i = 400$ K, the change in entropy of this body is

$$\Delta S_i = -Q/T_i = (-40 \text{ J})/(400 \text{ K}) = -0.10 \text{ J/K}.$$

Now the other metal piece gains 40 J at a temperature of 350 K, so its change in entropy is

$$\Delta S_o = Q/T_o = (40 \text{ J})/(350 \text{ K}) = 0.11 \text{ J/K}.$$

Therefore the overall change in entropy is

$$\Delta S = \Delta S_o + \Delta S_i = 0.01 \text{ J/K}.$$

This simple and obvious argument has breathtaking consequences. As we've seen, the entropy of a body can decrease: in our example, the hot body kept losing entropy. However, the *combined entropy* of both pieces increased. If energy can flow in or out of a body, its entropy can increase or decrease. But if the total energy of a system is fixed, then the entropy can only increase until it reaches a maximum value. At that point the system has reached thermal equilibrium, and nothing more will happen.

What we've seen now is that a rolling ball eventually coming to rest at its lowest point and two pieces of matter placed in contact eventually reaching the same temperature are both consequences of the law of increase of entropy. In both cases, of course, energy is conserved, but in neither instance does simple conservation of energy help predict the result that the entropy principle gives us. That is why Clausius chose the name *entropy*. Entropy is something that tags along with energy, keeping track of how useful or well organized the energy is. As time goes on, entropy tends to increase, meaning that energy tends to more random, disorganized, useless forms.

The association of entropy with usefulness of energy is obvious in our analysis of Galileo's experiment, and we can also see the connection in the example we just discussed. Before the combined body reached thermal equilibrium, there were still two pieces at temperatures T_i and T_o; instead of connecting a copper wire between them, we could have connected an engine, and thereby extracted work. In other words, a copper wire is just an example of the most inefficient possible engine, because it merely transforms heat from high to low temperature, doing no work at all. Once the process is finished, entropy has increased to the maximum value, everything is at the same temperature, and no work can be obtained. An increase in entropy means a loss of the ability to do work. Although energy is conserved, its usefulness is destroyed.

Also from these examples, we see that the entropy of a system can change in two different ways: Another form of energy may turn into heat, as in Galileo's experiment, or heat may flow from higher temperature to lower, as in the case of the two bodies connected by a copper wire. But whatever the cause, the entropy of the universe, that is, of a system and its surroundings, always increases up to a maximum value. The energy of a system becomes less organized, less able to do useful work.

The entropy of the universe increases with time as the energy of the universe becomes less useful. Ordered mechanical energy is eventually and inexorably converted into the unordered, random motion of atoms. This conclusion is a consequence of the second law of thermodynamics and allows the second law to be cast into another form:

The entropy of the universe always increases toward a maximum. As we shall see, this law makes a profound statement about the fate of the universe.

21.4 AN IMPLICATION OF THE ENTROPY PRINCIPLE

We know that entropy depends in some way on how many parts of a system have a share of the energy, and that when heat flows, an amount of entropy Q/T accompanies the flow. Entropy has one more very important quality: it is associated with the internal order or configuration of a body. For example, under otherwise identical conditions (total volume, temperature, etc.) we might imagine organizing a certain large number of molecules into either a liquid or a solid. The difference between these two states is that in the solid state, the molecules are arranged in a neat crystal lattice. If we form both states, the liquid will turn out to have more entropy than the solid.

We can see this clearly by noting what happens when an ice cube melts. At constant temperature (0°C) as heat flows into the ice from the warm drink it is immersed in, instead of warming up, the ice melts. An amount of heat Q flows in at temperature T, meaning that the entropy increases by Q/T. That entropy indicates the change in the internal structure of the ice from solid to liquid. In other words, the liquid has a higher entropy than the solid at the same temperature. So entropy is a measure not only of usefulness of energy, but also of disorder.

The question of why ice melts in the first place raises a paradox. If equilibrium is always a state of maximum entropy, and liquid is a state of higher entropy than solid, why do solids, like ice, ever exist in equilibrium at all? The crux of the problem is this: a body of a given *energy* reaches equilibrium when it has maximized its entropy. We don't know yet what constitutes the condition for equilibrium of a body of a given *temperature*. Let's now find the answer to that problem.

Consider something we'll call our system, which is divided into two parts. One part, which is very small compared to the system (but still large enough to be macroscopic) is called the sample. Everything else in the system is called the bath. The total energy of the system will always remain constant. We can therefore apply the entropy principle to the system: the entropy of the system tends to a maximum value. Moreover, the bath is always assumed to be in equilibrium; its entropy is always as large as it can be for the amount of energy it has. Being in equilibrium, the bath has a definite temperature T. Finally, there is the sample, which is not necessarily in equilibrium. In fact the sample could be in any state at all. Heat is free to flow either way between the bath and the sample. However, the sample is so small compared to the bath that whatever heat flows in or out doesn't appreciably affect the temperature of the bath. Therefore we know that when the sample finally reaches equilibrium, its temperature will be T. But we don't know whether it will be a solid or a liquid, or more importantly what decides its fate when it reaches equilibrium.

When we put our sample in contact with the bath, everything that happens tends to increase the entropy of the entire system. Now suppose some heat Q flows into the sample and increases the energy of the sample by $\Delta E = Q$. That action also decreases the entropy of the bath by

$$\Delta S_b = -\frac{Q}{T} = -\frac{\Delta E}{T}$$

where the minus sign reflects the decrease. Because the sample is not in equilibrium, we don't know what happens to its entropy, but the event must increase the entropy of the system, or at least leave it constant – that's the second law of thermodynamics. If we call S_s the entropy of the system and S the entropy of the sample

$$\Delta S_s = \Delta S_b + \Delta S \geq 0.$$

It follows that

$$\Delta S - \frac{\Delta E}{T} \geq 0.$$

In other words, the entropy of the sample must increase by at least as much or more than the entropy of the bath decreased. Thus, as time goes on, the quantity $\Delta S - \Delta E/T$

continually increases, until it reaches its maximum value. These quantities, the entropy, the energy, and final equilibrium temperature, are all properties of the sample. Therefore we can now ignore the bath and make a statement about the sample only.

Let's write the above result in a slightly different way by multiplying by T:

$$T \, \Delta S - \Delta E \geq 0$$

where all the changes refer to those that occurred when heat flowed into the sample. The temperature doesn't change, so we can write the left-hand side as $\Delta(TS - E)$, and because the left-hand side increases, its negative obeys

$$\Delta(E - TS) \leq 0. \tag{21.8}$$

In other words, whereas the entropy of the *system* increases to a maximum, the quantity $E - TS$ *of the sample* tends to a *minimum* when the sample is kept in a bath at constant temperature T.

The quantity $E - TS$ is called the Helmholtz free energy, or simply the free energy, and is written

$$\boxed{F = E - TS.} \tag{21.9}$$

The quantity F plays a role in thermodynamics somewhat analogous to that of potential energy in mechanics: the system is in a state of stable equilibrium whenever F is at a minimum. The crucial idea is that whenever a sample can lower its free energy, it does. Although that process might decrease the entropy of the sample, the action always increases the entropy of the universe.

Now we can understand why H_2O is sometimes solid and sometimes liquid. We need to compare the energies and entropies of the two states. And the state of the H_2O will be that for which the free energy is lower.

We already know that water, being a more disorganized state than ice, has more entropy at the same temperature. But what about its energy? The energy consists of the kinetic plus potential energies of all the atoms and molecules. At a given temperature, both states have the same kinetic energy. In a gas, liquid, or solid, each atom generally has $\frac{3}{2}kT$ of kinetic energy in equilibrium. The difference lies in the potential energy. In solids, unlike liquids, the molecules arrange themselves into stable configurations of the lowest possible potential energy. Of course, the arrangement that minimizes the potential energy of a bunch of molecules, fitting one together with another in just the right way, is the same for any small set of molecules, which is exactly why solids are made up of identical building blocks that repeat endlessly in a rigid lattice. The liquid, on the other hand, is disordered; it does not have all of its molecules in this optimal configuration. Therefore the liquid has higher potential energy than the solid.

Consequently, the H_2O molecule must find the arrangement that will minimize the free energy. For this purpose, the solid has smaller energy but also smaller entropy. The liquid, on the other hand, has larger E and also larger S. What combination wins? The decision is made by the temperature. When T is very small, the negative term TS is small and unimportant, and the state of lower E wins out: at low temperature H_2O is solid. But when the temperature is high, the term TS becomes more important, so the state of higher S wins and the H_2O is liquid. At some temperature in between, the winner switches from solid to liquid. That's the melting point and also why the ice cube would rather melt completely than just warm up a little bit.

21.5 A FINAL WORD

Entropy is a measure not only of uselessness, but also of disorder. As time goes on, entropy increases. Energy is degraded to more useless forms, and matter into less-oriented states. Of course, a given bit of matter might temporarily decrease its entropy, but this always means that something else nearby is increasing its entropy by at least as much, and usually more. Strictly speaking, the principle of increasing entropy only applies to systems of conserved total energy. The universe itself is such a system, so the universe appears to be headed for a state of thermal equilibrium, after which nothing else will happen. This cheerfully optimistic view of the future is generally referred to as the "heat death of the universe."

There is another equally extravagant extrapolation of the entropy principle which turns around the sentence, "As time goes on, entropy increases" to read "As entropy increases, time goes on." In other words, the increase in entropy is the very arrow of time; almost all other physical laws would work equally well if time ran backward instead of forward.*

We have seen that all systems, including the universe, tend to evolve in an irreversible way. Despite the action of men and women, the inexorable law of nature is for energy to become less useful and the universe more disorganized. This tendency of the flow of time was pointed out by an eleventh-century Persian poet-mathematician, Omar Khayyám, who wrote:

> The Moving Finger writes; and having writ,
> Moves on: not all thy Piety nor Wit
> Shall lure it back to cancel half a line,
> Nor all thy Tears Wash out a Word of it.

There have been many other statements of the second law of thermodynamics, but none so elegant.

Problems

Entropy Changes

1. What is the change in entropy of a ball that is strictly obeying the law of inertia?

2. Can you think of a process for which the entropy decreases?

3. Construct an argument explaining why a gas by itself never freely contracts.

*The expansion of the universe is another candidate for defining the arrow of time. In addition, a tiny forward–backward time asymmetry occurs in the force law governing the weak interactions (one of the four fundamental forces briefly introduced in Section 10.1), although this asymmetry seems too small to produce the observed differences between past and future according to our present understanding.

4. One mole ($N = 6.02 \times 10^{23}$) of an ideal gas expands reversibly and isothermally from a volume of 15 L to a volume of 45 L.

(a) Find the change in entropy of the gas.

(b) Determine the change in entropy of the universe for this process.

5. Suppose a vessel consists of two chambers, each of the same volume. In one chamber there is helium gas, and in the other there is argon gas at the same temperature and pressure. The partition between the chambers is suddenly removed. Does the entropy of the system increase? Explain your answer.

6. Calculate the change in entropy for the case of Problem 5.

7. A system absorbs 200 J from a heat bath at 300 K, does 50 J of work, and rejects 150 J of heat at a temperature T, then returns to its initial state.

(a) Calculate the change in entropy of the system for a complete cycle.

(b) If the cycle is reversible, what is the temperature T?

8. A 2.0-kg block is dropped from a height of 3.0 m above the ground, strikes the ground, and remains at rest. If the block, air, and ground are all initially at a temperature of 300 K, what is the change in entropy for the universe in this process?

9. A 1200-kg car traveling at 80 km/h crashes into a brick wall. If the temperature of the air is 25°C, calculate the entropy change of the universe.

10. Suppose 300 J of heat is conducted from one reservoir at a temperature of 500 K to another reservoir at a temperature T. Calculate the change in entropy of the system if T equals

(a) 100 K   (b) 200 K
(c) 400 K   (d) 490 K.

What can you conclude about the change in entropy as the reservoirs come closer in temperature?

11. One mole ($N = 6.02 \times 10^{23}$) of an ideal gas undergoes a free, adiabatic expansion from $V_1 = 12$ L, $T_1 = 400$ K to $V_2 = 24$ L, $T_2 = 400$ K. Afterward it is compressed isothermally back to its original state.

(a) Compute the change in entropy of the universe in this process.

(b) Show that the work made useless is given by $T \, \Delta S$.

12. Which process is more wasteful:

(a) a 1.0-kg ball starting from a height of 1.5 m and rolling down and up inclines until it comes to rest at its lowest point where the temperature of everything is 300 K, or

(b) the conduction of 150 J of heat from a reservoir at 350 K to one at 300 K?

Entropy and Disorder

13. Develop an argument about why different metals have different melting points.

14. Extend the discussion presented in the text to explain why H_2O is a gas and not a liquid above the vaporization temperature.

15. In Chapter 10 we discussed how nature tends to seek states of lowest possible potential energy. Does the behavior of H_2O at temperatures greater than the melting point contradict that idea? Explain.

16. A tree takes unorganized molecules and organizes them into branches and leaves. Do you think living organisms violate the second law of thermodynamics? Explain your reasoning.

CHAPTER 22

THE QUEST FOR LOW TEMPERATURE

We have seen that the gaseous and liquid states are only distant stages of the same condition of matter, and are capable of passing into one another by a process of continuous change. A problem of far greater difficulty yet remains to be solved, the possible continuity of the liquid and solid states of matter. The fine discovery made some years ago by James Thomson, of the influence of pressure on the temperature at which liquefaction occurs, and verified experimentally by Sir. W. Thomson, points, as it appears to me, to the direction this inquiry must take; and in the case at least of those bodies which expand in liquefying, and whose melting-points are raised by pressure, the transition may possibly be effected. But this must be a subject for future investigation; and for the present I will not venture to go beyond the conclusion I have already drawn from direct experiment, that the gaseous and liquid forms of matter may be transformed into one another by a series of continuous and unbroken changes.

Thomas Andrews, *Philosophical Transactions of 1869*, p. 575

22.1 COOLING OFF

How do you make something colder? Making something hotter is easy. For example, if you need to warm yourself on a chilly night, you can build a fire with little or no technology. But to cool yourself on a hot day is quite another matter. The difference between heating and cooling is reflected in our history: we've had the use of fire ever since Prometheus let the secret slip long before the dawn of history. But the secret of making things colder – refrigeration – is barely older than the oldest living person.

Nonetheless, some techniques of cooling are somewhat old. The ancient Sumerians used porous jugs for cooling household water. As some water seeped out it evaporated and cooled the rest, a technique used today in bottled water dispensers. Later in history, slaves were sent to bring snow from the mountains in summer. And later still, Leonardo da Vinci invented a form of air conditioning using air blown over a block of ice that had been stored since winter.

As we turn to the past to see how low temperatures were reached, we find that the quest for low temperatures is closely tied to the history of understanding the basic states of matter – solid, liquid, and gas.

As it turns out, the Sumerians had the right idea: evaporation, that is, the change of a liquid into its gaseous form, cools the liquid. That's not the only method, but it is powerful and convenient for most purposes. However, to make something very cold – temperatures in the range 10–100 K (-263 to $-173°C$) – further knowledge of the behavior of matter is needed. With the quest for low temperature came the discovery that all elements can exist in each of the basic states under the right conditions of temperature and pressure. We turn now to find what conditions govern the states of matter.

22.2   THE STATES OF MATTER

That all matter can exist in three basic states – solid, liquid, and gas – is a discovery less than a century old. But the idea of the states of matter can be traced back to ancient Greece. The elements of Aristotle's universe – earth, air, water, fire – also demonstrated the basic states of matter. The element earth reflected the solid appearance of matter, water generalized the liquid state, and fire and air represented the gaseous state. What was unknown to Aristotle and to others until the late nineteenth century was that any particular substance can exist in each of the three basic states. What conditions produce different forms of the same substance?

The physical state or phase of a substance depends on both its temperature and its pressure and can be illustrated visually by a *phase diagram* such as that in Fig. 22.1. For example, at low temperatures and high pressures a substance such as H_2O tends to be solid ice. At high temperatures and low pressures it tends to be gaseous, like steam or water vapor, and somewhere in the intermediate range of temperatures and pressures it tends to be liquid.

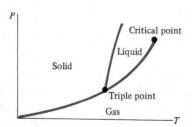

Figure 22.1 A typical pressure–temperature (phase) diagram illustrating regions in which a substance exists as either solid, liquid, or gas.

The curve that separates the solid and liquid regions on the phase diagram is called the *melting curve*. The curve between the liquid and gaseous states is the *vapor-pressure curve*, and that between the solid and gaseous states is the *sublimation curve*. Only on these curves can the respective phases coexist in equilibrium, and if two phases are in equilibrium, then specifying either the temperature or the pressure alone is sufficient to define the state. When a substance crosses one of these curves it is said to undergo a phase transition.

When the phase transition is from solid to liquid, the substance melts. The structure changes from an ordered, crystalline array of molecules and atoms to a less-ordered configuration. To achieve this change of phase a certain amount of heat must be added which goes entirely into changing the phase and does not raise the temperature. If heat Q is required to melt a substance of mass m, then the ratio Q/m is called L_f, the *latent heat of fusion*. Hence

$$Q = mL_f. \tag{22.1}$$

The latent heats of fusion for several substances are listed in Table 22.1.

Table 22.1 Phase Transitions and Their Temperatures and Latent Heats for Various Substances at 1 atm

Substance	Melting point (K)	L_f (kJ/kg)	Boiling point (K)	L_v (kJ/kg)
Hydrogen	14	59	20	452
Nitrogen	63	26	77	200
Oxygen	55	14	90	213
Ammonia	—	—	240	1369
Water	273	334	373	2257
Lead	600	25	2023	858

As the phase diagram indicates, a substance can go directly from solid to gas without passing through the liquid state. This process, known as *sublimation*, also is accompanied by a latent heat. The most common example is the sublimation of dry ice: solid carbon dioxide sublimes at atmospheric pressure. That's why dry ice is never wet.

When a substance crosses the vapor-pressure curve, boiling occurs as liquid is changed into vapor. In this case the structure of the substance does not change radically as in the solid–liquid transition. Instead the molecules and atoms increase their separation distance. Just as in the case of melting, once a substance is at the boiling point, a certain amount of heat must be added to change the phase from liquid to vapor. If Q is the amount of heat necessary to vaporize a mass m of liquid at the boiling point, the ratio Q/m is called L_v, the *latent heat of vaporization*. Thus we have

$$Q = mL_v. \tag{22.2}$$

The boiling points and latent heats of vaporization of a few substances are also listed in Table 22.1.

Two points of special interest on the phase diagram are the triple point and the critical point, which are shown in Fig. 22.1. The triple point is a unique pressure and temperature at which the three coexistence curves meet and at which all three phases can coexist. By international agreement, the triple point of water, occurring at 273.16 K and a pressure of 0.006 atm, is used as the fixed point in the absolute temperature scale. Triple-point temperatures and pressures for several other substances are listed in Table 22.2.

Table 22.2 Triple-Point Temperatures and Pressures for Several Gases

Substance	Temperature (K)	Pressure (10^5 Pa)
Hydrogen	13.84	0.070
Nitrogen	63.18	0.125
Oxygen	54.36	0.002
Ammonia	195.40	0.061
Carbon dioxide	216.55	5.17
Water	273.16	0.006

The critical point signals the end of the vapor-pressure curve and it also occurs for a particular pressure P_c and temperature T_c. At any higher pressure and temperature, there is no distinction between liquid and gas; there is only a homogeneous fluid phase.

A thorough study of the critical point of carbon dioxide was made by Thomas Andrews in the mid-nineteenth century. In an attempt to solve one of the great problems of his time – the liquefaction of gases – Andrews discovered what was to be a general property of all matter: above a critical temperature, no distinction between gas and liquid exists. This discovery, among others of the time, prompted physicists to investigate what distinguishes one phase from another. Table 22.3 lists the critical temperatures and pressures for several substances.

Table 22.3 Critical Temperatures and Pressures for Several Gases

Substance	Temperature (K)	Pressure (10^5 Pa)
Hydrogen	33.3	13.0
Nitrogen	126.2	33.9
Oxygen	154.8	50.8
Ammonia	405.5	112.8
Carbon dioxide	304.2	73.9
Water	647.4	221.2

In Fig. 22.2, the same phases as in Fig. 22.1 are shown on a PV diagram. Here we find the solid state at high pressure and low volume and the gas at low pressure and large volume. In the shaded regions of the graph more than one phase can coexist. The triple point of the pressure–temperature diagram corresponds to the single, horizontal isotherm labeled T_t. On the other hand, the critical point is also a single point on this diagram that occurs at the maximum of the gas–liquid coexistence curve, indicating that there is a critical volume as well as a critical temperature and pressure.

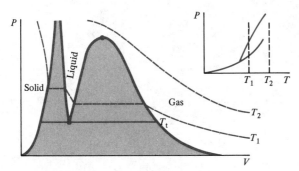

Figure 22.2 Phases of a given amount of a substance on a pressure–volume diagram.

The states of a substance in equilibrium depend on three quantities, pressure, temperature, and volume; the mathematical relationship between these variables can be extremely complicated. Nonetheless we can represent the state graphically by a surface on a pressure–volume–temperature plot. Such a representation is shown in Fig. 22.3. The region under the PV curve shown in Fig. 22.2 is a projection of this surface on a plane $T = $ const.

22.3 BEHAVIOR OF WATER

Let's consider the states of a substance without which life on Earth would not exist – water. The phase diagram for H_2O is illustrated in Fig. 22.4. Imagine an ice cube inside a sealed container that is much larger than the cube and from which the air has been evacuated. Even at very low temperature, sufficient water vapor exists inside the container to create the pressure called for by the sublimation curve. As we heat the container, the ice cube warms up, and the gas pressure increases slightly as water molecules evaporate from the surface. As this process occurs, the ice remains on the sublimation curve with the solid and gas states in coexistence.

As we continue to heat the container, the temperature eventually reaches that of the triple point. At this point no further increase in temperature occurs until all the ice has melted. The added heat goes entirely into changing the phase from ice to water.

Once all the ice has melted, the water can be heated further, thereby creating more water vapor. Because the water remains on the vapor-pressure curve, just enough water evaporates with every increase in temperature as called for by the curve (the amount

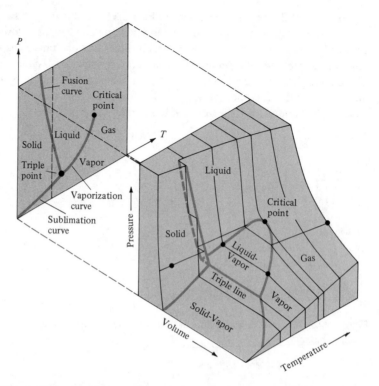

Figure 22.3 The state of a given amount of a substance can be represented by a surface in a plot of temperature, volume, and pressure.

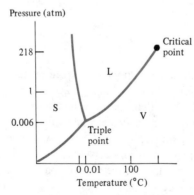

Figure 22.4 Phase diagram for H_2O.

depends on the size of the container). If we continue to warm the container, it eventually reaches 100°C, at which point the vapor pressure is equal to one atmosphere. If you opened the container at this point, the pressure of water vapor inside and the air outside would just balance. If you were to open the container and continue to heat it, you would have the same situation as a pot of water on a kitchen stove. Whereas before, the pressure could change because the can was sealed, now the pressure is kept constant and equal to that of the atmosphere. So heating will make the water evaporate without any change in temperature – that's boiling.

On the other hand, if you kept the container sealed, the water would pass through 100°C without any apparent effect and the pressure would continue to rise as well. This is what happens inside a pressure cooker. There is nothing special about the boiling point 100°C for water, except the accidental fact that at that point the vapor pressure of water happens to be equal to atmospheric pressure on the planet Earth, at sea level, on an average day. Table 22.4 lists the boiling point for water at various pressures.

Table 22.4 Vapor Pressures of Water for Various Temperatures

Temperature (°C)	Pressure (10^5 Pa)
0	0.006
20	0.023
40	0.073
60	0.199
80	0.473
100	1.01
120	1.99
140	3.61
160	6.17
180	10.0
200	15.0

The phase diagram for water is slightly different from the typical diagram we saw earlier in that the melting curve bends backward. This peculiarity of water makes it possible to ski or ice skate because, at a temperature below 0°C, ice can be melted by simply applying pressure. The force of the skis or ice skates melts the ice at that particular place and the liquid lubricates the path. Otherwise, skiing on snow would be like skiing on concrete.

22.4 LIQUEFACTION OF GASES

At the turn of the nineteenth century, a handful of scientists may have understood the phases of water, but even fewer saw a generalization to the states of matter. The idea crystalized through research into the liquefaction of gases. Earlier a Dutch scientist, Martin Van Marum, had accidentally liquefied ammonia gas under pressure while using it to test Boyle's law; however, he didn't grasp the significance of this first liquefaction. Until 1823, the liquefaction of gases remained tenuous.

In that year, at the Royal Institution of London, a young chemist made an explosive discovery. While heating a compound in a sealed glass tube, Michael Faraday produced an oily-looking substance. When he filed the tube open to investigate the substance, the tube promptly exploded. But Faraday realized that he had liquefied chlorine gas.

This great intuitive genius of the nineteenth century would later become famous for his researches in electromagnetism, but he revealed his talents in this early episode. Unlike Van Marum, Faraday understood perfectly the significance of his discovery, and immediately undertook a quest to liquefy other gases. Using an inverted U tube of glass, in one leg of which he could heat reagents to evolve the gas in question, while the other leg, under pressure, could be cooled, if necessary, to form the condenser, he succeeded in liquefying many gases. With that success, Faraday reached the plausible conclusion that all forms of matter could exist in each of the fundamental states, given the right conditions of temperature and pressure.

The next great advance in understanding the states of matter came in 1835 when the French scientist Thilorier managed to solidify carbon dioxide (CO_2). The phase diagram of carbon dioxide is shown in Fig. 22.5. Although the phase diagram appears like the ones we discussed earlier, peculiar things happen because of the relation between the characteristic temperatures and pressures of CO_2 and the temperature and pressure at which human life on earth flourishes.

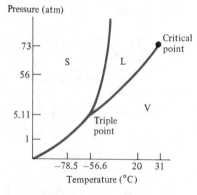

Figure 22.5 Phase diagram for carbon dioxide, CO_2.

At atmospheric pressure and temperature, carbon dioxide is commonly a gas. However, if it is compressed without allowing its temperature to rise above room temperature, carbon dioxide will condense into a liquid. On the phase diagram this compression corresponds to moving from room temperature and atmospheric pressure directly up the line to the coexistence curve, where liquid CO_2 coexists with gaseous CO_2 at high pressure.

Liquefying gases is what Faraday had already accomplished. The advance made by Thilorier comes when, starting from these conditions, the pressure is reduced too quickly for heat from the surroundings to keep the temperature constant. We can imagine opening a valve on the container of liquid plus gaseous CO_2 that leads directly into a room at atmospheric pressure. And whoosh, a high-pressure jet of CO_2 blows forth into the room.

Inside the container liquid CO_2 still coexists with gaseous CO_2, but now at reduced pressure. Because the two phases coexist, the carbon dioxide must be on the coexistence curve of the phase diagram. And to be on that curve at reduced pressure means that it must also be at reduced temperature. In other words, the liquid and gas in the container are *cooled* by evaporation. The process works for CO_2, or for that matter any other substance, just as it does for water.

We can summarize this observation as follows: To stay on the coexistence curve, temperature must decrease when pressure is allowed to decrease, just as the pressure must rise when a substance is heated. As gas escapes from a container, liquid evaporates in order to try to replace the gas. The result of the evaporation is cooling.

We can describe the same process in a more subtle way. Gas is a state of higher entropy than liquid. The entropy is higher not because the gas has a different internal structure from the liquid but because the gas is less dense. Each molecule of the gas has more space in which to move around, and that in itself is a form of disorder. In any case, when liquid evaporates to become a gas, the entropy per molecule increases. If this occurs at temperature T, then there is a corresponding transfer of heat equal to the heat of vaporization,

$$Q = mL_v,$$
(22.2)

and the change in entropy in going from liquid to gas is

$$\Delta S = \frac{Q}{T}.$$
(21.2)

Example 1
Compare the change in entropy for 0.1 kg of ice melting at 1 atm with that of the same mass of liquid H_2O vaporizing at 1 atm.

As Table 22.1 indicates, when ice melts it releases 334 kJ/kg of heat. Because $\Delta S = Q/T$, the change in entropy for this process is

$$\Delta S = (0.1 \text{ kg}) (334 \times 10^3 \text{ J/kg})/(273 \text{ K}) = 122 \text{ J/K}.$$

On the other hand, vaporization releases 2257 kJ/kg and the associated change in entropy is

$$\Delta S = (0.1 \text{ kg}) (2257 \times 10^3 \text{ J/kg})/(373 \text{ K}) = 605 \text{ J/K}.$$

The entropy change is much greater in the liquid–vapor transition than in the solid–liquid transition. This comparison indicates that for water the increase in distance between molecules accompanying the liquid–vapor transition is more effective in increasing the entropy than the change of form from crystalline array to liquid, which occurs at nearly constant density.

There is yet another way to describe this process. At any given temperature, the CO_2 molecules have the same kinetic energy, on the average, no matter whether they are in the liquid or the gas. According to Eq. (19.12) this energy is equal to $\frac{3}{2}kT$ for each atom making up each molecule. The molecules of liquid, however, have lower potential energy than those of the gas; it is precisely this potential energy that binds the molecules together, causing the liquid to form. Condensation is a consequence of the electric force of attraction between the molecules when they are close together in the liquid state.

Therefore, when molecules evaporate from liquid to gas, they are increasing their potential energy by escaping the attraction of their neighbors. If the gas plus liquid are isolated from contact with the rest of the world, so that overall energy must be conserved, the increase in potential energy must be balanced by a decrease in kinetic energy. But we've already seen that kinetic energy is equivalent to temperature, and so the temperature falls. On a molecular level this is why evaporation causes cooling of the gas. In the next section we'll return to this explanation.

Escape from the attractive forces of other molecules explains why the gas is cooled, but what cools the liquid? Inside the liquid, the average kinetic energy per molecule is $\frac{3}{2}kT$, but in fact some molecules have more kinetic energy than this and some less. The fastest-moving molecules are the ones most likely to escape into the gas. That leaves behind the slower, cooler molecules. That's why the liquid cools.

Of course, all of these different versions are not opposing views, but rather alternative descriptions of a single coherent understanding of the states of matter. The change in potential energy and the latent heat of fusion are really the same quantity viewed on different levels.

Any liquid, not just CO_2, can coexist with its own vapor. But carbon dioxide differs from most other substances in one peculiar respect: its vapor pressure at the triple point is above one atmosphere. When Thilorier opened his valve and allowed liquid CO_2 to cool by evaporation, as the pressure dropped, the mixture froze before it reached atmospheric pressure. CO_2 sublimes directly from the solid at a pressure of 1 atm. That's why dry ice never becomes a liquid under ordinary circumstances.

Thilorier's discovery led him to a method of producing even lower temperatures. By mixing ether with carbon dioxide and pumping away the vapor to reduce the pressure below one atmosphere, Thilorier produced a temperature of $-110°C$. Armed with this new advance in technology, Michael Faraday returned to the problem of liquefaction in 1844. Within a year he succeeded in liquefying all the known "permanent gases" except those he really wanted. Three elements – oxygen, nitrogen, and hydrogen – and three compounds – carbon monoxide, nitric oxide, and methane – failed to yield to his techniques.

The quest to liquefy these permanent gases now took on the character of an open scientific challenge. To succeed where the great Faraday had failed was no small matter. The story of that quest is filled with twists and ironies. For example, the liquefaction of oxygen was accomplished in 1877 by Louis Paul Cailletet and Raoul Pierre Pictet, who achieved success within a few days of each other 33 years after Faraday's failure set the stage. The basic process was a kind of cascade in which one bath was used to liquefy another bath under pressure. This second bath in turn was further cooled by released pressure, and so on. In the final step, the pressurized oxygen gas (Pictet achieved a pressure of 250 atm at a temperature of $-130°C$) did not liquefy. The critical point of oxygen hadn't been reached. However, when the pressure was subsequently released, both Pictet and Cailletet found that a cloud of partially liquefied oxygen was produced. The liquefaction had occurred because expansion of a gas by itself produces cooling.

22.5 THE JOULE–THOMSON EFFECT

Beginning in 1845 and culminating in 1862, Joule, and later Joule and Thomson (the same Thomson who later became Lord Kelvin), conducted a series of experiments to test the interchangeability of work and heat. Their investigations showed that under certain circumstances the expansion of a gas into a chamber held at constant pressure would produce cooling. The physical reason turns out to be the same one that causes a liquid to cool when it evaporates.

The cooling effect they discovered, known as the Joule–Thomson effect, provided experimental evidence for the existence of intermolecular forces between gas molecules. It is these forces that account for the structural changes in a substance as it passes through different phases as well as for the deviations of real gases from the ideal-gas law (which were discussed in Chapter 19). Further research revealed that intermolecular forces consist of two different interactions between pairs of molecules. One force is a weak, relatively long-range attractive force known as the *Van der Waals force*. The second force is a strong, short-range repulsive force which prevents molecules from coming too close together. The regions in which each force dominates are indicated in Fig. 22.6, where the potential energy of a molecule is plotted as a function of the distance between two interacting molecules. These forces are electrical in origin and depend on the structure of the particular molecule.

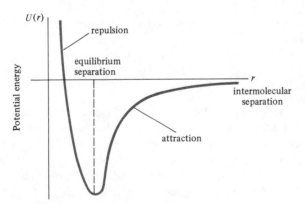

Figure 22.6 Potential-energy diagram for a molecule in a real gas. The force is repulsive at small r, where $u(r)$ slopes down, and attractive at larger r, where $u(r)$ slopes up.

In their experiments, Joule and Thomson kept a gas at constant pressure and allowed it to flow continuously through a porous plug of tightly packed cotton, as shown in Fig. 22.7. As the gas flows from the left chamber at pressure P_1 into the right chamber at lower pressure P_2, the pistons must move, changing the volume so as to keep the pressures constant. The porous plug, or nozzle, merely supports the pressure difference and the entire apparatus is isolated so that the process is adiabatic.

Let's discuss why the Joule–Thomson effect indicates the existence of intermolecular forces. As the gas is pushed from the left chamber of Fig. 22.7, decreasing its volume

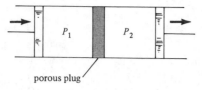

Figure 22.7 Schematic of the Joule–Thomson porous-plug apparatus.

from V_1 to zero, the work done *by* the gas is simply $W_1 = -P_1V_1$ because the pressure is held constant. On the other side of the nozzle the piston does an amount of work $W_2 = P_2V_2$ as the volume expands from zero to V_2. According to the first law of thermodynamics,

$$Q = W + \Delta U, \tag{20.3}$$

and because the process is adiabatic, $Q = 0$. Therefore the change in internal energy of the gas is equal to the negative of the net work done by the gas:

$$\Delta U = U_2 - U_1 = -W = -(-P_1V_1 + P_2V_2) = P_1V_1 - P_2V_2. \tag{22.3}$$

From this we see that there is a conserved quantity for the process:

$$U_1 + P_1V_1 = U_2 + P_2V_2. \tag{22.4}$$

Now if we assume an ideal gas, then both the internal energy and PV are functions only of temperature:

$$U = qNkT, \tag{19.13}$$

$$PV = NkT. \tag{19.9}$$

Consequently, the temperature of the gas should remain constant during the process. But the observation is that the Joule–Thomson process produces a cooling of the gas. This departure from ideal-gas behavior is attributed to intermolecular forces.

The intermolecular forces contribute potential energy, so the internal energy of the gas is not simply due to the kinetic energy of individual molecules. Let u denote the potential-energy contribution due to intermolecular forces. Then the internal energy of the gas is given by

$$U = qNkT + u. \tag{22.5}$$

According to Eq. (22.3), we should have

$$\Delta U = qNk\,\Delta T + \Delta u = P_1V_1 - P_2V_2. \tag{22.6}$$

When the gas cools, the right-hand side of this last result is found to be positive, that is,

$$P_1V_1 - P_2V_2 > 0,$$

which in turn implies

$$\Delta u > -qNk\,\Delta T. \tag{22.7}$$

Because ΔT is negative, this means Δu is positive.

As the gas expands, the potential energy of the molecules increases, because the molecules are pulling away from one another's attractive potentials. This is exactly what happens when a liquid evaporates. The increased potential energy is balanced by a decreased kinetic energy, or slower atoms, so the gas is cooler.

If the initial temperature and pressure are high enough, the repulsive part of the intermolecular force may be dominant. In this case, Δu is negative, so the gas should actually warm. Warming of the gas is not only expected, but also observed under these conditions.

Of course, intermolecular forces are responsible for change of phase as well. The difference between a gas and a liquid is that the molecules of a gas on the average are too far apart to have much influence on one another, while in a liquid, they are close enough for the Van der Waals forces between them to cause them to attract and condense. Expressed in energy terms, liquid molecules have more negative potential energy. This potential energy is the latent heat.

The Joule–Thomson effect creates a decrease in temperature that is found to be nearly proportional to the difference between the pressures of the two chambers and provides a practical method for liquefying gases. In 1895 the German chemist Karl von Linde adapted the process for the large-scale liquefaction of gases. His process did away with the cascade effect and makes use only of the gas to be liquefied, for example, nitrogen. Cooled, compressed nitrogen passes through an inside tube and through a Joule–Thomson expansion valve and back through an outer tube. The gas that is not liquefied is returned to the compressor by means of a heat exchanger, where the incoming compressed gas is cooled to a lower temperature by the outgoing cold product.

22.6   A FINAL WORD

After nitrogen and oxygen had been liquefied, only one permanent gas remained – hydrogen. Scientists knew that if it could be liquified at all, doing so would be a great achievement. And so two men set out in a race to liquefy hydrogen. They were Sir James Dewar, who was Faraday's successor as Professor at the Royal Institution in London, and Heike Kammerlingh-Onnes at the University of Leiden, in the Netherlands.

Each of these two great scientists spent decades pursuing the quest. In the end, Sir James Dewar won the race. He gained the lead by inventing the double-walled evacuated flask to retain cold liquids. It is known in scientific circles as a Dewar flask and in popular terms as a thermos bottle. However, in a final twist of irony, the rules of the race changed just before the finish line.

In 1895, just before Dewar succeeded in producing liquid hydrogen, his countryman William Ramsay isolated terrestrial helium, an element previously known to exist only in the sun. Kammerlingh-Onnes, knowing that he had lost the race for liquid hydrogen, immediately abandoned the pursuit and laid plans for an assault on helium. That siege lasted another thirteen years. To undertake it made sense only because of the work of one of his countrymen, Johannes Diderik van der Waals.

In the 1870s, Van der Waals developed a theory of the fluid state that took into account the fact that a compressed gas, especially above its critical point, could have potential energy. Such a gas, he argued, would exhibit deviations from the ideal gas law $PV = NkT$. He gave the modified law

$$(P + aN^2/V^2)\,(V - Nb) = NkT, \tag{19.10}$$

which we have described in Section 19.3, and concluded that by careful observation of the quantities a and b in his law it would be possible to deduce the strength of the forces between the molecules. But those forces were responsible for the Joule–Thomson cooling, and in fact, liquefaction itself. Thus, by careful measurements of the properties of helium gas, even well above its critical temperature, it would be possible to predict both the conditions under which the Joule–Thomson cooling would be effective, and the temperature and pressure that had to be reached in order to liquefy helium.

The results of this analysis, which Kammerlingh-Onnes carried out, gave him three indispensable pieces of information. First, helium did deviate from the ideal gas law like other gases, indicating that it could be liquefied. Although the effort was daunting, it would not be wasted. Second, when liquefied, the temperature of helium would be much lower than that of liquid hydrogen. So if he succeeded, he and not Dewar would be the winner of the quest in history's eyes. And finally, by giving him in advance the values of temperature and pressure he would have to attain, the analysis made it possible to undertake the job in a rational and orderly way.

Kammerlingh-Onnes and his research group succeeded in liquefying helium on July 10, 1908. The critical temperature turned out to be, as predicted, only 5 K – just five degrees above absolute zero. The last "permanent" gas had been liquefied. For his research into the properties of matter at extremely low temperature, Kammerlingh-Onnes was awarded the Nobel Prize in physics in 1913.

Today, liquid helium is made in huge commercial plants for use in scientific laboratories and industry, as are liquid nitrogen and oxygen. The cost is low: a liter of liquid nitrogen costs about as much as a liter of milk, and a liter of liquid helium costs about as much as a liter of vodka.

Problems

Cooling Off

1. Explain why you feel cool when you step out of the shower and into another room to dry yourself.

2. Explain why hot, humid weather is more uncomfortable than hot, dry weather.

3. When a frost is expected in Florida, why do citrus growers often spray their trees with water?

States of Matter

4. Can a liquid boil and freeze at the same time? Support your answer by reference to a phase diagram.

5. For the phase diagram shown below, explain what happens as the substance changes from (a) A to B, (b) C to D, (c) E to F, (d) G to H.

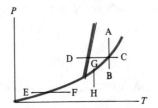

6. Can liquid become gas without passing through a phase transition? Can solid change into gas without passing through a phase transition? Explain your reasoning.

7. To raise the temperature of 1 kg of ice 1°C requires 2.0 kJ of heat; to raise the temperature of 1 kg of water 1°C requires 4.2 kJ of heat. Using this information, plot a graph of temperature versus heat added for 1 kg of ice taken from -10°C to vapor at 100°C.

Behavior of Water

8. In a car radiator the coolant water is under pressure. When the radiator cap is removed after the engine has been running for some time, why does the water boil and create a miniature geyser?

9. On hot, humid days why does a cold glass of lemonade "sweat"? That is, why do water droplets form on the outside of the glass?

10. A real estate developer proposes to create an artificial lake on the moon by filling a large crater with water. Would you invest in such a venture? Explain why or why not.

11. Can it be too cold to ice skate? Explain.

12. During extremely cold winters in places like Chicago, snow "vanishes" even though the temperature never rises above 0°C. What happens to the snow?

13. Use the phase diagram for water to explain how a snowball is formed.

14. In the "mile high" city of Denver (1 m = 1.6 km), a three-minute egg takes longer to cook. Explain why.

Liquefaction of Gases

15. Explain why the mist coming from an aerosol spray can is cool to the touch.

16. Formulate a simple theory of fog formation by citing the conditions that induce fog and the physical process that occurs.

17. Using the data in Table 22.1, calculate the entropy change associated with

(a) lead melting   (b) ice melting
(c) nitrogen vaporizing   (d) oxygen vaporizing.

18. Offer an explanation for the comparative difference in your answers to parts (c) and (d) in Problem 17.

The Joule–Thomson Effect

19. Use a potential-energy curve like the one in Fig. 22.6 to explain why a gas cools when it expands.

20. In the Joule–Thomson process, determine whether net work is done on the gas or by the gas if,

(a) the gas is ideal
(b) cooling occurs
(c) warming occurs.

What is the connection between net work and intermolecular forces?

APPENDIX

THE INTERNATIONAL SYSTEM OF UNITS

Basic Units	Definitions
Length	The meter (m) is currently defined as the distance that light travels in 1/299,792,458th of a second.
Time	The second (s) is the duration of 9,192,631,770 periods of the radiation emitted in a transition between two specified energy levels of the cesium-133 atom.

Mass   The kilogram (kg) is the mass of a particular cylinder of platinum–iridium alloy preserved in a vault at Sèvres, France.

Current   The ampere (A) is that current in two very long parallel wires 1 m apart that gives rise to a magnetic force per unit length of 2×10^{-7} N/m.

Temperature The kelvin (K) is 1/273.16 of the thermodynamic temperature of the triple point of water.

Names and Symbols for the SI Units

Quantity	Name of unit	Symbol	Definition
length	meter	m	
time	second	s	
mass	kilogram	kg	
current	ampere	A	
temperature	kelvin	K	
force	newton	N	$1\ N = 1\ kg\ m/s^2$
work, energy	joule	J	$1\ J = 1\ N\ m$
power	watt	W	$1\ W = 1\ J/s$
frequency	hertz	Hz	$1\ Hz = s^{-1}$
electric charge	coulomb	C	$1\ C = 1\ A\ s$
electric potential	volt	V	$1\ V = 1\ J/C$
electric field	volt per meter	V/m	$1\ V/m = V\ m^{-1}$
electric resistance	ohm	Ω	$1\ \Omega = 1\ V/A$
capacitance	farad	F	$1\ F = 1\ C/V$
inductance	henry	H	$1\ H = 1\ J/A^2$
magnetic field strength	tesla	T	
magnetic flux	weber	Wb	$1\ Wb = 1\ T\ m^2$
entropy	joule per kelvin	J/K	$1\ J/K = 1\ J\ K^{-1}$
specific heat	joule per kg kelvin	J/kg K	
pressure	pascal	Pa	$1\ Pa = 1\ N/m^2$

APPENDIX B

CONVERSION FACTORS

Length

    1 in. = 2.54 cm

    1 ft = 12 in. = 30.48 cm

    1 yd = 3 ft = 91.44 cm

$$1 \text{ km} = 0.6215 \text{ mi}$$

$$1 \text{ mi} = 1.609 \text{ km}$$

$$1 \text{ Å} = 0.1 \text{ nm}$$

Time

$$1 \text{ min} = 60 \text{ s}$$

$$1 \text{ h} = 60 \text{ min}$$

$$1 \text{ d} = 24 \text{ h} = 1440 \text{ min}$$

$$1 \text{ yr} = 365.24 \text{ d} \approx \pi \times 10^7 \text{ s}$$

Mass

$$1 \text{ kg} = 1000 \text{ g}$$

$$1 \text{ slug} = 14.59 \text{ kg}$$

$$1 \text{ kg} = 6.852 \times 10^{-2} \text{ slugs}$$

Area

$$1 \text{ m}^2 = 10^4 \text{ cm}^2$$

$$1 \text{ in.}^2 = 6.4516 \text{ cm}^2$$

$$1 \text{ m}^2 = 10.76 \text{ ft}^2$$

$$1 \text{ acre} = 43,560 \text{ ft}^2$$

Volume

$$1 \text{ m}^3 = 10^6 \text{ cm}^3$$

$$1 \text{ L} = 1000 \text{ cm}^3 = 10^{-3} \text{ m}^3$$

$$1 \text{ gal} = 3.786 \text{ L}$$

$$1 \text{ gal} = 4 \text{ qt} = 8 \text{ pt} = 128 \text{ oz} = 231 \text{ in.}^3$$

$$1 \text{ ft}^3 = 1728 \text{ in.}^3 = 28.32 \text{ L}$$

Force

$$1 \text{ N} = 0.2248 \text{ lb} = 10^5 \text{ dyn}$$

$$1 \text{ lb} = 4.4482 \text{ N}$$

Energy

$$1 \text{ ft lb} = 1.356 \text{ J}$$

$$1 \text{ cal} = 4.1840 \text{ J}$$

$$1 \text{ Cal} = 1000 \text{ cal}$$

$$1 \text{ Btu} = 778 \text{ ft lb} = 252 \text{ cal}$$

$$1 \text{ eV} = 1.602 \times 10^{-9} \text{ J}$$

$$1 \text{ erg} = 10^{-7} \text{ J}$$

$$1 \text{ J} = 1 \text{ W s}$$

Power

$$1 \text{ hp} = 550 \text{ ft lb/s} = 745.7 \text{ W}$$

$$1 \text{ W} = 1.341 \times 10^{-3} \text{ hp}$$

Pressure

$$1 \text{ atm} = 101.325 \text{ kPa}$$

$$1 \text{ atm} = 14.7 \text{ lb/in.}^2 = 1.01 \times 10^6 \text{ dyn/cm}^2 = 1.01 \times 10^6 \text{ erg/cm}^3$$

$$1 \text{ atm} = 760 \text{ mm Hg} = 38.8 \text{ ft H}_2\text{O}$$

$$1 \text{ lb/in.}^2 = 6.895 \text{ kPa}$$

$$1 \text{ torr} = 1 \text{ mm Hg} = 133.32 \text{ Pa}$$

$$1 \text{ bar} = 100 \text{ kPa}$$

$\frac{1}{100000}$ bar = Pa

Angles

$$\pi \text{ rad} = 180°$$

$$1 \text{ rad} = 57.30°$$

$$1° = 1.745 \times 10^{-2} \text{ rad}$$

FORMULAS FROM ALGEBRA, GEOMETRY, AND TRIGONOMETRY

Algebra

$$(a + b)^2 = a^2 + 2ab + b^2.$$

$$(a + b)^3 = a^3 + 3a^2b + 3ab^2 + b^3.$$

Quadratic formula: If $a \neq 0$ the roots of the quadratic equation $ax^2 + bx + c = 0$ are

$$x = \frac{-b \pm \sqrt{b^2 - 4ac}}{2a}.$$

Geometry

The slope of the straight line passing through points (x_1, y_1) and (x_2, y_2) with $x_1 \neq x_2$ is

$$\text{slope} = \frac{y_2 - y_1}{x_2 - x_1}.$$

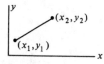

Equation of a straight line of slope m is

$$y = mx + b,$$

where b is the y intercept (the value of y when $x = 0$).

A circle of radius r has circumference $2\pi r$ and area πr^2.

A sphere of radius r has surface area $4\pi r^2$ and volume $\frac{4}{3}\pi r^3$.

Trigonometry

If r is the distance from the origin to a point (x, y) and θ is the polar coordinate angle, then

$$r = \sqrt{x^2 + y^2},$$

$$\cos\theta = x/r, \quad \sin\theta = y/r, \quad \tan\theta = y/x.$$

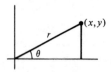

$$\csc\theta = 1/\sin\theta, \quad \sec\theta = 1/\cos\theta, \quad \cot\theta = 1/\tan\theta.$$

$$\sin^2\theta + \cos^2\theta = 1.$$

$$\sin(\theta \pm \alpha) = \sin\theta\cos\alpha \pm \cos\theta\sin\alpha,$$

$$\cos(\theta \pm \alpha) = \cos\theta\cos\alpha \mp \sin\theta\sin\alpha.$$

On a circular sector of radius r subtending an angle θ (measured in radians) the arc length of the circle is $s = r\theta$:

$$\theta\,(\text{in radians}) = \theta\,(\text{in degrees})\,\frac{\pi}{180},$$

$$\theta\,(\text{in degrees}) = \theta\,(\text{in radians})\,\frac{180}{\pi}.$$

APPENDIX

ASTRONOMICAL DATA

Earth

mass $\quad M_{\mathrm{E}} = 5.975 \times 10^{24} \mathrm{kg}$

radius $\quad 6371$ km

acceleration due to gravity $\quad 9.80665 \ \mathrm{m/s}^2$

Sun

    mass 1.987×10^{30} kg
    radius   696,500 km
    Earth–Sun mean distance 1.496×10^{11} m = 1 AU

Moon

    mass 7.343×10^{22} kg
    radius   1738 km
    Earth–Moon mean distance   384,400 km

Planet	Semimajor axis (10^6 km)	Orbital period (d)	Mass/M_E	Eccentricity
Mercury	57.9	87.96	0.055	0.2056
Venus	108.2	224.68	0.81	0.0068
Earth	149.6	365.24	1.00	0.0167
Mars	227.9	686.95	0.11	0.0934
Jupiter	778.3	4,337	318	0.0483
Saturn	1427.0	10,760	95	0.0560
Uranus	2871.0	30,700	15	0.0461
Neptune	4497.1	60,200	17	0.0100
Pluto	5983.5	90,780	0.0023	0.2484

APPENDIX

PHYSICAL CONSTANTS

Gravitational constant	G	6.672×10^{-11} N m^2/kg^2
Speed of light	c	2.997925×10^8 m/s
Electron's charge	e	1.60219×10^{-19} C
Coulomb constant	K_e	8.98755×10^9 N m^2/C^2
Permittivity of free space	ε_0	8.85419×10^{-12} C^2/N m^2
Magnetic constant	K_m	10^{-7} N/A^2

Permeability of free space	μ_0	$4\pi \times 10^{-7}$ N/A^2
Boltzmann's constant	k	1.3807×10^{-23} J/K
		8.617×10^{-5} eV/K
Avogadro's number	N_A	6.0220×10^{23} particles/mol
Gas constant	$R = N_A k$	8.314 J/mol K
		1.9872 cal/mol K
		8.206×10^{-2} L atm/mol K
Planck's constant	h	6.6262×10^{-34} J s
		4.1357×10^{-15} eV s
	$\hbar = h/2\pi$	1.05459×10^{-34} J s
		6.5822×10^{-16} eV s
Mass of the electron	m_e	9.1095×10^{-31} kg
		0.511 MeV/c^2
Mass of the proton	m_p	1.67265×10^{-27} kg
		938.28 MeV/c^2

SELECTED
BIBLIOGRAPHY

Apostol, T. M., *Calculus*, Vol. 1, Second Edition (John Wiley and Sons, New York, 1967).

Aristotle, *Works*, ed. by W. D. Ross, Vol. II, *De Caelo*, trans. by J. L. Stocks (Clarendon Press, Oxford, 1930).

Atallah, S., "Some Observations on the Great Fire of London, 1666," *Nature*, pp. 105–6, 2 July 1966.

Bekenstein, J. D., "Black Hole Thermodynamics," *Physics Today*, Vol. 3, pp. 24–31 (1980).

Boorstein, Daniel J., *The Discoverers* (Random House, New York, 1983).

Boscovich, Roger Joseph, *A Theory of Natural Philosophy* (MIT Press, Cambridge, Massachusetts, 1966).

Boyer, Carl, *The History of the Calculus* (Dover Publications, New York, 1949).

Boyer, Carl, *A History of Mathematics* (John Wiley and Sons, New York, 1968).

Casper, B. M., and Noyer, R. J., *Revolutions in Physics* (W. W. Norton and Co., New York, 1972).

Cohen, I. Bernard, *The Newtonian Revolution* (Cambridge University Press, Cambridge, 1980).

Cornford, Francis Macdonald, *Plato's Cosmology* (Routledge and Kegan Paul, London, 1966).

Crowe, Michael J., *A History of Vector Analysis* (University of Notre Dame Press, Notre Dame, Indiana, 1967).

Dijksterhuis, E. J., *The Mechanization of the World Picture*, trans. by C. Kikshoorn (Oxford University Press, Oxford, 1961).

Drake, Stillman, *Galileo Studies* (University of Michigan Press, Ann Arbor, 1970).

Drake, Stillman, and MacLachlan, James, "Galileo's Discovery of the Parabolic Trajectory," *Scientific American*, Vol. 232, pp. 102–110.

Epstein, L. C., and Hewitt, L. C., *Thinking Physics Part 1 and 2* (Insight Press, San Francisco, 1979).

French, A. P., *Newtonian Mechanics* (W. W. Norton, New York, 1971).

Galilei, Galileo, *Dialogues Concerning Two New Sciences*, trans. by Henry Crew and Alfonso de Salvio (Macmillan, New York, 1914).

Galilei, Galileo, *Dialogue Concerning The Two Chief World Systems*, trans. by Stillman Drake (University of California Press, Berkeley, 1953).

Galilei, Galileo, *Two New Sciences*, trans. by Stillman Drake (University of Wisconsin Press, Madison, 1974).

Goldreich, Peter, "Tides and the Earth-Moon System," *Scientific American*, Vol. 226, pp. 42–52, April 1972.

Goldreich, Peter, "Toward a Theory of the Uranian Rings," Nature, Vol. 277, p. 97, 1979.

Haldane, Elizabeth S., *Descartes His Life and Times* (John Murray, London, 1905).

Heilbron, J. L., *Elements of Early Modern Physics* (University of California Press, Berkeley, 1982).

Jespersen, James, *From Sundials to Atomic Clocks* (National Bureau of Standards, Monograph 155).

Joule, James Prescott, *The Scientific Papers of James Prescott Joule* (Taylor and Francis, London, 1884).

Kepler's Somnium, trans. by Edward Rosen (University of Wisconsin Press, Madison, 1967).

Koestler, Arthur, *The Sleepwalkers* (Grosset and Dunlap, New York, 1963).

Lawrence, E. N., "Meteorology and the Great Fire of London, 1666," *Nature*, pp. 168–9, 14 January 1967.

McCloskey, Michael, "Intuitive Physics," *Scientific American*, Vol. 248, pp. 122–130.

Magie, William Francis, *A Source Book in Physics* (McGraw-Hill, New York, 1935).

Maxwell, James Clerk, *The Scientific Papers of James Clerk Maxwell*, ed. by W. D. Niven (Dover Publications, New York, 1966).

Millikan, R. A., *Electrons* (University of Chicago Press, Chicago, 1927).

Millikan, R. A., *Physical Review*, Vol. 46, p. 1023 (1929).

Millikan, R.A., Roller, D., and Watson, E. A., *Mechanics, Molecular Physics, Heat, and Sound* (Ginn and Company, Boston, 1937).

Newton, Isaac, *Mathematical Principles*, trans. by Florian Cajori (University of California Press, Berkeley, 1934).

Peters, Philip C., "Black Holes: New Horizons in Gravitational Theory," *American Scientist*, Vol. 62, pp. 575–583, September–October 1974.

Rosenthal, Arthur, "The History of Calculus," *American Mathematical Monthly*, Vol. 58, pp. 75–86 (1951).

Sambursky, Shmuel, *Physical Thought from the Presocratics to the Quantum Physicists* (Pica Press, New York, 1975).

Settle, Thomas B., "An Experiment in the History of Science," *Science*, Vol. 133, No. 3445, pp. 19–23, January 6, 1961.

Sibulkin, M., "A Note on the Bathtub Vortex and the Earth's Rotation," *American Scientist*, Vol. 71, p. 352, July–August 1983.

Tea, P. L., Jr., and Falk, H., "Pumping on a Swing," *American Journal of Physics*, Vol. 36, p. 1165 (1968).

Turner, D. M., *Makers of Science: Electricity and Magnetism* (Oxford University Press, London, 1927).

Walker, Jearl, *The Flying Circus of Physics* (John Wiley and Sons, New York, 1975).

Walker, J., "The Physics of the Follow, the Draw, and the Massé (in billards and pool)," The Amateur Scientist, *Scientific American*, Vol. 248, No. 7, p. 124, July 1983.

Westfall, R. S., *Forces in Newton's Physics* (Macdonald, London, 1971).

Westfall, Richard S., "Newton and the Fudge Factor," *Science*, Vol. 179, pp. 751–758, 23 February 1973.

Westfall, R. S., *Never at Rest: A Biography of Isaac Newton* (Cambridge University Press, Cambridge, 1980).

Whitt, Lee, "The Standup Conic," Texas A & M University, 1981.

INDEX